Springer Series in Oper
and Financial Engineeri

Series Editors:
Thomas V. Mikosch
Sidney I. Resnick
Stephen M. Robinson

For further volumes:
http://www.springer.com/series/3182

David Simchi-Levi • Xin Chen • Julien Bramel

The Logic of Logistics

Theory, Algorithms, and Applications
for Logistics Management

Third Edition

 Springer

David Simchi-Levi
Massachusetts Institute of Technology
Cambridge, MA, USA

Xin Chen
University of Illinois at Urbana-Champaign
Urbana, IL, USA

Julien Bramel
Pine River Capital Management
New York, NY, USA

ISSN 1431-8598   ISSN 2197-1773 (electronic)
ISBN 978-1-4939-5021-8 ISBN 978-1-4614-9149-1 (eBook)
DOI 10.1007/978-1-4614-9149-1
Springer New York Heidelberg Dordrecht London

Springer is part of Springer Science+Business Media (www.springer.com)

To my wife, Yu, and my children, Jennifer, Jonathan and Hannah, for their love and support.

Xin Chen

In memory of Professor Uriel Rothblum, a friend, a colleague, and a scholar, whose research spans theory and applications.

David Simchi-Levi

Preface

We are pleased to introduce the third edition of the book, and we are thankful to those who used the book in research and practice and to those who sent us comments and feedback. As before, our objective is to present, in an easily accessible manner, logistics and supply chain models, algorithms, and tools. In this edition, we have attempted to build on the positive elements of the first two editions and to include what we have learned in the last few years, since the publication of the second edition.

In the last two decades, the academic community has focused on addressing many supply chain challenges. In some cases, the focus is on characterizing the structure of the optimal policy and identifying algorithms that generate the best possible policies. When this is not possible, the focus has been on an approach whose purpose is to ascertain characteristics of the problem or of an algorithm that are *independent of the specific problem data*. That is, the approach determines characteristics of the solution or the solution method that are intrinsic to the problem and not the data. This approach includes the so-called worst-case and average-case analyses, which, as illustrated in the book, help not only to understand characteristics of the problem or solution methodology, but also to provide specific guarantees of effectiveness. In many cases, the insights obtained from these analyses can then be used to develop practical and effective algorithms for specific complex logistics problems. Finally, game-theoretic approaches have been applied in the last few years to provide more insights to supply chain models involving competition and collaboration.

We have made several important changes to the third edition of this text. Many of these changes have been a result of new research or consulting engagement we have completed in the last few years. Our major changes include

- a new chapter on game theory, where we introduce the reader to key concepts and techniques (Chap. 3),

- a new chapter on supply chain competition and collaboration, where we extensively apply game theory (Chap. 11),

- a new chapter on process flexibility, where we explain the power of limited degree of flexibility (Chap. 13),

- a new section (Sect. 2.3) on discrete convex analysis,

- a new section (Sect. 9.6) on stochastic inventory models with positive lead times.

Additionally, we have extended the materials on integrated inventory and pricing models, including three new sections on the economic lot sizing model with pricing (Sect. 8.4), demand models (Sect. 10.2), and an alternative approach for deriving the structure of optimal policies (Sect. 10.5).

As before, this book is written for graduate students, researchers, and practitioners interested in the *mathematics of logistics and supply chain management.* We assume the reader is familiar with the basics of linear programming and probability theory and, in a number of sections, complexity theory and graph theory, although in many cases these can be skipped without loss of continuity.

Parts of this book are based on work we have done either together or with others. Indeed, some of the chapters originated from papers we have published in journals such as *Mathematics of Operations Research, Mathematical Programming, Operations Research,* and *IIE Transactions.* We rewrote most of these, trying to present the results in a simple yet general and unified way. However, a number of key results, proofs, and discussions are reprinted without substantial change. Of course, in each case this was done by providing the appropriate reference and by obtaining permission of the copyright owner. In the case of *Operations Research* and *Mathematics of Operations Research,* it is the Institute for Operations Research and the Management Sciences (INFORMS). Chapter 13 is heavily based on the paper by David Simchi-Levi and Yehua Wei, "Understanding the Performance of the Long Chain and Sparse Designs in Process Flexibility," which was published in *Operations Research* in 2012. Chapter 14 borrows extensively from "Supply Chain Design and Planning—Applications of Optimization Techniques for Strategic and Tactical Models," written by Ana Muriel and David Simchi-Levi and published in the *Handbooks in Operations Research and Management Science,* the volume on *Supply Chain Management,* S. Graves and A. G. Kok, eds., North-Holland, Amsterdam. Similarly, Chap. 20 borrows extensively from *Designing and Managing the Supply Chain,* written by David Simchi-Levi, Philip Kaminsky, and Edith Simchi-Levi and published by McGraw-Hill in 2007.

Cambridge, MA   David Simchi-Levi
Urbana-Champaign, IL   Xin Chen
New York, NY   Julien Bramel

Acknowledgments

It is our pleasure to acknowledge all those who helped us with the first, second, and third editions of this manuscript. First, we would like to acknowledge the contribution of our colleague Dr. Frank Chen, of the Chinese University of Hong Kong. Similarly, we are indebted to our colleague Professor Rafael Hassin, of Tel-Aviv University, and a number of referees, in particular, Professor James Ward of Purdue University, for carefully reading the manuscript and providing us with detailed comments and suggestions. We also would like to thank friends and colleagues Zhan Pang, Jin Qi, Diego Klabjan, Nir Halman, and Chenxi Zeng, whose inputs greatly improved the manuscript. In addition, we thank our former and current Ph.D. students Philip Kaminsky, Ana Muriel, Jennifer Ryan, Victor Martinez de Albeniz, Yehua Wei, Yan Zhao, Xiangyu Gao, Zhenyu Hu, and Limeng Pan, who read through and commented on various chapters or parts of earlier drafts. Our joint research and their comments and feedback were invaluable.

We would like to thank Edith Simchi-Levi, who is the main force behind the development of the network planning systems described in Chap. 20 and who carefully edited many parts of the book.

It is also a pleasure to acknowledge the support provided by the National Science Foundation, the Office of Naval Research, the Fund for the City of New York, Accenture, Bayer Business Services, DHL, Ford Motors Corporation, General Motors Corporation, Michelin, NASA, PwC, SAP, and Xerox. Their support made the development of some of the theory presented in the book possible.

Of course, we would like to thank our editors, Donna Chernyk and Achi Dosanjh, of Springer, who encouraged us throughout and helped us complete the project. Also, thanks to Ehrlich, Jamie, and the editorial staff at Springer in New York for their help.

Contents

List of Tables

List of Figures

1

Introduction

1.1 What Is Logistics Management?

For many companies, the ability to efficiently match demand and supply is key to their success. Failure to do so could lead to loss of revenue, reduced service levels, impacted reputation, and decline in the company's market share. Unfortunately, recent developments such as intense market competition, product proliferation, and the increase in the number of products with a short life cycle have created an environment where customer demand is volatile and unpredictable. In such an environment, traditional operations strategies such as building inventory, investing in capacity buffers, or increasing committed response time to consumers do not offer a competitive advantage. Therefore, many companies are looking for effective strategies to respond to market changes without significantly increasing cost, inventory, or response time. This has motivated a continuous evolution of the management of logistics systems.

In these systems, items are produced at one or more factories, shipped to warehouses and distribution centers for intermediate storage, and then shipped to retailers or customers. Consequently, to reduce cost and improve service levels, logistics strategies must take into account the interactions of these various levels in this *logistics network*, also referred to as the *supply chain*. This network consists of suppliers, manufacturing centers, warehouses, distribution centers, and retailer outlets, as well as raw materials, work-in-process inventory, and finished products that flow between the facilities; see Fig. 1.1.

D. Simchi-Levi et al., *The Logic of Logistics: Theory, Algorithms, and Applications for Logistics Management*, Springer Series in Operations Research and Financial Engineering, DOI 10.1007/978-1-4614-9149-1_1, © Springer Science+Business Media New York 2014

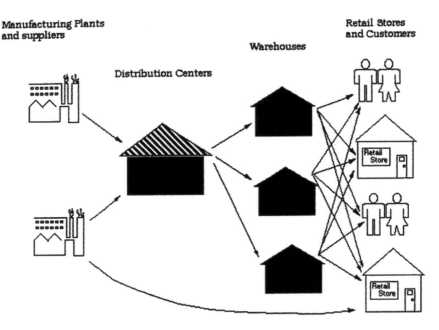

FIGURE 1.1. The logistics network

The goal of this book is to present the state-of-the-art in the *science of logistics management*. But what exactly is *logistics management*? According to the Council of Supply Chain Management Professionals (CSCMP), a nonprofit organization of business personnel, it is that part of the business that

> plans, implements, and controls the efficient, effective forward and reverses flow and storage of goods, services and related information between the point of origin and the point of consumption in order to meet customers' requirements.

This definition leads to several observations. First, logistics management takes into consideration every facility that has an impact on cost and plays a role in making the product conform to customer requirements: from supplier and manufacturing facilities, through warehouses and distribution centers, to retailers and stores. Indeed, in some supply chain analyses, it is necessary to account for the suppliers' suppliers and the customers' customers because they have an impact on supply chain performance.

Second, the objective of logistics management is to be efficient and cost-effective across the entire system; total systemwide costs, from transportation and distribution to inventories of raw materials, work-in-process, and finished goods, are to be minimized. Thus, the emphasis is not on simply minimizing transportation cost or reducing inventories but, rather, on taking a systems approach to logistics management.

Finally, because logistics management evolves around *planning, implementing, and controlling* the logistics network, it encompasses many of the firm's activities, from the strategic level through the tactical to the operational level.

Following Hax and Candea's (1984) treatment of production-inventory systems, logistical decisions are typically classified into three levels.

- The **strategic level** deals with decisions that have a long-lasting effect on the firm. This includes decisions regarding the number, location, and capacities of warehouses and manufacturing plants, or the flow of material through the logistics network.

- The **tactical level** typically includes decisions that are updated anywhere between once every week, month, or quarter. This includes purchasing and production decisions, inventory policies, and transportation strategies, including the frequency with which customers are visited.

- The **operational level** refers to day-to-day decisions such as scheduling, routing, and loading trucks.

Finally, what about supply chain management? What is the difference between supply chain management and logistics management? It is insightful to review the CSCMP definition. According to the CSCMP,

> Supply chain management encompasses the planning and management of all activities involved in sourcing and procurement, conversion, and all logistics management activities.

Thus, according to this definition, supply chain management includes logistics management as well as the coordination and collaboration with business partners such as suppliers, third-party service providers, and customers. In this book, we will focus on models important to both logistics as well as supply chain management.

1.2   Managing Cost and Uncertainty

What makes logistics, or supply chain management, difficult? Although we will discuss a variety of challenges throughout this text, they can all be related to one or both of the following observations:

1. It is challenging to design and operate a logistics system so that systemwide costs are minimized and systemwide service levels are maintained. Indeed, it is frequently difficult to operate a single facility so that costs are minimized and the service level is maintained. The difficulty increases significantly when an entire system is being considered.

2. Uncertainty is inherent in every logistics network; customer demand can never be forecast exactly, travel times will never be certain, and machines

and vehicles will break down. Logistics networks need to be designed to eliminate as much uncertainty as possible and to deal effectively with the uncertainty that remains.

One reason it is difficult to manage cost and uncertainty is due to **supply chain dynamics**. Indeed, in recent years, many suppliers and retailers have observed that while customer demand for specific products does not vary much, inventory and back-order levels fluctuate considerably across their supply chain. For instance, in examining the demand for Pampers disposal diapers, executives at Procter & Gamble noticed an interesting phenomenon.

As expected, retail sales of the product were fairly uniform; there is no particular day or month in which the demand is significantly higher or lower than in any other. However, the executives noticed that distributors' orders placed to the factory fluctuated much more than retail sales. In addition, P&G's orders to its suppliers fluctuated even more. This increase in variability as we travel up in the supply chain is referred to as the **bullwhip effect**.

Even when demand is known precisely (e.g., because of contractual agreements), the planning process needs to account for demand and cost parameters varying over time due to the impact of seasonal fluctuations, trends, advertising and promotions, competitors' pricing strategies, and so forth. These time-varying demand and cost parameters make it difficult to determine the most effective supply chain strategy, that is, the one that minimizes systemwide costs and conforms to customer requirements.

1.3 Examples

In this section, we introduce some of the logistics management issues that form the basis of the problems studied in the first four parts of the book. These issues span a large spectrum of logistics management decisions, at each of the three levels mentioned above. Our objective here is to briefly introduce the questions and the tradeoffs associated with these decisions.

Network Configuration

Consider the situation where several plants are producing products to serve a set of geographically dispersed retailers. The current set of facilities, that is, plants and warehouses, is deemed to be inappropriate, and management wants to reorganize or redesign the distribution network. This may be due, for example, to changing demand patterns or the termination of a leasing contract for a number of existing warehouses. In addition, changing demand patterns may entail a change in plant production levels, a selection of new suppliers, and, in general, a new flow pattern of goods throughout the distribution network. The goal is to choose a set of facility locations and capacities, to determine production levels for each product at each plant, and to set transportation flows between facilities, either from plant to warehouse or from warehouse to retailer, in such a way that total production, inventory, and transportation costs are minimized and various service-level requirements are satisfied.

Production Planning

A manufacturing facility must produce to meet demand for a product over a fixed finite horizon. In many real-world cases, it is appropriate to assume that demand is known over the horizon. This is possible, for example, if orders have been placed in advance or contracts have been signed specifying deliveries for the next few months. Production costs consist of a fixed amount, corresponding, say, to machine setup costs or times, and a variable amount, corresponding to the time it takes to produce one unit. A holding cost is incurred for each unit in inventory. The planner's objective is to satisfy demand for the product in each period and to minimize the total production and inventory costs over the fixed horizon. Obviously, this problem becomes more difficult as the number of products manufactured increases.

Inventory Control and Pricing Optimization

Consider a retailer that maintains an inventory of a particular product. Since customer demand is random, the retailer has information regarding the probabilistic distribution of demand only. The retailer's objective is to decide at what point to reorder a new batch of products, and how much to order. Typically, ordering costs consist of two parts: a fixed amount, independent of the size of the order, for example, the cost of sending a vehicle from the warehouse to the retailer; and a variable amount dependent on the number of products ordered. A linear inventory holding cost is incurred at a constant rate per unit of product per unit of time. The retailer must determine an optimal inventory policy to minimize the expected cost of ordering and holding inventory. In some situations, the price at which the product is sold to the end customer is also a decision variable. In this case, demand is not only random but is also affected by the selling price. The retailer's objective is thus to find an inventory policy and a pricing strategy maximizing expected profit over the finite, or infinite, horizon.

Procurement Strategies and Supply Contracts

In traditional logistics strategies, each party in the network focuses on its own profit and hence makes decisions with little regard to their impact on other partners. Relationships between suppliers and buyers are established by means of supply contracts that specify pricing and volume discounts, delivery lead times, quality, returns, and so forth. The question, of course, is whether supply contracts can also be used to replace the traditional strategy with one that optimizes the performance of the entire network. In particular, what is the impact of volume discount and revenue-sharing contracts on supply chain performance? Are there pricing strategies that can be applied by suppliers to incentivize buyers to order more products while at the same time increasing the supplier's profit? What are the risks associated with supply contracts, and how can these risks be minimized?

Process Flexibility

In the last few years, companies have been looking for new ways to respond to change in demand volume and mix without increasing inventory, capacity, or response time. One possible way to achieve that objective is to invest in process flexibility, where each plant is capable of producing multiple products. In this case, when the demand for one product is higher than expected while the demand for a different product is lower than expected, a flexible manufacturing system can quickly make adjustments by shifting production capacities appropriately. Unfortunately, flexibility does not come free, and hence the questions are how much flexibility is needed, how can one achieve flexibility, and what are the potential benefits of a (small) investment in flexibility?

Integration of Production, Inventory, and Transportation Decisions

Consider the problem faced by companies that rely on LTL (less than truckload) carriers for the distribution of products across their supply chain. Typically, these carriers offer volume discounts to encourage larger shipments; as a result, the transportation charges borne by the shipper are often piecewise linear and concave. In this case, the timing and routing of shipments need to be coordinated so as to minimize systemwide costs, including production, inventory, transportation, and shortage costs, by taking advantage of economies of scale offered by the carriers.

Vehicle Fleet Management

A warehouse supplies products to a set of retailers using a fleet of vehicles of limited capacity. A dispatcher is in charge of assigning loads to vehicles and determining vehicle routes. First, the dispatcher must decide how to partition the retailers into groups that can be feasibly served by a vehicle, that is, whose loads fit in a vehicle. Second, the dispatcher must decide what sequence to use so as to minimize cost. Typically, one of two cost functions is possible: In the first, the objective is to minimize the number of vehicles used, while in the second, the focus is on reducing the total distance traveled. The latter is an example of a single-depot capacitated vehicle routing rroblem (CVRP), where a set of *customers* has to be served by a fleet of vehicles of limited capacity. The vehicles are initially located at a *depot* (in this case, the warehouse) and the objective is to find a set of vehicle routes of minimal total length.

Truck Routing

Consider a truck that leaves a warehouse to deliver products to a set of retailers. The order in which the retailers are visited will determine how long the delivery will take and at what time the vehicle can return to the warehouse. Therefore, it is important that the vehicle follow an efficient route. The problem of finding the minimal length route, in either time or distance, from a warehouse through a set of retailers is an example of a traveling salesman problem (TSP). Clearly, truck routing is a subproblem of the fleet management example above.

Packing Problems

In many logistics applications, a collection of items must be packed into boxes, bins, or vehicles of limited size. The objective is to pack the items such that the number of bins used is as small as possible. This problem is referred to as the bin-packing problem (BPP). For example, it appears as a special case of the CVRP when the objective is to minimize the number of vehicles used to deliver the products. Bin-packing also appears in many other applications, including cutting standard-length wire or paper strips into specific customer order sizes. It also often appears as a subproblem in other combinatorial problems.

1.4 Modeling Logistics Problems

The reader observes that most of the problems and issues described in the previous section are fairly well defined mathematically. These are the types of issues, questions, and problems addressed in this book. Of course, many issues important to logistics or supply chain management are difficult to quantify and therefore to address mathematically; we will not cover these in this book. This includes topics related to information systems, outsourcing, third-party logistics, strategic partnering, etc. For a detailed analysis of these topics, we refer the reader to the book by Simchi-Levi et al. (2007) or the more recent book by Simchi-Levi (2010).

The fact that the examples provided in the previous section can be defined mathematically is, obviously, meaningless unless all required data are available. As we discuss in Part V of this book, finding, verifying, and tabulating the data are typically very problematic. Indeed, inventory holding costs, production costs, extra vehicle costs, and warehouse capacities are often difficult to determine. Furthermore, identifying the data relevant to a particular logistics or supply chain problem adds another layer of complexity to the data-gathering problem. Even when the data do exist, there are other difficulties related to modeling complex real-world problems. For example, in our analysis we ignore issues such as variations in travel times, variable yields in production, inventory shrinkage, forecasting, crew scheduling, and so on. These issues complicate logistics and supply chain practice considerably.

For most of this book, we assume that all relevant data, for example, production costs, production times, warehouse fixed costs, travel times, and holding costs, are given. As a result, each logistics or supply chain problem analyzed in Parts I–IV is well defined and thus *merely* a mathematical problem.

1.5   Logistics and Supply Chain in Practice

How are logistics and supply chain problems addressed in practice? That is, how are these difficult problems *solved* in the real world? In our experience, companies use several approaches. First and foremost, as in other aspects of life, people tend to repeat what has worked in the past. That is, if last year's safety stock level was enough to avoid backlogging demands, then the same level might be used this year. If last year's delivery routes were successful, that is, all retailers received their deliveries on time, then why change them? Second, there are so-called rules of thumb that are widely used and, at least on the surface, may be quite effective. For example, it is our experience that many logistics managers often use the "20/80 rule," which says that about 20 % of the products contribute to about 80 % of the total cost and therefore it is sufficient to concentrate efforts on these critical products. Logistics network design, to give another example, is an area where a variety of rules of thumb are used. One such rule might suggest that if your company serves the continental United States and it needs only one warehouse, then this warehouse should probably be located in the Chicago area; if two are required, then one in Los Angeles and one in Atlanta should suffice. Finally, some companies try to apply the experience and intuition of logistics experts and consultants, the idea being that what has worked well for a competitor should work well for itself.

Of course, while all these approaches are appealing and quite often result in logistics strategies that make sense, it is not clear how much is lost by not focusing on the *best* (or close to the best) strategy for the particular case at hand. Indeed, recently, with the advent of cheap computing power, it has become increasingly affordable for many firms, not just large ones, to acquire and use sophisticated *advance planning systems* (APS) to *optimize* their logistics and supply chain strategies. In these systems, data are entered, reviewed, and validated, various algorithms are executed, and a *suggested solution* is presented in a user-friendly way. Provided the data are correct and the system is solving the appropriate problem, these APS *can substantially reduce systemwide cost*. Also, generating a satisfactory solution that managers can implement is typically only arrived at after an iterative process in which the user evaluates various scenarios and assesses their impact on costs and service levels. Although this may not exactly be considered "optimization" in a strict sense, it usually serves as a useful tool for the system's user.

These planning systems have as their nucleus models and algorithms in some form or another. In some cases, the system may simply be a computerized version of the rules of thumb above. In more and more instances, however, these systems apply techniques that have been developed by the operations, management science, and computer science research communities.

In this book, we present the current state-of-the-art in mathematical research in the area of logistics. Some of the problems listed above represent difficult stochastic optimization problems that require concepts such as **convexity** and **supermodularity**, and their extensions for their analysis. Some problems require the use of methods from game theory in order to understand how different supply chain partners respond to various challenges. Other problems have at their core extremely difficult combinatorial problems in the class called $\mathcal{NP}$-Hard problems. This implies that it is very unlikely that one can construct an algorithm that will always find the optimal solution, or the best possible decision, in computation time that is polynomial in the "size" of the problem. The interested reader can refer to the excellent book by Garey and Johnson (1979) for details on computational complexity. Therefore, in many cases, an algorithm that consistently provides the optimal solution is not considered a reachable goal, and hence heuristic, or approximation, methods are employed.

1.6   Evaluation of Solution Techniques

A fundamental research question is how to evaluate heuristic or approximation methods. Such methods can range from simple "rules of thumb" to complex, computationally intensive, mathematical programming techniques. In general, these are methods that will find good solutions to the problem in a reasonable amount of time. Of course, the terms "good" and "reasonable" depend on the heuristic and on the problem instance. Also, what constitutes reasonable time may be highly dependent on the environment in which the heuristic will be used; that is, it depends on whether or not the algorithm needs to solve the logistics problem in *real time*.

Assessing and quantifying a heuristic's effectiveness is of prime concern. Traditionally, the following methods have been employed.

- **Empirical comparisons**: Here, a representative sample of problems is chosen and the performance of a variety of heuristics is compared. The comparison can be based on solution quality or computation time, or a combination of the two. This approach has one obvious drawback: deciding on a good set of test problems. The difficulty, of course, is that a heuristic may perform well on one set of problems but may perform poorly on the next. As pointed out by Fisher (1995), this lack of robustness forces practitioners to "patch up" the heuristic to fix the troublesome cases, leading to an algorithm with growing complexity. After considerable effort, a procedure may be created that works well for the situation at hand. Unfortunately, the resulting algorithm is usually extremely sensitive to changes in the data and may perform poorly when transported to other environments.

- **Worst-case analysis**: In this type of analysis, one tries to determine the maximum deviation from optimality, in terms of relative error, that a heuristic can incur on *any* problem instance. For example, a heuristic for the BPP might guarantee that any solution constructed by the heuristic uses at most 50% more bins than the optimal solution. Or a heuristic for the TSP might guarantee that the length of the route provided by the heuristic is at most twice the length of the optimal route. Using a heuristic with such a guarantee allays some of the fears of suboptimality, by guaranteeing that we are within a certain percentage of optimality. Of course, one of the main drawbacks of this approach is that a heuristic may perform very well on most instances that are likely to appear in a real-world application but may perform extremely poorly on some highly contrived instances. Hence, when we compare algorithms, it is not clear that a heuristic with a better worst-case performance guarantee is necessarily more effective in practice.

- **Average-case analysis**: Here, the purpose is to determine a heuristic's average performance. This is stated as the average relative error between the heuristic solution and the optimal solution under specific assumptions on the distribution of the problem data. This may include probabilistic assumptions on the depot location, demand size, item size, time windows, vehicle capacities, etc. As we shall see, while these probabilistic assumptions may be quite general, this approach also has its drawbacks. The most important includes the fact that an average-case analysis is usually only possible for large-size problems. For example, in the BPP, if the item sizes are uniformly distributed (between zero and the bin capacity), then a heuristic that will be "close to optimal" is one that first sorts the items in nonincreasing order and then, starting with the largest item, pairs each item with the largest item with which it fits. In what sense is it close to optimal? The analysis shows that as the problem size increases (the number of items increases), the relative error between the solution created by the heuristic and the optimal solution decreases to zero. Another drawback is that in order for an average-case analysis to be tractable, it is sometimes necessary to assume independent customer behavior. Finally, determining what probabilistic assumptions are appropriate in a particular real-world environment is not a trivial problem.

Because of the advantages and potential drawbacks of each of the approaches, we agree with Fisher (1980) that these should be treated as complementary approaches rather than competing ones. Indeed, it is our experience that the logistics algorithms that are most successfully applied in practice are those with good performance in at least two of the above measures.

We should also point out that characterizing the worst-case or average-case performance of a heuristic may be technically very difficult. Therefore, a heuristic may perform very well on average, or in the worst case, but *proving* this fact may be beyond our current abilities.

1.7   Additional Topics

We emphasize that due to space and time considerations, we have been obliged to omit some important and interesting results. These include results regarding yield management, machine scheduling, random yield in production, and dynamic and stochastic fleet management models, among others. We refer the reader to Graves et al. (1993), Ball et al. (1995), De Kok and Graves (2003), Simchi-Levi et al. (2004), and Özer and Phillips (2012) for excellent surveys of these and other related topics.

Also, while many elegant and strong results concerning approaches to certain logistics problems exist, there are still many areas where little, if anything, is known. This is, of course, partly due to the fact that as the models become more complex and integrate more and more issues that arise in practice, their analysis becomes more difficult.

Finally, we remark that it is our firmly held belief that logistics and supply chain management are one of the areas in which a rigorous mathematical analysis yields not only elegant results but, even more importantly, has had, and will continue to have, a significant impact on the practice of logistics and supply chains.

1.8   Book Overview

This book is meant as a survey of a variety of results covering most topics in the area of logistics. The reader should have a basic understanding of complexity theory, linear programming, probability theory, and graph theory. Of course, the book can be read easily without the reader's delving into the details of each proof.

The book is organized as follows. In Part I, we concentrate on performance analysis techniques. Specifically, in Chap. 2, we introduce the concepts, and associated properties, of convexity, supermodularity, and discrete convexity. In Chap. 3, we provide a concise introduction to some of the key concepts and results in game theory. In Chap. 4, we discuss some of the basic tools required to perform worst-case analysis, while in Chap. 5, we cover average-case analysis. Finally, in Chap. 6, we investigate the performance of mathematical programming-based approaches.

Part II concentrates on production and inventory problems. We start with lot sizing in two different deterministic environments, one with constant demand (Chap. 7) and the second with varying demand (Chap. 8). Chapter 9 focuses on stochastic inventory models, while Chap. 10 presents new results for the coordination of inventory and pricing decisions. The chapter distinguishes between models appropriate for risk-neutral and risk-averse decision makers.

Part III deals with supply chain design and coordination models. These include Chap. 11, which focuses on competition and collaboration models in supply chains, and Chap. 12 on effective supply contracts, such as buy back, revenue-sharing, and portfolio contracts. Chapter 13 deals with process flexibility, and Chap. 14 addresses models that integrate production, inventory, and transportation decisions across the supply chain. Finally, Chap. 15 analyzes distribution network configuration and facility location, also referred to as site selection, problems.

In Part IV, we consider vehicle routing problems, paying particular attention to heuristics with good worst-case or average-case performance. Chapter 16 contains an analysis of the single-depot capacitated vehicle routing problem when all customers have equal demands, while Chap. 17 analyzes the case of customers with unequal demands. In Chap. 18, we perform an average-case analysis of the vehicle routing problem with time window constraints. We also investigate set-partitioning-based approaches and column generation techniques in Chap. 19.

In Part V, we look at the practice of logistics management and in particular at issues related to the design, development, and implementation of APS. Specifically, in Chap. 20, we look at network planning issues from logistics network design, through inventory positioning, all the way to resource allocation. Finally, in Chap. 21, we report on the development of a decision support tool for school bus routing and scheduling in the City of New York.

Part I

Performance Analysis Techniques

2

Convexity and Supermodularity

The concepts of convexity and supermodularity are important in the optimization and economics literature. These concepts have been widely applied in the analysis of a variety of supply chain models, from stochastic, multi-period inventory problems to pricing models. Hence, in this chapter, we provide a brief introduction to convexity and supermodularity, focusing on materials most relevant to our context. We also briefly introduce some concepts and results from discrete convex analysis, which interestingly is an elegant combination of both convexity and submodularity. For more details, readers are referred to the three excellent books Rockafellar (1970) on convex analysis, Topkis (1998) on supermodularity, and Murota (2003) on discrete convex analysis.

2.1 Convex Analysis

2.1.1 Convex Sets and Convex Functions

Before we present the definition of convex sets, we introduce some notations that will be used. Throughout this book, we use $\Re^n$ to denote an n-dimensional Euclidean space, and "$\subseteq$" and "$\subset$" for set inclusion and strict set inclusion, respectively. For a set C, we write $x \in C$ if x is an element of C.

Definition 2.1.1 *A set $C \subseteq \Re^n$ is called convex if, for any $x, x' \in C$ and $\lambda \in [0, 1]$, $(1 - \lambda)x + \lambda x' \in C$.*

D. Simchi-Levi et al., *The Logic of Logistics: Theory, Algorithms, and Applications for Logistics Management*, Springer Series in Operations Research and Financial Engineering, DOI 10.1007/978-1-4614-9149-1_2, © Springer Science+Business Media New York 2014

Geometrically, a set is convex if and only if for any two points in the set, the line segment between these two points also lies in the set (Fig. 2.1). Here are some simple examples of convex sets: an interval in $\Re^1$; a disk and a square in $\Re^2$; a sphere and a cube in $\Re^3$. Also note that a set of solutions of a system of linear inequalities, that is, $\{x \in \Re^n : Ax \leq b\}$, is convex, where A is a linear mapping from $\Re^n$ to $\Re^m$ and b is a vector in $\Re^m$. Finally, the intersection of convex sets is also convex, and convexity is preserved under a linear transformation; namely, the set $AC + b := \{Ax + b | x \in C\}$ is still convex if C is.

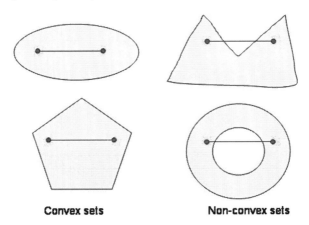

Convex sets   **Non-convex sets**

FIGURE 2.1. Examples of convex sets and nonconvex sets

Definition 2.1.2 *Given a convex set C in $\Re^n$, a function $f : C \to \Re$ is convex over set C if, for any $x, x' \in C$ and $\lambda \in [0, 1]$,*

$$f((1 - \lambda)x + \lambda x') \leq (1 - \lambda)f(x) + \lambda f(x'). \tag{2.1}$$

f is strictly convex if the inequality (2.1) holds strictly for any $x, x' \in C$ with $x \neq x'$ and $\lambda \in (0, 1)$. Finally, f is called (strictly) concave if $-f$ is (strictly) convex.

Remark: When f is (strictly) convex over $\Re^n$, we simply say that f is (strictly) convex. From now on, we mainly focus on the case when $C = \Re^n$ to simplify our presentation. In fact, almost all the results about convex functions defined over $\Re^n$ hold for convex functions defined over a convex subset of $\Re^n$, possibly with minor modification.

Sometimes it is convenient to work with functions that take the value of infinity. In this case, for a given convex set C in $\Re^n$, a function $f : C \to \bar{\Re}$ is convex over C in the extended sense if the inequality (2.1) still holds for f, where $\bar{\Re} = \Re \cup \{\infty\}$. The arithmetic convention here includes $\infty + \infty = \infty$, $0 \cdot \infty = 0$, and $\alpha \cdot \infty = \infty$ for $\alpha > 0$. Of course, usually we can restrict ourselves to the *effective domain* of function f, which is defined as follows:

$$\text{dom}(f) := \{x \in C \mid f(x) < \infty\}.$$

But on some occasions, it is more economical to use convex functions in the extended sense. Finally, for a convex function $f : C \to \Re$, define

$$\hat{f}(x) = \begin{cases} f(x), & \text{if } x \in C, \\ \infty, & \text{otherwise.} \end{cases}$$

It is easy to see that $\hat{f} : \Re^n \to \Re \cup \{\infty\}$ is convex in the extended sense.

It is also interesting to point out that if a function $f : \Re^n \to \Re$ is continuous, then the convexity of the function f is equivalent to the following midpoint convexity: For any $x, x' \in \Re^n$, $f((x + x')/2) \le 1/2(f(x) + f(x'))$. This is left as an exercise.

We now establish a connection between convex sets and convex functions through the epigraph mapping. The epigraph of a function $f : C \to \Re$ is defined as

$$\text{epi}(f) := \{(x, \alpha) \mid x \in C, \alpha \in \Re, f(x) \le \alpha\}.$$

It is easy to verify that a function f is convex on a convex set C if and only if its epigraph $\text{epi}(f)$ is convex. The epigraph mapping allows us to translate properties of convex sets into results about convex functions.

The graphical meaning of a convex function is clear; see Fig. 2.2 for an illustration. In fact, a function f is convex if and only if for any given x and x', the curve $((1-\lambda)x + \lambda x'), f((1-\lambda)x + \lambda x'))$ for $\lambda \in [0,1]$ always lies below the line segment connecting two points $(x, f(x))$ and $(x', f(x'))$. Obviously, a linear function is both convex and concave.

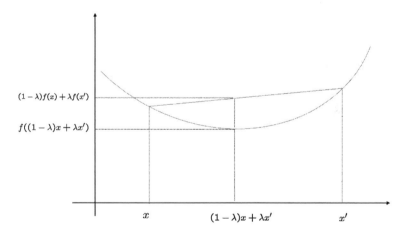

FIGURE 2.2. Illustration of the definition of convex function

In the following, we summarize some useful properties of convex functions. The proof for those properties is straightforward.

Proposition 2.1.3

(a) *Any nonnegative linear combination of convex functions is convex. That is, if $f_i : \Re^n \to \Re$ are convex for $i = 1, 2, \ldots, m$, then for any scalar $\alpha_i \ge 0$, $\sum_{i=1}^{m} \alpha_i f_i$ is also convex.*

(b) *A composition of a convex function and an affine function is still convex. That is, if $f : \Re^m \to \Re$ is convex, then for any linear mapping A from R^n to R^m and a vector b in $\Re^m$, $f(Ax + b)$ is also a convex function of $x \in \Re^n$.*

(c) *A composition of a nondecreasing convex function and a convex function is still convex. That is, if $f : \Re \to \Re$ is convex and nondecreasing and $g : \Re^n \to \Re$ is convex, then $f(g(x))$ is convex.*

(d) *If $f_k : \Re^n \to \Re$ is convex for $k = 1, 2, \ldots$ and $\lim_{k \to \infty} f_k(x) = f(x)$ for any $x \in \Re^n$, then $f(x)$ is convex.*

(e) *Assume that a function $f(\cdot, \cdot)$ is defined on the product space $\Re^n \times \Re^m$. If $f(\cdot, y)$ is convex for any given $y \in \Re^m$, then for a random vector ζ in $\Re^m$, $E_\zeta[f(x, \zeta)]$ is convex, provided it is well defined. As a special case, if $f : \Re^n \to \Re$ is convex, then $E_\zeta[f(x - \zeta)]$ is also convex.*

A weaker definition of convex functions is quasiconvex, which is also commonly used.

Definition 2.1.4 *A function $f : \Re^n \to \Re$ is called quasiconvex on a convex set C if, for any $x, x' \in \Re^n$ and $\lambda \in [0, 1]$,*

$$f((1 - \lambda)x + \lambda x') \leq \max\{f(x), f(x')\}.$$

The quasiconvexity of function $f : \Re^n \to \Re$ is equivalent to the fact that $-f$ is unimodal. That is, if x^* is a global maximizer of function $-f$, then for any $x \in \Re^n$, $-f((1 - \lambda)x + \lambda x^*)$ is a nondecreasing function of λ for $\lambda \in [0, 1]$.

For a function $f : C \to \Re$ and a given $\alpha \in \Re$, define the level set of f as

$$L_\alpha(f) := \{x \in C \mid f(x) \leq \alpha\}.$$

One can show, from the definition of quasiconvexity, that f is quasiconvex on a convex set C if and only if the level set $L_\alpha(f)$ is convex for any $\alpha \in \Re$.

2.1.2 Continuity and Differentiability Properties

In this section, we discuss the continuity and differentiability properties of convex functions. Before we proceed to prove the continuity of convex functions, observe that the convexity of a function $f : \Re^n \to \Re$ is equivalent to the following: For any $x^i \in \Re^n$, $\lambda_i \in \Re$ with $\lambda_i \geq 0$ $(i = 1, 2, \ldots, m)$ and $\sum_{i=1}^m \lambda_i = 1$,

$$f(\sum_{i=1}^m \lambda_i x^i) \leq \sum_{i=1}^m \lambda_i f(x^i). \tag{2.2}$$

This is a special case of Jensen's inequality.

Theorem 2.1.5 *If $f : \Re^n \to \Re$ is convex, then it is continuous.*

Proof. We only need to show, without loss of generality, that f is continuous at $x = 0$.

First, we argue that $f(x)$ is bounded above over the set $S = \{x \in \Re^n \mid \|x\|_1 = \sum_{i=1}^n |x_i| \le 1\}$. Let e_i $(i = 1, 2, \ldots, n)$ be a unit vector in $\Re^n$ with 1 at its ith component and 0 at other components, and let $e_{n+i} = -e_i$ $(i = 1, 2, \ldots, n)$. Then for any $x \in S$, there exists $\lambda_i \ge 0$ $(i = 1, 2, \ldots, 2n)$ with $\sum_{i=1}^{2n} \lambda_i = 1$ such that $x = \sum_{i=1}^{2n} \lambda_i e_i$. Therefore, (2.2) implies that $f(x) \le \max_{i=1,2,\ldots,2n} f(e_i)$.

We now show that for any sequence $x^k \in \Re^n$ $(k = 1, 2, \ldots,)$ convergent to $x = 0$, $f(x^k)$ converges to $f(0)$. Since x^k converges to 0, we can assume without loss of generality that $x^k \in S$ for all k. The definition of convex functions implies that

$$f(x^k) \le (1 - \|x^k\|_1)f(0) + \|x^k\|_1 f(x^k/\|x^k\|_1).$$

Letting k tend to infinity, we have

$$\varlimsup_{k \to \infty} f(x^k) \le f(0),$$

where $\varlimsup$ denotes the upper limit of a sequence. Also, observe that $0 = (1 - \frac{\|x^k\|_1}{1+\|x^k\|_1})x^k + \frac{\|x^k\|_1}{1+\|x^k\|_1}(-\frac{x^k}{\|x^k\|_1})$. Again, the definition of convex functions implies that

$$f(0) \le (1 - \frac{\|x^k\|_1}{1 + \|x^k\|_1})f(x^k) + \frac{\|x^k\|_1}{1 + \|x^k\|_1} f(-x^k/\|x^k\|_1).$$

When k goes to infinity, we have

$$f(0) \le \varliminf_{k \to \infty} f(x^k),$$

where $\varliminf$ denotes the lower limit of a sequence. Thus, $\lim_{k\to\infty} f(x^k) = f(0)$ for any sequence x^k convergent to $x = 0$, and therefore f is continuous at 0. ∎

It is appropriate to point out that if the domain of a convex function is not the whole space, it may not be continuous at the boundary points [for instance, $f : [0, 1] \to \Re$ with $f(x) = 0$ for $x \in (0, 1]$ and $f(0) = 1$]. However, it is always continuous at the interior points of its domain. A natural question is whether it is also differentiable at these interior points. Unfortunately, it is not always the case. For example, the absolute value function $|x|$ is convex while not differentiable at $x = 0$. Even though a convex function may not be differentiable, we will show in the following that for a convex function, its directional derivative always exists and possesses nice properties. Recall that for any $x, y \in \Re^n$, the directional derivative of a function $f : \Re^n \to \Re$ at x is defined as follows:

$$f'(x; y) := \lim_{t \downarrow 0} \frac{f(x + ty) - f(x)}{t}.$$

For a function f defined on one-dimensional space, let $f'_+(x) = f'(x;1)$ and $f'_-(x) = -f'(x;-1)$, and define

$$D_f(x,t) := \frac{f(x+t) - f(x)}{t}.$$

The following result, which has been widely applied, is helpful in establishing the monotonicity properties of f'_+ and f'_-.

Proposition 2.1.6 *Assume that a function* $f : \Re \to \Re$ *is convex. Then for any* $x, x', t, t' \in \Re$ *with* $x \le x'$ *and* $0 < t \le t'$ *or* $x < x + t \le x' < x' + t'$,

$$D_f(x,t) \le D_f(x',t'). \tag{2.3}$$

In particular, when $t = t'$, f *has increasing differences; that is, for any* $x \le x'$, $t \ge 0$,

$$f(x+t) - f(x) \le f(x'+t) - f(x').$$

Proof. Observe that $x \le x + t, x' \le x' + t$. There exist $\lambda, \lambda' \in [0,1]$ such that $x + t = (1-\lambda)x + \lambda(x'+t)$ and $x' = (1-\lambda')x + \lambda'(x'+t)$. The definitions of λ and λ' imply that $\lambda + \lambda' = 1$. From the convexity of f, we have

$$f(x+t) \le (1-\lambda)f(x) + \lambda f(x'+t)$$

and

$$f(x') \le (1-\lambda')f(x) + \lambda' f(x'+t).$$

Adding the two inequalities together and rearranging terms, we have that

$$f(x+t) - f(x) \le f(x'+t) - f(x'). \tag{2.4}$$

Thus, a convex function has increasing differences.

We now assume that $0 < t \le t'$. From the convexity of f, we have that

$$f(x'+t) \le \left(1 - \frac{t}{t'}\right)f(x') + \frac{t}{t'}f(x'+t'),$$

which immediately implies that

$$\frac{f(x'+t) - f(x')}{t} \le \frac{f(x'+t') - f(x')}{t'}.$$

This inequality, together with the inequality (2.4), implies the inequality (2.3).

Finally, assume that $x < x + t < x' < x' + t'$. Again, the convexity of f implies that

$$f(x+t) \le (1-\lambda)f(x) + \lambda f(x')$$

and

$$f(x') \le (1-\lambda')f(x+t) + \lambda' f(x'+t'),$$

where $\lambda = \frac{t}{x'-x}$ and $\lambda' = \frac{x'-(x+t)}{(x'+t')-(x+t)}$. The above inequalities are equivalent to the following:

$$\frac{f(x+t)-f(x)}{t} \leq \frac{f(x')-f(x+t)}{x'-(x+t)}$$

and

$$\frac{f(x')-f(x+t)}{x'-(x+t)} \leq \frac{f(x'+t')-f(x')}{t'}.$$

Therefore, the inequality (2.3) holds if $x < x+t < x' < x'+t'$. The continuity of convex functions implies that (2.3) is still true if $x < x+t \leq x' < x'+t'$.   ∎

Theorem 2.1.7 *Assume that $f : \Re \to \Re$ is convex. Then*

(a) *f'_+ and f'_- are well defined, and for any $x, x' \in \Re$ with $x < x'$, $f'_-(x) \leq f'_+(x) \leq f'_-(x')$.*

(b) *For any fixed $x \in \Re$,*

$$f(x+t) - f(x) \geq \xi t \ \forall \ t \in \Re$$

if and only if $\xi \in [f'_-(x), f'_+(x)]$.

Proof. From (2.3), we know that $D_f(x,t)$ is a nondecreasing function of t for $t > 0$ (simply let $x' = x$) and is bounded below by $D_f(x',t')$ for any $x < x'$ and $0 < t' < x' - x$. Therefore,

$$f'_+(x) = \inf_{t \downarrow 0} D_f(x,t).$$

Similarly, $D_f(x-t,t)$ is nonincreasing in t, and hence

$$f'_-(x) = \sup_{t \downarrow 0} D_f(x-t,t).$$

Indeed, define a new convex function $g(x) = f(-x)$ and the nonincreasing property of $D_f(x-t,t)$ follows from applying Proposition 2.1.6 to the function g.

Again from (2.3), it is easy to see that for any $x < x'$ and $0 < t < x' - x$,

$$D_f(x-t,t) \leq D_f(x,t) \leq D_f(x'-t,t).$$

Letting $t \downarrow 0$ yields $f'_-(x) \leq f'_+(x) \leq f'_-(x')$. Thus, part (a) is true.

Finally, part (b) is a direct consequence of the proof for part (a).   ∎

Theorem 2.1.7 part (b) implies that for any $\xi \in [f'_-(x), f'_+(x)]$, the function $f(x')$ always lies above the line $L_x = \{(x', f(x) + \xi(x' - x)) \mid x' \in \Re \}$ for any x; see Fig. 2.3 for an illustration. This result can be extended to convex functions defined in $\Re^n$. For this purpose, we introduce the concept of subgradient.

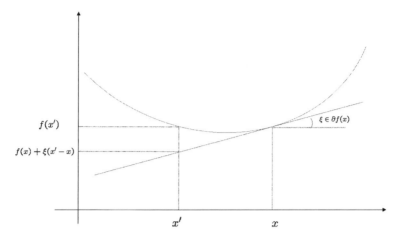

FIGURE 2.3. Illustration of the definition of subgradient

Definition 2.1.8 *Given a function $f : \Re^n \to \Re$, $\xi \in \Re^n$ is a subgradient of the function f at $x \in \Re^n$ if, for any $t \in \Re^n$,*

$$f(x + t) - f(x) \geq \langle \xi, t \rangle, \tag{2.5}$$

where $\langle \xi, t \rangle = \sum_{i=1}^{n} \xi_i t_i$ is the inner product between ξ and t. Let the subdifferential $\partial f(x)$ be the set of all subgradients of f at x.

The following theorem characterizes properties of subgradients. The proof of these properties is omitted since it is quite involved; see Rockafellar (1970) for more details.

Theorem 2.1.9 *Assume that a function $f : \Re^n \to \Re$ is convex. The following results hold.*

(a) For any $x \in \Re^n$, $\partial f(x)$ is nonempty, convex, and compact.

(b) For any $x, y \in \Re^n$, the directional derive of f can be expressed as follows:

$$f'(x; y) = \sup_{\xi \in \partial f(x)} \langle \xi, y \rangle.$$

(c) f is differentiable at $x \in \Re^n$ if and only if $\partial f(x) = \{\nabla f(x)\}$, where $\nabla f(x)$ denotes the gradient of f at x.

(d) For any compact set $C \subset \Re^n$, $\cup_{x \in C} \partial f(x)$ is compact.

Remark: If the domain of f is not the whole space, Theorem 2.1.9 may fail at the boundary points of its domain. For example, consider the convex function $f : [0, 1] \to \Re$ with $f(0) = 1$ and $f(x) = 0$ for $x \in (0, 1]$. Clearly, it is not continuous and has no subgradient at $x = 0$.

We now present the general form of Jensen's inequality.

Proposition 2.1.10 *Let f be a convex function over $\Re$ and ζ a random variable with finite expectation $E[\zeta]$. We have*

$$f(E[\zeta]) \leq E[f(\zeta)].$$

Proof. This proposition can be proven by using the special case of Jensen's inequality (2.2) and the continuity of convex functions as well as the definition of expectations. We present an alternative approach based on the properties of subgradients.

Choose any $\xi \in \partial f(E[\zeta])$. From the definition of subgradients, we have that

$$f(\zeta) - f(E[\zeta]) \geq \langle \xi, \zeta - E[\zeta] \rangle.$$

Taking expectations on both sides yields $f(E[\zeta]) \leq E[f(\zeta)]$.   ∎

2.1.3   Characterization of Convex Functions

The concept of convexity is widely used in optimization. However, identifying convex functions is not always simple. In this section, we give some sufficient and necessary conditions for a differentiable function to be convex.

Theorem 2.1.11 *Consider a function $f : \Re^n \to \Re$.*

(a) *If f is differentiable, then f is convex if and only if for any $x, x' \in \Re^n$,*

$$f(x') - f(x) \geq \langle \nabla f(x), x' - x \rangle, \tag{2.6}$$

where $\langle x, y \rangle = \sum_{i=1}^{n} x_i y_i$ is the inner product of $x, y \in \Re^n$.

(b) *If f is differentiable, then f is strictly convex if the inequality (2.6) holds strictly for any $x \neq x'$.*

(c) *If f is continuously differentiable, then f is convex if and only if ∇f is monotone; that is, for any $x, x' \in \Re^n$, $\langle \nabla f(x') - \nabla f(x), x' - x \rangle \geq 0$.*

(d) *If f is twice continuously differentiable, then f is convex if and only if $\nabla^2 f(x)$ is positive semidefinite for any $x \in \Re^n$.*

(e) *If f is twice continuously differentiable, then f is strictly convex if $\nabla^2 f(x)$ is positive definite for any $x \in \Re^n$.*

(f) *Assume that $f(x) = x^T Q x$ for a symmetric matrix Q of order n and $x \in \Re^n$. Then f is convex if and only if Q is positive semidefinite. And f is strictly convex if and only if Q is positive definite.*

Proof. Assume that f is differentiable. Pick any $x, x' \in \Re^n$ and define

$$\phi(\lambda) := f(x + \lambda(x' - x)).$$

First, notice that $f(x)$ is convex in x if and only if $\phi(\lambda)$ is convex for $\lambda \in [0, 1]$ for any picked $x, x' \in \Re^n$. Also, observe that

$$\phi'(\lambda) = \langle \nabla f(x + \lambda(x' - x)), x' - x \rangle.$$

If $\phi(\lambda)$ is convex in λ, then Theorem 2.1.7 part (b) implies that

$$f(x') - f(x) = \phi(1) - \phi(0) \geq \phi'(0) = \langle \nabla f(x), x' - x \rangle.$$

Hence, the inequality (2.6) holds for any $x, x' \in \Re^n$. On the other hand, if the inequality (2.6) is true for any $x, x' \in \Re^n$, then for any $z = (1 - \lambda)x + \lambda x'$ with $\lambda \in [0, 1]$, we have

$$f(x) - f(z) \geq \langle \nabla f(z), x - z \rangle$$

and

$$f(x') - f(z) \geq \langle \nabla f(z), x' - z \rangle.$$

Multiplying the first inequality by $(1 - \lambda)$ and the second inequality by λ and summing them up, we end up with

$$f((1 - \lambda)x + \lambda x') \leq (1 - \lambda)f(x) + \lambda f(x').$$

Thus, f is convex and part (a) is true. Obviously, from the above argument, one can see that f is strictly convex if the inequality (2.6) holds strictly. Therefore, part (b) holds.

For part (c), notice that if $\phi(\lambda)$ is convex, then Theorem 2.1.7 part (a) implies that $\phi'(0) \leq \phi'(1)$. Thus,

$$\langle \nabla f(x') - \nabla f(x), x' - x \rangle = \phi'(1) - \phi'(0) \geq 0.$$

On the other hand, if ∇f is monotone, then for any $\lambda' \geq \lambda \geq 0$,

$$\phi'(\lambda') = \langle \nabla f(x + \lambda'(x' - x)), x' - x \rangle \geq \langle \nabla f(x + \lambda(x' - x)), x' - x \rangle = \phi'(\lambda).$$

Therefore, $\phi'(\lambda)$ is nondecreasing, and hence

$$\phi(\lambda) = \phi(0) + \int_0^\lambda \phi'(\xi)d\xi \leq \phi(0) + \lambda \phi'(\lambda) \tag{2.7}$$

and

$$\phi(\lambda) = \phi(1) - \int_\lambda^1 \phi'(\xi)d\xi \leq \phi(1) - (1 - \lambda)\phi'(\lambda). \tag{2.8}$$

Multiplying the first inequality by $(1 - \lambda)$ and the second inequality by λ and summing them up, we end up with

$$\phi(\lambda) \leq (1 - \lambda)\phi(0) + \lambda\phi(1); \tag{2.9}$$

that is, ϕ is convex for $\lambda \in [0,1]$. Thus, f is convex. Also, notice that from the above proof, ∇f is monotone if and only if $\phi'(\lambda)$ is nondecreasing for $\lambda \in [0,1]$.

We now assume that f is twice continuously differentiable. In this case,

$$\phi''(\lambda) = \langle x' - x, \nabla^2 f(x + \lambda(x' - x))(x' - x)\rangle.$$

Notice that for any $0 \leq \lambda \leq \lambda' \leq 1$,

$$\phi'(\lambda') - \phi'(\lambda) = \int_\lambda^{\lambda'} \phi''(\xi)d\xi.$$

Therefore, if $\nabla^2 f(x)$ is positive semidefinite for any $x \in \Re^n$, then $\phi''(\xi) \geq 0$ for any ξ, and hence $\phi'(\lambda)$ is nondecreasing, which in turn implies that f is convex as we already proved for part (c). On the other hand, the convexity of f implies that ϕ' is nondecreasing, which in turn implies that $\phi''(\lambda) \geq 0$ for any $\lambda \in [0,1]$. In particular, we have

$$0 \leq \phi''(0) = \langle x' - x, \nabla^2 f(x)(x' - x)\rangle.$$

Since $x' \in \Re^n$ is arbitrary, we have that $\nabla^2 f(x)$ is positive semidefinite. This proves part (d).

If $\nabla^2 f(x)$ is positive definite for any $x \in \Re^n$, then for $x \neq x'$, $\phi'(\lambda)$ is strictly increasing for $\lambda \in [0,1]$ and at least one of the inequalities (2.7) and (2.8) holds as a strict inequality. Hence, the inequality (2.9) holds strictly. This implies that ϕ, and therefore f is strictly convex. Thus, part (e) holds.

Finally, for part (f), we only need to prove that if f is strictly convex, then Q is positive definite. The remaining results are special cases of parts (d) and (e). If f is strictly convex, then part (d) implies that Q is positive semidefinite. Assume to the contrary that Q is not positive definite. There exists a nonzero vector $z \in \Re^n$ such that $f(z) = \langle z, Qz\rangle = 0$. Therefore, for any $x \in \Re^n$, $f(x + \lambda z) = f(x) + 2\lambda\langle x, Qz\rangle$, which is a linear function of λ. Thus, f is not strictly convex, a contradiction. Hence, Q is positive definite if f is strictly convex.   ∎

2.1.4   Convexity and Optimization

Convexity plays an important role in optimization theory. In particular, we show in the following that a local minimizer of a convex minimization problem, namely, the problem of minimizing a convex function, is a global minimizer of this problem, and the first-order optimality condition is both sufficient and necessary for a point to be a global minimizer. As we shall see, this result has implications both for optimization theory and for algorithms.

Theorem 2.1.12 *Assume that $f : \Re^n \to \Re$ is convex. If x^* is a local minimizer of f, then x^* is a global minimizer of f. Furthermore, x^* is a global minimizer of f if and only if $0 \in \partial f(x^*)$.*

Proof. If x^* is a local minimizer, then there exists a ball $B_\epsilon(x^*) = \{x \in \Re^n \mid \|x - x^*\|_2 < \epsilon\}$ for some $\epsilon > 0$ such that $f(x) \geq f(x^*)$ for any $x \in B_\epsilon(x^*)$. Moreover, for any $x \in \Re^n$, there exists $\lambda \in (0, 1)$ such that $(1 - \lambda)x^* + \lambda x \in B_\epsilon$. From the definition of convexity, we have

$$f(x^*) \leq f((1 - \lambda)x^* + \lambda x) \leq (1 - \lambda)f(x^*) + \lambda f(x).$$

The inequality implies that $f(x^*) \leq f(x)$ for any $x \in \Re^n$. Hence, x^* is a global minimizer of f, and from the definition of a subgradient, we have $0 \in \partial f(x^*)$. Finally, if $0 \in \partial f(x^*)$, then the definition of a subgradient implies that $f(x) \geq f(x^*)$ for any $x \in \Re^n$. In other words, x^* is a global minimizer of the function f. Therefore, x^* is a global minimizer of f if and only if $0 \in \partial f(x^*)$.   ∎

The following result is a straightforward consequence of the definition of strictly convex functions.

Theorem 2.1.13 *A strictly convex function $f : \Re^n \to \Re$ has at most one local and global minimizer.*

We now consider the convex function maximization problem. If $f : \Re \to \Re$ is convex, then from the definition of convexity, one can see that either a or b is an optimal solution for the problem $\max_{x \in [a,b]} f(x)$. More generally, we have the following result regarding the convex function maximization problem.

Theorem 2.1.14 *Assume that a set $C \subset \Re^n$ is compact and $f : \Re^n \to \Re$ is convex. Then $\max_{x \in C} f(x)$ achieves maximization at an extreme point x^* of C. That is, there exists no $x, x' \in C$ with $x \neq x'$ such that $x^* = (x + x')/2$.*

We provide some intuition to the theorem instead of a formal proof. Assume that a maximizer of $f(x)$ over the set C, x^*, is not an extreme point of C. Then there exist $x, x' \in C$ with $x \neq x'$ such that $x^* = (x + x')/2$. Let $L = \{x^* + t(x' - x) \in C \mid t \in \Re\}$ be a line segment in C. It is clear, from the definition of convexity, that all points on the line segment L are maximizers of the function $f(x)$. Let $\hat{x}$ be one of the endpoints of L. If $\hat{x}$ is an extreme point of C, we are done; otherwise, we can repeat the above process and the theorem follows since such a process cannot proceed an infinite number of times.

The following proposition shows that under some conditions, convexity is preserved under optimization operations.

Proposition 2.1.15 *Given a function $f(\cdot, \cdot)$ defined on the product space $\Re^n \times \Re^m$,*

(a) *If $f(\cdot, y)$ is convex for any given $y \in \Re^m$, then for any set $C \subset \Re^m$, the function $h : \Re^n \to \Re \cup \{\infty\}$ defined by*

$$h(x) := \sup_{y \in C} f(x, y)$$

is also convex (possibly in the extended sense).

(b) *Assume that for any $x \in \Re^n$, there is an associated set $C(x) \subset \Re^m$ and $C := \{(x, y) \mid y \in C(x), x \in \Re^n\}$ is convex. If f is convex and the function*

$$g(x) := \inf_{y \in C(x)} f(x, y)$$

is well defined, then g is also convex over $\Re^n$.

Proof. To prove part (a), observe that for any given $y \in C$, the set

$$\text{epi}(f(\cdot, y)) = \{(x, \alpha) \mid x \in \Re^n, \alpha \in \Re, f(x, y) \leq \alpha\}$$

is convex. Therefore, $\text{epi}(h) = \cap_{y \in C}\text{epi}(f(\cdot, y))$ is a convex set, which implies that h is convex.

For part (b), let us fix $x, x' \in \Re^n$ and $\lambda \in [0, 1]$. From the definition of an infimum, there exists, for any given $\epsilon > 0$, $y, y' \in C$ such that

$$f(x, y) \leq g(x) + \epsilon \text{ and } f(x', y') \leq g(x') + \epsilon. \tag{2.10}$$

Since C is convex, we have that $((1 - \lambda)x + \lambda x', (1 - \lambda)y + \lambda y') \in C$. Thus, $(1 - \lambda)y + \lambda y' \in C((1 - \lambda)x + \lambda x')$ and

$$\begin{aligned} g((1 - \lambda)x + \lambda x') &\leq f((1 - \lambda)x + \lambda x', (1 - \lambda)y + \lambda y') \\ &\leq (1 - \lambda)f(x, y) + \lambda f(x', y') \\ &\leq (1 - \lambda)g(x) + \lambda g(x') + \epsilon, \end{aligned}$$

where the second inequality holds since $f(\cdot, \cdot)$ is convex and the last inequality follows from (2.10). Since ϵ is arbitrary, we conclude that g is convex. ∎

2.2 Supermodularity

In this section, we introduce the concept of supermodularity. Though this concept can be defined on general partially ordered sets, for our purpose we focus on the Euclidean space $\Re^n$ with the standard partial order $\leq$; that is, for any $x, x' \in \Re^n$, $x \leq x'$ if and only if $x_i \leq x'_i$ for $i = 1, 2, \ldots, n$.

To present the definition of supermodularity, we first introduce two operations, *join* and *meet* operations, in $\Re^n$. For any two points $x = (x_1, x_2, \ldots, x_n)$ and $x' = (x'_1, x'_2, \ldots, x'_n)$ in $\Re^n$, define their *join* as

$$x \vee x' = (\max\{x_1, x'_1\}, \max\{x_2, x'_2\}, \ldots, \max\{x_n, x'_n\}),$$

and their *meet* as

$$x \wedge x' = (\min\{x_1, x'_1\}, \min\{x_2, x'_2\}, \ldots, \min\{x_n, x'_n\}).$$

Of course, if $x \leq x'$, namely, $x_i \leq x'_i$ for $i = 1, 2, \ldots, n$, then $x \vee x' = x'$ and $x \wedge x' = x$. A set $X \subseteq \Re^n$ is called a *lattice* if for any $x, x' \in X$, $x \vee x', x \wedge x' \in X$. Note that X is called a *sublattice* of $\Re^n$ in some literature as it inherits the infimum and supremum from $\Re^n$.

Definition 2.2.1 *Suppose X is a lattice in $\Re^n$ and a function $f : X \to \Re$. The function f is supermodular on the set X if, for any $x, x' \in X$,*

$$f(x) + f(x') \le f(x \vee x') + f(x \wedge x'). \tag{2.11}$$

f is strictly supermodular if the inequality (2.11) holds strictly for unordered pairs x and x'; that is, none of $x \le x'$ nor $x \ge x'$ is true. A function f is (strictly) submodular if $-f$ is (strictly) supermodular.

A closely related and more intuitive concept is *(strictly) increasing differences*. Given $X \subseteq \Re^n$ and $T \subseteq \Re^m$, a function $f : X \times T \to \Re$ has *(strictly) increasing differences* if for any $t, t' \in T$ with $t \le t'$ $(t < t')$, $f(x,t') - f(x,t)$ is (strictly) increasing in $x \in X$. Notice that a function $g : X \to \Re$ is called increasing if for any $x, x' \in X$ with $x \le x'$, $g(x) \le g(x')$.

This concept can be extended to functions defined on sets with a product structure $\Pi_{i=1}^n X_i$, where X_i is a subset of some Euclidean space. For this purpose, we define, for a function $f : \Pi_{i=1}^n X_i \to \Re$, any pair of indexes $i, j \in \{1, 2, \ldots, n\}$ and any vector

$$\hat{x}_{ij} = (x_1, \ldots, x_{i-1}, x_{i+1}, \ldots, x_{j-1}, x_{j+1}, \ldots, x_n) \in \Pi_{l=1:n, l \neq i,j} X_l,$$

a function

$$f_{\hat{x}_{ij}}(x_i, x_j) = f(x_1, \ldots, x_{i-1}, x_i, x_{i+1}, \ldots, x_{j-1}, x_j, x_{j+1}, \ldots, x_n).$$

The function $f : \Pi_{i=1}^n X_i \to \Re$ has *(strictly) increasing differences* if the function $f_{\hat{x}_{ij}}(x_i, x_j)$ has *(strictly) increasing differences* for any pair of distinct indexes $i, j \in \{1, 2, \ldots, n\}$ and any vector $\hat{x}_{ij} \in \Pi_{l=1:n, l \neq i,j} X_l$.

The following result shows that for functions defined on $\Re^n$, the concept of supermodularity is equivalent to the property of increasing differences.

Theorem 2.2.2 *A function $f : \Re^n \to \Re$ is (strictly) supermodular if and only if f has (strictly) increasing differences when $\Re^n$ is regarded as the product of n one-dimensional real space.*

Proof. Assume that f has increasing differences. Then, for any $x, x' \in \Re^n$, we have

$$
\begin{aligned}
f(x) - f(x \wedge x') &= \sum_{i=1}^n (f(x_1, \ldots, x_i, x_{i+1} \wedge x'_{i+1}, \ldots, x_n \wedge x'_n) \\
&\quad - f(x_1, \ldots, x_{i-1}, x_i \wedge x'_i, x_{i+1} \wedge x'_{i+1}, \ldots, x_n \wedge x'_n)) \\
&\le \sum_{i=1}^n (f(x_1 \vee x'_1, \ldots, x_{i-1} \vee x'_{i-1}, x_i, x_{i+1}, \ldots, x'_n) \\
&\quad - f(x_1 \vee x'_1, \ldots, x_{i-1} \vee x'_{i-1}, x_i \wedge x'_i, x_{i+1}, \ldots, x'_n)) \\
&= \sum_{i=1}^n (f(x_1 \vee x'_1, \ldots, x_{i-1} \vee x'_{i-1}, x_i \vee x'_i, x'_{i+1}, \ldots, x'_n) \\
&\quad - f(x_1 \vee x'_1, \ldots, x_{i-1} \vee x'_{i-1}, x'_i, x'_{i+1}, \ldots, x'_n)) \\
&= f(x \vee x') - f(x'),
\end{aligned}
$$

where the inequality holds since f has increasing differences and the second equality holds since $\{x_i, x'_i\} = \{x_i \vee x'_i, x_i \wedge x'_i\}$. Hence, f is supermodular.

Assume now that f is supermodular. For any pair of distinct indexes $i, j \in \{1, 2, \ldots, n\}$, any vector

$$\hat{x}_{ij} = (x_1, \ldots, x_{i-1}, x_{i+1}, \ldots, x_{j-1}, x_{j+1}, \ldots, x_n) \in \Re^{n-2}$$

and $x_i, x'_i, x_j, x'_j \in \Re$ with $x_i \leq x'_i$ and $x_j \geq x'_j$, let

$$x = (x_1, \ldots, x_{i-1}, x_i, x_{i+1}, \ldots, x_{j-1}, x_j, x_{j+1}, \ldots, x_n)$$

and

$$x' = (x_1, \ldots, x_{i-1}, x'_i, x_{i+1}, \ldots, x_{j-1}, x'_j, x_{j+1}, \ldots, x_n).$$

The supermodularity of f implies that

$$
\begin{aligned}
f_{\hat{x}_{ij}}(x_i, x_j) - f_{\hat{x}_{ij}}(x_i, x'_j) &= f(x) - f(x \wedge x') \\
&\leq f(x \vee x') - f(x') \\
&= f_{\hat{x}_{ij}}(x'_i, x_j) - f_{\hat{x}_{ij}}(x'_i, x'_j).
\end{aligned}
$$

Thus, f has increasing differences.

Finally, the equivalence of the strict supermodularity and the strictly increasing differences can be established by following a similar argument.   ∎

If a function $f : \Re^n \to \Re$ is differentiable, it is easy to verify that f is supermodular if and only if the partial derivative $\frac{\partial f(x)}{\partial x_i}$ is nondecreasing in x_j for all distinct indexes i and j and for any $x \in \Re^n$. Furthermore, if f is twice differentiable, then f is supermodular if and only if $\frac{\partial^2 f(x)}{\partial x_i \partial x_j} \geq 0$ for any distinct indexes i and j and for any $x \in \Re^n$.

From the definition of supermodularity, we can easily conclude that a separable function $f : \Re^n \to \Re$ is both supermodular and submodular. In fact, the reverse is essentially true.

Theorem 2.2.3 *A function $f : \Re^n \to \Re$ is both supermodular and submodular if and only if f is separable; that is, there exist functions $f_i : \Re \to \Re$ $(i = 1, 2, \ldots, n)$ such that $f(x) = \sum_{i=1}^n f_i(x_i)$ for any $x = (x_1, x_2, \ldots, x_n) \in \Re^n$.*

Proof. The "if" part is obvious, since as we already pointed out, a function defined on $\Re$ is both supermodular and submodular. Hence, we focus on the "only if" part.

Assume that $f : \Re^n \to \Re$ is both supermodular and submodular. Then

$$
\begin{aligned}
f(x) &= f(0) + \sum_{i=1}^n (f(x_1, \ldots, x_{i-1}, x_i, 0, \ldots, 0) - f(x_1, \ldots, x_{i-1}, 0, 0, \ldots, 0)) \\
&= f(0) + \sum_{i=1}^n (f(0, \ldots, 0, x_i, 0, \ldots, 0) - f(0)),
\end{aligned}
$$

where the second equality holds since from Theorem 2.2.2, f has both increasing differences and decreasing differences. Therefore, f is separable.   ∎

In the following, we present some examples of supermodular functions, whose proof is left as an exercise.

Theorem 2.2.4

(a) *The function $f(x,z) = \sum_{i=1}^{n} x_i z_i$ is supermodular on the product space $\Re^n \times \Re^n$, where $x, z \in \Re^n$.*

(b) *The* **Cobb–Douglas** *function $f(x) = x_1^{\alpha_1} x_2^{\alpha_2} \cdots x_n^{\alpha_n}$ for $\alpha_i \geq 0$ is supermodular on the set $\{x | x = (x_1, x_2, \ldots, x_n) \geq 0\}$.*

(c) *The function $f(x,z) = -\sum_{i=1}^{n} |x_i - z_i|^p$ is supermodular on $(x,z) \in \Re^{2n}$ for any $p \geq 1$.*

(d) *If $f_i(z)$ is increasing (decreasing) on $\Re$ for $i = 1, 2, \ldots, n$, then the function $f(x) = \min_{i \in \{1,2,\ldots,n\}} f_i(x_i)$ is supermodular on $\Re^n$.*

We now list below some useful properties about supermodular functions. Some of these properties are similar to Proposition 2.1.3, which deals with convex functions.

Proposition 2.2.5

(a) *Any nonnegative linear combination of supermodular functions is supermodular. That is, if $f_i : \Re^n \to \Re$ $(i = 1, 2, \ldots, m)$ are supermodular, then for any scalar $\alpha_i \geq 0$, $\sum_{i=1}^{m} \alpha_i f_i$ is still supermodular.*

(b) *If f_k is supermodular for $k = 1, 2, \ldots$ and $\lim_{k \to \infty} f_k(x) = f(x)$ for any $x \in \Re^n$, then $f(x)$ is supermodular.*

(c) *A composition of an increasing (decreasing) convex function and an increasing supermodular (submodular) function is still supermodular. That is, if $f : \Re \to \Re$ is convex and nondecreasing (nonincreasing) and $g : \Re^n \to \Re$ is increasing and supermodular (submodular), then $f(g(x))$ is supermodular.*

(d) *Assume that a function $f(\cdot, \cdot)$ is defined in the product space $\Re^n \times \Re^m$. If $f(\cdot, y)$ is supermodular for any given $y \in \Re^m$, then for a random vector ζ in $\Re^m$, $E_\zeta[f(x, \zeta)]$ is supermodular, provided it is well defined.*

(e) *Assume that A is a lattice in $\Re^n \times \Re^m$ and a function $f(\cdot, \cdot) : A \to \Re$ is supermodular. For any $x \in \Re^n$, let $S(x) = \{y \in \Re^m \mid (x,y) \in A\}$ and $\Pi_y(A) = \{x \in \Re^n \mid S(x) \neq \emptyset\}$. Then $\Pi_y(A)$ is a lattice. If the function*

$$g(x) = \sup_{y \in S(x)} f(x,y)$$

is finite on $\Pi_y(A)$, then g is supermodular over $\Pi_y(A)$.

Proof. Parts (a), (b), and (d) follow directly from the definition of supermodular functions.

We now prove part (c). Assume that g is increasing and supermodular and f is convex and nondecreasing. For any $x, x' \in \Re^n$, since g is increasing, we conclude that $g(x \wedge x') \leq g(x), g(x') \leq g(x \vee x')$. Therefore, there exist $\lambda, \lambda' \in [0, 1]$ such that

$$g(x) = (1 - \lambda)g(x \wedge x') + \lambda g(x \vee x') \text{ and } g(x') = (1 - \lambda')g(x \wedge x') + \lambda' g(x \vee x').$$

Adding the two equalities together gives us

$$g(x) + g(x') = (2 - \lambda - \lambda')g(x \wedge x') + (\lambda + \lambda')g(x \vee x').$$

Since g is supermodular, we must have either $g(x \wedge x') = g(x \vee x')$ or $\lambda + \lambda' \leq 1$. In the first case, clearly

$$f(g(x)) + f(g(x')) = f(g(x \wedge x')) + f(g(x \vee x')).$$

In the second case,

$$
\begin{aligned}
f(g(x)) + f(g(x')) &\leq f(g(x \wedge x')) + f(g(x \vee x')) \\
&+ (1 - \lambda - \lambda')(f(g(x \vee x')) - f(g(x \wedge x'))) \\
&\leq f(g(x \wedge x')) + f(g(x \vee x')),
\end{aligned}
$$

where the first inequality follows from the convexity of function f and the second inequality holds since $\lambda + \lambda' \leq 1$, f is nondecreasing, and g is increasing. Hence, $f(g(x))$ is supermodular. Obviously, the above argument holds true when f is convex and nonincreasing and g is increasing and submodular.

For part (e), for any $x, x' \in \Pi_y(A)$ and any $y \in S(x), y' \in S(x')$, we have that

$$(x \wedge x', y \wedge y') \in A \text{ and } (x \vee x', y \vee y') \in A.$$

Thus, $y \wedge y' \in S(x \wedge x')$ and $y \vee y' \in S(x \vee x')$, which imply that $x \wedge x', x \vee x' \in \Pi_y(A)$ and $\Pi_y(A)$ is a lattice. From the supermodularity of f, we have that

$$
\begin{aligned}
f(x, y) + f(x', y') &\leq f(x \wedge x', y \wedge y') + f(x \vee x', y \vee y') \\
&\leq g(x \wedge x') + g(x \vee x').
\end{aligned}
$$

Finally, taking the supremum for $y \in S(x)$ and $y' \in S(x')$ in the left-hand side of the above inequality, we have

$$g(x) + g(x') \leq g(x \wedge x') + g(x \vee x').$$

Thus, g is supermodular on $\Pi_y(A)$.   ∎

The following result establishes some connections between convexity and supermodularity.

Theorem 2.2.6 *Let X be a lattice in $\Re^n$ and $a_i \in \Re$ $(i = 1, 2, \ldots, n)$. For a function $f : \Re \to \Re$, define $g : \Re^n \to \Re$ with $g(x) := f(\sum_{i=1}^n a_i x_i)$ for any $x = (x_1, x_2, \ldots, x_n) \in \Re^n$. We have the following.*

(a) If $a_i \geq 0$ for $i = 1, 2, \ldots, n$, and f is convex, then g is supermodular on X.

(b) *If $n = 2$, $a_1 > 0$, $a_2 < 0$, and f is concave, then g is supermodular on X.*

Suppose, in addition, that for any x, x' with $x \leq x'$, $x \in X$ implies $x' \in X$, $\Re = \{\sum_{i=1}^{n} a_i x_i \mid x \in X\}$, and f is continuous.

(c) *If $n \geq 2$, $a_1 > 0$, $a_2 > 0$, and g is supermodular on X, then f is convex.*

(d) *If $n \geq 2$, $a_1 > 0$, $a_2 < 0$, and g is supermodular on X, then f is concave.*

(e) *If $n \geq 3$, $a_1 > 0$, $a_2 > 0$, $a_3 < 0$, and g is supermodular on X, then f is a linear function.*

Proof. First, observe that for any $x, x' \in \Re^n$,

$$\sum_{i=1}^{n} a_i x_i - \sum_{i=1}^{n} a_i(x_i \wedge x_i') = \sum_{i=1}^{n} a_i(x_i \vee x_i') - \sum_{i=1}^{n} a_i x_i'. \tag{2.12}$$

We now prove part (a). Since $a_i \geq 0$, we have

$$\sum_{i=1}^{n} a_i(x_i \wedge x_i') \leq \sum_{i=1}^{n} a_i x_i, \quad \sum_{i=1}^{n} a_i x_i' \leq \sum_{i=1}^{n} a_i(x_i \vee x_i').$$

Therefore, there exist $\lambda, \lambda' \in [0,1]$ such that

$$\sum_{i=1}^{n} a_i x_i = (1 - \lambda) \sum_{i=1}^{n} a_i(x_i \wedge x_i') + \lambda \sum_{i=1}^{n} a_i(x_i \vee x_i')$$

and

$$\sum_{i=1}^{n} a_i x_i' = (1 - \lambda') \sum_{i=1}^{n} a_i(x_i \wedge x_i') + \lambda' \sum_{i=1}^{n} a_i(x_i \vee x_i').$$

Moreover, (2.12) implies $\lambda + \lambda' = 1$. Thus, from the convexity of f, we have that

$$
\begin{aligned}
g(x) + g(x') &= f(\textstyle\sum_{i=1}^{n} a_i x_i) + f(\textstyle\sum_{i=1}^{n} a_i x_i') \\
&\leq f(\textstyle\sum_{i=1}^{n} a_i(x_i \wedge x_i')) + f(\textstyle\sum_{i=1}^{n} a_i(x_i \vee x_i')) \\
&= g(x \wedge x') + g(x \vee x').
\end{aligned}
$$

Hence, g is supermodular on X.

For part (b), we argue that for any $x = (x_1, x_2)$ and $x' = (x_1', x_2')$,

$$
\begin{aligned}
f(a_1 x_1 + a_2 x_2) &- f(a_1(x_1 \wedge x_1') + a_2(x_2 \wedge x_2')) \\
&\leq f(a_1(x_1 \vee x_1') + a_2(x_2 \vee x_2')) - f(a_1 x_1' + a_2 x_2').
\end{aligned} \tag{2.13}
$$

If $x \leq x'$ or $x \geq x'$, it is obvious that the inequality (2.13) holds true. Otherwise, assume, without loss of generality, that $x_1 = x_1 \vee x_1'$ and $x_2 = x_2 \wedge x_2'$. We have, from (2.12), that

$$\sum_{i=1}^{2} a_i x_i - \sum_{i=1}^{2} a_i(x_i \wedge x_i') = \sum_{i=1}^{2} a_i(x_i \vee x_i') - \sum_{i=1}^{2} a_i x_i' \geq 0.$$

It is also easy to verify that

$$a_1x_1 + a_2x_2 \geq a_1(x_1 \vee x_1') + a_2(x_2 \vee x_2').$$

Hence, Proposition 2.1.6, together with the above inequality, implies the inequality (2.13), and thus g is supermodular.

To prove part (c), fix any $z, z' \in \Re$ with $z \leq z'$. Choose $x \in \Re^n$ such that $\sum_{i=1}^n a_i x_i = z$. Let $\epsilon = (z' - z)/(2a_1) \geq 0$ and $\delta = (z' - z)/(2a_2) \geq 0$. Also, let $y = x + \epsilon e_1$, $y' = x + \delta e_2$, and $x' = x + \epsilon e_1 + \delta e_2$, where e_i is the unit vector with 0 at all components except 1 at its ith component. From the definitions of ϵ and δ, we have that $\sum_{i=1}^n a_i y_i = \sum_{i=1}^n a_i y_i' = (z + z')/2$ and $\sum_{i=1}^n a_i x_i' = z'$. Hence, the supermodularity of g implies that

$$f((z + z')/2) - 1/2(f(z) + f(z')) = 1/2(g(y) + g(y') - g(x) - g(x')) \leq 0,$$

since $x = y \wedge y'$ and $x' = y \vee y'$. Thus, the convexity of the function f follows from this inequality and the continuity of f.

We now prove part (d). Fix any $z, z' \in \Re$ with $z \leq z'$. Choose $x \in \Re^n$ such that $\sum_{i=1}^n a_i x_i = (z + z')/2$. Let $\epsilon = (z' - z)/(2a_1) \geq 0$ and $\delta = -(z' - z)/(2a_2) \geq 0$. Also, let $y = x + \epsilon e_1$, $y' = x + \delta e_2$, and $x' = x + \epsilon e_1 + \delta e_2$. From the definitions of ϵ and δ, we have that $\sum_{i=1}^n a_i y_i = z'$, $\sum_{i=1}^n a_i y_i' = z$, and $\sum_{i=1}^n a_i x_i' = (z + z')/2$. Hence, the supermodularity of g implies that

$$f((z + z')/2) - 1/2(f(z) + f(z')) = 1/2(g(x) + g(x') - g(y) - g(y')) \geq 0,$$

since $x = y \wedge y'$ and $x' = y \vee y'$. Thus, the concavity of the function f follows from this inequality and the continuity of f.

Finally, if $n \geq 3$, $a_1 > 0$, $a_2 > 0$, and $a_3 < 0$, the proof for parts (c) and (d) implies that f is both convex and concave, and hence f is linear. ∎

One of the most widely used properties associated with supermodularity is on the monotonicity of the sets of optimal solutions of a class of parametric optimization problems. To present this property, define a new concept of *increasing set function*. Let $S(t)$ be a set function in $\Re^n$ parameterized by $t \in T \subset \Re^m$; that is, for a parameter $t \in T$, $S(t)$ is a subset of $\Re^n$. The set function $S(t)$ is called increasing in t if for any $t, t' \in T$ with $t \leq t'$, $x \in S(t)$, and $x' \in S(t')$, we have that $x \wedge x' \in S(t)$ and $x \vee x' \in S(t')$.

The concept of increasing set functions is different from set inclusion. To see this, notice that $S(t) = [t, +\infty)$ is an increasing set function for $t \in \Re$, but $S(t') \subset S(t)$ for $t < t'$. However, for an increasing set function S, it is straightforward to show that given $t, t' \in T$ with $t \leq t'$, for any $x \in S(t)$, there exists a point $x' \in S(t')$ and for any $x' \in S(t')$, there exists a point $x \in S(t)$ such that $x \leq x'$.

Proposition 2.2.7 *Let $S(t)$ be an increasing set function in $\Re^n$ parameterized by $t \in T \subset \Re^m$. We have that $S(t)$ is a lattice of $\Re^n$ for any $t \in T$. If, in addition, $S(t)$ is nonempty and compact for any $t \in T$, then $S(t)$ has a largest element and a smallest element, which are increasing in t, respectively.*

Proof. It is straightforward to see from the definition of increasing set function that $S(t)$ is a lattice of $\Re^n$ for any $t \in T$. We now show that $S(t)$ has a largest and a smallest element if $S(t)$ is compact. Let $x^* = (x_1^*, x_2^*, \ldots, x_n^*) \in \Re^n$ with $x_i^* = \sup_{x \in S(t)} x_i$, $i = 1, 2, \ldots, n$. Since $S(t)$ is compact, there exists $x^i \in S(t)$ such that $x_i^i = x_i^*$, $i = 1, 2, \ldots, n$. Hence, $x^* = x^1 \vee x^2 \vee \ldots \vee x^n \in S(t)$ since $S(t)$ is a lattice. Obviously, x^* is the largest element of $S(t)$.

To show that the largest element of $S(t)$, denoted as $\bar{x}(t)$, is increasing, note that for any $t, t' \in T$ with $t \leq t'$, $\bar{x}(t') \leq \bar{x}(t) \vee \bar{x}(t') \in S(t')$. Since $\bar{x}(t')$ is the largest element of $S(t')$, we must have $\bar{x}(t') = \bar{x}(t) \vee \bar{x}(t') \geq \bar{x}(t)$.

The properties regarding the smallest element of $S(t)$ can be established similarly. ∎

Under some supermodularity assumptions, the sets of optimal solutions for a collection of optimization problems parameterized by a parameter are increasing in the parameter. In addition, for a given parameter, there exist a largest and a smallest optimal solution, which are increasing in the parameter as well. Consider the parametric optimization problem

$$\max_{x \in S(t)} g(x, t).$$

Let $A := \{(x, t) \mid t \in T, x \in S(t)\} \subset \Re^n \times \Re^m$, and $S^*(t) := \operatorname{argmax}_{x \in S(t)} g(x, t)$.

Theorem 2.2.8 *Assume that $S(t)$ is increasing in $t \in T$, and $g(x, t) : A \to \Re$ is supermodular in x for any fixed $t \in T$ and has increasing differences in (x, t).*

(a) $S^*(t)$ is increasing in t on $\{t \in T \mid S^*(t) \neq \emptyset\}$.

(b) Assume, in addition, that $S(t)$ is a nonempty and compact set of $\Re^n$ for any $t \in T$, and $g(x, t)$ is continuous in x on $S(t)$ for any $t \in T$. Then $S^*(t)$ is a nonempty and compact lattice and thus there exist $\bar{x}(t), \underline{x}(t) \in S^*(t)$ such that for any $x \in S^*(t)$, $\underline{x}(t) \leq x \leq \bar{x}(t)$. Furthermore, $\bar{x}(t)$ and $\underline{x}(t)$ are increasing.

Proof. To prove part (a), pick any $t, t' \in T$ with $t \leq t'$ such that $S^*(t)$ and $S^*(t')$ are nonempty. For any $x \in S^*(t)$ and $x' \in S^*(t')$, $x \wedge x' \in S(t)$ and $x \vee x' \in S(t')$, since $S(t)$ is increasing in t. We have that

$$
\begin{aligned}
0 &\geq g(x \vee x', t') - g(x', t') \\
&\geq g(x, t') - g(x \wedge x', t') \\
&\geq g(x, t) - g(x \wedge x', t) \\
&\geq 0,
\end{aligned}
$$

where the first and last inequalities follow from the optimality of x and x' for t and t', respectively, the second inequality from the supermodularity of g in x for any fixed t, and the third inequality from the increasing differences property of

g in (x, t). Thus, $x \wedge x' \in S^*(t)$ and $x \vee x' \in S^*(t')$, which implies that $S^*(t)$ is increasing in t on $\{t \in T \mid S^*(t) \neq \emptyset\}$.

If $g(x, t)$ is continuous in x and $S(t)$ is nonempty and compact, the set of optimal solutions $S^*(t)$ is also nonempty and compact for any $t \in T$. Part (b) follows Proposition 2.2.7 directly.   ∎

To conclude this subsection, we present an extension of a fixed-point theorem due to Tarski (1955), which will be useful to prove the existence of a Nash equilibrium of supermodular games in the next section.

Theorem 2.2.9 *Let C be a nonempty and compact lattice of $\Re^n$. If S is an increasing set function mapping any point $x \in C$ to a nonempty and compact lattice that itself is a subset of C, then the set of fixed points $E = \{x \in C : x \in S(x)\}$ is nonempty and has a largest and a smallest fixed point. If, in addition, S depends on a parameter $t \in T \subset \Re^m$ and for fixed $x \in C$, $S(x, t)$ is increasing in t, then the largest and the smallest fixed points increase in t.*

Proof. Define a set

$$\bar{C} = \{x \in C : \exists y \in S(x) \text{ such that } y \geq x\}.$$

The set $\bar{C}$ is nonempty since it includes the smallest element of C, which is guaranteed to exist by Proposition 2.2.7.

Let $\bar{x} = \sup\{x : x \in \bar{C}\}$. Since C is a compact set, $\bar{x}$ is well defined. We show that $\bar{x}$ is a fixed point. Note that for $x \in \bar{C}$, there exists a $y_x \in S(x)$ such that $y_x \geq x$. Since S is increasing and $x \leq \bar{x}$ for any $x \in \bar{C}$, there exists a point $\bar{y}_x \in S(\bar{x})$ such that $\bar{y}_x \geq y_x$. Therefore, $\bar{y} = \sup\{\bar{y}_x : x \in \bar{C}\} \in S(\bar{x})$ since $S(\bar{x})$ is a compact lattice. Because $y_x \geq x$ for any $x \in \bar{C}$, we have that $\bar{y} \geq \bar{x}$. Thus, there exists a point $z \in S(\bar{y})$ such that $z \geq \bar{y}$ and $\bar{y} \in \bar{C}$. However, $\bar{x}$ is the least upper bound of $\bar{C}$ and $\bar{y} \geq \bar{x}$. We must have that $\bar{x} = \bar{y} \in S(\bar{x})$. Thus, E is nonempty and $\bar{x}$ is its largest element. Similarly, we can show that $\underline{x} = \inf\{x \in C : \exists y \in S(x) \text{ such that } y \leq x\}$ is the smallest fixed point in E.

If S also depends on a parameter $t \in T$, define for a given $t \in T$

$$\bar{C}(t) = \{x \in C : \exists y \in S(x, t) \text{ such that } y \geq x\}.$$

From the argument in the previous paragraph, it suffices to show that $\bar{C}(t)$ is increasing in t if $S(x, t)$ is increasing in t for any fixed $x \in C$. To see this, pick two points $t, t' \in T$ with $t \leq t'$ and $x \in \bar{C}(t)$ and $x' \in \bar{C}(t')$. By definition, there exist $y \in S(x, t)$ and $y' \in S(x', t')$ such that $y \geq x$ and $y' \geq x'$. Since $S(x, t)$ is increasing in t for any fixed $x \in C$, there exists a point $z \in S(x, t')$ such that $z \geq y \geq x$. Therefore, $x \vee x' \leq z \vee y' \in S(x \vee x', t')$ and $x \vee x' \in \bar{C}(t')$. Similarly, we can show that $x \wedge x' \in \bar{C}(t)$. Thus, $\bar{C}(t)$ is increasing in t.   ∎

2.3 Discrete Convex Analysis

In this section, we give a brief introduction to discrete convex analysis. Some of the concepts and results there have been demonstrated useful to inventory models, and we expect they can find more applications in supply chain management.

Discrete convex analysis can be treated as an extension of convex analysis to combinatorial structures. It aims at building a unified theoretical framework for tractable discrete optimization problems. Two key concepts, $L^\natural$-convexity and $M^\natural$-convexity defined on either real variables or integer variables, play prominent roles (here L and M stand for "lattice" and "matroid," respectively). In the following, we use the notation $\mathcal{F}$ to denote either the real space $\Re$ or the set with all integers $\mathcal{Z}$ and $\mathcal{F}_+$ to denote the set of nonnegative elements in $\mathcal{F}$. Recall $\bar{\Re} = \Re \cup +\infty$.

2.3.1 $L^\natural$-Convexity

Definition 2.3.1 ($L^\natural$-Convexity) *A function $f : \mathcal{F}^n \to \bar{\Re}$ is $L^\natural$-convex if for any $x, x' \in \mathcal{F}^n$, $\alpha \in \mathcal{F}_+$,*

$$f(x) + f(x') \geq f((x + \alpha e) \wedge x') + f(x \vee (x' - \alpha e)), \qquad (2.14)$$

where e is the n-dimensional all-ones vector. A function f is $L^\natural$-concave if $-f$ is $L^\natural$-convex.

Inequality (2.14) is called *translation submodular*. Notice that in the above definition, if $f(x) = +\infty$ or $f(x') = +\infty$, inequality (2.14) is assumed to hold automatically. Thus, for an $L^\natural$-convex function f, its effective domain $\mathcal{V} = \text{dom}(f) = \{x \in \mathcal{F}^n | f(x) < +\infty\}$ is an $L^\natural$-convex set; that is, it satisfies the following condition:

$$\forall\, x, x' \in \mathcal{V} \text{ and } \alpha \in \mathcal{F}_+, (x + \alpha e) \wedge x' \in \mathcal{V} \text{ and } x \vee (x' - \alpha e) \in \mathcal{V}.$$

Let $\delta_\mathcal{V} : \mathcal{F}^n \to \bar{\Re}$ be the indicator function of a given set $\mathcal{V} \subset \mathcal{F}^n$; that is, $\delta_\mathcal{V}(x) = 0$ if $x \in \mathcal{V}$ and $+\infty$ otherwise. It is easy to verify that the set $\mathcal{V}$ is $L^\natural$-convex if and only if its indicator function $\delta_\mathcal{V}$ is $L^\natural$-convex.

We sometimes say a function f is $L^\natural$-convex on a set $\mathcal{V} \subset \mathcal{F}^n$ with the understanding that $\mathcal{V}$ is an $L^\natural$-convex set and the extension of f to the whole space $\mathcal{F}^n$ by defining $f(x) = +\infty$ for $x \notin \mathcal{V}$ is $L^\natural$-convex. It is also straightforward to show that an $L^\natural$-convex function f restricted to an $L^\natural$-convex set $\mathcal{V}$ by defining $f(x) = +\infty$ for $x \notin \mathcal{V}$ is also $L^\natural$-convex. From the definition of $L^\natural$-convexity, it is clear that if $f : \Re^n \to \bar{\Re}$ is $L^\natural$-convex, then $f_\mathcal{Z} : \mathcal{Z}^n \to \bar{\Re}$ is also $L^\natural$-convex, where $f_\mathcal{Z}(x) = f(x)$ for any $x \in \mathcal{Z}^n$.

We have the following equivalent definition of $L^\natural$-convexity.

Proposition 2.3.2 *A function $f : \mathcal{F}^n \to \bar{\Re}$ is $L^\natural$-convex if and only if $g(x, \xi) := f(x - \xi e)$ is submodular on $(x, \xi) \in \mathcal{F}^n \times \mathcal{S}$, where $\mathcal{S}$ is the intersection of $\mathcal{F}$ and any unbounded interval in $\Re$.*

Proof. Assume that f is $L^\natural$-convex. Consider any two vectors $(x, \xi) \in \mathcal{F}^n \times \mathcal{S}$ and $(x', \xi') \in \mathcal{F}^n \times \mathcal{S}$. Without loss of generality, assume that $\xi \geq \xi'$. Let $\alpha = \xi - \xi'$. We have that $\alpha \in \mathcal{F}_+$ and

$$
\begin{aligned}
g(x, \xi) + g(x', \xi') &= f(x - \xi e) + f(x' - \xi' e) \\
&\geq f((x - \xi e + \alpha e) \wedge (x' - \xi' e)) \\
&\quad + f((x - \xi e) \vee (x' - \xi' e - \alpha e)) \\
&= f(x \wedge x' - \xi' e) + f(x \vee x' - \xi e) \\
&= g(x \wedge x', \xi \wedge \xi') + g(x \vee x', \xi \vee \xi'),
\end{aligned}
$$

where the inequality follows from the $L^\natural$-convexity of f and the second equality from the definition of α. Thus, g is submodular.

On the other hand, assume that g is submodular. For any $x, x' \in \mathcal{F}^n$ and $\alpha \in \mathcal{F}_+$, since $\mathcal{S}$ is the intersection of $\mathcal{F}$ and an unbounded interval in $\Re$, it is clear that we can find a pair $\xi \in \mathcal{S}$ and $\xi' \in \mathcal{S}$ such that $\xi = \alpha + \xi'$. We have that

$$
\begin{aligned}
f(x) + f(x') &= g(x{+}\xi e, \xi){+}g(x'{+}\xi'e, \xi') \\
&\geq g((x{+}\xi e) \wedge (x'{+}\xi'e), \xi \wedge \xi'){+}g((x{+}\xi e) \vee (x'{+}\xi'e), \xi \vee \xi') \\
&= g((x{+}\xi'e{+}\alpha e) \wedge (x'{+}\xi'e), \xi'){+}g((x{+}\xi e) \vee (x'{+}\xi e - \alpha e), \xi) \\
&= f((x + \alpha e) \wedge x') + f(x \vee (x' - \alpha e)),
\end{aligned}
$$

where the inequality follows from the submodularity of g. Hence, f is $L^\natural$-convex. ∎

In the following we present some examples of $L^\natural$-convex functions and $L^\natural$-convex sets, whose proof is left as an exercise.

Proposition 2.3.3

(a) *Given any univariate convex functions $g_i : \Re \to \bar{\Re}$ $(i = 1, \cdots, n)$ and $h_{ij} : \Re \to \bar{\Re}$ $(i \neq j)$, the function $f : \Re^n \to \bar{\Re}$ defined by*

$$
f(x) := \sum_{i=1}^n g_i(x_i) + \sum_{i \neq j} h_{ij}(x_i - x_j)
$$

is $L^\natural$-convex. As a special case, any linear function is $L^\natural$-convex.

(b) *A quadratic function $f : \Re^n \to \Re$ defined by $f(x) = \sum_{i,j=1}^n a_{ij} x_i x_j$ with $a_{ij} = a_{ji} \in \Re$ is $L^\natural$-convex if and only if the matrix A with its ijth component being a_{ij} is a diagonally dominated M-matrix; that is,*

$$
a_{ij} \leq 0, \forall\, i \neq j, a_{ii} \geq 0, \ \text{and} \ \sum_{j=1}^n a_{ij} \geq 0, \forall\, i.
$$

(c) *A twice continuously differentiable function $f : \Re^n \to \Re$ is $L^\natural$-convex if and only if its Hessian is a diagonally dominated M-matrix.*

(d) *For a given vector $a \in \Re^n$ and a nondecreasing univariate function $f : \Re \to \Re$, the function $g : \Re^n \to \Re$ defined by $g(x) = f(\max_{i=1:n}\{a_i + x_i\})$ is $L^\natural$-convex.*

(e) *A set with a representation $\{x \in \mathcal{F}^n : l \leq x \leq u, x_i - x_j \leq v_{ij}, \forall\, i \neq j\}$, is $L^\natural$-convex in the space $\mathcal{F}^n$, where $l, u \in \mathcal{F}^n$ and $v_{ij} \in \mathcal{F}$ $(i \neq j)$. In fact, any closed $L^\natural$-convex set in the space $\mathcal{F}^n$ can have such a representation.*

We now list below some useful preservation properties about $L^\natural$-convex functions. Some of these properties are similar to but more restrictive compared with those in Proposition 2.2.5, which deals with supermodular functions.

Proposition 2.3.4

(a) *Any nonnegative linear combination of $L^\natural$-convex functions is $L^\natural$-convex. That is, if $f_i : \mathcal{F}^n \to \bar{\Re}$ ($i = 1, 2, \ldots, m$) are $L^\natural$-convex, then for any scalar $\alpha_i \geq 0$, $\sum_{i=1}^m \alpha_i f_i$ is also $L^\natural$-convex.*

(b) *If f_k is $L^\natural$-convex for $k = 1, 2, \ldots$ and $\lim_{k\to\infty} f_k(x) = f(x)$ for any $x \in \mathcal{F}^n$, then $f(x)$ is $L^\natural$-convex.*

(c) *Assume that a function $f(\cdot, \cdot)$ is defined on the product space $\mathcal{F}^n \times \mathcal{F}^m$. If $f(\cdot, y)$ is $L^\natural$-convex for any given $y \in \mathcal{F}^m$, then for a random vector ζ in $\mathcal{F}^m$, $E_\zeta[f(x, \zeta)]$ is $L^\natural$-convex, provided it is well defined.*

(d) *If $f : \mathcal{F}^n \to \bar{\Re}$ is an $L^\natural$-convex function, then $g : \mathcal{F}^n \times \mathcal{F} \to \bar{\Re}$ defined by $g(x, \xi) = f(x - \xi e)$ is also $L^\natural$-convex.*

(e) *Assume that A is an $L^\natural$-convex set of $\mathcal{F}^n \times \mathcal{F}^m$ and $f(\cdot, \cdot) : \mathcal{F}^n \times \mathcal{F}^m \to \bar{\Re}$ is an $L^\natural$-convex function. Then the function*

$$g(x) = \inf_{y:(x,y)\in A} f(x, y)$$

is $L^\natural$-convex over $\mathcal{F}^n$ if $g(x) \neq -\infty$ for any $x \in \mathcal{F}^n$.

Proof. Parts (a), (b), and (c) are straightforward; thus, we only prove parts (d) and (e). We first prove part (d). To show that $g(x, \xi)$ is $L^\natural$-convex on $\mathcal{F}^n \times \mathcal{F}$, notice that for any $(x, \xi, \zeta) \in \mathcal{F}^n \times \mathcal{F} \times \mathcal{F}$, $g(x - \zeta e, \xi - \zeta) = g(x, \xi)$, which is independent of ζ and submodular in (x, ξ). Therefore, from Proposition 2.3.2, $g(x, \xi)$ is $L^\natural$-convex.

To prove part (e), we assume without loss of generality that $A = \mathcal{F}^n \times \mathcal{F}^m$; otherwise, we can focus on the restriction of f on A. From the definition of g, we have

$$g(x - \xi e) = \inf_{y \in \mathcal{F}^m} f(x - \xi e, y) = \inf_{y \in \mathcal{F}^m} f(x - \xi e, y - \xi e).$$

Since f is $L^\natural$-convex, from Proposition 2.3.2, the function $f(x - \xi e, y - \xi e)$ is submodular in $(x, y, \xi) \in \mathcal{F}^n \times \mathcal{F}^m \times \mathcal{F}$. The preservation of the supermodularity property in Proposition 2.2.5 part (e) then implies that $g(x - \xi e)$ is submodular for $(x, \xi) \in \mathcal{F}^n \times \mathcal{F}$, and thus from Proposition 2.3.2, g is $L^\natural$-convex. ∎

Since an $L^\natural$-convex function is submodular, we can prove monotonicity properties of optimal solutions similar to the ones in Theorem 2.2.8. In a simple setting presented below, one can also show that the optimal solutions have bounded sensitivities. The result was first established by Zipkin (2008) to analyze the structural properties of stochastic inventory models with lost sales.

Lemma 2.3.5 *Let $g(x, \xi) : \mathcal{F}^n \times \mathcal{F} \to \bar{\Re}$ be $L^\natural$-convex, and let $\xi(x)$ be the largest optimal solution (assuming existence) of the optimization problem $f(x) = \min_{\xi \in \mathcal{F}} g(x, \xi)$ for any $x \in dom(f)$. Then $\xi(x)$ is nondecreasing in $x \in dom(f)$, but $\xi(x + \omega e) \leq \xi(x) + \omega$ for any $\omega > 0$ with $\omega \in \mathcal{F}$ and $\xi(x) + \omega \in dom(f)$.*

Proof. The statement that $\xi(x)$ is nondecreasing in $x \in \text{dom}(f)$ follows directly from Theorem 2.2.8. We now prove that $\xi(x + \omega e) \leq \xi(x) + \omega$. Define

$$g^e(x, \xi, \omega) = g(x - \omega e, \xi - \omega).$$

The $L^\natural$-convexity of function $g(x, \xi)$, together with Proposition 2.3.2, implies the submodularity of function $g^e(x, \xi, \omega)$. For any given $\xi' \in \Re$ with $\xi' > \xi(x) + \omega$, consider two vectors $(x, \xi' - \omega, -\omega)$ and $(x, \xi(x), 0)$. Since $\omega > 0$ and $\xi' - \omega > \xi(x)$,

$$(x, \xi' - \omega, -\omega) \wedge (x, \xi(x), 0) = (x, \xi(x), -\omega), \quad (x, \xi' - \omega, -\omega) \vee (x, \xi(x), 0) = (x, \xi' - \omega, 0),$$

and we have that for any $\xi' > \xi(x) + \omega$,

$$
\begin{aligned}
g(x + \omega e, \xi') - g(x + \omega e, \xi(x) + \omega) &= g^e(x, \xi' - \omega, -\omega) - g^e(x, \xi(x), -\omega) \\
&\geq g^e(x, \xi' - \omega, 0) - g^e(x, \xi(x), 0) \\
&= g(x, \xi' - \omega) - g(x, \xi(x)) \\
&> 0,
\end{aligned}
$$

where the first inequality follows from the submodularity of $g^e(x, \xi, \omega)$ and the last inequality from the assumption of $\xi(x)$ and ξ'. The above inequalities imply that ξ' cannot be optimal for the optimization problem $\min_{\xi \in \mathcal{F}} g(x + \omega e, \xi)$ for any $\xi' > \xi(x) + \omega$. ∎

2.3.2 $M^\natural$-Convexity

We now introduce the $M^\natural$-convexity. For a given $x \in \mathcal{F}^n$, define its positive support set

$$\text{supp}^+(x) = \{ i \mid x_i > 0 \}.$$

Let $e_i \in \mathcal{F}^n$ be the unit vector with 1 at its ith component for $i = 1, 2, \ldots, n$ and e_0 be the vector zero.

Definition 2.3.6 ($M^\natural$-Convexity) *A function $f : \mathcal{F}^n \to \bar{\Re}$ is $M^\natural$-convex if the exchange property holds; that is, for any $x, x' \in \text{dom}(f)$, $i \in \text{supp}^+(x - x')$, there exist an index $j \in \text{supp}^+(x' - x) \cup \{0\}$ and a positive number $\alpha_0 \in \mathcal{F}$ such that*

$$f(x) + f(x') \geq f(x - \alpha(e_i - e_j)) + f(x' + \alpha(e_i - e_j)) \tag{2.15}$$

for any $\alpha \in \mathcal{F}$ with $0 \leq \alpha \leq \alpha_0$. A function f is $M^\natural$-concave if $-f$ is $M^\natural$-convex.

Similar to $L^\natural$-convexity, a set $\mathcal{V} \in \mathcal{F}^n$ is called an $M^\natural$-convex set if its indicator function $\delta_\mathcal{V}$ is $M^\natural$-convex. It is clear that the effective domain of an $M^\natural$-convex function is $M^\natural$-convex. We can also prove that $M^\natural$-convexity is preserved when taking the limit of a sequence of $M^\natural$-convex functions.

Unfortunately, unlike $L^\natural$-convexity, $M^\natural$-convexity is less amicable for analysis. In fact, an $M^\natural$-convex function restricted to an $M^\natural$-convex set is not necessarily $M^\natural$-convex, and neither is the sum of two $M^\natural$-convex functions.

A subclass of $M^\natural$-convex functions, relatively easier to deal with, is the so-called laminar convex functions, defined as the sum of univariate convex functions indexed by a laminar family.

Definition 2.3.7 *A nonempty set $\mathcal{L}$ consisting of subsets of $\{1, 2, \ldots, n\}$ is called a laminar family if for any $X, Y \in \mathcal{L}$, $X \cap Y = \emptyset$ or $X \subseteq Y$ or $X \supseteq Y$. A function $f : \mathcal{F}^n \to \bar{\Re}$ is a laminar convex function if it can be represented as*

$$f(x) = \sum_{S \in \mathcal{L}} f_S(\sum_{i \in S} x_i),$$

where f_S are univariate convex functions and $\mathcal{L}$ is a laminar family.

We now present a few facts about $M^\natural$-convex functions and laminar convex functions, whose proof is left as an exercise.

Theorem 2.3.8

(a) *A laminar convex function $f : \mathcal{F}^n \to \bar{\Re}$ is $M^\natural$-convex.*

(b) *Any separable convex function, defined as the sum of univariate convex functions, is laminar convex and thus $M^\natural$-convex. As a special case, any linear function is laminar convex and $M^\natural$-convex.*

(c) *The intersection of the sets of $L^\natural$-convex and $M^\natural$-convex functions is exactly the set of separable convex functions.*

(d) *A quadratic function $f : \Re^n \to \Re$ defined by $f(x) = \sum_{i,j=1}^n a_{ij} x_i x_j$ with $a_{ij} = a_{ji} \in \Re$ is $M^\natural$-convex in $\mathcal{Z}^n$ if and only if it is laminar convex, which is also equivalent to the following condition:*

$$a_{ij} \geq 0, \forall\, i, j = 1, \ldots, n, \text{ and } a_{ij} \geq \min\{a_{ik}, a_{jk}\} \text{ for } k \notin \{i, j\}. \quad (2.16)$$

(e) *A quadratic function $f : \Re^n \to \Re$ defined by $f(x) = \sum_{i,j=1}^n a_{ij} x_i x_j$ with $a_{ij} = a_{ji} \in \Re$ is $M^\natural$-convex in $\Re^n$ if and only if for any $\gamma > 0$, $A + \gamma I$ is nonsingular and $(A + \gamma I)^{-1}$ is a diagonally dominated M-matrix, where A is a matrix with a_{ij} at its ijth component.*

(f) *The infimal convolution of two $M^\natural$-convex functions $f_1, f_2 : \mathcal{F}^n \to \bar{\Re}$, defined as*

$$g(x) = \min_{u,v \in \mathcal{F}^n, u+v=x} (f_1(u) + f_2(v)),$$

remains $M^\natural$-convex in $\mathcal{F}^n$ if $g(x) > -\infty$ for any $x \in \mathcal{F}^n$.

$L^\natural$-convex functions and $M^\natural$-convex functions enjoy nice properties for optimization. For example, as we present below, these functions defined on $\mathcal{Z}^n$ are convex extendable (its proof is not trivial; readers are referred to Murota 2003), and similar to Theorem 2.1.12, a local minimizer of an $L^\natural$-convex function or an $M^\natural$-convex function is guaranteed to be a global minimizer. We refer to Murota (2003) for efficient algorithms for the minimization of $L^\natural$-convex functions and $M^\natural$-convex functions and elegant duality properties between the two classes of functions.

Theorem 2.3.9 $L^\natural$-convex functions and $M^\natural$-convex functions defined on $\mathcal{Z}^n$ are convex extendable. That is, for any $L^\natural$-convex or $M^\natural$-convex function $f : \mathcal{Z}^n \to \bar{\Re}$, there exists a convex function $f^e : \Re^n \to \bar{\Re}$ such that $f^e(x) = f(x)$ for any $x \in \mathcal{Z}^n$.

Theorem 2.3.10

(a) Assume that $f : \mathcal{Z}^n \to \bar{\Re}$ is $L^\natural$-convex. A vector $x^* \in dom(f)$ is a global minimizer of f if and only if it is a local minimizer in the sense that

$$f(x^*) \le \min\{f(x^* + v), f(x^* - v)\}, \text{ for any } v \in \{0,1\}^n.$$

(b) Assume that $f : \mathcal{Z}^n \to \bar{\Re}$ is $M^\natural$-convex. A vector $x^* \in dom(f)$ is a global minimizer of f if and only if it is a local minimizer in the sense that

$$f(x^*) \le \min\{f(x^* - e_i + e_j)\}, \text{ for any } i,j \in \{0,1,\dots,n\}.$$

Proof. The "if" direction in both parts (a) and (b) is straightforward. We now prove the "only if" direction of part (a) by induction on $\|x - x^*\|_1$, where for any given $x \in \Re^n$, $\|x\|_1 = \sum_{i=1}^n |x_i|$. For any $x \in \mathcal{Z}^n$ with $\|x - x^*\|_1 = 1$, we have from the definition of local minimizers that $f(x^*) \le f(x)$ since $x = x^* + e_i$ or $x = x^* - e_i$ for some i. Assume that $f(x^*) \le f(x)$ for any $x \in \mathcal{Z}^n$ with $\|x - x^*\|_1 \le k$. Consider any $x \in \mathcal{Z}^n$ with $\|x - x^*\|_1 = k+1$. If x^* and x are unordered, it is easy to verify that

$$\|x \wedge x^* - x^*\|_1 < \|x - x^*\|_1 \text{ and } \|x \vee x^* - x^*\|_1 < \|x - x^*\|_1,$$

and by the induction assumption, $f(x \wedge x^*) \ge f(x^*)$ and $f(x \vee x^*) \ge f(x^*)$, which together with the definition of $L^\natural$-convexity imply that

$$f(x) \ge f(x \wedge x^*) + f(x \vee x^*) - f(x^*) \ge f(x^*).$$

If x^* and x are ordered, without loss of generality assume that $x \le x^*$. It is clear that $x \le (x + e) \wedge x^* \le x^*$ and $x \le x \vee (x^* - e) \le x^*$. If $x = x \vee (x^* - e)$, we immediately have from the definition of local minimizers that $f(x^*) \le f(x)$. If $x \ne x \vee (x^* - e)$, we have that

$$\|(x + e) \wedge x^* - x^*\|_1 < \|x - x^*\|_1 \text{ and } \|x \vee (x^* - e) - x^*\|_1 < \|x - x^*\|_1,$$

and by the induction assumption,

$$f((x + e) \wedge x^*) \ge f(x^*) \text{ and } f(x \vee (x^* - e)) \ge f(x^*).$$

Again from the definition of $L^\natural$-convexity, we have that

$$f(x) \ge f((x + e) \wedge x^*) + f(x \vee (x^* - e)) - f(x^*) \ge f(x^*).$$

We now prove the "only if" direction of part (b) by induction on $\|x - x^*\|_1$. For any $x \in \mathcal{Z}^n$ with $\|x - x^*\|_1 = 1$, it is clear from the definition of local minimizers that $f(x^*) \le f(x)$ since $x = x^* + e_i$ or $x = x^* - e_i$ for some i. Assume that

$f(x^*) \leq f(x)$ for any $x \in \mathcal{Z}^n$ with $\|x - x^*\|_1 \leq k$. Consider any $x \in \mathcal{Z}^n$ with $\|x - x^*\|_1 = k + 1$. We have that there exist some $i \in \text{supp}(x - x^*) \cup \{0\}$ and $j \in \text{supp}(x^* - x) \cup \{0\}$ with $i \neq j$ such that $\|x - (e_i - e_j) - x^*\|_1 < \|x - x^*\|_1$, and thus,

$$f(x) \geq f(x - (e_i - e_j)) + f(x^* + (e_i - e_j)) - f(x^*) \geq f(x^*),$$

where the first inequality follows from the definition of $M^\natural$-convexity and the second inequality from the induction assumption and the definition of local minimizers. ∎

2.4   Exercises

Exercise 2.1. Assume that a function $f : \mathfrak{R}^n \to \mathfrak{R}$ is continuous.

(a) Prove that f is convex if and only if it satisfies the *midpoint convexity*; namely, for any $x, x' \in \mathfrak{R}^n$,

$$f\left(\frac{x + x'}{2}\right) \leq \frac{1}{2}(f(x) + f(x')).$$

(b) Prove that f is convex if and only if it satisfies the *equidistance convexity*; that is, for any $x, x' \in \mathfrak{R}^n$ and $\alpha \in [0, 1]$,

$$f(x) + f(x') \geq f(x - \alpha(x - x')) + f(x' + \alpha(x - x')).$$

Exercise 2.2. (Private Communication with Peng Sun) Assume that $f : \mathfrak{R} \to \mathfrak{R}$ is convex, and random variables $X_1, X_2, \ldots,$ are nonnegative and independently and identically distributed. Prove that $E[f(\sum_{i=1}^n X_i)]$ is convex on the set of natural numbers.

Exercise 2.3. Prove Theorem 2.2.4.

Exercise 2.4. Assume that a function $f(\cdot, \cdot)$ is defined on the product space $\mathfrak{R}^n \times \mathfrak{R}^m$ and $f(\cdot, y)$ is convex for any given $y \in \mathfrak{R}^m$. Let ζ be a random vector in $\mathfrak{R}^m$. Prove the following:

(a) The exponential function $\exp(x)$ is strictly increasing and convex.

(b) The function $E_\zeta[\exp(w + f(x, \zeta))]$ is jointly convex in x and w.

(c) The function $\ln(E_\zeta[\exp(f(x, \zeta))])$ is convex.

Exercise 2.5. If $f : \mathfrak{R}^n \times \mathfrak{R}^m \to \mathfrak{R}$ is quasiconvex, show that $g(x) = \min_y f(x, y)$ is also quasiconvex provided that g is well defined.

Exercise 2.6. Assume that A is a lattice in the product space $\Re^n \times \Re^m$. Prove that the set function $S(t) = \{x \in \Re^n | (x, t) \in A\}$ is increasing on the set $\{t \in \Re^m | S(t) \neq \emptyset\}$.

Exercise 2.7. Let Z be the set of all integers in $\Re$. Prove that a function f is convex on Z if and only if either of the following two conditions holds:

(a) $\Delta f(x)$ is nondecreasing, where $\Delta f(x) = f(x+1) - f(x)$.

(b) There exists a convex function g on $\Re$ such that $g(x) = f(x)$ for all $x \in Z$. In other words, g is a convex extension of f.

Exercise 2.8. Prove that a function $f : \mathcal{Z}^n \to \bar{\Re}$ is $L^\natural$-convex if and only if for any $x, x' \in \mathcal{Z}^n$, the following discrete midpoint convexity holds:

$$f(x) + f(x') \geq f\left(\left\lfloor \frac{x+x'}{2} \right\rfloor\right) + f\left(\left\lceil \frac{x+x'}{2} \right\rceil\right), \qquad (2.17)$$

where for any $x \in \Re^n$, $\lfloor x \rfloor$ is an n-dimensional integer vector derived by rounding down each component of x to its nearest integer, and $\lceil x \rceil$ is an n-dimensional integer vector derived by rounding up each component of x to its nearest integer.

Exercise 2.9. Prove Theorem 2.3.3.

Exercise 2.10. Prove that the set of minimizers of an $L^\natural$-convex ($M^\natural$-convex) function is $L^\natural$-convex ($M^\natural$-convex).

Exercise 2.11. Prove that any $L^\natural$-convex ($M^\natural$-convex) set $S \subset \mathcal{Z}^n$ is hole-free; that is, S is exactly the intersection of the convex hull of S and $\mathcal{Z}^n$.

Exercise 2.12. Prove Theorem 2.3.8.

Exercise 2.13. Prove the following statements by providing counterexamples: (a) a quadratic $M^\natural$-convex function in $\Re^n$ may not be laminar convex; (b) the condition in (2.16) is sufficient but not necessary for a quadratic function to be $M^\natural$-convex in $\Re^n$.

Exercise 2.14. Prove or disprove the following statement: The infimal convolution of two laminar convex functions defined on the same laminar family remains laminar convex.

Exercise 2.15. Prove Theorem 2.3.9.

Exercise 2.16. The definitions of $L^\natural$-convex functions and $M^\natural$-convex functions on $\Re^n$ are different from the original ones in Murota and Shioura (2004), which explicitly impose the convexity condition. You are asked to show that they are equivalent under the continuity assumption; that is, continuous $L^\natural$-convex functions and $M^\natural$-convex functions defined on $\Re^n$ are convex. Is the continuity assumption dispensable?

Exercise 2.17. Prove that an $M^\natural$-convex function $f : \mathcal{Z}^n \to \bar{\Re}$ is supermodular in $\mathcal{Z}^n$.

Exercise 2.18. (Chen et al. 2012b) Consider the following optimization problem parameterized by two-dimensional vectors $x \in S = \{Ay : y \in D\}$:

$$
\begin{aligned}
f(x) \quad = \quad &\max \quad g(y) \\
&\text{s.t.} \quad Ay = x, \\
&\qquad y \in D,
\end{aligned}
$$

where A is a nonnegative $2 \times n$ matrix, D is a closed convex lattice of $\Re^n$, and $g : D \to \Re$. Show that if g is concave and supermodular on D, then so is f on S.

Exercise 2.19. (Hu 2011) For $i = 1, 2, \ldots, m$, let D_i be a subset of $\Re^2$ and $g_i : D_i \to \Re$. Define

$$
\begin{aligned}
f(x) \quad = \quad &\max \quad \sum_{i=1}^{m} g_i(x_i) \\
&\text{s.t.} \quad \sum_{i=1}^{m} x_i = x, \\
&\qquad x_i \in D_i \; \forall \; i = 1, 2, \ldots, m.
\end{aligned}
$$

Show that if D_i are $L^\natural$-convex sets and g_i are $L^\natural$-concave functions, then f is $L^\natural$-concave.

3

Game Theory

Game theory provides a powerful mathematical framework for modeling and analyzing systems with multiple decision makers, referred to as players, with possibly conflicting objectives. A game studied in game theory consists of a set of players, a set of strategies (or moves) available to the players, and their payoffs (or utilities) for each combination of their strategies. Depending on whether the players can sign enforceable binding agreements, game theory consists of two branches: noncooperative game theory and cooperative game theory. Noncooperative game theory provides concepts and tools to study the behaviors of the players when they make their decisions independently. Cooperative game theory, on the other hand, assumes that it is possible for the players to sign enforceable binding agreements and provides concepts describing basic principles these binding agreements should follow. Both noncooperative game theory and cooperative game theory have been widely used in many disciplines, such as economics, political science, social science, as well as biology and computer science, among others. They have also received considerable attention in supply chain management literature in recent years. In this chapter, we provide a concise introduction to some of the key concepts and results that are most relevant in our context. We refer to Osborne (2003) and Myerson (1997) for both noncooperative game theory and cooperative game theory, Fudenberg and Tirole (1991) and Başar and Olsder (1999) for noncooperative game theory, Vives (2000) on oligopoly pricing from the perspective of noncooperative game theory, and Peleg and Sudhölter (2007) for cooperative game theory, respectively.

D. Simchi-Levi et al., *The Logic of Logistics: Theory, Algorithms, and Applications for Logistics Management*, Springer Series in Operations Research and Financial Engineering, DOI 10.1007/978-1-4614-9149-1_3, © Springer Science+Business Media New York 2014

3.1 Noncooperative Game Theory

Noncooperative games can be represented in both extensive form and normal form. A game in extensive form specifies the sequence of moves of the players and the associated information structure based on which the players decide their moves, while a game in normal form is a simultaneous move game in which the players move only once and at the same time. Whether to represent a game in either extensive form or normal form depends on which form is more convenient for a specific application. For example, a Stackelberg game in which one player, referred to as the leader, makes its move first followed by the moves of other players, referred to as the followers, after observing the leader's move, is naturally described in extensive form. However, mathematically, any extensive form representation of a game can be equivalently translated into a game in normal form, and a game in normal form can be treated as an extensive form game with simultaneous moves. For our purpose, we will restrict our attention to games in normal form here.

A game in normal form is specified by a triple $(N, \{S_i\}_{i \in N}, \{u_i\}_{i \in N})$, where $N = \{1, 2 \ldots, n\}$ is the set of players, S_i is the set of strategies of player i, $u_i : S \to \Re$ is player i's payoff as a function of the composition of all players' strategies, referred to as a strategy profile, and $S = \Pi_{i \in N} S_i$ is the set of all strategy profiles. All these elements are common knowledge to the players. That is, the players know these elements, they know the other players know these elements, they know the other players know the other players know these elements, and so on.

A simple example of a noncooperative game is the famous prisoner's dilemma. In this game, two criminals, referred to as Players 1 and 2, are arrested and imprisoned. They have to decide independently whether to betray the other or keep silent. If both criminals keep silent, each of them will serve one year in prison; if both betray, each serves two years; if one betrays and the other keeps silent, the one who betrays is set free and the other will serve three years.

The normal form of the prisoner's dilemma is specified by $(N, \{S_i\}_{i \in N}, \{u_i\}_{i \in N})$, where $N = \{1, 2\}$, $S_1 = S_2 = \{\text{Silent}, \text{Betray}\}$, and the payoff of each player is the negative of the number of years served in prison by the player. The payoffs of the game can be described in Table 3.1. In the table, the first column specifies the strategies of Player 1 and the first row specifies the strategies of Player 2. In each cell, the first number is the payoff of Player 1 and the second number is the payoff of Player 2, given their corresponding strategies.

	Silent	Betray
Silent	$-1, \quad -1$	$-3, \quad 0$
Betray	$0, \quad -3$	$-2, \quad -2$

TABLE 3.1. The prisoner's dilemma

3.1.1 Definition and Existence of Nash Equilibrium

Given the specification of a game, one is interested in predicating its outcome—which strategies the players will take and what payoff each of them will receive. In this respect, the Nash equilibrium is one of the most important concepts in noncooperative game theory.

Definition 3.1.1 *Given a game* $(N, \{S_i\}_{i \in N}, \{u_i\}_{i \in N})$, *a strategy profile* $s^* = (s_1^*, \dots, s_n^*) \in S$ *is a Nash equilibrium (in pure strategies) if for any* $i \in N$,

$$u_i(s^*) \geq u_i(s_i, s_{-i}^*) \ \forall \ s_i \in S_i,$$

where for a given $s \in S$, $s_{-i} \in \Pi_{j \neq i} S_j$ *is the strategy profile of all players except player* i.

The above definition says that in a Nash equilibrium, any player's strategy maximizes its own payoff assuming that the other players' strategies are specified in the Nash equilibrium; that is, for any $i \in N$,

$$s_i^* \in \operatorname{argmax}_{s_i \in S_i} u_i(s_i, \mathbf{s}_{-i}^*).$$

Clearly, a Nash equilibrium specifies basic requirements rational players would respect and is often used as an appropriate predication of their behaviors. In fact, when playing the game, a player may conjecture the strategies of other players. In an equilibrium, the conjectures of the players are correct and the players, if rational, would respectively maximize their payoffs given their conjectures.

Consider the prisoner's dilemma. For Player 1, it is clear that its best strategy is to betray Player 2 no matter what strategy Player 2 takes. Similarly, Player 2's best strategy is to betray Player 1. Thus, the strategy profile (Betray, Betray) is a Nash equilibrium. It is interesting to observe that both players can be better off by keeping silent. However, the strategy profile (Silent, Silent) is not a Nash equilibrium.

Despite its conceptual appealing, the concept of Nash equilibrium faces several theoretical and practical difficulties. One of the main difficulties is that a game may not admit any Nash equilibrium, as demonstrated in the following matching penny game. In this game, two players, Player 1 and Player 2, each of whom has a penny, need to decide independently which side to show. If the sides of the pennies match, namely, both players show either heads or tails, Player 1 collects both pennies and thus has a gain of one penny; otherwise, Player 2 collects both pennies and has a gain of one penny. Formally, we face a game $(N, \{S_i\}_{i \in N}, \{u_i\}_{i \in N})$, with $N = \{1, 2\}$, $S_1 = S_2 = \{\text{Head}, \text{Tail}\}$, and (u_1, u_2) described in Table 3.2.

	Head	Tail
Head	+1, −1	−1, +1
Tail	−1, +1	+1, −1

TABLE 3.2. Payoffs of matching penny game

The matching penny game is a zero-sum game in which the total payoff of the two players is zero for any strategy profile. It does not have a Nash equilibrium. To see this, notice that for any strategy chosen by Player 1, Player 2's best response is to pick a different strategy, while for any strategy chosen by Player 2, Player 1's best response is to follow the same strategy. Thus, any strategy profile is not a Nash equilibrium.

We first present a set of conditions under which the existence of a Nash equilibrium is guaranteed.

Theorem 3.1.2 *Given a game* $(N, \{S_i\}_{i \in N}, \{u_i\}_{i \in N})$, *if for any player* $i \in N$, *its strategy set* S_i *is a nonempty, convex, and compact set in a Euclidean space, its payoff function* u_i *is continuous in* $s \in S$ *and quasiconcave in* $s_i \in S_i$, *then the game has a Nash equilibrium. If, in addition, the game is symmetric, namely, for any* $i, j \in N$, $S_i = S_j$, *and* $u_i(s_i, s_{-i}) = u_j(s'_j, s'_{-j})$ *if* $s_i = s'_j$ *and* s_{-i} *and* s'_{-j} *are the same when ignoring the identities of the players, there is a symmetric Nash equilibrium.*

The proof is an application of the well-known Kakutani fixed-point theorem, which we state in the following without proof.

Theorem 3.1.3 *Let* C *be a nonempty, compact, and convex subset of* $\Re^n$. *Let* T *be a set-valued operator that maps any* $x \in C$ *to a nonempty and convex subset of* C. *If the set* $\{(x, y) : x \in C, y \in T(x)\}$, *referred to the graph of* T, *is closed, then* T *has a fixed point. That is, there exists an* $x \in C$ *such that* $x \in T(x)$.

To gain some intuition, let's look at a simple case with a function $f : [0, 1] \to [0, 1]$. If f is continuous, then one can see directly from a graph that the curve $y = f(x)$ must intersect the line $y = x$ at some $x \in [0, 1]$; that is, a fixed point exists.

Theorem 3.1.2 is an immediate consequence of the Kakutani fixed-point theorem. To see this, denote for any $s \in S$, $T_i(s) = \text{argmax}_{s'_i \in S_i} u_i(s'_i, s_{-i})$, the best response of player i given the strategies of the other players, and note that a Nash equilibrium is exactly a fixed point of the best response operator $T = (T_1, \ldots, T_n)$, which maps a strategy profile in S to a subset of S. The continuity of u_i together with the compactness of S_i implies that T has a closed graph, and the quasiconcavity of each player's payoff in its own strategy together with the convexity of the strategy set implies that $T(x)$ is convex for any $x \in S$. Thus, T has a fixed point and the game has a Nash equilibrium. To show that a symmetric game admits a symmetric equilibrium, denote for any $\tilde{s} \in S_1$, $\tilde{T}(\tilde{s}) = \text{argmax}_{s'_1 \in S_i} u_1(s'_1, \tilde{s}, \ldots, \tilde{s})$. Again, the conditions in Theorem 3.1.2 imply that $\tilde{T}$ has a fixed point $\tilde{s}^*$, and therefore the symmetric strategy profile $(\tilde{s}^*, \tilde{s}^*, \ldots, \tilde{s}^*)$ is a Nash equilibrium. We emphasize that $\tilde{s}^*$ is a fixed point of $\tilde{T}$. In most cases, it is not a global maximizer of $u_1(s, s, \ldots, s)$. A common mistake we have seen is to claim that if $(s^0, s^0, \ldots, s^0)$ is a symmetric equilibrium, then s^0 is a global maximizer of $u_1(s, s, \ldots, s)$, or vice versa.

A different set of conditions to guarantee the existence of a Nash equilibrium without quasiconcavity of the payoff functions is based on supermodularity. Specifically, a game $(N, \{S_i\}_{i \in N}, \{u_i\}_{i \in N})$ is called (strictly) supermodular if, for each $i \in N$, S_i is a lattice, and u_i is (strictly) supermodular in $s_i \in S_i$ for any fixed $s_{-i} \in \Pi_{j \neq i} S_j$ and has (strictly) increasing differences in (s_i, s_{-i}) on $S_i \times \Pi_{j \neq i} S_j$. For simplicity, we restrict our discussion to cases in which S_i is a lattice in a Euclidean space. We have the following theorem about a Nash equilibrium for supermodular games.

Theorem 3.1.4 *Given a supermodular game $(N, \{S_i\}_{i \in N}, \{u_i\}_{i \in N})$, if for $i \in N$, S_i is nonempty and compact, u_i is continuous in $s_i \in S_i$ for any fixed $s_{-i} \in \Pi_{j \neq i} S_j$, then the set of Nash equilibria is nonempty and has a largest and a smallest element. If, in addition, the payoff functions u_i are parameterized by $t \in T \subseteq \Re^m$ and have increasing differences in (s_i, t) for any fixed $s_{-i} \in \Pi_{j \neq i} S_j$, the largest and the smallest Nash equilibria are increasing in t.*

Proof. We prove this theorem using the generalized Tarski fixed-point theorem, Theorem 2.2.9. Again, for any $s \in S$, let $T_i(s) = \text{argmax}_{s_i' \in S_i} u_i(s_i', s_{-i})$ and $T = (T_1, \ldots, T_n)$. Our assumption on $u_i(s_i, s_{-i})$ together with Theorem 2.2.8 implies that $T_i(s)$, as a subset of S_i, is a nonempty and compact lattice for any $s \in S$ and T_i is increasing. Thus, $T(s)$, as a subset of S, is a nonempty and compact lattice for any $s \in S$ and T is increasing. From Theorem 2.2.9, the set of fixed points of T, or equivalently the set of Nash equilibria, is nonempty and has a largest and a smallest element.

If, in addition, the payoff functions u_i ($i \in N$) are parameterized by $t \in T \subseteq \Re^m$ and have increasing differences in (s_i, t) for any fixed $s_{-i} \in \Pi_{j \neq i} S_j$, from Theorem 2.2.8, the set $T_i(s, t) = \text{argmax}_{s_i' \in S_i} u_i(s_i', s_{-i}, t)$ is increasing in t as well. The remaining argument is similar to the one in the previous paragraph using the second part of Theorem 2.2.9. ∎

One remedy for games without Nash equilibria in pure strategies is to allow mixed strategies. Specifically, a mixed strategy for a player i with a strategy set S_i is a probability distribution defined over S_i. Let Σ_i be the set of all mixed strategies of player i and $\Sigma = \Sigma_1 \times \ldots \times \Sigma_n$. We now define a Nash equilibrium in mixed strategies.

Definition 3.1.5 *Given a game $(N, \{S_i\}_{i \in N}, \{u_i\}_{i \in N})$, a strategy profile $\sigma^* = (\sigma_1^*, \ldots, \sigma_n^*) \in \Sigma$ is a Nash equilibrium in mixed strategies if, for any $i \in N$,*

$$u_i(\sigma^*) \geq u_i(s_i, \sigma_{-i}^*) \ \forall \ s_i \in S_i,$$

where, for a given σ, σ_{-i} is the strategy profile of all players except player i and $u_i(\sigma)$ is the expected payoff.

Unlike the case with a Nash equilibrium in pure strategies, one can show using Theorem 3.1.2 that a Nash equilibrium in mixed strategies always exists for games with finite (pure) strategies. Indeed, using mixed strategies can be regarded as a way to convexify the payoffs and the strategy sets of a game. Come back to the matching penny game. We now show it has a unique Nash equilibrium in mixed strategies. For this purpose, let $\sigma_1 = (x_1, x_2) \in \Sigma_1 = \{(x_1, x_2) : x_1 + x_2 = 1, x_1 \geq 0, x_2 \geq 0\}$, which specifies a mixed strategy of choosing "Head" and "Tail" with probabilities x_1 and x_2, respectively. Similarly, let $\sigma_2 = (y_1, y_2) \in \Sigma_2(= \Sigma_1)$. We have $u_1(\sigma) = (x_1 - x_2)(y_1 - y_2)$ and $u_2(\sigma_2) = -u_1(\sigma)$. The definition of a Nash equilibrium in mixed strategies implies that

$$u_1(\sigma^*) = (x_1^* - x_2^*)(y_1^* - y_2^*) \geq u_1((1,0), \sigma_2^*) = y_1^* - y_2^*,$$

$$u_1(\sigma^*) = (x_1^* - x_2^*)(y_1^* - y_2^*) \geq u_1((0,1), \sigma_2^*) = -(y_1^* - y_2^*).$$

Thus,

$$|y_1^* - y_2^*| \leq u_1(\sigma^*) \leq |x_1^* - x_2^*||y_1^* - y_2^*| \leq |y_1^* - y_2^*|.$$

If $y_1^* \neq y_2^*$, we must have $|x_1^* - x_2^*| = 1$; that is, σ_1^* must be a pure strategy. Symmetrically, if $x_1^* \neq x_2^*$, σ_2^* must be a pure strategy. However, as demonstrated earlier, the matching penny game does not have a Nash equilibrium in pure strategies. Thus, we must have $\sigma_1^* = \sigma_2^* = (1/2, 1/2)$. That is, both players choose their pure strategies with equal probability, resulting in zero expected payoffs for each of the players.

Though the existence of a Nash equilibrium in mixed strategies can be guaranteed even when a Nash equilibrium in pure strategies does not exist, it is not clear how mixed strategies would be implemented in supply chain management settings. Thus, we mainly focus on a Nash equilibrium in pure strategies and simply refer to it as a Nash equilibrium when there is no confusion.

3.1.2 Uniqueness of Nash Equilibrium

Another difficulty with the concept of Nash equilibrium is that a game may have multiple equilibria. In this case, it often depends on factors that are not included in the formal description of the game to determine which equilibrium would be an appropriate outcome. Consider a game in which two friends, again referred to as Player 1 and Player 2, decide independently between two locations, A and B, for lunch. If it happens that they pick the same location, they enjoy each other's company and get a payoff of 1 each; otherwise, they eat alone and get a payoff of 0 (Table 3.3). This is a coordination game without communication. Similar to the matching penny game, we can present the payoffs of the players in a table format. One can easily check that the game has two Nash equilibria, (A, A) and (B, B). From the description of the game, it is not clear how one player can make a right guess of the move of the other player and whether any equilibrium could be realized.

Thus, in many applications, it is desirable to establish the uniqueness of a Nash equilibrium. There are three basic approaches for this purpose, which are termed the "contraction," "univalence," and "index theory" approaches, respectively.

	A	B
A	+1, +1	−1, −1
B	−1, −1	+1, +1

TABLE 3.3. Payoffs of coordination game without communication

The first approach is based on the contraction principle, which says that a contraction operator in a Euclidean space (or, more generally, complete space) has a unique fixed point. Specifically, given a subset C of a Euclidean space, an operator $T : C \to C$ is a contraction if there exists a constant α with $0 < \alpha < 1$ such that for any $x, x' \in C$, $\|T(x) - T(x')\| \leq \alpha \|x - x'\|$, where $\|\cdot\|$ is a norm (for example, the commonly used 1-norm, 2-norm, or ∞-norm). If C is closed, then T has a unique fixed point in C. If, in addition, C is compact, it suffices to require $\|T(x) - T(x')\| < \|x - x'\|$ for any $x, x' \in C$. A sufficient condition for the best response operator T to be a contraction is that the ∞-norm of the Jacobian of T is less than 1.

Under the assumptions in Theorem 3.1.2, if one can show that the best response operator is single-valued and a contraction, then we have a unique Nash equilibrium. One sufficient condition for the case in which the strategy sets S_i are one-dimensional is the diagonal dominance of the second-order derivatives of the payoffs

$$\frac{\partial^2 u_i(s)}{\partial^2 s_i} + \sum_{j \neq i} \left| \frac{\partial^2 u_i(s)}{\partial s_i \partial s_j} \right| < 0, \forall s \in S, i \in N, \tag{3.1}$$

which implies that the ∞-norm of the Jacobian of T is less than 1. To see this, observe that if all equilibria are interior points of the strategy sets and the payoff functions are differentiable, we have the first-order optimality condition $\nabla u_i(s_i', s_{-i})|_{s_i' = T_i(s)} = 0$ for $i \in N$, where $\nabla_i u_i(s)$ is the gradient of u_i with respect to s_i. From the implicit function theorem, $\frac{\partial T_i(s)}{\partial s_i} = 0$ and for $j \neq i$,

$$\frac{\partial T_i(s)}{\partial s_j} = - \left(\frac{\partial^2 u_i(s_i', s_{-i})}{\partial^2 s_i} \right)^{-1} \frac{\partial^2 u_i(s_i', s_{-i})}{\partial s_i \partial s_j} \Big|_{s_i' = T_i(s)}.$$

The ∞-norm of the Jacobian of T is given by

$$\max_{i \in N} \sum_{j \in N} \left| \frac{\partial T_i(s)}{\partial s_j} \right|.$$

Similar conditions can be derived when the strategy sets S_i are multidimensional.

The second approach is based on analyzing the first-order optimality conditions of all players' maximization problems. A key condition for the uniqueness of a Nash equilibrium is the diagonally strict concavity of the payoff functions. Define $\nabla u(s) = [\nabla_1 u_1(s), \ldots, \nabla_n u_n(s)]^T$.

Definition 3.1.6 *The payoff functions* $\{u_i\}_{i \in N}$ *are diagonally strictly concave on* S *if for any* $s, s' \in S$,

$$(s - s')^T (\nabla u(s) - \nabla u(s')) < 0.$$

When the strategy sets S_i are one-dimensional, the condition (3.1) implies that the payoff functions are diagonally strictly concave.

Denote $G(s)$ the Jacobian of $\nabla u(s)$. It is straightforward to show that if $G(s) + G(s)^T$ is negative definite for any $s \in S$, then the payoff functions are diagonally strictly concave. Some additional technical conditions are also needed. Specifically, we assume that the strategy sets have explicit algebraic representations:

$$S_i = \{s_i \in \Re^{n_i} | h_{ij}(s_i) \geq 0, j = 1, \ldots, \tau_i\}, \tag{3.2}$$

where n_i and τ_i are positive integers and $h_{ij} : \Re^{n_i} \to \Re$.

The following uniqueness result is due to Rosen (1965).

Theorem 3.1.7 *Consider a game* $(N, \{S_i\}_{i \in N}, \{u_i\}_{i \in N})$. *Assume that* S_i *is compact and has the representation (3.2), where* h_{ij} *is concave, and there exists some* $x_i^0 \in \Re^{n_i}$ *such that* $h_{ij}(x_i^0) > 0$ *for any nonlinear function* h_{ij}. *If the payoff functions are diagonally strictly concave on* S, *then the game has a unique Nash equilibrium.*

The existence of a Nash equilibrium follows from Theorem 3.1.2. The conditions on h_{ij} are imposed so that we can write down the first-order optimality conditions, commonly referred to as the Karush–Kuhn–Tucker (KKT) conditions, and the diagonally strict concavity of the payoff functions is sufficient for the uniqueness of a solution to the KKT conditions.

The third approach is based on the Poincaré–Hopf index theorem. Consider a function $g : C \to \Re^m$, where C is a nonempty compact set in $\Re^m$. An implication of the Poincaré–Hopf index theorem is that under some boundary condition on C, $g(a) = 0$ has a unique solution if the determinant of $-G(a)$ is positive whenever $g(a) = 0$, where $G(a)$ is the Jacobian of g. To gain some intuition, consider a differentiable function $g : [0, 1] \to \Re$ with $g(0) > 0$ and $g(1) < 0$. Clearly, $g(a) = 0$ has at least one solution. It cannot have multiple solutions. To see this, let a^* be any solution of $g(a) = 0$; then for $a > a^*$, $g(a)$ is less than zero until it reaches zero for the first time at some a'. However, it is impossible since g has a negative derivative at a'.

If all equilibria are interior points of the strategy sets and the payoff functions are differentiable, then any equilibrium s^* would be a solution of the equation system $\nabla u(s) = 0$. The index theory thus provides an approach to check the uniqueness of a Nash equilibrium. That is, if the determinant of the Jacobian of $-\nabla u(s)$ is positive at the equilibria, then we have a unique Nash equilibrium. This approach is less restrictive but harder to check than the other two approaches.

3.2   Cooperative Game Theory

Cooperative games can be represented in either coalitional form or strategic form. Cooperative strategic games assume that the players can make binding agreements on the choice of strategies, while cooperative coalitional games assume that they can make binding agreements on the distribution of payoffs. Since many applications can be naturally formulated in coalitional form and the players focus more on the choice of stable payoffs rather than on the choice of stable strategies, as any combination of strategies can be supported by a binding agreement, we will only consider cooperative games in coalitional form.

Cooperative coalitional games can be divided into two categories: transferrable utilities and nontransferrable utilities. A coalitional game with transferrable utilities is a pair (N, V), where $N = \{1, 2, \ldots, n\}$ is the set of all players and referred to as the grand coalition, and V is the characteristic function that maps any subset S of N, called a coalition, to a real number with $V(\emptyset) = 0$. If a coalition S forms, its value $V(S)$ can be allocated in any possible way among its members by allowing side payments. That is, a feasible payoff vector x of the players in S satisfies the condition $x(S) \leq V(S)$, where we use the notation $x(S)$ to denote $\sum_{i \in S} x_i$ and x_i is the payoff to player i. A coalitional game with nontransferrable utilities is a pair (N, V), where again $N = \{1, 2, \ldots, n\}$ is the set of all players, and V is a mapping that associates a set in a Euclidean space with each coalition S. For simplicity, we focus on coalitional games with transferrable utilities and we simply refer to them as coalitional games or cooperative games in coalition form.

Depending on the properties of the characteristic functions, we can define monotonic, superadditive, and convex games. Specifically, a coalition game (N, V) is called monotonic if

$$V(S) \leq V(T), \forall\, S \subseteq T \subseteq N;$$

that is, the larger a coalition in terms of set inclusion, the higher the worth of the coalition. A coalition game (N, V) is called superadditive if

$$V(S \cup T) \geq V(S) + V(T), \forall\, S, T \subseteq N, S \cap T = \emptyset;$$

namely, combining two independent coalitions leads to a higher value. A coalition game (N, V) is called convex if

$$V(S) + V(T) \leq V(S \cup T) + V(S \cap T), \forall\, S, T \subseteq N.$$

A convex game is also referred to as a supermodular game since v is supermodular as a set function. One can show that a game (N, V) is convex if and only if

$$V(S \cup \{i\}) - V(S) \leq V(T \cup \{i\}) - V(T), \forall\, S \subseteq T \subseteq N, i \notin T; \tag{3.3}$$

that is, the marginal contribution of a player is increasing with respect to set inclusion. Since $V(\emptyset) = 0$, a convex game is superadditive.

Given a cooperative game (N, V), one would like to know how the coalition value should be distributed among the players. We will introduce several solution concepts: the core, the nucleolus, and the Shapley value. A solution concept specifies for each game a set of payoff vectors in $\Re^n$ whose ith components represent the values allocated to player i, under the assumption that the grand coalition forms. If the players do not form the grand coalition, we can restrict these solution concepts to whatever coalitions the players form. In cooperative game theory, most solution concepts can be derived from a list of axioms, which stipulate properties a reasonable allocation should respect. We list some of the properties desirable for a solution concept $\sigma(N, V)$.

- Efficiency: Any payoff vector $x \in \sigma(N, V)$ is efficient; namely, $x(N) = V(N)$, or, equivalently, the grand coalition value $V(N)$ is fully allocated, where for any set $S \subseteq N$ and any vector $z \in \Re^n$, $z(S) = \sum_{i \in S} z_i$.

- Individual rationality: Any payoff vector $x \in \sigma(N, V)$ is individually rational; namely, $x_i \geq V(\{i\})$ for all $i \in N$, or, equivalently, no player receives less than what he can get on his own.

- Group rationality: Any payoff vector $x \in \sigma(N, V)$ is group rational; namely, $x(S) \geq V(S)$ for all $S \subseteq N$, or, equivalently, no group of players receives less than what the group can get on its own.

- Symmetry: If the marginal contributions of two players i and j are the same for all $S \subseteq N \setminus \{i, j\}$, then any payoff vector $x \in \sigma(N, V)$ should assign the same value to i and j; that is, $x_i = x_j$.

- Anonymity: For a game (N, V) and any permutation π of N,

$$\sigma(N, V^\pi) = \{x^\pi : x \in \sigma(N, V)\},$$

where $V^\pi(S) = V(\{\pi(i)\}_{i \in S})$ and for any $x \in \Re^n$, x^π is a vector such that $x_i^\pi = x_{\pi(i)}$. That is, the names of the players would not affect their payoffs.

- Superadditivity: Given two games (N, V) and (N, V'), $\sigma(N, V) + \sigma(N, V') \subseteq \sigma(N, V + V')$, where $(N, V + V')$ is a game whose characteristic function is the sum of those of (N, V) and (N, V').

- Additivity: Given two games (N, V) and (N, V'), $\sigma(N, V + V') = \sigma(N, V) + \sigma(N, V')$.

- Covariance under strategic equivalence: Given a game (N, V), a scalar $\alpha > 0$ and $\beta \in \Re^n$, $\sigma(N, \alpha V + \beta) = \alpha \sigma(N, V) + \beta$, where in the game $(N, \alpha V + \beta)$, for any $S \subseteq N$, $(\alpha v + \beta)(S) = \alpha V(S) + \beta(S)$.

- Null player: For any $x \in \sigma(N, V)$, $x_i = 0$ if player i has zero marginal contribution added to any S; that is, $V(S \cup \{i\}) = V(S)$.

The efficiency property is sometimes referred to as the Pareto optimality. Group rationality implies individual rationality. Additivity implies superadditivity and covariance under strategic equivalence.

Another desirable property for a solution concept is consistency. To present it, we need to define reduced games. Given a game (N, V), for a nonempty set $S \subset N$ and a payoff vector $x \in \Re^n$, define a reduced game $(S, V_{x,S})$ as follows: $V_{x,S}(\emptyset) = 0$, $V_{x,S}(S) = V(N) - x(N \setminus S)$, and

$$V_{x,S}(T) = \max_{T' \subseteq N \setminus S} V(T \cup T') - x(T') \ \forall \emptyset \neq T \subset S.$$

A solution concept $\sigma(N, V)$ is consistent if for any $x \in \sigma(N, V)$, $x_S \in \sigma(S, V_{x,S})$, where $x_S = (x_i)_{i \in S}$. It is called weakly consistent if for any $x \in \sigma(N, V)$ and $1 \leq |S| \leq 2$, $x_S \in \sigma(S, V_{x,S})$, where $|S|$ is the cardinality of S. The consistency properties basically say that if a payoff is reasonable in the grand coalition, any coalition should accept that the payoffs allocated to them are reasonable.

Here are some monotonicity properties that a solution concept $\sigma(N, V)$ may want to satisfy. Among them, the coalitional monotonicity and strong monotonicity are defined for a single-valued solution concept. In this case, we simply use $\sigma(N, V)$ to denote the payoff vector when there is no confusion.

- Aggregate monotonicity: For two games (N, V) and (N, V'), if $V(N) \geq V'(N)$, $V(S) = V'(S)$ for all $S \subset N$, then for any $x \in \sigma(N, V')$, there exists $y \in \sigma(N, V)$ such that $y_i \geq x_i$ for all $i \in N$.

- Coalitional monotonicity: For two games (N, V) and (N, V'), if $V(T) \geq V'(T)$ for some $T \subseteq N$ and $V(S) = V'(S)$ for all S with $T \neq S \subseteq N$, then $\sigma(N, V)_i \geq \sigma(N, V')_i$ for any $i \in T$.

- Strong monotonicity: For two games (N, V) and (N, V') and any $i \in N$, $\sigma(N, V)_i \geq \sigma(N, V')_i$ if

$$V(S \cup \{i\}) - V(S) \geq V'(S \cup \{i\}) - V'(S) \ \forall \ S \subseteq N.$$

In the following, we introduce the concepts of the core, the nucleolus, and the Shapley value and present their properties.

3.2.1   Core

The core of a game (N, V), denoted by $\mathcal{C}(N, V)$, is the set of efficient payoff vectors that are group rational. That is,

$$\mathcal{C}(N, V) = \{x \in \Re^n : x(N) = V(N), x(S) \geq V(S) \ \forall \ S \subset N\}.$$

Clearly, the core is a polyhedral set that can be defined by 2^n linear inequalities. A payoff vector x in the core, called a core allocation, implies that based on this allocation x, no group of players has incentive to deviate from the grand coalition and thus the grand coalition is stable. It also implies that the allocation is fair in the sense that no group would subsidize its complement since for any $S \subset N$,

$$x(N \setminus S) = x(N) - x(S) \leq V(N) - V(S);$$

that is, the payoff to the group $N \setminus S$ is no more than its added value to the grand coalition. The core is attractive because it satisfies several plausible properties.

Theorem 3.2.1 *The core $\mathcal{C}(N, V)$ if nonempty satisfies the following properties: (a) efficiency, (b) individual rationality, (c) group rationality, (d) anonymity, (e) superadditivity, (f) covariance under strategic equivalence, (g) null player, (h) consistency, and (i) aggregate monotonicity.*

Proof. We only prove parts (g) and (h). The other parts are straightforward. To prove part (g), note that for a null player, $V(S \cup \{i\}) = V(S)$ for any $S \subset N$. Specifically, $V(\{i\}) = 0$ and $V(N \setminus \{i\}) = V(N)$. For any $x \in \mathcal{C}(N, V)$, we have that $x_i \geq V(\{i\})$. On the other hand,

$$x_i = x(N) - x(N \setminus \{i\}) \leq V(N) - V(N \setminus \{i\}) = 0.$$

Thus, $x_i = 0$.

We now show that the core is consistent. To see this, let $x \in \mathcal{C}(N, V)$. For any nonempty sets $S \subset N$ and $T \subset S$, $x_S(S) = x(N) - x(N \setminus S) = V_{x,S}(S)$, and there exists $T' \subseteq N \setminus S$ such that

$$V_{x,S}(T) = V(T \cup T') - x(T') = V(T \cup T') - x(T \cup T') + x(T) \leq x_S(T).$$

Thus, x_S is in the core of the reduced game $(S, V_{x,S})$.   ∎

Unfortunately, the core of a game may be empty, and even if it is nonempty, finding an allocation in the core is usually computationally challenging. In fact, determining whether a given vector is in the core or not can be challenging as well since the core is defined by an exponential number (in n) of linear inequalities.

In the following, we present a sufficient and necessary condition for the existence of a nonempty core credited to Bondareva (1963) and Shapley (1967). For this purpose, define for any $S \subseteq N$, the characteristic vector χ_S as a vector in $\Re^n$ with

$$\chi_S^i = \begin{cases} 1, & \text{if } i \in S \\ 0, & \text{if } i \notin S. \end{cases}$$

A collection $\mathcal{B}$ of subsets of N is called balanced if there exist positive numbers δ_S, $S \in \mathcal{B}$, such that

$$\sum_{S \in \mathcal{B}} \delta_S \chi_S = \chi_N.$$

The collection $\{\delta_S\}_{S \in \mathcal{B}}$ is called a system of balancing weights associated with $\mathcal{B}$. A game is called *balanced* if for any balanced collection $\mathcal{B}$ and any associated system of balancing weights $\{\delta_S\}_{S \in \mathcal{B}}$,

$$\sum_{S \in \mathcal{B}} \delta_S V(S) \leq V(N).$$

Theorem 3.2.2 *(The Bondareva–Shapley theorem) A cooperative coalitional game (N, V) has a nonempty core if and only if it is balanced.*

Proof. Notice that the core of (N, V) is nonempty if and only if $V(N)$ is the optimal objective value of the linear program

$$\begin{array}{ll} \max & x(N) \\ \text{s.t.} & x(S) \geq V(S), \forall\, S \subseteq N. \end{array}$$

Let $y(S)$ be the dual variable associated with the inequality indexed by S. The dual of the above linear program is

$$\min \quad \sum_{S \subseteq N} V(S)y(S)$$
$$\text{s.t.} \quad \sum_{S:i \in S} y(S) = 1, \forall\, i \in N,$$
$$y(S) \geq 0.$$

Since $V(N)$ is the objective value of the dual for the feasible solution $\{y^0(S)\}_{S \subseteq N}$ with $y^0(N) = 1$ and $y^0(S) = 0$ for $S \subset N$, from the strong duality theorem, the core is nonempty if and only if for any feasible solution $\{y(S)\}_{S \subseteq N}$ of the dual,

$$V(N) \leq \sum_{S \subseteq N} V(S)y(S).$$

For any feasible solution $\{y(S)\}_{S \subseteq N}$ of the dual, we can define a balanced collection $\mathcal{B} = \{S : y(S) > 0\}$ and its associated system of balanced weights $\delta_S = y(S)$ for $S \in \mathcal{B}$. On the other hand, any system of balancing weights $\{\delta_S\}_{S \in \mathcal{B}}$ associated with a balanced collection $\mathcal{B}$ can be extended to be a feasible solution $\{y(S)\}_{S \subseteq N}$ for the dual by defining $y(S) = \delta_S$ if $S \in \mathcal{B}$ and 0 otherwise. Therefore, the core is nonempty if and only if for any balanced collection $\mathcal{B}$ and its associated system of balancing weights $\{\delta_S\}_{S \in \mathcal{B}}$,

$$V(N) \leq \sum_{S \subseteq N} V(S)y(S) = \sum_{S \in \mathcal{B}} \delta_S V(S);$$

that is, the game is balanced. ∎

Though general cooperative games may have empty cores, the core of a convex game is always nonempty and has a nice characterization. We will show that the following greedy algorithm will construct an extreme point of the core. Given a permutation π of $\{1, 2, \ldots, n\}$, let $S_i^\pi = \{j : \pi(j) \leq i\}$ for any $i = 1, \ldots, n$ and $S_0^\pi = \emptyset$. Define for $i = 1, \ldots, N$,

$$x_{\pi(i)}^\pi = V(S_i^\pi) - V(S_{i-1}^\pi);$$

that is, the payoff the greedy algorithm assigns to player $\pi(i)$ is the marginal contribution added to the players in S_{i-1}^π.

Theorem 3.2.3 *For a convex game (N, V), the set of extreme points of the core is exactly $\{x^\pi : \pi$ is a permutation$\}$.*

Proof. It is sufficient to prove that the payoff vector, denoted as x^*, associated with the identity permutation π with $\pi(i) = i$ for any $i \in N$ is an extreme point of the core. According to its definition, for $i \in N$,

$$x_i^* = V(S_i) - V(S_{i-1}), \forall\, i \in N,$$

and $x = (x_1, \ldots, x_n)$, where $S_i = \{1, 2, \ldots, i\}$.

We first show that x^* is in the core. Notice that

$$x^*(N) = \sum_{i \in N}(V(S_i) - V(S_{i-1})) = V(N).$$

For any $S \subset N$, let $S = \{i_1, \ldots, i_k\}$ with $i_1 < i_2 \ldots < i_k$, and define for any $j \leq k$, $\hat{S}_j = \{i_1, \ldots, i_j\}$. We have for any $j = 1, \ldots, k$,

$$\begin{aligned}
x^*_{i_j} &= V(S_{i_j}) - V(S_{i_j-1}) \\
&= V(S_{i_j-1} \cup \{i_j\}) - V(S_{i_j-1}) \\
&\geq V(\{S_{i_j-1} \cap S\} \cup \{i_j\}) - V(S_{i_j-1} \cap S) \\
&= V(\hat{S}_j) - V(\hat{S}_{j-1}),
\end{aligned}$$

where the inequality follows from the supermodularity of the characteristic function V. Therefore,

$$x^*(S) = \sum_{j=1}^{k} x^*_{i_j} \geq \sum_{j=1}^{k}(V(\hat{S}_j) - V(\hat{S}_{j-1})) = V(S),$$

which implies that x is in the core.

To show that x^* is an extreme point of the core, consider the following optimization problem:

$$\begin{aligned}
\min \quad & \sum_{i \in N} f_i x_i \\
\text{s.t.} \quad & x \in \mathcal{C}(N, V),
\end{aligned} \tag{3.4}$$

where $f = (f_1, \ldots, f_n)$ is a given vector in $\Re^n$ with $f_1 > f_2 > \ldots > f_n$. It suffices to show that x^* is a unique optimal solution of the above optimization problem. For this purpose, note that for any given payoff vector x in the core,

$$\begin{aligned}
\sum_{i \in N} f_i x_i &= \sum_{i \in N} f_i(x(S_i) - x(S_{i-1})) \\
&= \sum_{i \in N \setminus \{n\}}(f_i - f_{i-1})x(S_i) + f_n x(S_n) \\
&\geq \sum_{i \in N \setminus \{n\}}(f_i - f_{i-1})V(S_i) + f_n V(N) \\
&= \sum_{i \in N} f_i(V(S_i) - V(S_{i-1})) \\
&= \sum_{i \in N} f_i x^*_i,
\end{aligned}$$

where the inequality holds since $x \in \mathcal{C}(N, V)$, and the last inequality follows from the definition of x^*. Since f_i is strictly decreasing in i, the inequality would hold as a strict inequality if $x(S_i) \neq V(S_i)$ for some $i \in N \setminus \{n\}$. If $x(S_i) = V(S_i)$ for all $i \in N$, $x = x^*$. Hence, x^* is the unique optimal solution of problem (3.4) and therefore an extreme point of the core.

To show that no extreme point exists other than those in $\{x^\pi : \pi$ is a permutation$\}$, it suffices to show that for any given vector f, the optimization problem (3.4) has an optimal solution in $\{x^\pi : \pi$ is a permutation$\}$. In fact, without loss of generality, assume that

$$f_1 \geq f_2 \geq \ldots \geq f_n.$$

Following an analysis similar the one in the above paragraph, we can show that x^* is optimal to problem (3.4). ∎

Of course, it is possible that $x^\pi = x^{\pi'}$ for different permutations π and π'. However, if the characteristic function V is strictly supermodular, namely,

$$V(S) + V(T) < V(S \cup T) + V(S \cap T) \ \forall \ S, T \subseteq N \text{ with } S \not\subseteq T, T \not\subseteq S,$$

then one can show that $x^\pi \neq x^{\pi'}$ (see Exercise 3.5).

For cooperative games with empty cores, an alternative solution concept is the ϵ-core. For a given real number ϵ, the ϵ-core C_ϵ of a game (N, V) is given as

$$C_\epsilon = \{x \in \Re^n : x(N) = V(N), x(S) \geq V(S) - \epsilon \ \forall \ \emptyset \neq S \subset N\}.$$

That is, a payoff vector x is in the ϵ-core if it is efficient and a group of players won't be better off if it deviates from the grand coalition to form a subcoalition by paying a cost ϵ. Clearly, $C_\epsilon \subseteq C_{\epsilon'}$ if $\epsilon \leq \epsilon'$ and C_ϵ is nonempty for large ϵ.

The least-core of a game (N, V) is defined as the intersection of all nonempty ϵ-cores of (N, V), or equivalently C_{ϵ_0}, where ϵ_0 is the smallest ϵ such that C_ϵ is nonempty.

3.2.2 Nucleolus

As shown in the previous subsection, the core may be empty and even if nonempty, it may not be unique. In contrast, the nucleolus exists and is unique. Since the nucleolus is a singleton, it is often referred to as the unique element in it. To present its definition, we first define the lexicographic order. A vector $x \in \Re^n$ is smaller than a vector $y \in \Re^n$ in the lexicographic order if there exists an index $k \in \{1, \ldots, n-1\}$ such that

$$x_i = y_i \ \forall i \leq k, x_{k+1} < y_{k+1}.$$

We use $x <_{lex} y$ and $x \leq_{lex} y$ if x is smaller than y and x is smaller than or equal to y in the lexicographic order, respectively. Consider a game (N, V). For a given payoff vector $x \in \Re^n$, define the excess of coalition S under payoff x as

$$e_x(S) = V(S) - x(S) \ \forall \ \emptyset \neq S \subset N,$$

which is the difference of the value of and the proposed payoff to coalition S. Let $\theta(e_x)$ be a vector in $\Re^{2^n-2}$ whose components consist of the excesses $e_x(S)$ in decreasing order. That is, its first element is the largest $e_x(S)$, its second component is the second-largest $e_x(S)$, and so on. It is straightforward to show that for $i = 1, \ldots, 2^n - 2$, $\theta_i(e_x)$ is a continuous function of x. In addition, if x is in the core of (N, V), then $\theta_i(e_x) \leq 0$ for all $i = 1, \ldots, 2^n - 2$.

The nucleolus of the game (N, V), denoted by $\mathcal{N}(N, V)$, consists of optimal solutions minimizing $\theta(e_x)$ over all efficient payoff vectors in terms of the lexicographic order. That is, given a payoff vector $x \in \mathcal{N}(N, V)$,

$$\theta(e_x) \leq_{lex} \theta(e_y) \text{ for any efficient payoff vector } y.$$

One can argue that the nucleolus is the most equitable allocation because it sequentially minimizes the excess of the worst-treated coalitions.

Theorem 3.2.4 *For a given game (N, V), we have the following:*

(a) The nucleolus $\mathcal{N}(N, V)$ is always nonempty and is a singleton.

(b) If (N, V) is superadditive, the nucleolus is individually rational.

(c) The nucleolus belongs to the core if the core is nonempty.

Proof. To prove part (a), we first describe a procedure to find an element of the nucleolus. Given any efficient payoff vector y^0, let

$$\delta = \max_{\emptyset \neq S \subset N} e_{y^0}(S).$$

To search for the nucleolus, it suffices to focus on payoff vectors in the set

$$I_0 = \{y \in \Re^n : y(N) = y^0(N), y_i \geq V(\{i\}) - \delta\},$$

which is nonempty since $y^0 \in I_0$ and compact. Since $\theta_i(\theta_y)$ is continuous in y, for $i = 1, \ldots, 2^n - 2$, we can recursively show that

$$I_i = \text{argmin}_{y \in I_{i-1}} \theta_i(e_y)$$

is nonempty and compact as well. The set $I_{2^n - 2}$ is exactly the nucleolus, which is clearly nonempty.

We now show that the nucleolus is a singleton. Assume to the contrary that $x, y \in I_{2^n - 2}$ with $x \neq y$. The above procedure implies that $\theta(e_x) = \theta(e_y)$. Let $S_1, S_2, \ldots, S_{2^n - 2}$ be an ordered sequence of all nonempty proper subsets of N so that $\theta_i(e_x) = e_x(S_i)$ and let k be the smallest index such that $\theta_k(e_y) \neq e_y(S_k)$. We assume that

$$\text{either } e_x(S_l) < e_x(S_k) \text{ or } e_y(S_l) < \theta_k(e_y) = e_x(S_k) \; \forall l > k.$$

This assumption can be made without loss of generality. In fact, if the assumption does not hold, there exists l with $l > k$ such that $e_x(S_l) = e_x(S_k) = e_y(S_l)$. We can simply consider a new sequence in which the positions of S_k and S_l are switched. Repeat the process until we end up with a sequence satisfying the assumption. Define $z = \frac{x+y}{2}$. For any $l \leq k$, since $e_x(S_l) = e_y(S_l)$, we have that $\theta_l(z) = \theta_l(x) = e_z(S_l)$, and for any $l \geq k$,

$$e_z(S_l) = (e_x(S_l) + e_y(S_l))/2 < e_x(S_k) = \theta_k(e_x).$$

Therefore, $\theta_k(e_z) < \theta_k(e_x)$, which contradicts the assumption that x has the minimum $\theta_k(e_x)$ in terms of the lexicographic order. Thus, the nucleolus is unique.

To prove part (b), let x be the nucleolus and $i \in \text{argmax}_{j \in N}(V(\{j\}) - x_j)$. Assume to the contrary that $V(\{i\}) > x_i$. We claim $i \in S$ for any $S \subset N$ such that $\theta_1(e_x) = e_x(S)$; that is, $S \in \text{argmax}_{\hat{S}} e_x(\hat{S})$. In fact, if $i \notin S$, from the superadditive property of V, e_x is superadditive and

$$e_x(S \cup \{i\}) \geq e_x(S) + e_x(\{i\}) = e_x(S) + V(\{i\}) - x_i > e_x(S),$$

which contradicts the definition of S.

Define a new payoff vector y such that $y_i = x_i + \epsilon$ and $y_j = x_j - \frac{\epsilon}{n-1}$ for $j \neq i$ for some small $\epsilon > 0$. Pick any $S \in \operatorname{argmax}_{\hat{S}} e_x(\hat{S})$. Since $i \in S$, we have that

$$e_y(S) = e_x(S) - \epsilon \frac{n - |S|}{n - 1} < e_x(S) = \theta_1(e_x).$$

For any $\hat{S}$ with $e_x(\hat{S}) < \theta_1(e_x)$, as long as ϵ is sufficiently small, $y(S) - x(S)$ is small and

$$e_y(\hat{S}) = e_x(\hat{S}) - (y(\hat{S}) - x(\hat{S})) < \theta_1(e_x) - \epsilon \frac{n - |S|}{n - 1} = e_y(S).$$

Thus, $S \in \operatorname{argmax}_{\hat{S}} e_x(\hat{S})$ or, equivalently, $e_y(S) = \theta_1(e_y)$. However, $\theta_1(e_y) < \theta_1(e_x)$, which contradicts the definition of the nucleolus. Thus, the nucleolus is individually rational.

Finally, if the core is nonempty, pick any core allocation z. Again, let x be the nucleolus. By definition,

$$V(S) - x(S) = e_x(S) \leq \theta_1(e_x) \leq \theta_1(e_z) \leq 0 \,\forall\, \emptyset \neq S \subset N.$$

Thus, the nucleolus is in the core. ∎

Loosely speaking, if the core is nonempty, its nucleolus occupies a center position in the sense that the minimum distance to any boundary of the core is as large as possible.

In addition to what we proved in the above theorem, the nucleolus satisfies the following properties.

Theorem 3.2.5 *The nucleolus $\mathcal{N}(N,V)$ satisfies the following properties: (a) efficiency, (b) anonymity, (c) symmetry, (d) covariance under strategic equivalence, (e) null player, and (f) consistency.*

Parts (a) and (b) of the above theorem are obvious. However, the proof of the remaining parts is involved and is left as an exercise.

3.2.3 Shapley Value

The Shapley value is another single-valued solution concept. For a cooperative game (N,V), the Shapley value $\phi \in \Re^n$ is given by

$$\phi_i = \sum_{S \subseteq N \setminus \{i\}} \frac{|S|!(n - |S| - 1)!}{n!} (V(S \cup \{i\}) - V(S)) \,\forall\, i \in N.$$

The Shapley value can be interpreted as follows. The players arrive one by one to form the grand coalition, and all possible arrival sequences are equally likely.

When Player i arrives and finds a set of players S has arrived, he receives a payoff $V(S \cup \{i\}) - V(S)$, his marginal contribution added to S. The Shapley value of Player i is his expected marginal contribution. To see this, note that there are $n!$ different arrival sequences in total and each arrival sequence, corresponding to a permutation of $\{1, 2, \ldots, n\}$, is equally likely. In addition, the number of different arrival sequences in which the players in S are exactly those who arrive before i is $|S|!(n - |S| - 1)!$, the product of the number of different arrival sequences of the first $|S|$ players before i and that of the remaining $n - |S| - 1$ players after i.

The Shapley value is widely used because it is unique and enjoy several plausible properties.

Theorem 3.2.6 *The Shapley value satisfies the following properties: (a) efficiency, (b) symmetry, (c) anonymity, (d) additivity, (e) null player, (f) (strong) aggregate monotonicity, (g) coalitional monotonicity, and (h) strong monotonicity.*

To show that the Shapley value is efficient, recall that the Shapley value of player i can be interpreted as Player i's expected marginal contribution added to its predecessors when any arrival sequence is equally likely. Thus, for any given arrival sequence, the total marginal contribution added to their predecessors of all players is exactly $V(N)$. Thus, $\sum_{i \in N} \phi_i = V(N)$. It is straightforward to show that the Shapley value satisfies the remaining properties.

Interestingly, the Shapley value is a unique solution concept that satisfies the above properties.

Theorem 3.2.7 *The Shapley value is a unique solution concept that satisfies the efficiency, symmetry, additivity, and null player properties. It is also a unique solution concept that satisfies the efficiency, symmetry, and strong monotonicity properties.*

The proof of the theorem is left as an exercise.

For a convex game, we have shown that for any permutation of $\{1, 2, \ldots, n\}$, π, the greedy algorithm gives a payoff vector x^π in the core, which is exactly the vector of marginal contributions added to their predecessors of the players if the arrival sequence is $(\pi(1), \pi(2), \ldots, \pi(n))$, with $p(i)$ denoting the i arrival. Therefore, the Shapley value is the convex combination of payoff vectors x^π for all permutations π and thus is in the core. Actually, it is in the center of gravity of the core since the weights of the payoff vectors x^π, the extreme points of the core, are $1/n!$.

Theorem 3.2.8 *For a convex game (N, V), the Shapley value is in the core.*

Unfortunately, for a nonconvex game, the Shapley value may not be core allocations even when the core is nonempty.

3.3   Exercises

Exercise 3.1. Consider the generalized form of the prisoner's dilemma in which the two prisoners' payoffs are specified in general values (Table 3.4). Find the

TABLE 3.4. The generalized form of the prisoner's dilemma

	Silent		Betray	
Silent	A,	A	C,	B
Betray	B,	C	D,	D

equilibria of the game for all possible values of A, B, C, and D.

Exercise   3.2. Consider   a   game $(N, \{S_i\}_{i \in N}, \{u_i\}_{i \in N})$.   If S_i is   one-dimensional and u_i is smooth for all $i \in N$, show that the diagonally dominant conditions (3.1) on the second derivatives of u_i are sufficient for the best response operator to be a contraction and thus a unique equilibrium exists. Extend the diagonal dominance conditions to multidimensional case.

Exercise 3.3. Provide the detailed proof of Theorem 3.1.7.

Exercise 3.4. Prove that a game (N, V) is convex if and only if (3.3) holds.

Exercise 3.5. Show that for a cooperative game with a strictly supermodular characteristic function, a different permutation π of $\{1, 2, \ldots, n\}$ leads to a different extreme point x^π of its core in Sect. 3.2.1.

Exercise 3.6. Prove that the nucleolus is a unique solution concept for the class of all cooperative coalitional games that satisfies the efficiency, consistency, covariance under strategic equivalence, and anonymity properties.

Exercise 3.7. Prove Theorem 3.2.5.

Exercise 3.8. Prove Theorem 3.2.7.

Exercise 3.9. Consider the cooperative coalitional game (N, V) with $N = \{1, 2, 3\}$ and the characteristic function given by

$$V(\emptyset) = 0, V(\{i\}) = 0 \; \forall \; i \in N, V(S) = \alpha \; \forall \; |S| = 2, V(N) = 1,$$

where α is parameter. Compute the core, the nucleolus, and the Shapley value.

Exercise 3.10. Consider the cooperative production game in which k players collaborate to produce n products using m resources. Assume that player l ($l = 1, \ldots, n$) has a resource vector $b^l \in \Re^m$. One unit of product j ($j = 1, \ldots, n$)

can be sold at a profit r_j and takes a_{ij} units of resource i $(i = 1, \ldots, m)$ to produce. To determine how the profit should be distributed among the players if they decide to cooperate, we can formulate this as a production game (K, V) with $K = \{1, 2, \ldots, k\}$ and for any coalition S, $V(S)$ being the maximum profit the coalition can generate by pooling their resources together. Is the game convex? Show that it has a nonempty core. Is the Shapley value in the core?

4
Worst-Case Analysis

4.1 Introduction

Since most complicated logistics problems, for example, the bin-packing problem and the traveling salesman problem, are $\mathcal{NP}$-Hard, it is unlikely that polynomial-time algorithms will be developed for their optimal solutions. Consequently, a great deal of work has been devoted to the development and analyses of heuristics. In this chapter, we demonstrate one important tool, referred to as *worst-case performance analysis*, which establishes the maximum deviation from optimality that can occur for a given heuristic algorithm. We will characterize the worst-case performance of a variety of algorithms for the bin-packing problem and the traveling salesman problem. The results obtained here serve as important building blocks in the analysis of algorithms for vehicle routing problems.

Worst-case effectiveness is essentially measured in two different ways. Take a generic problem, and let I be a particular instance. Let $Z^*(I)$ be the total cost of the optimal solution, for instance, I. Let $Z^H(I)$ be the total cost of the solution provided by the heuristic H on instance I. Then, the *absolute performance ratio of heuristic H* is defined as

$$R^H \doteq \inf \left\{ r \geq 1 \mid \frac{Z^H(I)}{Z^*(I)} \leq r, \text{ for all } I \right\}.$$

This measure, of course, is specific to the particular problem. The absolute performance ratio is often achieved for very small problem instances. It is therefore

D. Simchi-Levi et al., *The Logic of Logistics: Theory, Algorithms, and Applications for Logistics Management*, Springer Series in Operations Research and Financial Engineering, DOI 10.1007/978-1-4614-9149-1_4, © Springer Science+Business Media New York 2014

desirable to have a measure that takes into account problems of large size only. This measure is the *asymptotic performance ratio*. For a heuristic H, this ratio is defined as

$$R_\infty^H \doteq \inf \left\{ r \geq 1 \mid \exists n \text{ such that } \frac{Z^H(I)}{Z^*(I)} \leq r, \text{ for all } I \text{ with } Z^*(I) \geq n \right\}.$$

This measure sometimes gives a more accurate picture of a heuristic's performance. Note that $R_\infty^H \leq R^H$.

In general, it is important also to show that no better worst-case bound (for a given heuristic) is possible. This is usually achieved by providing an example, or family of examples, where the bound is tight, or arbitrarily close to tight.

In this chapter, we will analyze several heuristics for two difficult problems, the bin-packing problem and the traveling salesman problem, along with their worst-case performance bounds.

4.2   The Bin-Packing Problem

The bin-packing problem (BPP) can be stated as follows: Given a list of n real numbers $L = (w_1, w_2, \ldots, w_n)$, where we call $w_i \in (0, 1]$ the size of item i, the problem is to assign each item to a bin such that the sum of the item sizes in a bin does not exceed 1, while minimizing the number of bins used. For simplicity, we also use L as a set, but this should cause no confusion. In this case, we write $i \in L$ to mean $w_i \in L$.

Many heuristics have been developed for this problem since the early 1970s. Some of the more popular ones are first-fit (FF), best-fit (BF), first-fit decreasing (FFD), and best-fit decreasing (BFD) analyzed by Johnson et al. (1974). First-fit and best-fit assign items to bins according to the order they appear in the list without using any knowledge of subsequent items in the list; these are *online* algorithms. First-fit can be described as follows: Place item 1 in bin 1. Suppose we are packing item j; place item j in the lowest indexed bin whose current content does not exceed $1 - w_j$. The BF heuristic is similar to FF except that it places item j in the bin whose current content is the *largest* but does not exceed $1 - w_j$. In contrast to these heuristics, FFD first sorts the items in nonincreasing order of their size and then performs FF. Similarly, BFD first sorts the items in nonincreasing order of their size and then performs BF. These are called *offline* algorithms.

Let $b^H(L)$ be the number of bins produced by a heuristic H on list L. Similarly, let $b^*(L)$ be the minimum number of bins required to pack the items in list L; that is, $b^*(L)$ is the optimal solution to the bin-packing problem defined in list L.

The best asymptotic performance bounds for the FF and BF heuristics are given in Garey et al. (1976), where they show that

$$b^{FF}(L) \leq \left\lceil \frac{17}{10} b^*(L) \right\rceil$$

and

$$b^{\text{BF}}(L) \leq \left\lceil \frac{17}{10} b^*(L) \right\rceil.$$

Here $\lceil x \rceil$ is defined as the smallest integer greater than or equal to x.

The best asymptotic performance bounds for FFD and BFD have been obtained by Baker (1985), who shows that

$$b^{\text{FFD}}(L) \leq \frac{11}{9} b^*(L) + 3$$

and

$$b^{\text{BFD}}(L) \leq \frac{11}{9} b^*(L) + 3.$$

Johnson et al. (1974) provides instances with arbitrarily large values of $b^*(L)$ such that the ratios $\frac{b^{\text{FF}}(L)}{b^*(L)}$ and $\frac{b^{\text{BF}}(L)}{b^*(L)}$ approach $\frac{17}{10}$ and instances where $\frac{b^{\text{FFD}}(L)}{b^*(L)}$ and $\frac{b^{\text{BFD}}(L)}{b^*(L)}$ approach $\frac{11}{9}$. Thus, the maximum deviation from optimality for all lists that are sufficiently "large" is no more than 70 % times the minimal number of bins in the case of FF and BF, and 22.2 % in the case of FFD and BFD.

We now show that by using simple arguments, one can characterize the absolute performance ratio for each of the four heuristics. We start, however, by demonstrating that in general we cannot expect to find a polynomial-time heuristic with absolute performance ratio less than $\frac{3}{2}$.

Lemma 4.2.1 *Suppose there exists a polynomial-time heuristic H for the BPP with $R^H < 3/2$; then $\mathcal{P} = \mathcal{NP}$.*

Proof. We show that if such a heuristic exists, then we can solve the $\mathcal{NP}$-Complete 2-partition problem in polynomial time. This problem is defined as follows: Given a set $A = \{a_1, a_2, \ldots, a_n\}$, does there exist an $A_1 \subset A$ such that $\sum_{a_i \in A_1} a_i = \sum_{a_i \in A \setminus A_1} a_i$?

For a given instance A of 2-partition, we construct an instance L of the bin-packing problem with items sizes a_i and bins of capacity $\frac{1}{2} \sum_A a_i$ Observe that if there exists an A_1 such that $\sum_{A_1} a_i = \sum_{A \setminus A_1} a_i = \frac{1}{2} \sum_A a_i$, then the heuristic H must find a solution such that $b^{\text{H}}(L) = 2$. On the other hand, if there is no such A_1 in the 2-partition problem, then the corresponding bin-packing problem has no solution with fewer than three bins, and hence $b^{\text{H}}(L) \geq 3$.

Consequently, to solve the 2-partition problem, apply the heuristic H to the corresponding bin-packing problem. If $b^{\text{H}}(L) \geq 3$, there is no subset A_1 with the desired property. Otherwise, there is one. Since 2-partition is $\mathcal{NP}$-Complete, this implies $\mathcal{P} = \mathcal{NP}$. ∎

Let XF be either FF or BF, and let XFD be either FFD or BFD. In this section, we prove the following result due to Simchi-Levi (1994).

Theorem 4.2.2 *For all lists L,*

$$\frac{b^{\text{XF}}(L)}{b^*(L)} \leq \frac{7}{4}.$$

and

$$\frac{b^{\mathrm{XFD}}(L)}{b^*(L)} \leq \frac{3}{2}.$$

In view of Lemma 4.2.1, it is clear that FFD and BFD have the best possible absolute performance ratios for the bin-backing problem among all polynomial-time heuristics. As Garey and Johnson (1979, p. 128) point out, it is easy to construct examples in which an optimal solution uses two bins while FFD or BFD uses three bins. Similarly, Johnson et al. give examples in which an optimal solution uses 10 bins while FF and BF use 17 bins. Thus, the absolute performance ratio for FFD and BFD is exactly $\frac{3}{2}$, while it is at least 1.7 and no more than $\frac{7}{4}$ for FF and BF.

We now define the following terms, which will be used throughout this section. An item is called *large* if its size is (strictly) greater than 0.5; otherwise, it is called *small*. Define a bin to be of type I if it has only small items and of type II if it is not a type I bin; that is, it has at least one large item in it. A bin is called *feasible* if the sum of the item sizes in the bin does not exceed 1. An item is said to *fit* in a bin if the bin resulting from the insertion of this item is a feasible bin. In addition, a bin is said to be *opened* when an item is placed in a bin that was previously empty.

4.2.1   First-Fit and Best-Fit

The proof of the worst-case bounds for FF and BF, the first part of Theorem 4.2.2, is based on the following observation. Recall XF = FF or BF.

Lemma 4.2.3 *Consider the jth bin opened by XF ($j \geq 2$). Any item that was assigned to it before it was more than half-full does not fit in any bin opened by XF prior to bin j.*

Proof. The property is clearly true for FF, and in fact holds for any item assigned to the jth bin, $j \geq 2$, not necessarily to items assigned to it before it was more than half-full. To prove the property for BF, suppose by contradiction that item i was assigned to the jth bin before it was more than half-full, and this item fits in one of the previously opened bins, say the kth bin. Clearly, in that case, i cannot be the first item assigned to the jth bin since BF would not have opened a new bin if i fits in one of the previously opened bins. Let the levels of bins k and j, just before the time item i was packed by BF, be α_k and α_j, and let item h be the first item in bin j. Hence, $w_h \leq \alpha_j \leq \frac{1}{2}$ by the hypothesis. Since BF assigns an item to the bin where it fits with the largest content, and item i would have fit in bin k, we have $\alpha_j > \alpha_k$. Thus, $\alpha_k < \frac{1}{2}$, meaning that item H would have fit in bin k, a contradiction. ∎

We use Lemma 4.2.3 to construct a lower bound on the minimum number of bins. For this purpose, we introduce the following procedure. For a given integer v, $2 \leq v \leq b^{\mathrm{XF}}(L)$, select v bins from those produced by XF. Index the v bins in the order they are opened, starting with 1 and ending with v. Let X_j be the set of items assigned by XF to the jth bin before it was more than half-full, $j = 1, 2, \ldots, v$. Let S_j be the set of items assigned by XF to the jth bin, $j = 1, 2, \ldots, v$. Observe that $X_j \subseteq S_j$ for all $j = 1, 2, \ldots, v$.

Procedure LBBP (Lower-Bound Bin-Packing)

Step 1: Let $X'_i = X_i$, $i = 1, 2, \ldots, v$.

Step 2: For $i = 1$ to $v - 1$, do
  Let $j = \max\{k : X'_k \neq \emptyset\}$.
  If $j \leq i$, stop.
  Else, let u be the smallest item in X'_j.
  Set $S_i \leftarrow S_i \cup \{u\}$ and $X'_j \leftarrow X'_j \setminus \{u\}$.

In view of Lemma 4.2.3, it is clear that Procedure LBBP generates nonempty subsets $S_1, S_2, \ldots, S_m$, for some $m \leq v$, such that $\sum_{i \in S_j} w_i > 1$ for $j \leq m - 1$ and possibly for $j = m$. This is true since by Lemma 4.2.3, item u (as defined in the LBBP procedure), originally assigned to bin j before it was more than half-full, does not fit in any bin i with $i < j$. Then the following must hold.

Lemma 4.2.4 $\max \left\{ |\bigcup_{j=m+1}^{v} X_j|, \; m - 1 \right\} < \sum_{j=1}^{v} \sum_{i \in S_j} w_i$.

Proof. Since bins $1, 2, \ldots, m - 1$ generated by Procedure LBBP are not feasible, we have $\sum_{j=1}^{v} \sum_{i \in S_j} w_i > m - 1$. Note that every item in $\bigcup_{j=m+1}^{v} X_j$ is moved by Procedure LBBP to exactly one S_j, $j = 1, 2, \ldots, m - 1$, and possibly to S_m. Thus, if S_m is feasible, that is, no (additional) item is assigned by Procedure LBBP to S_m, then $|\bigcup_{j=m+1}^{v} X_j| \leq m - 1 < \sum_{j=1}^{v} \sum_{i \in S_j} w_i$. On the other hand, if an item is assigned by Procedure LBBP to S_m, then none of the subsets S_j, $j = 1, 2, \ldots, m$, is feasible, and therefore, $m = |\bigcup_{j=m+1}^{v} X_j| < \sum_{j=1}^{v} \sum_{i \in S_j} w_i$. ∎

We are now ready to prove the first part of Theorem 4.2.2, that is, establish the upper bound on the absolute performance ratio of the XF heuristic. Let c be the number of large items in the list L. Without loss of generality, assume $b^{\mathrm{XF}}(L) > c$ since otherwise the solution produced by XF is optimal. So $b^{\mathrm{XF}}(L) - c > 0$ is the number of type I bins produced by XF. We consider the following two cases.

Case 1: c is even. In this case, we partition the bins produced by XF into two sets. The first set includes only type I bins, while the second set includes the remaining bins produced by XF, that is, all the type II bins. Index the bins in the first set in the order they are opened, from 1 to $b^{\mathrm{XF}}(L) - c$. Let $v = b^{\mathrm{XF}}(L) - c$, and apply Procedure LBBP to the set of type I bins, producing m bins out of which at least $m - 1$ are infeasible. Then

Lemma 4.2.5 *If c is even,*

$$\max \left\{ \frac{c}{2} + m, \; 2(b^{\mathrm{XF}}(L) - m) - \frac{3c}{2} \right\} \leq b^*(L).$$

Proof. Combining Lemma 4.2.4 with the fact that no two large items fit in the same bin, we have $\sum_{i\in L} w_i > m-1+\frac{c}{2}$. On the other hand, every bin in an optimal solution is feasible, and therefore, $\sum_{i\in L} w_i \leq b^*(L)$. Since c is even, $m+\frac{c}{2} \leq b^*(L)$. Since we applied Procedure LBBP only to the type I bins produced by XF, each one of these bins has at least two items except possibly one that may have only one item. Hence, $2(b^{\mathrm{XF}}(L) - m - c - 1) + 1 \leq |\bigcup_{j=m+1}^{v} X_j|$ and therefore, using Lemma 4.2.4,

$$2(b^{\mathrm{XF}}(L) - m - c - 1) + \frac{c}{2} + 1 < \sum_{i\in L} w_i \leq b^*(L)$$

or

$$2(b^{\mathrm{XF}}(L) - m - c - 1) + \frac{c}{2} + 2 \leq b^*(L).$$

Rearranging the left-hand side gives the second lower bound. ∎

Theorem 4.2.6 *If c is even,*

$$b^{\mathrm{XF}}(L) \leq \frac{7}{4}b^*(L).$$

Proof. From Lemma 4.2.5, we have $2(b^{\mathrm{XF}}(L) - m) - \frac{3c}{2} \leq b^*(L)$. Hence,

$$b^{\mathrm{XF}}(L) \leq \frac{b^*(L)}{2} + \frac{3c}{4} + m$$

$$= \frac{b^*(L)}{2} + (m + \frac{c}{2}) + \frac{c}{4}$$

$$\leq \frac{7}{4}b^*(L)$$

since $m + \frac{c}{2}$, $b^*(L)$ and c are lower bounds. ∎

Case 2: c is odd. In this case, we partition the set of all bins generated by the XF heuristic in a slightly different way. The first set of bins, called B_1, comprises all the type I bins except the last type I bin opened by XF. The second set is made up of the remaining bins; that is, these are all the type II bins together with the type I bin not included in B_1. We now apply Procedure LBBP to the bins in B_1 [with $v = b^{\mathrm{XF}}(L) - c - 1$], producing m bins out of which at least $m - 1$ bins are not feasible.

Lemma 4.2.7 *If c is odd,*

$$\max\left\{\frac{c}{2} + m + \frac{1}{2},\ 2(b^{\mathrm{XF}}(L) - m) - \frac{3c}{2} - \frac{1}{2}\right\} \leq b^*(L).$$

Proof. Take one of the type II bins and "match" it with the only type I bin not in B_1; the total weight of these two bins is more than 1. Thus, using Property 4.2.4, we have $\frac{c-1}{2} + 1 + (m-1) < \sum_{i \in L} w_i \leq b^*(L)$, which proves the first lower bound. To prove the second lower bound, we use the fact that every bin in B_1 has at least two items, and therefore, $2(b^{\mathrm{XF}}(L) - m - c - 1) \leq |\bigcup_{j=m+1}^{v} X_j|$. Using Property 2.2, we get

$$2(b^{\mathrm{XF}}(L) - m - c - 1) + \frac{c-1}{2} + 1 < \sum_{i \in L} w_i \leq b^*(L)$$

or

$$2(b^{\mathrm{XF}}(L) - m - c - 1) + \frac{c-1}{2} + 2 \leq b^*(L).$$

Rearranging the left-hand side gives the second lower bound. ∎

Theorem 4.2.8 *If c is odd,*

$$b^{\mathrm{XF}}(L) \leq \frac{7}{4}b^*(L) - \frac{1}{4}.$$

Proof. From Lemma 4.2.7, we have $2(b^{\mathrm{XF}}(L) - m) - \frac{3c}{2} - \frac{1}{2} \leq b^*(L)$. Hence,

$$
\begin{aligned}
b^{\mathrm{XF}}(L) &\leq \frac{b^*(L)}{2} + m + \frac{3c}{4} + \frac{1}{4} \\
&= \frac{b^*(L)}{2} + \left(m + \frac{c}{2} + \frac{1}{2}\right) + \frac{c}{4} - \frac{1}{4} \\
&\leq \frac{7}{4}b^*(L) - \frac{1}{4}.
\end{aligned}
$$

∎

4.2.2 First-Fit Decreasing and Best-Fit Decreasing

The proof of the worst-case bounds for FFD and BFD is based on Lemma 4.2.3. This lemma states that if a bin produced by these heuristics contains only items of size at most $\frac{1}{2}$, then the first two items assigned to the bin cannot fit in any bin opened prior to it.

Let XFD denote either FFD or BFD. Index the bins produced by XFD in the order they are opened. We consider three cases. First, suppose $b^{\mathrm{XFD}}(L) = 3p$ for some integer $p \geq 1$. Consider the bin with index $2p + 1$. If this bin contains a large item, we are done since in that case $b^*(L) > 2p = \frac{2}{3}b^{\mathrm{XFD}}(L)$. Otherwise, bins $2p + 1$ through $3p$ must contain at least $2p - 1$ small items, none of which can fit in the first $2p$ bins. Hence, the total sum of the item sizes exceeds $2p - 1$, meaning that $b^*(L) \geq 2p = \frac{2}{3}b^{\mathrm{XFD}}(L)$.

Suppose $b^{\mathrm{XFD}}(L) = 3p + 1$. If bin $2p + 1$ contains a large item, we are done. Otherwise, bins $2p + 1$ through $3p + 1$ contain at least $2p + 1$ small items, none of which can fit in the first $2p$ bins, implying that the total sum of the item sizes exceeds $2p$ and hence $b^*(L) \geq 2p + 1 > \frac{2}{3}b^{\mathrm{XFD}}(L)$.

Similarly, suppose $b^{XFD}(L) = 3p + 2$. If bin $2p + 2$ contains a large item, we are done. Otherwise, bins $2p + 2$ through $3p + 2$ contain at least $2p + 1$ small items, none of which can fit in the first $2p + 1$ bins, implying the sum of the item sizes exceeds $2p + 1$, and hence $b^*(L) \geq 2p + 2 > \frac{2}{3}b^{XFD}(L)$.

4.3   The Traveling Salesman Problem

Interesting worst-case results have been obtained for another combinatorial problem that plays an important role in the analysis of logistics systems: the traveling salesman problem (TSP). The problem can be defined as follows: Let $G = (V, E)$ be a complete undirected graph with vertices V, $|V| = n$, and edges E, and let d_{ij} be the *length* of edge (i, j). [We use the term *length* to designate the "cost" of using edge (i, j). The most general formulation of the TSP allows for completely arbitrary "lengths," and, in fact, in many applications the physical distance is irrelevant and the d_{ij} simply represents the *cost* of sequencing j immediately after i.] The objective in the TSP is to find a tour that visits each vertex exactly once and whose total length is as small as possible. The problem has been analyzed extensively in the last three decades; see Lawler et al. (1985) for an excellent survey and, in particular, the chapter written by Johnson and Papadimitriou (1985), which includes some of the worst-case results presented here.

We shall examine a variety of heuristics for the TSP and show that, for an important special case of this problem, heuristics with strong worst-case bounds exist. We start, however, with a negative result, due to Sahni and Gonzalez (1976), which states that, in general, finding a heuristic for the TSP with a constant worst-case bound is as hard as solving any $\mathcal{NP}$-Complete problem, no matter what the bound.

To present the result, let I be an instance of the TSP. Let $L^*(I)$ be the length of the optimal traveling salesman tour through V. Given a heuristic H, let $L^H(I)$ be the length of the tour generated by H.

Theorem 4.3.1 *Suppose there exist a polynomial-time heuristic H for the TSP and a constant R^H such that for all instances I,*

$$\frac{L^H(I)}{L^*(I)} \leq R^H;$$

then $\mathcal{P} = \mathcal{NP}$.

Proof. The proof is in the same spirit as the proof of Lemma 4.2.1. Suppose such a heuristic exists. We will use it to solve the $\mathcal{NP}$-Complete Hamiltonian cycle problem in polynomial time. The Hamiltonian cycle problem is defined as follows. Given a graph $G = (V, E)$, does there exist a *simple cycle* (a cycle that does not visit a point more than once) in G that includes all of V? To answer this question, we construct an instance I of the TSP and apply H to it; the length of the tour generated by H will tell us whether G has a Hamiltonian cycle.

The instance I is defined on a complete graph whose set of vertices is V and the length of each edge $\{i, j\}$ is

$$d_{ij} = \begin{cases} 1, & \text{if } \{i, j\} \in E; \\ |V|R^{\mathrm{H}}, & \text{otherwise.} \end{cases}$$

We distinguish between two cases depending on whether G contains a Hamiltonian cycle. If G does not contain a Hamiltonian cycle, then any traveling salesman tour in I must contain at least one edge with length $|V|R^{\mathrm{H}}$, and hence the length of the tour generated by H is at least $|V|R^{\mathrm{H}} + |V| - 1$.

On the other hand, if G has a Hamiltonian cycle, then I must have a tour of length $|V|$. This is true since we can use the Hamiltonian cycle as a traveling salesman tour for the instance I in which the vertices appear on the traveling salesman tour in the same order they appear in the Hamiltonian cycle. Thus, if G has a Hamiltonian cycle, heuristic H applied to I must provide a tour of length no more than $|V|R^{\mathrm{H}}$.

Consequently, we have a method for solving the Hamiltonian cycle problem: Apply H to the TSP defined on the instance I. If $L^{\mathrm{H}}(I) \leq |V|R^{\mathrm{H}}$, then there exists a Hamiltonian cycle in G. Otherwise, there is no such cycle in G. Finally, since H is assumed to be polynomial, we conclude that $\mathcal{P} = \mathcal{NP}$.   ∎

The theorem thus implies that it is very unlikely that a polynomial-time heuristic for the TSP with a constant absolute worst-case bound exists. However, there is an important version of the traveling salesman problem that excludes the above negative result. This is when the distance matrix $\{d_{ij}\}$ satisfies *the triangle inequality assumption*.

Definition 4.3.2 *A distance matrix satisfies the triangle inequality assumption if for all $i, j, k \in V$, we have $d_{ij} \leq d_{ik} + d_{kj}$.*

In many logistics environments, the triangle inequality assumption is not a very restrictive one. It merely states that traveling directly from point (vertex) i to point (vertex) j is at most the cost of traveling from i to j through the point k.

In the next four sections, we describe and analyze different heuristics developed for the TSP. To simplify the presentation in what follows, we write L^* instead of $L^*(I)$; this should cause no confusion.

4.3.1   A Minimum Spanning Tree-Based Heuristic

The following algorithm provides a simple example of how a fixed worst-case bound is possible for the TSP when the distance matrix satisfies the triangle inequality assumption. In this case, the bound is 2; that is, the heuristic provides a solution with total length at most $100\,\%$ above the length of an optimal tour.

A *spanning tree* of a graph $G = (V, E)$ is a connected subgraph with $|V| - 1$ edges spanning all of V. The cost (or weight) of a tree is the sum of the length of the edges in the tree. A minimum spanning tree (MST) is a spanning tree with minimum cost. It is well known and easy to show that a minimum spanning tree can be found in polynomial time [see, for example, Papadimitriou and Stieglitz (1982)]. If W^* denotes the weight (cost) of the minimum spanning tree, then we must have $W^* \leq L^*$ since deleting any edge from the optimal tour results in a spanning tree.

The minimum spanning tree can be used to find a feasible traveling salesman tour in polynomial time. The idea is to perform a depth-first search [see Aho et al. (1974)] over the minimum spanning tree and then to do simple improvements on this solution. Formally, this is done as follows (Johnson and Papadimitriou 1985).

A Minimum Spanning Tree-Based Heuristic

Step 1: Construct a minimum spanning tree and color its edges white, and all other edges black.

Step 2: Let the *current vertex* (denoted v) be an arbitrary vertex.

Step 3: If one of the edges adjacent to v in the MST is white, color it black and proceed to the vertex at the other end of this edge. Else (all edges from v are black), go back along the edge by which the current vertex was originally reached.

Step 4: Let this vertex be v. Stop if v is the vertex you started with and all edges of MST are black. Otherwise, go to *step 3*.

Observe that the above strategy produces a tour that starts and ends at one of the vertices and visits all other vertices in the graph covering each arc twice. This is not a very efficient tour since some vertices may be visited more than once. To improve on this tour, we can modify the above strategy as follows: Instead of going back to a visited vertex, we can use a *shortcut* strategy in which we skip this vertex, and go directly to the next unvisited vertex. The triangle inequality assumption implies that the above modification will not increase the length of the tour and, in fact, may reduce it.

Let L^{MST} be the length of the traveling salesman tour generated by the above strategy. We clearly have

$$L^{\mathrm{MST}} \leq 2W^* \leq 2L^*,$$

where the first inequality follows since without shortcuts, the length of the tour is exactly $2W^*$. This proves that the worst-case bound of the algorithm is at most 2. It remains to verify that the worst-case bound of this heuristic cannot be improved. For this purpose, consider Fig. 4.1, the example constructed by Johnson and Papadimitriou (1985). Here, $W^* = \frac{n}{3} + \frac{n}{3}(1-\epsilon) + 2\epsilon - 1$, $L^{\mathrm{MST}} \approx \frac{2n}{3} + \frac{2n}{3}(1-\epsilon)$, and $L^* = \frac{2n}{3}$.

4.3.2 The Nearest-Insertion Heuristic

Before describing this heuristic, we consider the following intuitively appealing strategy, called the *nearest-neighbor heuristic*. Given an instance I of the TSP, start with an arbitrary vertex and find the vertex not yet visited that is closest to the current vertex. Travel to this vertex. Repeat this until all vertices are visited; then go back to the starting vertex.

Unfortunately, Rosenkrantz et al. (1977) show the existence of a family of instances for the TSP with arbitrary n with the following property. The length of the tour generated by the nearest-neighbor heuristic on each instance in the family is $O(\log n)$ times the length of the optimal tour. Thus, the nearest-neighbor heuristic does not have a bounded worst-case performance.

This comes as no surprise since the algorithm obviously suffers from one major weakness. This "greedy" strategy tends to begin well, inserting very short arcs into the path, but ultimately it ends with arcs that are quite long. For instance, the last edge added, the one connecting the last node to the starting node, may be very long due to the fact that at no point does the heuristic consider the location of the starting vertex and possible ending vertices.

One way to improve the performance of the nearest-neighbor heuristic is presented in the following variant, called the *nearest-insertion* (NI) heuristic, developed and analyzed by Rosenkrantz et al. Informally, the heuristic works as follows: At each iteration of the heuristic, a Hamiltonian cycle containing a subset of the vertices is constructed. The heuristic then selects a new vertex not yet in the cycle that is "closest" in a specific sense and inserts it between two adjacent vertices in

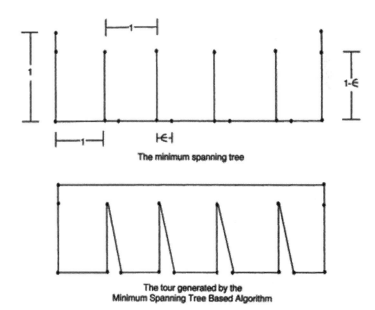

FIGURE 4.1. An example for the minimum spanning tree-based algorithm with $n = 18$

the cycle. The process stops when all vertices are in the cycle. Formally, this is done as follows.

The Nearest-Insertion Heuristic

Step 1: Choose an arbitrary node v and let the cycle C consist of only v.

Step 2: Find a node outside C closest to a node in C; call it k.

Step 3: Find an edge $\{i, j\}$ in C such that $d_{ik} + d_{kj} - d_{ij}$ is minimal.

Step 4: Construct a new cycle C by replacing $\{i, j\}$ with $\{i, k\}$ and $\{k, j\}$.

Step 5: If the current cycle C contains all the vertices, stop. Otherwise, go to *step 2*.

Let L^{NI} be the length of the solution obtained by the nearest-insertion heuristic. Then

Theorem 4.3.3 *For all instances of the TSP satisfying the triangle inequality,*

$$L^{\mathrm{NI}} \leq 2L^*.$$

We start by proving the following interesting result. Let T be a spanning tree of G and let $W(T)$ be the weight (cost) of that tree; that is, $W(T)$ is the sum of the length of all edges in the tree T. Then

Lemma 4.3.4 *For every spanning tree T,*

$$L^{\mathrm{NI}} \leq 2W(T).$$

Proof. We prove the lemma by matching each vertex we insert during the execution of the algorithm with a single edge of the given tree T. To do that, we describe a procedure that will be carried out in parallel to the nearest-insertion heuristic.

The Dual Nearest-Insertion Procedure

Step 1: Start with a family $\mathcal{T}$ of trees that, at first, consists of only the tree T.

Step 2: Given k (the vertex selected in *step 2* of NI), find the unique tree in $\mathcal{T}$ containing k. Let this tree be T_k.

Step 3: Let ℓ be the unique vertex in T_k that is in the current cycle.

Step 4: Let h be the vertex adjacent to ℓ on the unique path from ℓ to k. Replace T_k in $\mathcal{T}$ with two trees obtained from T_k by deleting edge $\{\ell, h\}$.

Step 5: If $\mathcal{T}$ contains n trees, stop. Otherwise, go to *step 2*.

The dual nearest-insertion procedure is carried out in parallel to the nearest-insertion heuristic in the sense that each time *step 1* is performed in the latter procedure, *step 1* is performed in the former procedure. Each time *step 2* is performed in the latter, *step 2* is performed in the former, etc.

Observe that each time *step 4* of the dual nearest-insertion procedure is performed, the set of trees $\mathcal{T}$ is updated so that each tree in $\mathcal{T}$ has exactly one vertex from the current cycle and each vertex of the current cycle belongs to exactly one tree. This is true since when edge $\{\ell, h\}$ is deleted, two subtrees are constructed, one containing the vertex ℓ and the other containing the vertex k. Edge $\{\ell, h\}$ is the one we associate with the insertion of vertex k.

Let m be the vertex in the current cycle to which vertex k (not in the cycle) was closest. That is, m is the vertex such that d_{km} is the smallest among all d_{uv}, where u is in the cycle and v is outside the cycle. Let $m+1$ be one of the vertices on the cycle adjacent to m. Finally, let edge $\{i, j\}$ be the edge deleted from the current cycle. Clearly, inserting k into the current cycle increases the length of the tour by

$$d_{ik} + d_{kj} - d_{ij} \leq d_{mk} + d_{k,m+1} - d_{m,m+1} \leq 2d_{mk},$$

where the left-hand inequality holds because of *step 3* of the nearest-insertion heuristic and the right-hand inequality holds in view of the triangle inequality assumption. Of course, this is true only when the cycle contains at least two vertices. When it contains exactly one vertex, that is, when the nearest-insertion algorithm enters *step 2* for the first time, inserting k into the current cycle increases the length of the tour by exactly $2d_{mk}$.

Since ℓ is in the current cycle and h is not, $d_{mk} \leq d_{\ell h}$. Hence, the increase in the cost of the current cycle is no more than $2d_{\ell h}$. Finally, since this relationship holds for every edge of T and the corresponding inserted vertex, we have

$$L^{\mathrm{NI}} \leq 2W(T).$$

∎

To finish the proof of Theorem 4.3.3, apply Theorem 4.3.4 with T^*; thus,

$$W^* = W(T^*) < L^* \leq L^{\mathrm{NI}} \leq 2W(T^*).$$

This completes the proof of the theorem.

To see that the bound is tight, consider the example (constructed by Rosenkrantz et al. 1977) depicted in Fig. 4.2. In this example, the length of every edge connecting two consecutive vertices on the perimeter is 1 while all other edges have length 2. Thus, the optimal traveling salesman tour visits the vertices according to their appearance on the circle, and therefore, $L^* = n$. It is easy to see that the nearest-insertion heuristic generates the tour depicted in Fig. 4.2b with cost $L^{\mathrm{NI}} = 2n - 2$.

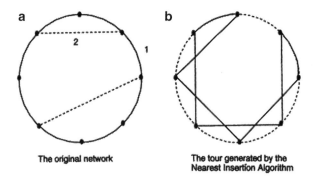

FIGURE 4.2. An example for the nearest-insertion algorithm with $n = 8$

4.3.3 Christofides' Heuristic

In 1976, Christofides presented a very simple algorithm that currently has the best-known worst-case performance bound for the TSP. To present the algorithm, we need to state several properties of graphs.

Lemma 4.3.5 *Given a graph with at least two vertices, the number of vertices with odd degree is even.*

Definition 4.3.6 *An Eulerian tour is a tour that traverses all edges of a graph exactly once.*

Definition 4.3.7 *An Eulerian graph is a graph that has an Eulerian tour.*

Then it is a simple exercise to show the following.

Lemma 4.3.8 *A connected graph is Eulerian if and only if the degree of each vertex is even.*

Christofides' algorithm starts with a minimum spanning tree. Of course, this tree (as any other tree) is not Eulerian, since some of the vertices have odd degree. We can augment the graph (by adding suitably chosen arcs) so that it becomes Eulerian. In fact, we would like to add a number of arcs connecting odd-degree vertices so that they then have even degree. To do this, we will find a *minimum weight matching* among the odd-degree vertices.

Given a graph with an even number of vertices, a *matching* is a subset of edges with the property that every vertex is the endpoint of exactly one edge of the subset. In the minimum-weight matching problem, the objective is to find a matching whose total length of all its edges is minimum. This problem can be solved in $O(n^3)$, where n is the number of vertices in the graph [see Lawler (1976)].

Lemma 4.3.5 tells us that the number of vertices with odd degree in the MST is even. Thus, adding the edges of a matching defined on those odd-degree vertices clearly increases the degree of each of these vertices by one. The resulting graph is Eulerian, by Lemma 4.3.8. Of course, to minimize the total cost, we would like to select the edges of a minimum-weight matching. Finally, the Eulerian tour generated is transformed into a traveling salesman tour using shortcuts, similarly to what was done in the minimum spanning tree-based heuristic of Sect. 4.3.1.

Let L^C be the length of the tour generated by Christofides' heuristic. We prove the following:

Theorem 4.3.9 *For all instances of the TSP satisfying the triangle inequality, we have*

$$L^C \leq \frac{3}{2} L^*.$$

Proof. Recall that $W^* \doteq W(T^*)$ is the cost of the MST and let $W(M^*)$ be the weight of the minimum-weight matching, that is, the sum of edge lengths of all edges in the optimal matching. Because of the triangle inequality assumption,

$$L^C \leq W(T^*) + W(M^*).$$

We already know that $W(T^*) \leq L^*$. It remains to show that $W(M^*) \leq \frac{1}{2} L^*$. For this purpose, index the vertices of odd degree in the minimum spanning tree $i_1, i_2, \ldots, i_{2k}$ according to their appearance on an optimal traveling salesman tour. Consider two feasible solutions for the minimum-weight matching problem defined on these vertices. The first matching, denoted M^1, consists of edges $\{i_1, i_2\}, \{i_3, i_4\}, \ldots, \{i_{2k-1}, i_{2k}\}$. The second matching, denoted M^2, consists of edges $\{i_2, i_3\}, \{i_4, i_5\}, \ldots, \{i_{2k}, i_1\}$.

We clearly have $W(M^*) \leq \frac{1}{2}[W(M^1) + W(M^2)]$. The triangle inequality assumption tells us that $W(M^1) + W(M^2) \leq L^*$; see Fig. 4.3. Hence, $W(M^*) \leq \frac{1}{2} L^*$ and, consequently,

$$L^* \leq W(T^*) + W(M^*) \leq \frac{3}{2} L^*.$$

∎

As in the two previous heuristics, this bound is tight. Consider the example depicted in Fig. 4.4 for which $L^* = n$ while $L^C = n - 1 + \frac{n-1}{2}$.

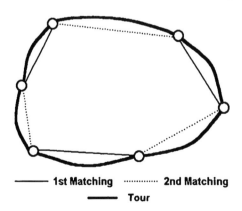

FIGURE 4.3. The matching and the optimal traveling salesman tour

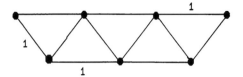

FIGURE 4.4. An example for Christofides' algorithm with $n = 7$

4.3.4   Local Search Heuristics

Some of the oldest and, by far, the most extensively used heuristics developed for the traveling salesman problem are the so-called k-opt procedures ($k \geq 2$). These heuristics, part of the extensive class of *local search* procedures, can be described as follows. Given a traveling salesman tour through the set of vertices V, say the sequence

$$\{i_1, i_2 \ldots, i_{u_1}, i_{u_2}, \ldots, i_{v_1}, i_{v_2}, \ldots, i_n\},$$

an *ℓ-exchange* is a procedure that replaces ℓ edges currently in the tour by ℓ new edges so that the result is again a traveling salesman tour. For instance, a 2-exchange procedure replaces edges $\{i_{u_1}, i_{u_2}\}$ and $\{i_{v_1}, i_{v_2}\}$ with $\{i_{u_1}, i_{v_1}\}$ and $\{i_{u_2}, i_{v_2}\}$ and results in a new tour:

$$\{i_1, i_2 \ldots, i_{u_1}, i_{v_1}, i_{v_1-1}, \ldots, i_{u_2}, i_{v_2}, i_{v_2+1}, \ldots, i_n\}.$$

An *improving ℓ-exchange* is an ℓ-exchange that results in a tour whose total length (cost) is smaller than the cost of the original tour.

A k-opt procedure starts from an arbitrary traveling salesman tour and, using improving ℓ-exchanges, for $\ell \leq k$, successively generates tours of smaller and smaller length. The procedure terminates when no improving ℓ-exchange is found for all $\ell \leq k$. Let $L^{\mathrm{OPT}(k)}$ be the length of the tour generated by a k-opt heuristic, for $k \geq 2$.

Recently, Chandra et al. (1999) obtained interesting results on the worst-case performance of the k-opt heuristic. They show

Theorem 4.3.10 *For all instances of the TSP satisfying the triangle inequality, we have*

$$\frac{L^{\mathrm{OPT}(2)}}{L^*} \leq 4\sqrt{n}.$$

In addition, there exists an infinitely large family of TSP instances satisfying the triangle inequality assumption for which

$$\frac{L^{\mathrm{OPT}(2)}}{L^*} \geq \frac{1}{4}\sqrt{n}.$$

They also provide a lower bound on the worst-case performance of k-opt for all $k \geq 3$.

Theorem 4.3.11 *There exists an infinitely large family of TSP instances satis-fying the triangle inequality assumption with*

$$\frac{L^{\mathrm{OPT}(k)}}{L^*} \geq \frac{1}{4} n^{\frac{1}{2k}}$$

for any $k \geq 2$.

Thus, the above results indicate that the worst-case performances of k-opt heuristics are quite poor. By contrast, many researchers and practitioners have reported that k-opt heuristics can be highly effective; see, for instance, Golden and Stewart (1985).

This raises a fundamental dilemma. Although worst-case analysis provides a rigid guarantee on a heuristic's performance, it suffers from being highly deter-mined by certain pathological examples. Is there a more appropriate measure to assess the effectiveness of a particular heuristic, one that would assess the effec-tiveness on an *average* or realistic example? We will try to address this question in the next chapter.

4.4   Exercises

Exercise 4.1. Prove Lemma 4.3.8.

Exercise 4.2. The 2-TSP is the problem of designing two tours that together visit each of the customers and use the same starting point. Show that any algo-rithm for the TSP can solve this problem as well.

Exercise 4.3. (Papadimitriou and Stieglitz 1982) Consider the n-city TSP in which the triangle inequality assumption holds. Let $c^* > 0$ be the length of an optimal tour, and let c' be the length of the second-best tour. Prove $(c' - c^*)/c^* \leq \frac{2}{n}$.

Exercise 4.4. Prove that in every completely connected directed graph (a graph in which between every pair of vertices there is a directed edge in one of the two possible directions), there is a directed Hamiltonian path.

Exercise 4.5. Let Z^{C} be the length of the tour provided by Christofides' heuris-tic, and let Z^* be the length of the optimal tour. Construct an example with $Z^{\mathrm{C}} = \frac{3}{2} Z^*$.

Exercise 4.6. Prove that for any graph G, there exists an even number of nodes with odd degree.

Exercise 4.7. Let G be a tree with $n \geq 2$ nodes. Show that

(a) There exist at least two nodes with degree 1.

(b) The number of arcs is $n - 1$.

Exercise 4.8. Consider the n-city TSP defined with distances d_{ij}. Assume that there exist $a, b \in \mathbb{R}^n$ such that for each i and j, $d_{ij} = a_i + b_j$. What is the length of the optimal traveling salesman tour? Explain your solution.

Exercise 4.9. Consider the TSP with the triangle inequality assumption and two prespecified nodes s and t. Assume that the traveling salesman tour has to include edge (s, t) (that is, the salesman has to travel from s directly to t). Modify Christofides' heuristic for this model and show that the worst-case bound is $\frac{3}{2}$.

Exercise 4.10. Show that a minimum spanning tree T satisfies the following property. When T is compared with any other spanning tree T', the kth-shortest edge of T is no longer than the kth-shortest edge of T', for $k = 1, 2, \ldots, n - 1$.

Exercise 4.11. (Papadimitriou and Stieglitz 1982) The wandering salesman problem (WSP) is a traveling salesman problem except that the salesman can start wherever he or she wishes and does not have to return to the starting city after visiting all cities.

(a) Describe a heuristic for the WSP with worst-case bound $\frac{3}{2}$.

(b) Show that the same bound can be obtained for the problem when one of the endpoints of the path is specified in advance.

Exercise 4.12. (Papadimitriou and Stieglitz 1982) Which of the following problems remain essentially unchanged (complexity-wise) when they are transformed from minimization to maximization problems? Why?

(a) Traveling salesman problem

(b) Shortest path from s to t

(c) Minimum-weight matching

(d) Minimum spanning tree

Exercise 4.13. Suppose there are n jobs that require processing on m machines. Each job must be processed by machine 1, then by machine 2, ..., and finally by machine m. Each machine can work on at most one job at a time, and once it begins work on a job it must work on it until completion, without interruption. The amount of time machine j must process job i is denoted $p_{ij} \geq 0$ (for $i = 1, 2, \ldots, n$ and $j = 1, 2, \ldots, m$). Further suppose that once the processing of a job is completed on machine j, its processing must begin *immediately* on machine $j+1$ (for $j \leq m - 1$). This is a *flow shop* with no wait-in-process.

Show that the problem of sequencing the jobs so that the last job is completed as early as possible can be formulated as an $(n + 1)$-city TSP. Specifically, show how the d_{ij} values for the TSP can be expressed in terms of the p_{ij} values.

Exercise 4.14. Consider the bin-packing problem with items of size w_i, $i = 1, 2, \ldots, n$, such that $0 < w_i \leq 1$. The objective is to find the minimum number of unit-size bins b^* needed to pack all the items without violating the capacity constraint.

(a) Show that $\sum_{i=1}^{n} w_i$ is a lower bound on b^*.

(b) Define a *locally optimal* solution to be one where no two bins can be feasibly combined into one. Show that any locally optimal solution uses no more than twice the minimum number of bins, that is, no more than $2b^*$ bins.

(c) The next-fit heuristic is the following. Start by packing the first item in bin 1. Then each subsequent item is packed in the last-opened bin if possible, or else a new bin is opened and it is placed there. Show that the next-fit heuristic produces a solution with at most $2b^*$ bins.

Exercise 4.15. (Anily et al. 1994) Consider the bin-packing problem and the next-fit increasing heuristic. In this strategy, items are ordered in a nondecreasing order according to their size. Start by packing the first item in bin 1. Then each subsequent item is packed in the last-opened bin if possible, or else a new bin is opened and it is placed there. Show that the number of bins produced by this strategy is no more than $\frac{7}{4}$ times the optimal number of bins. For this purpose, consider the following two steps.

(a) Consider the following procedure. First, order the items in nondecreasing order of their size. When packing bin $i \geq 1$, follow the packing rule: If the bin is currently feasible (i.e., the total load is no more than 1), then assign the next item to this bin; otherwise, close this bin, open bin $i+1$, and put this item in bin $i + 1$. Show that the number of bins generated by this procedure is a lower bound on the minimal number of bins needed.

(b) Relate this lower-bounding procedure to the number of bins produced by the next-fit increasing heuristic.

Exercise 4.16. Given a network $G = (V, E)$, and edge length l_e for every $e \in E$, assume that edge (u, v) has a variable length x. Find an expression for the length of the shortest path from s to t as a function of x.

Exercise 4.17. A complete directed network $G = (V, A)$ is a directed graph such that for every pair of vertices $u, v \in V$, there are arcs $u \to v$ and $v \to u$ in A with nonnegative arc lengths $d(u, v)$ and $d(v, u)$, respectively. The network $G = (V, A)$ satisfies the triangle inequality if, for all $u, v, w \in V$, $d(u, v) + d(v, w) \geq d(u, w)$.

A directed cycle is a sequence of vertices $v_1 \to v_2 \to \cdots \to v_\ell \to v_1$ without any repeated vertex other than the first and last ones. If the cycle contains all the vertices in G, then it is said to be a *directed Hamiltonian cycle*. To keep notation simple, let $d_{ij} \doteq d(v_i, v_j)$.

A directed cycle containing exactly k vertices is called a k-cycle. The length of a cycle is defined as the sum of arc lengths used in the cycle. A directed network $G = (V, A)$ with $|V| \geq k$ is said to be k-symmetric if, for every k-cycle $v_1 \to v_2 \to \cdots \to v_k \to v_1$ in G,

$$d_{12} + d_{23} + \cdots + d_{k-1,k} + d_{k1} = d_{1k} + d_{k,k-1} + \cdots + d_{32} + d_{21}.$$

In other words, a k-symmetric network is a directed network in which the length of every k-cycle remains unchanged if its orientation is reversed.

(a) Show that the asymmetric traveling salesman problem on a $|V|$-symmetric network (satisfying the triangle inequality) can be solved via solving a corresponding symmetric traveling salesman problem. In particular, show that any heuristic with a fixed worst-case bound for the symmetric traveling salesman problem can be used for the asymmetric traveling salesman problem on a $|V|$-symmetric network to obtain a result with the same worst-case bound.

(b) Prove that any 3-symmetric network is k-symmetric for $k = 4, 5, \ldots, |V|$.

Thus, part (a) can be used if we have a 3-symmetric network. Argue that a 3-symmetric network can be identified in polynomial time.

5

Average-Case Analysis

5.1 Introduction

Worst-case performance analysis is one method of characterizing the effectiveness of a heuristic. It provides a guarantee on the maximum relative difference between the solution generated by the heuristic and the optimal solution for any possible problem instance, even those that are not likely to appear in practice. Thus, a heuristic that works well in practice may have a weak worst-case performance, if, for example, it provides very bad solutions for one (or more) pathological instance(s).

To overcome this important drawback, researchers have recently focused on *probabilistic analysis* of algorithms with the objective of characterizing the average performance of a heuristic under specific assumptions on the distribution of the problem data. As pointed out, for example, by Coffman and Lueker (1991), probabilistic analysis is frequently quite difficult and even the analysis of simple heuristics can often present a substantial challenge. Therefore, usually the analysis is *asymptotic*. That is, the average performance of a heuristic can only be quantified when the problem size is extremely large.

As we demonstrate in Parts II and IV, an asymptotic probabilistic analysis is useful for several reasons:

D. Simchi-Levi et al., *The Logic of Logistics: Theory, Algorithms, and Applications for Logistics Management*, Springer Series in Operations Research and Financial Engineering, DOI 10.1007/978-1-4614-9149-1_5, © Springer Science+Business Media New York 2014

1. It can foster new insights into which algorithmic approaches will be effective for solving large problems. That is, the analysis provides a framework where one can analyze and compare the performance of heuristics on large problems.

2. For problems with fast rates of convergence, the analysis can sometimes explain the observed empirical behavior of heuristics for more reasonable-size problems.

3. The approximations derived from the analysis can be used in other models and may lead to a better understanding of the tradeoffs in more complex problems integrating vehicle routing with other issues important to the firm, such as inventory control.

In this chapter, we present some of the basic tools used in the analysis of the average performance of heuristics. Again, we use the bin-packing problem and the traveling salesman problem as the "raw materials" on which to present them.

5.2   The Bin-Packing Problem

The bin-packing problem provides a very well-studied example for which to demonstrate the benefits of a probabilistic analysis.

Without loss of generality, we scale the bin capacity q so that it is 1. Consider the item sizes $w_1, w_2, w_3 \ldots$ to be independently and identically distributed on $(0, 1]$ according to some general distribution Φ. In this section, we demonstrate two elegant and powerful techniques that can be used in the analysis of b_n^*, the random variable representing the optimal solution value on the items $w_1, w_2, \ldots, w_n$. The first is the theory of *subadditive processes* and the second is the theory of *martingale inequalities*.

Subadditive Processes

Let $\{a_n\}$, $n \geq 1$, be a sequence of positive real numbers. We say that the sequence is *subadditive* if for all n and m, we have $a_n + a_m \geq a_{n+m}$. The following important result was proved by Kingman (1976) and Steele (1990), whose proof we follow.

Theorem 5.2.1 *If the sequence $\{a_n\}$, $n \geq 1$ is subadditive, then there exists a constant γ such that*

$$\lim_{n \to \infty} \frac{a_n}{n} = \gamma.$$

Proof. Let $\gamma = \underline{\lim}_{n \to \infty} \frac{a_n}{n}$. For a given ϵ, select n such that $\frac{a_n}{n} \leq \gamma + \epsilon$. Since the sequence $\{a_n\}$ is subadditive, we have

$$a_{nm} \leq a_n + a_{n(m-1)}.$$

Making a repeated use of this inequality, we get $a_{nm} \leq m a_n$, which implies

$$\frac{a_{nm}}{nm} \leq \gamma + \epsilon.$$

For any k, $0 \leq k \leq n$, define $\ell \doteq nm + k$. Using subadditivity again, we have

$$a_\ell = a_{nm+k} \leq a_{nm+k-1} + a_1$$
$$\leq a_{nm} + k a_1$$
$$\leq a_{nm} + n a_1,$$

where the second inequality is obtained by repeating the first one k times. Thus,

$$\frac{a_\ell}{\ell} = \frac{a_{nm+k}}{nm+k} \leq \frac{a_{nm} + n a_1}{nm+k} \leq \frac{a_{nm}}{nm} + \frac{a_1}{m} \leq \gamma + \epsilon + \frac{a_1}{m}.$$

Taking the limit with respect to m, we have

$$\overline{\lim_{\ell \to \infty}} \frac{a_\ell}{\ell} \leq \gamma + \epsilon + \overline{\lim_{m \to \infty}} \frac{a_1}{m} = \gamma + \epsilon.$$

The proof is therefore complete since ϵ was chosen arbitrarily.   ∎

It is clear that the optimal solution of the bin-packing problem possesses a subadditivity-like property; that is, for any sets $S, T \subseteq N$:

$$b^*(S \cup T) \leq b^*(S) + b^*(T),$$

where $b^*(S)$ denotes the optimal solution to the bin-packing problem on a set $S \subseteq N$. Using similar arguments as in the above analysis shows that there exists a constant γ such that the optimal solution to the bin-packing problem b_n^* satisfies

$$\lim_{n \to \infty} \frac{b_n^*}{n} = \gamma \qquad (a.s.).$$

In addition, γ is dependent only on the item size distribution Φ.

The Uniform Model

To illustrate the concepts just developed, consider the case where Φ is the uniform distribution on $[0, 1]$. In order to pack a set of n items drawn randomly from this distribution, we use the following sliced interval partitioning heuristic with parameter r $(SIP(r))$. It works as follows. For any *fixed* positive integer $r \geq 1$, the set of items N is partitioned into the following $2r$ disjoint subsets, some of which may be empty:

$$N_j = \left\{ k \in N \,\Big|\, \frac{1}{2}\left(1 - \frac{j+1}{r}\right) < w_k \leq \frac{1}{2}\left(1 - \frac{j}{r}\right) \right\}, \qquad j = 1, 2, \ldots, r-1,$$

and

$$N^j = \left\{ k \in N \,\Big|\, \frac{1}{2}\left(1 + \frac{j-1}{r}\right) < w_k \leq \frac{1}{2}\left(1 + \frac{j}{r}\right) \right\}, \qquad j = 1, 2, \ldots, r-1.$$

Also,

$$N_0 = \left\{ k \in N \middle| \frac{1}{2}\left(1 - \frac{1}{r}\right) < w_k \le \frac{1}{2} \right\}$$

and

$$N^r = \left\{ k \in N \middle| \frac{1}{2}\left(1 + \frac{r-1}{r}\right) < w_k \right\}.$$

The number of items in each N_j (respectively, N^j) is denoted by n_j (respectively, n^j) for all possible values of j.

Note that for any $j = 1, 2, \ldots, r - 1$, one bin can hold an item from N_j together with exactly one item from N^j. The $SIP(r)$ heuristic generates pairs of items, one item from N_j and one from N^j, for every $j = 1, 2, \ldots, r - 1$. The items in $N_0 \cup N^r$ are put in individual bins; one bin is assigned to each of these items.

For any $j = 1, 2, \ldots, r - 1$, we arbitrarily match one item from N_j with exactly one item from N^j; one bin holds each such pair. If $n_j = n^j$, then all the items in $N_j \cup N^j$ are matched. If, however, $n_j \ne n^j$, then we can match exactly $\min\{n_j, n^j\}$ pairs of items. The remaining $|n_j - n^j|$ items in $N_j \cup N^j$ that have not yet been matched are put one per bin. Thus, the total number of bins used is

$$n_0 + n^r + \sum_{j=1}^{r-1} \max\{n_j, n^j\}.$$

The heuristic clearly generates a feasible solution to the bin-packing problem. Since

$$\lim_{n \to \infty} \frac{n_j}{n} = \lim_{n \to \infty} \frac{n^j}{n} = \frac{1}{2r} \qquad (a.s.) \qquad \text{for all } j = 1, 2, \ldots, r,$$

we have

$$\gamma = \lim_{n \to \infty} \frac{b_n^*}{n} \le \lim_{n \to \infty} \frac{1}{n}\left[n_0 + n^r + \sum_{j=1}^{r-1} \max\{n_j, n^j\}\right] = \frac{1}{2} + \frac{1}{2r} \qquad (a.s.).$$

Since this holds for any $r > 1$, we see that $\gamma \le \frac{1}{2}$. Since $\gamma \ge E(w)$ (see Exercise 5.4), then $\gamma \ge \frac{1}{2}$ and we conclude that $\gamma = \frac{1}{2}$ for the uniform distribution on $[0, 1]$.

Using this idea, we can actually devise an *asymptotically optimal* heuristic for instances where the item sizes are uniformly distributed on $[0, 1]$. To formally define this property, let Z_n^* be the cost of the optimal solution to the problem on a problem of size n, and let Z_n^H be the cost of the solution provided by a heuristic H. Let the relative error of a heuristic H on a particular instance of n points be

$$e_n^H = \frac{Z_n^H - Z_n^*}{Z_n^*}.$$

Definition 5.2.2 *Let Ψ be a probability measure on the set of instances $\mathcal{I}$. A heuristic H is asymptotically optimal for Ψ if almost surely*

$$\lim_{n \to \infty} e_n^H = 0,$$

where the problem data are generated randomly from Ψ.

That is, under certain assumptions on the distribution of the data, H generates solutions whose relative error tends to zero as n, the number of points, tends to infinity. The above $SIP(r)$ heuristic is not asymptotically optimal since for any fixed r, the relative error converges to $\frac{1}{r}$.

A truly asymptotically optimal heuristic can easily be constructed. The following heuristic is called MATCH. First, sort the items in nonincreasing order of the item sizes. Then take the largest item, say item i, and match it with the largest item with which it will fit. If no such item exists, then put item i in a bin alone. Otherwise, put item i and the item it was matched with in a bin together. Now repeat this until all items are packed. The proof of asymptotic optimality is given as an exercise (Exercise 5.11).

An additional use for the bin-packing constant γ is as an approximation for the number of bins needed. When n is large, the number of bins required to pack n random items from Φ is very close to $n\gamma$. How close the random variable representing the number of bins is to $n\gamma$ is discussed next.

Martingale Inequalities

Consider the stochastic processes $\{X_n\}$ and $\{Y_n\}$ with $n \geq 0$. We say that the stochastic process $\{X_n\}$ is a *martingale* with respect to $\{Y_n\}$ if, for every $n \geq 0$, we have

$$(i) \quad E[X_n] < +\infty, \text{ and}$$

$$(ii) \quad E[X_{n+1}|Y_1, \ldots, Y_n] = X_n.$$

To get some insight into the definition of a martingale, consider someone playing a sequence of fair games. Let $X_n = Y_n$ be the amount of money the player has at the end of the nth game. If $\{X_n\}$ is a martingale with respect to $\{Y_n\}$, then this says that the expected amount of money the player will have at the end of the $(n + 1)$st game is equal to what the player had at the beginning of that game X_n, regardless of the game's history prior to state n. See Karlin and Taylor (1975) for details.

Consider now the random variable

$$D_n \doteq E[X_{n+1}|Y_1, \ldots, Y_n] - E[X_{n+1}|Y_1, \ldots, Y_{n-1}].$$

The sequence $\{D_n\}$ is called a *martingale difference sequence* if $E[D_n] = 0$ for every $n \geq 0$. Azuma (1967) developed the following interesting inequality for martingale difference sequences; see also Stout (1974) or Rhee and Talagrand (1987).

Lemma 5.2.3 *Let $\{D_i\}$, $i = 1, 2, \ldots, n$, be a martingale difference sequence. Then for every $t > 0$, we have*

$$Pr\left\{\left|\sum_{i \leq n} D_i\right| > t\right\} \leq 2\exp\left\{-t^2 / \left(2\sum_{i \leq n}\|D_i\|_\infty^2\right)\right\},$$

where $\|D_i\|_\infty$ is a uniform upper bound on the D_is.

The lemma can be used to establish upper bounds on the probable deviations of both

- b_n^* from its mean $E[b_n^*]$, and

- $\frac{b_n^*}{n}$ from its asymptotic value γ.

For this purpose, define

$$
D_i = \begin{cases} E[b_n^*|w_1,\ldots,w_i] - E[b_n^*|w_1,\ldots,w_{i-1}], & \text{if } i \geq 2; \\ E[b_n^*|w_1] - E[b_n^*|\emptyset], & \text{if } i = 1, \end{cases}
$$

where $E[b_n^*|w_1,\ldots,w_i]$ is the random variable that represents the expected optimal solution value of the bin-packing problem obtained by fixing the sizes of the first i items and averaging on all other item sizes. Clearly, $E[b_n^*|w_1,\ldots,w_n] = b_n^*$, while $E[b_n^*|\emptyset] = E[b_n^*]$. Hence, $\sum_{i=1}^{n} D_i = b_n^* - E[b_n^*]$. Furthermore, the sequence D_i defines a martingale difference sequence with the property that $D_i \leq 1$ for every $i \geq 1$.

Applying Lemma 5.2.3, we obtain the following upper bound:

$$
Pr\left\{ |b_n^* - E[b_n^*]| > t \right\} = Pr\left\{ \left| \sum_{i=1}^{n} D_i \right| > t \right\} \leq 2\exp\left\{ -t^2/(2n) \right\}.
$$

This bound can now be used to construct an upper bound on the likelihood that $\frac{b_n^*}{n}$ differs from its asymptotic value by more than some fixed amount.

Theorem 5.2.4 *For every $\epsilon > 0$, there exists an integer n_0 such that for all $n \geq n_0$,*

$$
Pr\left\{ \left| \frac{b_n^*}{n} - \gamma \right| > \epsilon \right\} < 2\exp\left(-\frac{n\epsilon^2}{2} \right).
$$

Proof. Theorem 5.2.1 implies that $\lim_{n \to \infty} E[\frac{b_n^*}{n}] = \gamma$, and therefore, for every $\epsilon > 0$ and $k \geq 2$, there exists n_0 such that for all $n \geq n_0$, we have

$$
\left| E\left[\frac{b_n^*}{n} \right] - \gamma \right| < \frac{\epsilon}{k}.
$$

Consequently,

$$
Pr\left\{ \left| \frac{b_n^*}{n} - \gamma \right| > \epsilon \right\} \leq Pr\left\{ \left| \frac{b_n^*}{n} - \frac{E[b_n^*]}{n} \right| + \left| \frac{E[b_n^*]}{n} - \gamma \right| > \epsilon \right\}
$$

$$
\leq Pr\left\{ \left| \frac{b_n^*}{n} - \frac{E[b_n^*]}{n} \right| + \frac{\epsilon}{k} > \epsilon \right\}
$$

$$
\leq Pr\left\{ \left| b_n^* - E[b_n^*] \right| > \frac{n\epsilon(k-1)}{k} \right\}
$$

$$
\leq 2\exp\left\{ -\frac{n\epsilon^2(k-1)^2}{2k^2} \right\}.
$$

Since this last inequality holds for arbitrary $k \geq 2$, this completes the proof.   ∎

These results demonstrate that b_n^* is, in fact, very close to $n\gamma$, and this is true for any distribution of the item sizes. Therefore, it suggests that $n\gamma$ may serve as a good approximation for b_n^* in other, more complex, combinatorial problems.

5.3   The Traveling Salesman Problem

In this section, we demonstrate an important use for the tools presented above. Our objective is to show how probabilistic analysis can be used to construct effective algorithms with certain attractive theoretical properties.

Let $x_1, x_2, \ldots, x_n$ be a sequence of points in the Euclidean plane ($\mathbb{R}^2$), and let L_n^* be the length of the optimal traveling salesman tour through these n points. We start with a deterministic upper bound on L_n^* developed by Few (1955). We follow Jaillet's (1985) presentation.

Theorem 5.3.1 *Let $a \times b$ be the size of the smallest rectangle that contains $x_1, x_2 \ldots, x_n$; then*

$$L_n^* \leq \sqrt{2(n-2)ab} + 2(a+b).$$

Proof. For an integer m (to be determined), partition the rectangle of size $a \times b$ (where a is the length and b is the height) into $2m$ equal-width horizontal strips. This creates $2m+1$ horizontal lines and two vertical lines (counting the boundaries of the rectangle). Label the horizontal lines $1, 2, \ldots, 2m+1$ moving downward. Now temporarily delete all horizontal lines with an even label. Connect each point x_i, $i = 1, 2 \ldots, n$, with two vertical segments, to the closest (odd-labeled) horizontal line. A path through $x_1, \ldots, x_n$ can now be constructed by proceeding from, say, the upper left-hand corner of the $a \times b$ rectangle and moving from left to right on the first horizontal line, picking up all points that are connected (with the two vertical segments) to this line. Then we proceed downward and cover the third horizontal line from right to left. This continues until we reach the end of the $2m+1$st line. This path can be extended to a traveling salesman tour by returning from the last point to the first by adding at most one vertical and one horizontal line (we avoid diagonal movements for the sake of simplicity). Now repeat this procedure with the even-labeled horizontal lines and, in a similar manner, create a path through all the customers. Extend this path to a traveling salesman tour by adding one horizontal line and one vertical segment of length $b - \frac{b}{m}$. See Fig. 5.1.

Clearly, the sum of the length of the two traveling salesman tours is

$$a(2m+1) + \frac{nb}{m} + 2b + a + 2\left(b - \frac{b}{m}\right).$$

Since L_n^* is no larger than either of these two tours, we have

$$L_n^* \leq a + 2b + ma + (n-2)\frac{b}{2m}.$$

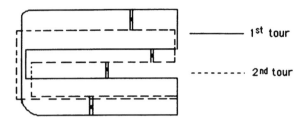

FIGURE 5.1. The two traveling salesman tours constructed by the partitioning algorithm

The right-hand side is convex in m; hence, we minimize on m. That is, we choose

$$m^* = \left\lceil \sqrt{\frac{b(n-2)}{2a}} \right\rceil;$$

then

$$L_n^* \leq a + 2b + m^*a + \frac{b(n-2)}{2m^*}$$

$$\leq a + 2b + a\left(\sqrt{\frac{b(n-2)}{2a}} + 1\right) + \frac{b(n-2)}{2}\sqrt{\frac{2a}{(n-2)b}}$$

$$= \sqrt{2(n-2)ab} + 2(a+b).$$

∎

The above result implies that the length of the optimal traveling salesman tour is at most $O(\sqrt{n})$. In 1959, Beardwood et al. showed that the rate of growth of L^*, when customer locations are independent and identically distributed, is $\Theta(\sqrt{n})$. Specifically, they prove the following result.

Theorem 5.3.2 *Let $x_1, x_2, \ldots, x_n$ be a sequence of independent random variables having a distribution μ with compact support in $\mathbb{R}^2$. Then there exists a constant $\beta > 0$, independent of the distribution μ, such that with probability 1,*

$$\lim_{n \to \infty} \frac{L_n^*}{\sqrt{n}} = \beta \int_{\mathbb{R}^2} f^{1/2}(x)dx,$$

where f is the density of the absolutely continuous part of the distribution μ.

Since Beardwood et al. proved this result, many researchers have proved it using a variety of techniques. One of these methods is based on the concept of Euclidean subadditive processes (Steele 1981), which is a generalization of the concept of subadditive processes described earlier.

In this subsection, we are not going to prove the result, but rather concentrate on its algorithmic implications. Specifically, we will describe the following polynomial-time algorithm, which is *asymptotically optimal*. The heuristic was suggested by Karp (1977) although we have modified it in several places for the purpose of clarifying the presentation.

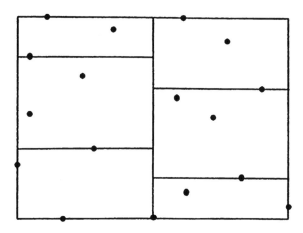

FIGURE 5.2. Region-partitioning example with $n = 17$, $q = 3$, $h = 2$, and $t = 1$

A Region-Partitioning Heuristic

In the region-partitioning heuristic, the region containing the points is sub-divided into subregions such that each nonempty subregion contains exactly q customers (except possibly for one) and where q is to be determined later. The heuristic then constructs an optimal traveling salesman tour on the set of points within or bordering each subregion and then connects these tours to form a trav-eling salesman tour through all the points.

To generate subregions each with exactly q points, except for possibly one subre-gion where there may be fewer points, we use the following strategy: The smallest rectangle with sides a and b containing the set of points $x_1, x_2, \ldots, x_n$ is partitioned by means of horizontal and vertical lines. First, the region is divided by t vertical lines such that each subregion contains exactly $(h + 1)q$ points except possibly the last one. This is done precisely as follows: Temporarily index the customers in increasing order of their horizontal coordinate. Place the vertical lines so that the jth vertical line (for $j \leq t$) goes through the customer with index $j(h + 1)q$. Each of these $t + 1$ subregions is then partitioned by means of h horizontal lines into $h + 1$ smaller subregions such that each contains exactly q points except possibly the last one. More precisely, this is done as follows: In each vertical strip, index the customers in increasing order of their vertical coordinates. Place the horizontal lines so that the jth horizontal line (for $j \leq h$) goes through the customer with index jq. See Fig. 5.2 for an example.

To solve the traveling salesman problems within each subregion, we use a dy-namic programming algorithm developed by Held and Karp (1962). It finds an opti-mal traveling salesman tour through m points in running time, which is $O(m^2 2^m)$. If we choose $q = \lceil \log n \rceil$, then solving the traveling salesman problem for a sin-gle region takes $O(n \log^2 n)$, and since the number of subregions is no more than $1 + n/\log n$, the total time spent solving these traveling salesman problems is $O(n^2 \log n)$.

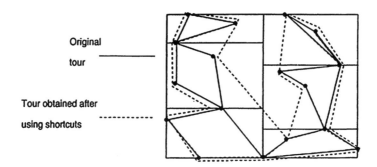

FIGURE 5.3. The tour generated by the region-partitioning algorithm

After finding optimal traveling salesman tours within each subregion, observe that this collection of traveling salesman tours can be easily transformed into a single traveling salesman tour through all the points. This is true since this collection of tours, along with the lines added as above, defines an Eulerian graph, where the degree of each point (node) is either two or four (a point that is on the boundary of two subregions will have degree 4). Thus, this tour can be transformed into a single traveling salesman tour, and using shortcuts can further reduce its length. See Fig. 5.3.

To guarantee that each nonempty subregion has exactly q points, except for maybe one, h and t must satisfy

$$t = \left\lceil \frac{n}{(h+1)q} \right\rceil - 1$$

and

$$t(h+1)q < n \leq (t+1)(h+1)q.$$

This is achieved by choosing $h = \lceil \sqrt{\frac{n}{q}} - 1 \rceil$.

Let L^{RP} be the length of the tour generated by the above region-partitioning heuristic. To establish the quality of the heuristic, we need to find an upper bound on L^{RP}; this is provided by the following.

Lemma 5.3.3

$$L^{\mathrm{RP}} \leq L^* + \frac{3}{2} P^{\mathrm{RP}},$$

where P^{RP} is the sum of the perimeters of all subregions generated by the region-partitioning heuristic.

Proof. Let L_j be the length of the optimal traveling salesman tour in subregion $j = 1, 2, \ldots, \lceil \frac{n}{q} \rceil$. Similarly, let L_j^* be the sum of the lengths of all segments of the optimal traveling salesman tour through all n customers that are contained in the jth subregion, for $j \geq 1$. Since the collection of tours and lines constructed above defines an Eulerian graph, we have $L^{RP} \leq \sum_j L_j$. Also, by definition, we have $L^* = \sum_j L_j^*$. Thus, it is sufficient to show that

$$L_j \leq L_j^* + \frac{3}{2}P_j, \tag{5.1}$$

where P_j is the perimeter of subregion j.

To prove inequality (5.1), assume there are exactly k continuous segments $S_1, \ldots, S_k$, of the globally optimal traveling salesman tour, in subregion j; see Fig. 5.4. Let the $2k$ endpoints of these segments be $y_1, y_2, \ldots, y_{2k}$ ordered consecutively around the boundary of subregion j. Without loss of generality, we assume that

$$\ell(y_1y_2) + \ell(y_3y_4) + \cdots + \ell(y_{2k}y_{2k-1}) \leq \ell(y_2y_3) + \ell(y_4y_5) + \cdots + \ell(y_{2k}y_1),$$

where $\ell(y_iy_{i+1})$ is the distance between points y_i and y_{i+1} along the perimeter of the jth subregion. We construct a feasible solution for the traveling salesman problem defined by the points x_i that are in the jth subregion. The tour is based on (i) the segments $S_1, \ldots, S_k$, (ii) two copies of each segment y_1y_2, y_3y_4, $\ldots$, $y_{2k-1}y_{2k}$, and (iii) one copy of each segment y_2y_3, y_4y_5, $\ldots$, $y_{2k}y_1$.

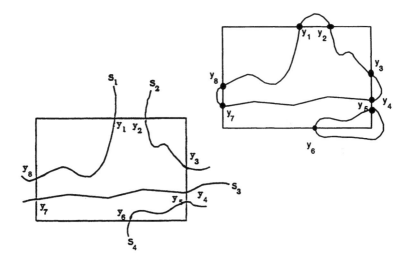

FIGURE 5.4. The segments $S_1, \ldots, S_k$ and the corresponding Eulerian graph

Observe (Fig. 5.4) that the above three components define an Eulerian graph whose set of vertices is the points x_i that belong to the jth subregion plus all the points y_i, for $i = 1, 2, \ldots, 2k$. This implies that the graph has an Eulerian tour whose cost is no more than

$$L_j^* + \frac{3}{2}P_j.$$

This tour can be converted into a traveling salesman tour, using shortcuts, and therefore,

$$L_j \le L_j^* + \frac{3}{2}P_j.$$

Summing these up on j completes the proof. ∎

We can now prove the following result due to Karp.

Theorem 5.3.4 *Under the conditions of Theorem 5.3.2, with probability 1,*

$$\lim_{n\to\infty} \frac{L^*}{\sqrt{n}} = \lim_{n\to\infty} \frac{L^{\mathrm{RP}}}{\sqrt{n}}.$$

Proof. Lemma 5.3.3 implies

$$L^* \le L^{\mathrm{RP}} \le L^* + \frac{3}{2}P^{\mathrm{RP}}.$$

Hence, we need to evaluate the quantity P^{RP}. Note that the number of vertical lines added in the construction of the subregions is $t \le \sqrt{\frac{n}{q}}$. Each of these lines is counted twice in the quantity P^{RP}.

In the second step of the RP heuristic, we add h horizontal lines, where $h \le \sqrt{\frac{n}{q}}$. These horizontal lines are also counted twice in P^{RP}. It follows that

$$P^{\mathrm{RP}} \le 2\sqrt{\frac{n}{q}}(a + b) + 2(a + b) \le 2\sqrt{\frac{n}{\log n}}(a + b) + 2(a + b),$$

where the right-side inequality is justified by the definition of q.

Consequently,

$$\frac{L^{\mathrm{RP}}}{\sqrt{n}} \le \frac{L^*}{\sqrt{n}} + \frac{3}{2}\frac{P^{\mathrm{RP}}}{\sqrt{n}}$$

$$\le \frac{L^*}{\sqrt{n}} + \frac{3(a + b)}{\sqrt{\log n}} + \frac{3(a + b)}{\sqrt{n}}.$$

Taking the limit as n goes to infinity proves the theorem. ∎

5.4   Exercises

Exercise 5.1. *A lower bound on β.* Let $X(n) = \{x_1, x_2, \ldots, x_n\}$ be a set of points uniformly and independently distributed in the unit square. Let ℓ_j be the distance from $x_j \in X(n)$ to the nearest point in $X(n) \setminus x_j$. Let $L(X(n))$ be the length of the optimal traveling salesman tour through $X(n)$. Clearly, $E(L(X(n))) \ge nE(\ell_1)$. We evaluate a lower bound on β in the following way.

(a) Find $\Pr(\ell_1 \geq \ell)$.

(b) Use (a) to calculate a lower bound on $E(\ell_1) = \int_0^\infty \Pr(\ell_1 \geq \ell)d\ell$.

(c) Use Stirling's formula to approximate the bound when n is large.

(d) Show that $\frac{1}{2}$ is a lower bound on β.

Exercise 5.2. *An upper bound on β.* (Karp and Steele 1985) The *strips method* for constructing a tour through n random points in the unit square dissects the square into $\frac{1}{\Delta}$ horizontal strips of width Δ, and then follows a zigzag path, visiting the points in the first strip in left-to-right order, then the points in the second strip in right-to-left order, etc., finally returning to the initial point from the final point of the last strip. Prove that when Δ is suitably chosen, the expected length of the tour produced by the strips method is at most $1.16\sqrt{n}$.

Exercise 5.3. Consider the TSP defined on a set of points N indexed $1, 2, \ldots, n$. Let Z^* be the length of the optimal tour. Consider now the following strategy: Starting with point 1, the salesman moves to the closest point in the set $N \setminus \{1\}$, say point 2. The salesman then constructs an optimal traveling salesman tour defined on this set of $n - 1$ points $(N \setminus \{1\})$ and then returns to point 1 through point 2. Show that the length of this tour is no larger than $3Z^*/2$. Is the bound tight?

Exercise 5.4. Prove that the bin-packing constant γ satisfies $1 \leq \gamma/E(w) \leq 2$, where $E(w)$ is the expected item size.

Exercise 5.5. The harmonic heuristic with parameter M, denoted $H(M)$, is the following. For each $k = 1, 2, \ldots, M - 1$, items of size $\frac{1}{k+1} < w_i \leq \frac{1}{k}$ are packed separately, at most k items per bin. That is, items of size greater than $\frac{1}{2}$ are packed one per bin, items of size $\frac{1}{3} < w_i \leq \frac{1}{2}$ are packed two per bin, and so forth. Finally, items of size $w_i \leq \frac{1}{M}$ are packed separately from the rest using first-fit.

Given n items drawn randomly from the uniform distribution on $(\frac{1}{6}, 0]$, what is the asymptotic number of bins used by $H(5)$?

Exercise 5.6. Suggest a method to pack n items drawn randomly from the uniform distribution on $[\frac{1}{3}, 1]$. Can you prove that your method is asymptotically optimal? What is the bin-packing constant (γ) for this distribution?

Exercise 5.7. Suggest a method to pack n items drawn randomly from the uniform distribution on $[0, \frac{5}{12}]$. Can you prove that your method is asymptotically optimal? What is the bin-packing constant (γ) for this distribution?

Exercise 5.8. Suggest a method to pack n items drawn randomly from the uniform distribution on $[\frac{1}{40}, \frac{59}{120}]$. Can you prove that your method is asymptotically optimal? What is the bin-packing constant (γ) for this distribution?

Exercise 5.9. (Dreyfus and Law 1977) The following is a dynamic programming procedure to solve the TSP. Let city 1 be an arbitrary city. Define the following function:

$$f_i(j, S) = \text{the length of the shortest path from city 1 to}$$
$$\text{city } j \text{ visiting cities in the set } S, \text{ where } |S| = i.$$

Determine the recursive formula and solve the following instance.

The distances between cities

d_{ij}	1	2	3	4	5
1	0	3	1	5	4
2	1	0	5	4	3
3	5	4	0	2	1
4	3	1	3	0	3
5	5	2	4	1	0

Exercise 5.10. What is the complexity of the dynamic program developed in the previous exercise?

Exercise 5.11. (Coffman and Lueker 1991) Consider flipping a fair coin n times in succession. Let X_n represent the random variable denoting the maximum excess of the number of heads over tails at any point in the sequence of n flips. It is known that $E(X_n)$ is $\Theta(\sqrt{n})$. From this, argue that

$$E[Z_n^{\text{MATCH}}] = \frac{n}{2} + \Theta(\sqrt{n}).$$

Exercise 5.12. Assume n cities are uniformly distributed in the unit disc. Consider the following heuristic for the n-city TSP. Let d_i be the distance from city i to the depot. Order the points so that $d_1 \leq d_2 \leq \cdots \leq d_n$. For each $i = 1, 2, \ldots, n$, draw a circle of radius d_i centered at the depot; call this circle i. Starting at the depot, travel directly to city 1. From city 1, travel to circle 2 in a direction along the ray through city 1 and the depot. When circle 2 is reached, follow circle 2 in the direction (clockwise or counterclockwise) that results in a shorter route to city 2. Repeat this same step until city n is reached; then return to the depot. Let Z_n^H be the length of this traveling salesman tour. What is the asymptotic rate of growth of Z_n^H? Is this heuristic asymptotically optimal?

6

Mathematical Programming-Based Bounds

6.1 Introduction

An important method of assessing the effectiveness of any heuristic is to compare it to the value of a lower bound on the cost of an optimal solution. In many cases, this is not an easy task; constructing strong lower bounds on the optimal solution may be as difficult as solving the problem. An attractive approach for generating a lower bound on the optimal solution to an $\mathcal{NP}$-Complete problem is the following mathematical programming approach. First, formulate the problem as an integer program; then relax the integrality constraint and solve the resulting linear program.

What problems do we encounter when we try to use this approach? One difficulty is deciding on an integer programming formulation. There are myriad possible formulations from which to choose. Another difficulty may be that in order to formulate the problem as an integer program, a large (sometimes exponential) number of variables are required. That is, the resulting linear program may be very large, so that it is not possible to use standard linear programming solvers. The third problem is that it is not clear how tight the lower bound provided by the linear relaxation will be. This depends on the problem and the formulation.

In the sections below, we demonstrate how a general class of formulations can provide tight lower bounds on the original integer program. In later chapters, we show that these and similar linear programs can be solved effectively and implemented in algorithms that solve logistics problems to optimality or near optimality.

D. Simchi-Levi et al., *The Logic of Logistics: Theory, Algorithms, and Applications for Logistics Management*, Springer Series in Operations Research and Financial Engineering, DOI 10.1007/978-1-4614-9149-1_6, © Springer Science+Business Media New York 2014

6.2   An Asymptotically Tight Linear Program

Again, consider the bin-packing problem. There are many ways to formulate the problem as an integer program. The one we use here is based on formulating it as a set-partitioning problem. The idea is as follows. Let F be the collection of all sets of items that can be feasibly packed into one bin; that is,

$$F \doteq \{S \subseteq N : \sum_{i \in S} w_i \le 1\}.$$

For any $i \in N$ and $S \in F$, let

$$\alpha_{iS} = \begin{cases} 1, & \text{if } i \in S, \\ 0, & \text{otherwise.} \end{cases}$$

Let

$$y_S = \begin{cases} 1, & \text{if the set of items } S \text{ are placed in a single bin,} \\ 0, & \text{otherwise.} \end{cases}$$

Then the set-partitioning formulation of the bin-packing problem is the following integer program: /

$$\text{Problem } P: \quad Min \sum_{S \in F} y_S$$

$$\text{s.t.}$$

$$\sum_{S \in F} \alpha_{iS} y_S = 1, \quad \forall i \in N, \tag{6.1}$$

$$y_S \in \{0, 1\}, \quad \forall S \in F.$$

In this section, we prove that the relative difference between the optimal solution of the linear relaxation of Problem P and the optimal solution of Problem P (the integer solution) tends to zero as $|N| = n$, the number of items, increases. First, we need the following definition.

Definition 6.2.1 *A function ϕ is Lipschitz continuous of order q on a set $A \subseteq \mathbb{R}$ if there exists a constant K such that*

$$|\phi(x) - \phi(y)| \le K|x - y|^q, \quad \forall x, y \in A.$$

Our first result of this section is the following.

Theorem 6.2.2 *Let the item sizes be independently and identically distributed according to a distribution Φ, which is Lipschitz continuous of order $q \ge 1$ on $[0, 1]$. Let b_n^{LP} be the value of the optimal solution to the linear relaxation of P, and let b_n^* be the value of the optimal integer solution to P, that is, the value of the optimal solution to the bin-packing problem. Then, with probability 1,*

$$\lim_{n \to \infty} \frac{1}{n} b_n^{LP} = \lim_{n \to \infty} \frac{1}{n} b_n^*.$$

To prove the theorem, we consider a related model. Consider a *discretized* bin-packing problem in which there are a finite number W of item sizes. Each different size defines an *item type*. Let n_i be the number of items of type i, for $i = 1, 2, \ldots, W$, and let $n = \sum_{i=1}^{W} n_i$ be the total number of items. Clearly, this discretized bin-packing problem can be solved by formulating it as the set-partitioning problem P. To obtain some intuition about the linear relaxation of P, we first introduce another formulation closely related to P.

Let a *bin assignment* be a vector $(a_1, a_2, \ldots, a_W)$, where $a_i \geq 0$ are integers, and such that a single bin can contain a_1 items of type 1, along with a_2 items of type 2, $\ldots$, along with a_W items of type W, without violating the capacity constraint. Index all the possible bin assignments $1, 2, \ldots, R$, and note that R is independent of n. The bin-packing problem can be formulated as follows. Let

$$A_{ir} = \text{number of items of type } i \text{ in bin assignment } r,$$

for each $i = 1, 2, \ldots, W$ and $r = 1, 2, \ldots, R$. Let

$$y_r = \text{number of times bin assignment } r \text{ is used in the optimal solution.}$$

The new formulation of the discretized bin-packing problem is

$$\text{Problem } P_D : \quad Min \ \sum_{r=1}^{R} y_r$$

$$\text{s.t.}$$

$$\sum_{r=1}^{R} y_r A_{ir} \geq n_i, \quad \forall i = 1, 2, \ldots, W,$$

$$y_r \geq 0 \text{ and integer}, \quad \forall r = 1, 2, \ldots, R.$$

Let b_D^* be the value of the optimal solution to Problem P_D and let b_D^{LP} be the optimal solution to the linear relaxation of Problem P_D. Clearly, Problem P and Problem P_D have the same optimal solution values; that is, $b^* = b_D^*$. On the other hand, b^{LP} is not necessarily equal to b_D^{LP}. However, it is easy to see that any feasible solution to the linear relaxation of Problem P can be used to construct a feasible solution to the linear relaxation of Problem P_D, and therefore,

$$b^{\text{LP}} \geq b_D^{\text{LP}}. \tag{6.2}$$

The following is the crucial lemma needed to prove Theorem 6.2.2.

Lemma 6.2.3

$$b^{\text{LP}} \leq b^* \leq b_D^{\text{LP}} + W \leq b^{\text{LP}} + W.$$

Proof. The leftmost inequality is trivial, while the rightmost inequality is due to (6.2). To prove the central inequality, note that in Problem P_D there are W constraints, one for each item type. Let $\bar{y}_r$, for $r = 1, 2, \ldots, R$, be an optimal solution to the linear relaxation of Problem P_D and observe that there exists such an optimal solution with at most W positive variables, one for each constraint. We construct a feasible solution to Problem P_D by rounding the linear solution up; that is, for each $r = 1, 2, \ldots, R$ with $\bar{y}_r > 0$, we make $y_r = \lceil \bar{y}_r \rceil$, and for each $r = 1, 2, \ldots, R$ with $\bar{y}_r = 0$, we make $y_r = 0$. Hence, the increase in the objective function is no more than W. ∎

Observe that the upper bound on b^* obtained in Lemma 6.2.3 consists of two terms. The first, b^{LP}, is a lower bound on b^*, which clearly grows with the number of items n. The second term (W) is independent of n. Therefore, the upper bound on b^* of Lemma 6.2.3 is dominated by b^{LP}; consequently, we see that for large n, $b^* \approx b^{\mathrm{LP}}$, exactly what is implied by Theorem 6.2.2.

We can now use the intuition developed in the above analysis of the discrete bin-packing problem to prove Theorem 6.2.2.

Proof. It is clear that $b^{\mathrm{LP}} \leq b^*$, and therefore, $\overline{\lim}_{n \to \infty} b^{\mathrm{LP}}/n \leq \lim_{n \to \infty} b^*/n$. To prove the upper bound, partition the interval $(0, 1]$ into $k \geq 2$ subintervals of equal length. Let N_j be the set of items whose size w satisfies $\frac{j-1}{k} < w \leq \frac{j}{k}$, and let $|N_j| = n_j$, $j = 1, 2, \ldots, k$. We construct a new bin-packing problem where item sizes take only the values $\frac{j}{k}$, $j = 1, 2, \ldots, k-1$ and where the number of items of size $\frac{j}{k}$ is $\min\{n_j, n_{j+1}\}$, $j = 1, 2, \ldots, k-1$. We refer to this instance of the bin-packing problem as the *reduced instance*. For this reduced instance, define $\underline{b}^*$, $\underline{b}^{\mathrm{LP}}$, and $\underline{b}_D^{\mathrm{LP}}$ to be the obvious quantities.

It is easy to see that we can always construct a feasible solution to the original bin-packing problem by solving the bin-packing problem defined on the reduced instance and then assigning each of the remaining items to a single bin. This results in

$$b^* \leq \underline{b}^* + \sum_{j=1}^{k-1} |n_j - n_{j+1}| + n_k$$

$$\leq \underline{b}_D^{\mathrm{LP}} + k + \sum_{j=1}^{k-1} |n_j - n_{j+1}| + n_k \qquad \text{(using Lemma 6.2.3)}$$

$$\leq \underline{b}^{\mathrm{LP}} + k + \sum_{j=1}^{k-1} |n_j - n_{j+1}| + n_k.$$

We now argue that $\underline{b}^{\mathrm{LP}} \leq b^{\mathrm{LP}}$. This must be true since every item in the reduced instance can be associated with a unique item in the original instance whose size is at least as large. Thus, every feasible solution to the linear relaxation of the set-partitioning problem defined on the original instance is feasible for the same problem on the reduced instance. Hence,

$$b^* \leq b^{\mathrm{LP}} + k + \sum_{j=1}^{k-1} |n_j - n_{j+1}| + n_k.$$

The strong law of large numbers and the mean value theorem imply that for a given $j = 1, \ldots, k$, there exists s_j such that

$$\lim_{n \to \infty} \frac{n_j}{n} = \frac{1}{k} \phi(s_j),$$

where ϕ is the density of item sizes. Hence,

$$\lim_{n \to \infty} \frac{1}{n} |n_j - n_{j+1}| = \frac{1}{k} |\phi(s_j) - \phi(s_{j+1})|$$

$$\leq \frac{1}{k} K(s_{j+1} - s_j)^q \qquad \text{(by Lipschitz continuity)}$$

$$\leq \frac{2}{k^{q+1}} K \qquad \left(\text{since } s_{j+1} - s_j \leq \frac{2}{k} \right)$$

$$\leq \frac{2}{k^2} K \qquad \text{(since } q \geq 1 \text{)}.$$

Consequently,

$$\lim_{n \to \infty} \frac{b^*}{n} \leq \lim_{n \to \infty} \frac{b^{\text{LP}}}{n} + \frac{K(2k-1)}{k^2}.$$

Since this holds for arbitrary k, this completes the proof. ∎

In fact, it appears that the linear relaxation of the set-partitioning formulation may be extremely close to the optimal solution in the case of the bin-packing problem. Chan et al. (1998) show that the worst-case effectiveness of the set-partitioning lower bound (the linear relaxation), that is, the maximum ratio of the optimal integer solution (b^*) to the optimal linear relaxation b^{LP}, is $\frac{4}{3}$. They also provide an example that achieves this bound. That is, for any number of items and any set of item weights, the linear program is at least $75\,\%$ of the optimal solution.

6.3   Lagrangian Relaxation

In 1971, Held and Karp applied a mathematical technique known as *Lagrangian relaxation* to generate a lower bound on a general integer (linear) program. Our discussion of the method follows the elegant presentation of Fisher (1981). We start with the following integer program:

$$\text{Problem } P: \quad Z = Min \qquad cx$$

$$\text{s.t.}$$

$$Ax = b, \qquad\qquad (6.3)$$

$$Dx \leq e, \qquad\qquad (6.4)$$

$$x \geq 0 \ \text{ and integer,}$$

where x is an n-vector, b is an m-vector, e is a k-vector, A is an $m \times n$ matrix, and D is a $k \times n$ matrix. Let the optimal solution to the linear relaxation of Problem P be Z_{LP}. The Lagrangian relaxation of constraints (6.3) with multipliers $u \in \mathbb{R}^m$ is

$$\text{Problem } LR_u : \quad Z_D(u) = Min \quad cx + u(Ax - b)$$

$$\text{s.t.}$$

$$Dx \leq e, \tag{6.5}$$

$$x \geq 0 \qquad \text{and integer.}$$

The following is a simple observation.

Lemma 6.3.1 *For all $u \in \mathbb{R}^m$, $Z_D(u) \leq Z$.*

Proof. Let x be any feasible solution to Problem P. Clearly, x is also feasible for LR_u, and since $Z_D(u)$ is its optimal solution value, we get

$$Z_D(u) \leq cx + u(Ax - b) = cx.$$

Consequently, $Z_D(u) \leq Z$. ∎

Remark: If the constraints $Ax = b$ in Problem P are replaced with the constraints $Ax \leq b$, then Lemma 6.3.1 holds for $u \in \mathbb{R}^m_+$.

Since $Z_D(u) \leq Z$ holds for all u, we are interested in the vector u that provides the largest possible lower bound. This is achieved by solving Problem D, called the *Lagrangian dual*, defined as follows:

$$\text{Problem D}: \ Z_D = max_u Z_D(u).$$

Problem D has a number of important and interesting properties.

Lemma 6.3.2 *The function $Z_D(u)$ is a piecewise linear concave function of u.*

This implies that $Z_D(u)$ attains its maximum at a nondifferentiable point. This maximal point can be found using a technique called *subgradient optimization*, which can be described as follows: Given an initial vector u^0, the method generates a sequence of vectors $\{u^k\}$ defined by

$$u^{k+1} = u^k + t_k(Ax^k - b), \tag{6.6}$$

where x^k is an optimal solution to Problem LR_{u^k} and t_k is a positive scalar called the *step size*. Polyak (1967) shows that if the step sizes $t_1, t_2, \ldots$, are chosen such that $\lim_{k\to\infty} t_k = 0$ and $\sum_{k \geq 0} t_k$ is unbounded, then $Z_D(u^k)$ converges to Z_D.

The step size commonly used in practice is

$$t_k = \frac{\lambda_k(UB - Z_D(u^k))}{\sum_{i=1}^n (a_i x^k - b_i)^2},$$

where UB is an upper bound on the optimal integer solution value (found using a heuristic), $a_i x^k - b_i$ is the difference between the left-hand side and the right-hand side of the ith constraint in $Ax^k \leq b$, and λ_k is a scalar satisfying $0 < \lambda_k \leq 2$. Usually, one starts with $\lambda_0 = 2$ and cuts it in half every time $Z_D(u)$ fails to increase after a number of iterations.

It is now interesting to compare the Lagrangian relaxation lower bound (Z_D) to the lower bound achieved by solving the linear relaxation of the set-partitioning formulation (Z_{LP}).

Theorem 6.3.3

$$Z_{\mathrm{LP}} \leq Z_{\mathrm{D}}.$$

Proof.

$$Z_{\mathrm{D}} = \max_u \left\{ \min_x cx + u(Ax - b) \,\Big|\, Dx \leq e, x \geq 0 \text{ and integer} \right\}$$

$$\geq \max_u \left\{ \min_x cx + u(Ax - b) \,\Big|\, Dx \leq e, x \geq 0 \right\}$$

$$= \max_u \max_v \left\{ ve - ub \,\Big|\, vD \leq c + uA, v \leq 0 \right\} \qquad \text{(by strong duality)}$$

$$= \max_{u,v} \left\{ ve - ub \,\Big|\, vD \leq c + uA, v \leq 0 \right\}$$

$$= \min_y \left\{ cy \,\Big|\, Ay = b, Dy \leq e, y \geq 0 \right\} \qquad \text{(by strong duality)}$$

$$= Z_{\mathrm{LP}}.$$

∎

We say a mathematical Program P possesses the *integrality property* if the solution to the linear relaxation of P always provides an integer solution. An inspection of the above proof reveals the following corollary.

Corollary 6.3.4 *If Problem LR_u possesses the integrality property, then $Z_{\mathrm{D}}{=}Z_{\mathrm{LP}}$.*

6.4 Lagrangian Relaxation and the Traveling Salesman Problem

Held and Karp (1970, 1971) developed the Lagrangian relaxation technique in the context of the traveling salesman problem. They show some interesting relationships between this method and a graph-theoretic problem called the *minimum-weight 1-tree problem*.

6.4.1   The 1-Tree Lower Bound

We start by defining a 1-tree. For a given choice of vertex, say vertex 1, a 1-tree is a tree having vertex set $\{2, 3, \ldots, n\}$ together with two distinct edges connected to vertex 1. Therefore, a 1-tree is a graph with exactly one cycle. Define the weight of a 1-tree to be the sum of the costs of all its edges. In the minimum-weight 1-tree problem, the objective is to find a 1-tree of minimum weight. Such a 1-tree can be constructed by finding a minimum spanning tree on the entire network excluding vertex 1 and its corresponding edges and by adding to the minimum spanning tree the two edges incident to vertex 1 of minimum cost.

We observe that any traveling salesman tour is a 1-tree tour in which each vertex has a degree 2. Moreover, if a minimum-weight 1-tree is a tour, then it is an optimal traveling salesman tour. Thus, the minimum-weight 1-tree provides a lower bound on the length of the optimal traveling salesman tour.

Unfortunately, this bound can be quite weak. However, there are ways to improve it. For this purpose, consider the vector $\pi = \{\pi_1, \pi_2, \ldots, \pi_n\}$ and the following transformation of the distances $\{d_{ij}\}$:

$$d'_{ij} \doteq d_{ij} + \pi_i + \pi_j.$$

Let L^* be the length of the optimal tour with respect to the distance matrix $\{d_{ij}\}$. It is clear that the same tour is also optimal with respect to the distance matrix $\{d'_{ij}\}$. To see that, observe that any traveling salesman tour S of cost L with respect to $\{d_{ij}\}$ has a cost $L + 2\sum_{i=1}^{n} \pi_i$ with respect to $\{d'_{ij}\}$. Thus, the difference between the length of any traveling salesman tour in $\{d_{ij}\}$ and $\{d'_{ij}\}$ is constant, independent of the tour.

Observe also that the above transformation of the distances *does* change the minimum 1-tree. How can this idea be used? First, enumerate all possible 1-trees and let d_i^k be the degree of vertex i in the kth 1-tree. Let T_k be the weight (cost) of that 1-tree (before transforming the distances). This implies that the cost of that 1-tree after the transformation is exactly

$$T_k + \sum_{i \in V} d_i^k \pi_i.$$

Thus, the minimum-weight 1-tree on the transformed distance matrix is obtained by solving

$$\min_k \left\{ T_k + \sum_{i \in V} d_i^k \pi_i \right\}.$$

Since, in the transformed distance matrix, the optimal traveling salesman tour does not change while the 1-tree provides a lower bound, we have

$$L^* + 2\sum_{i \in V} \pi_i \geq \min_k \left\{ T_k + \sum_{i \in V} d_i^k \pi_i \right\},$$

which implies

$$L^* \geq \min_k \left\{ T_k + \sum_{i \in V} (d_i^k - 2)\pi_i \right\} \doteq w(\pi).$$

Consequently, the best lower bound is obtained by maximizing the function $w(\pi)$ over all possible values of π. How can we find the best value of π? Held and Karp (1970, 1971) use the subgradient method described in the previous section. That is, starting with some arbitrary vector π^0, in step k the method updates the vector π^k according to

$$\pi_i^{k+1} = \pi_i^k + t_k(d_i^k - 2),$$

where π_i^k is the ith element in the vector π^k, and t_k, the step size, equals

$$t_k = \frac{\lambda_k(UB - w(\pi^k))}{\sum_{i=1}^n (d_i^k - 2)^2}.$$

6.4.2 The 1-Tree Lower Bound and Lagrangian Relaxation

We now relate the 1-tree lower bound to a Lagrangian relaxation associated with the following formulation of the traveling salesman problem. For every $e \in E$, let d_e be the cost of the edge and let x_e be a variable that takes on the value 1 if the optimal tour includes the edge and the value 0 otherwise. Given a subset $S \subset V$, let $E(S)$ be the set of edges from E such that each edge has its two endpoints in S. Let $\delta(S)$ be the collection of edges from E in the cut separating S from $V \backslash S$. The traveling salesman problem can be formulated as follows:

$$\text{Problem } P' : \quad Z^* = Min \sum_{e \in E} d_e x_e$$

s.t.

$$\sum_{e \in \delta(i)} x_e = 2, \quad \forall i = 1, 2, \ldots, n, \tag{6.7}$$

$$\sum_{e \in E(S)} x_e \le |S| - 1, \quad \forall S \subseteq V \backslash \{1\}, S \neq \emptyset, \tag{6.8}$$

$$0 \le x_e \le 1, \quad \forall e \in E, \tag{6.9}$$

$$x_e \quad \text{integer}, \quad \forall e \in E. \tag{6.10}$$

Constraints (6.7) ensure that each vertex has an edge going in and an edge going out. Constraints (6.8), called *subtour elimination* constraints, forbid integral solutions consisting of a set of disjoint cycles.

Observe that constraints (6.7) can be replaced by the following constraints:

$$\sum_{e \in \delta(i)} x_e = 2, \quad \forall i = 1, \ldots, n-1, \tag{6.11}$$

$$\sum_{e \in E} x_e = n. \tag{6.12}$$

This is true since constraints (6.11) are exactly constraints (6.7) for $i = 1, \ldots, n-1$. The only missing constraint is $\sum_{e \in \delta(n)} x_e = 2$. Therefore, it is sufficient to show that (6.12) holds if and only if this one holds. To see this, calculate

$$\sum_{e \in E} x_e = \frac{1}{2} \sum_{i=1}^{n} \sum_{e \in \delta(i)} x_e$$

$$= \frac{1}{2} \sum_{i=1}^{n-1} \sum_{e \in \delta(i)} x_e + \frac{1}{2} \sum_{e \in \delta(n)} x_e$$

$$= (n-1) + \frac{1}{2} \sum_{e \in \delta(n)} x_e.$$

Thus, $\sum_{e \in E} x_e = n$ if and only if $\sum_{e \in \delta(n)} x_e = 2$.

The resulting formulation of the traveling salesman problem is

$$\left\{ Min \sum_{e \in E} d_e x_e \,\middle|\, (6.8), (6.9), (6.10), (6.11), \text{ and } (6.12) \right\}.$$

We can now use the Lagrangian relaxation technique described in Sect. 6.3 and get the following lower bound on the length of the optimal tour:

$$\max_u \left\{ \min_x \sum_{i,j \in V} (d_{ij} + u_i + u_j) x_{ij} \,\middle|\, (6.8), (6.9), (6.10), \text{ and } (6.12) \right\}.$$

Interestingly enough, Edmonds (1971) showed that the extreme points of the polyhedron defined by constraints (6.8)–(6.10) and (6.12) is the set of all 1-trees; that is, the optimal solution to a linear program defined on these constraints must be integral. Thus, we can apply Corollary 6.3.4 to see that the lower bound obtained from the 1-tree approach is the same as the linear relaxation of Problem P'.

6.5   The Worst-Case Effectiveness of the 1-Tree Lower Bound

We conclude this chapter by demonstrating that the Held and Karp (1970, 1971) 1-tree relaxation provides a lower bound that is not far from the length of the optimal tour. For this purpose, we show that the Held and Karp lower bound can be written as follows:

$$\text{Problem HK} \quad : \quad Z_{LP} = Min \sum_{e \in E} d_e x_e$$

$$\text{s.t.}$$

$$\sum_{e \in \delta(i)} x_e = 2, \quad \forall i = 1, 2, \ldots, n, \qquad (6.13)$$

$$\sum_{e\in\delta(S)} x_e \geq 2, \ \forall S \subseteq V\setminus\{1\}, S\neq\emptyset, \qquad (6.14)$$

$$0 \leq x_e \leq 1, \quad \forall e \in E. \qquad (6.15)$$

Lemma 6.5.1 *The linear relaxation of Problem P' is equivalent to Problem HK.*

Proof. We first show that any feasible solution $\bar{x}$ to the linear relaxation of Problem P' is feasible for Problem HK. Since $\sum_{e\in S}\bar{x}_e \leq |S|-1$, $\sum_{e\in E(V\setminus S)}\bar{x}_e \leq n-|S|-1$, and $\sum_{e\in E(V)} x_e = n$ (why?), we get $\sum_{e\in\delta(S)}\bar{x}_e \geq 2$.

Similarly, we show that any feasible solution $\tilde{x}$ to Problem HK is feasible for the linear relaxation of Problem P'. The feasibility of $\tilde{x}$ in Problem HK implies that $\sum_{i\in S}\sum_{e\in\delta(i)}\tilde{x}_e = 2|S|$. However,

$$\sum_{i\in S}\sum_{e\in\delta(i)}\tilde{x}_e = 2\sum_{e\in E(S)}\tilde{x}_e + \sum_{e\in\delta(S)}\tilde{x}_e = 2|S|,$$

and since $\sum_{e\in\delta(S)}\tilde{x}_e \geq 2$, we get $\sum_{e\in E(S)}\tilde{x}_e \leq |S|-1$. ∎

Shmoys and Williamson (1990) have shown that the Held and Karp lower bound (Problem HK) has a particular monotonicity property, and as a consequence, they obtain a new proof of an old result from Wolsey (1980), who showed the following:

Theorem 6.5.2 *For every instance of the TSP for which the distance matrix satisfies the triangle inequality, we have $Z^* \leq \frac{3}{2}Z_{LP}$.*

The proof presented here is based on the monotonicity property established by Shmoys and Williamson (1990) . However, we use a powerful tool discovered by Goemans and Bertsimas (1993), called the *parsimonious property*. This is a property that holds for a general class of network design problems.

To present the property, consider the following linear program defined on the complete graph $G = (V, E)$. Associated with each vertex $i \in V$ is a given number r_i, which is either 0 or 2. Let $V_2 = \{i \in V | r_i = 2\}$.

We will analyze the following linear program (here ND stands for network design).

$$\text{Problem ND} : \quad Min \quad \sum_{e\in E} d_e x_e$$

$$\text{s.t.}$$

$$\sum_{e\in\delta(i)} x_e = r_i, \quad \forall i = 1, 2, \ldots, n, \qquad (6.16)$$

$$\sum_{e\in\delta(S)} x_e \geq 2, \quad \forall S \subset V, V_2 \cap S \neq \emptyset,$$

$$V_2 \cap (V\setminus S) \neq \emptyset, \qquad (6.17)$$

$$0 \leq x_e \leq 1, \quad \forall e \in E. \qquad (6.18)$$

It is easy to see that when $V_2 = V$, this linear program is equivalent to the linear program Problem HK. We now provide a short proof of the following result.

Lemma 6.5.3 *The optimal solution value to Problem ND is unchanged if we omit constraint (6.16).*

Our proof is similar to the proof presented in Bienstock and Simchi-Levi (1993); see also Bienstock et al. (1993a), which uses a result of Lovasz (1979). In his book of problems (Exercise 6.51), Lovasz presents the following result, together with a short proof. But first, we need a definition.

Definition 6.5.4 *An undirected graph G is k-connected between two vertices i and j if there are k (node) disjoint paths between i and j.*

Lemma 6.5.5 *Let G be an Eulerian multigraph and $s \in V(G)$, such that G is k-connected between any two vertices different from s. Then, for any neighbor u of s, there exists another neighbor w of s, such that the multigraph obtained from G by removing $\{s, u\}$ and $\{s, w\}$ and adding a new edge $\{u, w\}$ (the splitting-off operation) is also k-connected between any two vertices different from s.*

Lovasz's proof of Lemma 6.5.5 can be easily modified to yield the following.

Lemma 6.5.6 *Let G be an Eulerian multigraph, $Y \subseteq V(G)$ and $s \in V(G)$, such that G is k-connected between any two vertices of Y different from s. Then, for any neighbor u of s, there exists another neighbor w of s, such that the multigraph obtained from G by removing $\{s, u\}$ and $\{s, w\}$ and adding a new edge $\{u, w\}$ is also k-connected between any two vertices of Y different from s.*

We can now prove Lemma 6.5.3.

Proof. Let $V_0 = V \setminus V_2$; that is, $V_0 = \{i \in V | r_i = 0\}$. Let Problem ND$'$ be Problem ND without (6.16). Finally, let $\tilde{x}$ be a rational vector feasible for Problem ND$'$, chosen such that (i) $\tilde{x}$ is optimal for Problem ND$'$, and (ii) subject to (i), $\sum_{e \in E} \tilde{x}_e$ is minimized.

Let M be a positive integer, large enough so that $\tilde{v} = 2M\tilde{x}$ is a vector of even integers. We may regard $\tilde{v}$ (with a slight abuse of notation) as the incidence vector of the edge-set $\tilde{E}$ of a multigraph $\tilde{G}$ with vertex set V. Clearly, $\tilde{G}$ is Eulerian, and by (6.17), it is $4M$-connected between any two elements of V_2.

Now suppose that for some vertex s, $\sum_{e \in \delta(\{s\})} \tilde{x}_e > r_s$ (i.e., s has a degree larger than $2Mr_s$ in $\tilde{G}$). Let us apply Lemma 6.5.6 to s and any neighbor u of s (where $Y = V_2$), and let $\tilde{H}$ be the resulting multigraph, with incidence vector $\tilde{z}$.
Clearly,

$$\sum_{e \in E} d_e \tilde{z}_e \leq \sum_{e \in E} d_e \tilde{v}_e,$$

and so

$$\sum_{e \in E} d_e \frac{\tilde{z}_e}{2M} \leq \sum_{e \in E} d_e \tilde{x}_e.$$

Moreover,

$$\sum_{e \in E} \frac{\tilde{z}_e}{2M} = \sum_{e \in E} \tilde{x}_e - \frac{1}{2M}.$$

Hence, by the choice of $\tilde{x}$, $z = \frac{\tilde{z}}{2M}$ cannot be feasible for Problem ND$'$.

If $s \in V_0$, then by Lemma 6.5.6, z is feasible for Problem ND$'$. Thus, we must have $s \in V_2$ and, in fact, $\sum_{e \in \delta(\{t\})} \tilde{x}_e = 0$ for all $t \in V_0$. In other words, $\tilde{E}$ spans precisely V_2, $\tilde{G}$ is $4M$-connected, and $\sum_{e \in \delta(\{s\})} \geq 4M + 2$. But we claim now that the multigraph $\tilde{H}$ is $4M$-connected. For by Lemma 6.5.6, it could only fail to be $4M$-connected between s and some other vertex, but the only possible cut of size less than $4M$ is the one separating s from $V \setminus \{s\}$. Since this cut has at least $4M$ edges, the claim is proved. Consequently, again we obtain that z is feasible for Problem ND$'$, a contradiction. In other words, $\sum_{e \in E} \tilde{v}_e = 2Mr_i$ for all i; that is, (6.16) holds.   ∎

An immediate consequence of Lemma 6.5.3 is that in Problem HK, one can ignore constraint (6.13) without changing the value of its optimal solution. This new formulation reveals the following monotonocity property of the Held and Karp lower bound: Let $A \subseteq V$ and consider the Held and Karp lower bound on the length of the optimal traveling salesman tour through the vertices in A; that is,

$$\text{Problem HK}(A) : \quad Z_{LP}(A) = Min \sum_{e \in E} d_e x_e$$

s.t.

$$\sum_{e \in \delta(S)} x_e \geq 2, \quad \forall S \subset A, \qquad (6.19)$$

$$0 \leq x_e \leq 1, \quad \forall e \in E. \qquad (6.20)$$

Since any feasible solution to Problem HK(V) is feasible for Problem HK(A), the cost of this linear program is monotone with respect to the set of nodes A.

We are ready to prove Theorem 6.5.2.

Proof. Section 4.3.3 presents and analyzes the heuristic developed by Christofides for the TSP which is based on constructing a minimum spanning tree plus a matching on the nodes of odd degree. Observe that a similar heuristic can be obtained if we start from a 1-tree instead of a minimum spanning tree. Thus, the length of the optimal tour is bounded by $W(T_1^*) + W(M^*(A))$, where $W(T_1^*)$ is the weight (cost) of the best 1-tree and $W(M^*(A))$ is the weight of the optimal weighted matching defined on the set of odd-degree nodes in the best 1-tree, denoted by A.

We argue that $W(M^*(A)) \leq \frac{1}{2} Z_{LP}(A)$. Let $\bar{x}$ be an optimal solution to Problem HK(A). It is easy to see that the vector $\frac{1}{2}\bar{x}$ is feasible for the following constraints:

$$\sum_{e \in \delta(i)} x_e = 1, \quad \forall i \in A, \qquad (6.21)$$

$$\sum_{e \in E(S)} x_e \le \frac{1}{2}(|S| - 1), \quad \forall S \subset A, S \ne \emptyset, \ |S| \ge 3, \ |S| \text{ is odd}, \quad (6.22)$$

$$0 \le x_e \le 1, \quad \forall e \in E. \quad (6.23)$$

A beautiful result of Edmonds (1965) tells us that these constraints are sufficient to formulate the matching problem as a linear program. Consequently,

$$W(M^*(A)) \le \frac{1}{2} Z_{LP}(A) \le \frac{1}{2} Z_{LP}(V) = \frac{1}{2} Z_{LP},$$

and therefore,

$$L^* \le W(T_1^*) + W(M^*(A))$$

$$\le Z_{LP} + \frac{1}{2} Z_{LP}$$

$$\le \frac{3}{2} Z_{LP}.$$

∎

6.6   Exercises

Exercise 6.1.   Prove Lemma 6.3.2.

Exercise 6.2.   Show that a lower bound on the cost of the optimal traveling salesman tour can be given by

$$\frac{2}{|N|} \max_{i \in N} \sum_{j \in N} d_{ij},$$

where N is the set of cities and d_{ij} is the distance from city i to city j.

Exercise 6.3.   Consider an instance of the bin-packing problem where there are m_j items of size $w_j \in (0, 1]$ for $j = 1, 2, \ldots, n$. Define a *bin configuration* to be a vector $\bar{c} = (c_1, c_2, \ldots, c_n)$ with the property that $c_i \ge 0$ for $i = 1, 2, \ldots, n$ and $\sum_{j=1}^{n} c_j w_j \le 1$. Enumerate all possible bin configurations. Let there be M such configurations. Define C_{jk} to be the number of items of size w_j in bin configuration k, for $k = 1, 2, \ldots, M$ and $j = 1, 2, \ldots, n$.

Formulate an integer program to solve this bin-packing problem using the following variables: x_k is the number of times configuration k is used, for $k = 1, 2, \ldots, M$.

Exercise 6.4.   A function $u : [0, 1] \to [0, 1]$ is *dual-feasible* if, for any sets of numbers $w_1, w_2, \ldots, w_k$, we have

$$\sum_{i=1}^{k} w_i \le 1 \ \Rightarrow \ \sum_{i=1}^{k} u(w_i) \le 1.$$

(a) Given an instance of the bin-packing problem with item sizes $w_1, w_2, \ldots, w_n$ and a dual-feasible function u, prove that $\sum_{i=1}^{n} u(w_i) \leq b^*$.

(b) Assume n is even. Let half of the items be of size $\frac{2}{3}$ and the other half of size $\frac{1}{2}$. Find a dual-feasible function u that satisfies

$$\sum_{i=1}^{n} u(w_i) = b^*.$$

Exercise 6.5. Consider a list L of n items of sizes in $(\frac{1}{3}, \frac{1}{2}]$. Let b^{LP} be the optimal fractional solution to the set-partitioning formulation of the bin-packing problem, and let b^* be the optimal integer solution to the same formulation. Prove that

$$b^* \leq b^{\text{LP}} + 1.$$

Exercise 6.6. Prove that if a graph has exactly $2k$ vertices of odd degree, then the set of edges can be partitioned into k paths such that each edge is used exactly once.

Part II

Inventory Models

7

Economic Lot Size Models with Constant Demands

7.1 Introduction

Production planning is also an area where difficult combinatorial problems appear in day-to-day logistics operations. In this chapter, we analyze problems related to lot sizing when demands are constant and known in advance. Lot sizing in this deterministic setting is essentially the problem of balancing the fixed costs of ordering with the costs of holding inventory. In this chapter, we look at several different models of deterministic lot sizing. First, we consider the most basic single-item model, the economic lot size model. Then we look at coordinating the ordering of several items with a warehouse of limited capacity. Finally, we look at a one-warehouse multiretailer system.

7.1.1 The Economic Lot Size Model

The classical economic lot size model, introduced by Harris (1915) (see Erlenkotter 1990 for an interesting historical discussion), is a framework where we can see the simple tradeoffs between ordering and storage costs. Consider a facility, possibly a warehouse or a retailer, that faces a constant demand for a *single* item and places orders for the item from another facility in the distribution network, which is assumed to have an unlimited quantity of the product. The model assumes the following.

D. Simchi-Levi et al., *The Logic of Logistics: Theory, Algorithms, and Applications for Logistics Management*, Springer Series in Operations Research and Financial Engineering, DOI 10.1007/978-1-4614-9149-1_7, © Springer Science+Business Media New York 2014

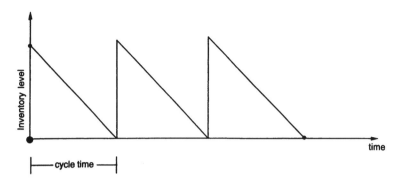

FIGURE 7.1. Inventory level as a function of time

- Demand is constant at a rate of D items per unit time.

- Order quantities are fixed at Q items per order.

- A fixed setup cost K is incurred every time the warehouse places an order.

- A linear inventory carrying cost h, also referred to as the *holding cost*, is accrued for every unit held in inventory per unit time.

- The lead time, that is, the time that elapses between the placement of an order and its receipt, is zero.

- The initial inventory is zero.

- The planning horizon is infinite.

The objective is to find the optimal ordering policy minimizing the total purchasing and carrying cost per unit of time without shortage.

Like all models, this is a simplified version of what might actually occur in practice. The assumption of a known fixed demand over the infinite horizon is clearly unrealistic. The lead time is most likely positive, and the requirement of a fixed order quantity is restrictive. As we shall see, all these assumptions can be easily relaxed while maintaining a relatively simple optimal policy. For the purposes of understanding the basic tradeoffs in the model, we keep the assumptions listed above.

It is easy to see that an optimal ordering policy must satisfy the *zero-inventory-ordering property*, which says that every order is received precisely when the inventory level drops to zero. This can be seen by considering the case where an order is placed when the inventory level is not zero. In that case, the cost is not increased if we simply wait until the inventory is zero to order.

To find the optimal ordering policy in the economic lot size model, we consider the inventory level as a function of time (see Fig. 7.1). This is the so-called saw-toothed inventory pattern. We refer to the time between two successive replenishments as a *cycle* time. Thus, the total inventory cost in a cycle of length

T is

$$K + \frac{hTQ}{2},$$

and since $Q = TD$, the average total cost per unit of time is

$$\frac{KD}{Q} + \frac{hQ}{2}.$$

Hence, the optimal order quantity is

$$Q^* \doteq \sqrt{\frac{2KD}{h}}.$$

This quantity is referred to as the economic order quantity (EOQ); it is the quantity at which the inventory setup cost per unit of time ($\frac{KD}{Q}$) equals the inventory holding cost per unit of time ($\frac{hQ}{2}$).

We now see how some of our assumptions can be relaxed, without losing any of the model's simplicity. Consider the case in which the initial inventory is positive, say at level I_0; then the first order for Q^* items is simply delayed until time $\frac{I_0}{D}$. Further, the assumption of zero lead time can also be easily relaxed. In fact, the model can handle any deterministic lead time L. To do this, simply place an order for Q^* items when the inventory level is DL. On the other hand, relaxing the assumptions of fixed demands and infinite planning horizon requires significant changes to the above solution.

7.1.2 The Finite-Horizon Model

To make the model more realistic, we now introduce a finite horizon, say t. For instance, in the retail apparel industry, such a horizon may represent an 8–12-week period, for example, the "winter season," in which demand for the product might be assumed to be constant and known. We also relax the assumption that the order quantities are fixed. We seek an inventory policy on the interval $[0, t]$ that minimizes the ordering and carrying costs.

For this purpose, consider any inventory policy, say, $\mathcal{P}$, that places $m \geq 1$ orders in the interval $[0, t]$. Clearly, the first order must be placed at time zero and the last must be placed so that the inventory at time t is zero. For any $i, 1 \leq i \leq m-1$, let T_i be the time between the placement of the ith order and the $(i+1)$st order and let T_m be the time between the placement of the last order and t. Thus, by definition, $t = \sum_{i=1}^{m} T_i$, and $\mathcal{P}$ places the jth order at time $\sum_{i=1}^{j} T_i$, for $1 \leq j \leq m$. Again, it is clear that the policy $\mathcal{P}$ must satisfy the zero-inventory-ordering property. Figure 7.2 illustrates the inventory level of policy $\mathcal{P}$.

For policy $\mathcal{P}$, let $I(\tau)$ be the inventory level at time $\tau \in [0, t]$. Thus, the total cost per unit of time associated with $\mathcal{P}$ is

$$\frac{1}{t} \left[Km + h \int_0^t I(\tau) d\tau \right].$$

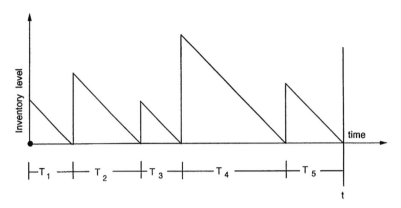

FIGURE 7.2. Inventory level as a function of time under policy $\mathcal{P}$

The only thing we know about the function $I(\tau)$ is that it decreases at a rate of D (a slope of $-D$) between orders and reaches zero exactly m times. Thus, we can express the total inventory up to time t as a function of the time between orders $\{T_i\}_{i=1,\ldots,m}$ as follows:

$$\sum_{i=1}^{m} \frac{T_i \cdot DT_i}{2} = \frac{D}{2} \sum_{i=1}^{m} T_i^2.$$

Consequently, if m orders are placed, we can find the best times to place them by solving

$$Min \left\{ \sum_{i=1}^{m} T_i^2 \;\middle|\; \sum_{i=1}^{m} T_i = t, \; T_i \geq 0, \; \forall i = 1, 2, \ldots, m \right\}.$$

The optimal solution to this convex optimization problem is $T_i = \frac{t}{m}$ for each $i = 1, 2, \ldots, m$. Hence, an optimal policy must have the following property.

Property 7.1.1 *For a problem with one product over the interval $[0, t]$, the inventory policy with minimum cost that places m orders is achieved by placing orders of equal size at equally spaced points in time.*

The property thus implies that the total purchasing and carrying cost per unit time associated with $\mathcal{P}$ is at least

$$\frac{Km}{t} + \frac{hDt}{2m}.$$

Consequently, by selecting the value of m that minimizes this value, we can construct a policy of minimal cost. Let

$$\alpha = t\sqrt{\frac{hD}{2K}},$$

and thus the best value of m is either $\lfloor \alpha \rfloor$ or $\lceil \alpha \rceil$, depending on which yields the smaller cost. Thus, our policy in the finite-horizon case is, in fact, very similar to the infinite-horizon case. Orders are placed at regularly spaced intervals of time, and, of course, the orders are of the same size each time.

7.1.3   Power-of-Two Policies

Consider the infinite-horizon model described in Sect. 7.1. For this model, we know that the average total cost per unit of time is

$$\frac{KD}{Q} + \frac{hQ}{2} = \frac{K}{T} + \frac{hTD}{2} \doteq f(T),$$

where T is the time between orders. In this subsection, following Muckstadt and Roundy (1993), we introduce a new class of policies called power-of-two policies.

To simplify the analysis, and in accordance with the notation used in the literature (see Roundy 1985 and Muckstadt and Roundy 1993), let $g \doteq \frac{hD}{2}$, and hence

$$f(T) = \frac{K}{T} + gT.$$

Observe that the function $f(T)$ motivates another interpretation of the model. We can consider the problem to be an economic lot size model with unit demand rate, that is, $D = 1$, and inventory holding cost $2g$. The optimal reorder interval is $T^* = \sqrt{\frac{K}{g}}$, and the total cost per unit time is $f(T^*) = 2\sqrt{Kg}$.

One difficulty with the economic lot size model is that the optimal reorder interval T^* may take on any value and thus might lead to highly impractical optimal policies. For instance, reorder intervals of $\sqrt{3}$ days, or $\sqrt{\pi}$ weeks, would not be easy to implement. That is, the model might specify that orders be placed on Monday of one week, Thursday of the next, Tuesday of the next week, and so forth, a schedule of orders that may not have an easily recognizable pattern. Therefore, it is natural to consider policies where the reorder interval T is restricted to values that would entail easily implementable policies. One such restriction is termed the *power-of-two* restriction. In this case, T is restricted to be a power-of-two multiple of some fixed base planning period T_B; that is,

$$T = T_B 2^k, \quad k \in \{0, 1, 2, 3, \ldots\}. \tag{7.1}$$

Such a policy is called a *power-of-two policy*. The base planning period T_B may represent a day, week, or month, for example, and is usually fixed beforehand. It represents the minimum possible reorder interval.

Restricting ourselves to power-of-two policies requires addressing the following issues.

- How does one find the best power-of-two policy, the one minimizing the cost over all possible power-of-two policies?

- How far from optimal is the best policy of this type?

We start by answering the first question. Let $T^* = \sqrt{\frac{K}{g}}$ be the optimal (unrestricted) reorder interval, and let T be the optimal power-of-two reorder interval. Since f is convex, the optimal k in (7.1) is the smallest integer k satisfying

$$f(T_B 2^k) \leq f(T_B 2^{k+1})$$

or

$$\frac{K}{T_B 2^k} + g T_B 2^k \le \frac{K}{T_B 2^{k+1}} + g T_B 2^{k+1}.$$

Hence, k is the smallest integer such that

$$\sqrt{\frac{K}{2g}} = \frac{1}{\sqrt{2}} T^* \le T_B 2^k = T.$$

Thus, finding the optimal power-of-two policy is straightforward.

Observe that by the definition of the optimal k, it must also be true that

$$T = T_B 2^k \le \sqrt{\frac{2K}{g}} = \sqrt{2} T^*,$$

and hence the optimal power-of-two policy, for a given base planning period T_B, must be in the interval $[\frac{1}{\sqrt{2}} T^*, \sqrt{2} T^*]$. It is easy to verify that

$$f\left(\frac{1}{\sqrt{2}} T^*\right) = f(\sqrt{2} T^*) = \frac{1}{2}\left(\frac{1}{\sqrt{2}} + \sqrt{2}\right) f(T^*),$$

and hence, since f is convex, we have

$$\frac{f(T)}{f(T^*)} \le \frac{1}{2}\left(\frac{1}{\sqrt{2}} + \sqrt{2}\right) \approx 1.06.$$

Consequently, the average inventory purchasing and carrying cost of the best power-of-two policy is guaranteed to be within 6% of the average cost of the overall minimum policy. The reader can see that this property is a result of the "flatness" of the function f around its minimum.

This restriction, to power-of-two multiples of the base planning period, will also prove to be quite useful later in a more general setting.

7.2   Multi-Item Inventory Models

7.2.1   Introduction

The previous models established optimal inventory policies for single-item models. It is simple to show that without the presence of joint order costs, a problem with several items each facing a constant demand can be handled by solving each item's replenishment problem separately. In reality, the management of a single warehouse inventory system involves coordinating inventory orders to minimize cost without exceeding the *warehouse capacity*. The warehouse capacity limits the total volume held by the warehouse at any point in time. This constraint ties together the different items and necessitates careful coordination (or scheduling) of the orders. That is, it is important to know not only how often an item is ordered, but also exactly the point in time at which each order takes place. This problem is called the economic warehouse lot scheduling problem (EWLSP). The scheduling part, hereafter called the *staggering* problem, is exactly the problem of time-phasing the placement of the orders to satisfy the warehouse capacity constraint. Unfortunately, this problem has no easy solution, and consequently it has attracted a considerable amount of attention in the last three decades.

The earliest known reference to the problem appears in Churchman et al. (1957) and subsequently in Holt (1958) and Hadley and Whitin (1963). These authors were concerned with determining lot sizes that made an overall schedule satisfy the capacity constraint, and not with the possibility of phasing the orders to avoid holding the maximum volume of each item at the same time. Thus, they only considered what are called *independent solutions,* wherein every item is replenished without any regard for coordination with other items.

Several authors considered another class of policies called *rotation cycle* policies wherein all items share the same order interval. Homer (1966) showed how to optimally time-phase (stagger) the orders to satisfy the warehouse constraint for a given common order interval. Page and Paul (1976), Zoller (1977), and Hall (1988) independently rediscovered Homer's result. At the end of his paper devoted to rotation-cycle policies, Zoller indicates the possibility of partitioning the items into disjoint subsets, or *clusters,* if the assumption of a rotation policy "proves to be too restrictive." This is precisely Page and Paul's partitioning heuristic. In their heuristic, all the items in a cluster share a common order interval. The orders are then optimally staggered within each cluster, but no attempt is made to time-phase the orders of different clusters. Goyal (1978) argued that such a time-phasing across the different clusters may lead to a further reduction in warehouse space requirements. Hartley and Thomas (1982) and Thomas and Hartley (1983) considered the two-item case in detail.

A number of studies have been concerned with the *strategic* version of the EWLSP in which the warehouse capacity is not a constraint but rather a decision variable. These include Hodgson and Howe (1982), Park and Yun (1985), Hall (1988), Rosenblatt and Rothblum (1990), and Anily (1991). In this model, the inventory carrying cost consists of two parts; one part is proportional to the average inventory, while the second part is proportional to the peak inventory. A component of the latter cost, discussed in Silver and Peterson (1985), is the cost of *leasing* the storage space. This cost is typically proportional to the size of the warehouse, and not to the inventories actually stored in it.

Define a policy to be a *stationary order size* policy if all replenishments of an item are of the same size. Likewise, a *stationary order intervals* policy has all orders for an item equally spaced in time. It is easily verified that an optimal stationary order size (respectively, stationary order interval) policy is also a stationary order interval (respectively, a stationary order size) policy if every order of an item is received precisely when the inventory of that item drops to zero; that is, it also satisfies the zero-inventory-ordering property. Thus, it is natural to consider policies that have all three properties: stationary order size, stationary order interval, and zero inventory ordering. We call such policies *stationary order size and interval* policies, in short, *SOSI policies*. Two "extreme" cases of SOSI policies are the independent solutions and the rotation-cycle policies defined above. All the authors cited above considered SOSI policies exclusively. Zoller claims that SOSI policies are the only rational alternative, and most authors agree that SOSI policies are much easier to implement in practice. In his Ph.D. thesis, however, Hariga (1988) investigated both time-variant and stationary order sizes. He was motivated to study time-variant order sizes by their successful application in resolving the feasibility issue in the economic lot scheduling problem (ELSP) (see Dobson 1987).

Anily's paper departs from earlier work on the EWLSP in its focus on the worst-case performance of heuristics. In her paper, Anily restricts herself to the class of SOSI policies for the strategic model. She proves lower bounds on the minimum-required warehouse size and on the total cost for this class of policies. She presents a partitioning heuristic of which the best independent solution and the best rotation-cycle policies are special cases. This partitioning heuristic is similar to the one proposed by Page and Paul for the tactical model although the precise methods for finding the partition are different. Anily proves that the ratio of the cost of the best independent solution to her lower bound is at most $\sqrt{2}$. She also provides a data-dependent bound for the best rotation cycle, derived from Jones and Inman's (1989) work on the economic lot size problem. As a result, her partitioning heuristic is at least as good as either special case and thus has a worst-case bound of $\sqrt{2}$ relative to SOSI policies.

In this section, we determine easily computable lower bounds on the cost of the EWLSP as well as some simple heuristics for the problem. These bounds are used to determine the worst-case performance of these heuristics on different versions of the problem. First, in Sect. 7.2.2, we introduce notation, state assumptions, and formally define the strategic and tactical versions of the EWLSP. In Sect. 7.2.3, we establish the worst-case results. The discussion in this section is based on the work of Gallego et al. (1996).

7.2.2 Notation and Assumptions

Let $N = \{1, 2, \ldots, n\}$ be a set of n items each facing a constant *unit* demand rate (this can be done without loss of generality). An ordering cost K_i is incurred each time an order for item i is placed. A linear holding cost $2h_i$ is accrued for each unit of item i held in inventory per unit of time. Demand for each item must be met over an infinite horizon without shortages or backlogging.

The volume of inventory of item i held at a given point in time is the product of its inventory level at that time and the volume usage rate of item i, denoted by $\gamma_i > 0$. The volume usage rate is defined as the volume displaced by one unit of item i. Without loss of generality, we select the unit of volume so that $\sum_{i=1}^{n} \gamma_i = 1$.

The objective in the strategic version of the EWLSP is to minimize the long-run average inventory carrying and ordering cost plus a cost proportional to the maximum volume held by the warehouse at any point in time. Formally, for any inventory policy $\mathcal{P}$, let $V(\mathcal{P})$ denote the maximum inventory volume held by the warehouse, and let $C(\mathcal{P})$ be the long-run average inventory carrying and holding cost incurred by this policy. Then the objective is to find a policy $\mathcal{P}$ minimizing

$$Z(\mathcal{P}) \doteq C(\mathcal{P}) + V(\mathcal{P}).$$

The tactical version of the EWLSP has also received much attention in the literature. There, the objective is to find a policy $\mathcal{P}$ minimizing the long-run average inventory carrying and holding costs subject to the inventory always being less than the warehouse capacity. Hence, the tactical version can be formulated as follows: Find a policy $\mathcal{P}$ minimizing $C(\mathcal{P})$ subject to $V(\mathcal{P}) \leq v$, where v denotes the available warehouse volume.

7.2.3   Worst-Case Analyses

Preliminaries

We present here two simple results that are used in subsequent analyses.

Given a SOSI policy, let $T = \{T_1, T_2, \ldots, T_n\}$ be the vector of reorder intervals, where T_i is the reorder interval of item i. For any such vector T, let $V(T)$ denote the maximum volume of inventory held by the warehouse over all points in time. The following provides a simple upper bound on $V(T)$.

Lemma 7.2.1 *For any vector* $T = \{T_1, T_2, \ldots, T_n\}$, *we have*

$$V(T) \leq \sum_{i=1}^{n} \gamma_i T_i.$$

Proof. Clearly, the inventory level of item i, at any moment in time, is no more than T_i (recall demand is 1 for all i). ∎

For the next result, we need some additional notation. Consider any inventory policy $\mathcal{P}$ and any time interval $[0, t]$. Let $V(\mathcal{P}, t)$ be the maximum inventory held by the warehouse in policy $\mathcal{P}$ over the interval $[0, t]$, and let $C(\mathcal{P}, t)$ be the average inventory holding and carrying cost incurred over $[0, t]$. Let m_i be the number of times the warehouse places an order for item i over the interval $[0, t]$. For $\tau \in [0, t]$, let $I_i(\tau)$ be the inventory level of item i at time τ. Let $v_i(\tau)$ be the volume of inventory held by item i at time τ; that is, $v_i(\tau) = \gamma_i I_i(\tau)$. Also, let $v(\tau) = \sum_{i=1}^{n} v_i(\tau)$ be the volume of inventory held by the warehouse at time τ.

Lemma 7.2.2 *For any inventory policy* $\mathcal{P}$ *and time interval* $[0, t]$, *we have*

$$\frac{1}{2} \sum_{i=1}^{n} \frac{\gamma_i t}{m_i} \leq \sum_{i=1}^{n} \frac{1}{t} \gamma_i \int_{\tau=0}^{t} I_i(\tau) d\tau \leq V(\mathcal{P}, t).$$

Proof. Clearly, $v(\tau) \leq V(\mathcal{P}, t)$ for all $\tau \leq t$. Taking the integral up to time $t > 0$ gives

$$V(\mathcal{P}, t) \geq \frac{1}{t} \int_{\tau=0}^{t} \sum_i v_i(\tau) d\tau$$

$$= \frac{1}{t} \int_{\tau=0}^{t} \sum_i \gamma_i I_i(\tau) d\tau$$

$$= \sum_i \frac{1}{t} \gamma_i \int_{\tau=0}^{t} I_i(\tau) d\tau$$

$$\geq \sum_i \frac{1}{2} \frac{\gamma_i t}{m_i},$$

where the last inequality follows from Property 7.1.1, which states that when m_i orders for a single item are placed over the interval $[0, t]$, the average inventory level is minimized by placing equal orders at equally spaced points in time. ∎

The Strategic Model

Consider the following heuristic for the strategic version of the EWLSP. Use the vector of reorder intervals T that solves

$$Z^H = \min_T \left\{ \sum_i \left(\frac{K_i}{T_i} + h_i T_i \right) + \sum_i \gamma_i T_i \right\}.$$

Clearly, the vector T can be found in $O(n)$ time by solving n separate economic lot scheduling models, and

$$Z^H = 2 \sum_i \sqrt{K_i (h_i + \gamma_i)}. \tag{7.2}$$

By Lemma 7.2.1, Z^H must provide an upper bound on the optimal solution value of the strategic model.

We now construct a lower bound on the optimal solution value over all possible inventory policies. The lower bound is the cost of the optimal policy if the warehouse cost were based on average inventory rather than maximum inventory. This bound will be used to prove the worst-case result.

Lemma 7.2.3 *A lower bound on the optimal solution value over all possible inventory strategies is given by*

$$Z^{LB} = 2 \sum_i \sqrt{K_i (h_i + \gamma_i/2)}. \tag{7.3}$$

Proof. We show that $Z^{LB} \leq C(\mathcal{P}, t) + V(\mathcal{P}, t)$ for all possible inventory policies $\mathcal{P}$ and for all $t > 0$. Given an inventory policy $\mathcal{P}$, where m_i orders for item i are placed over a time interval $[0, t]$, then

$$C(\mathcal{P}, t) = \frac{1}{t} \sum_i \left(m_i K_i + 2h_i \int_{\tau=0}^t I_i(\tau) d\tau \right).$$

Combining this cost with the lower bound obtained in Lemma 7.2.2 on $V(\mathcal{P}, t)$ yields the following lower bound on $C(\mathcal{P}, t) + V(\mathcal{P}, t)$:

$$C(\mathcal{P}, t) + V(\mathcal{P}, t) \geq \frac{1}{t} \sum_i \left[m_i K_i + 2h_i \int_{\tau=0}^t I_i(\tau) d\tau \right] + \frac{1}{t} \sum_i \gamma_i \int_{\tau=0}^t I_i(\tau) d\tau$$

$$= \frac{1}{t} \sum_i \left[m_i K_i + (2h_i + \gamma_i) \int_{\tau=0}^t I_i(\tau) d\tau \right]$$

$$\geq \sum_i \left[K_i \left(\frac{m_i}{t} \right) + \frac{(2h_i + \gamma_i)}{2} \left(\frac{t}{m_i} \right) \right].$$

The last inequality again follows from Property 7.1.1. Minimizing the last expression with respect to $\frac{t}{m_i}$ for each $i \in N$ proves the result. ∎

We now show that this heuristic is effective in terms of worst-case performance.

Theorem 7.2.4

$$\frac{Z^H}{Z^{LB}} \leq \sqrt{2}.$$

Proof. Combining (7.2) and (7.3), we get

$$\frac{Z^H}{Z^{LB}} = \frac{2 \sum_i \sqrt{K_i(h_i + \gamma_i)}}{2 \sum_i \sqrt{K_i(h_i + \gamma_i/2)}} \leq \sqrt{2}.$$

∎

Can this bound be improved? The following example shows that the bound is tight as the number of items grows to infinity. Consider an example n items with $K_i = K$, $h_i = 0$ and $\gamma_i = \gamma = \frac{1}{n}$ for all $i \in N$. Clearly,

$$Z^H = 2n\sqrt{K\gamma}.$$

We now construct a feasible solution whose cost approaches the lower bound Z^{LB} as n goes to infinity. Consider a feasible policy $\mathcal{P}$ with identical reorder intervals denoted by $\tilde{T}$. To reduce the maximum volume $V(\tilde{T})$, we stagger the orders such that item i is ordered at times $\tilde{T}[\frac{(i-1)}{n} + k]$ for $k \geq 0$. Then the maximum volume of inventory is $\frac{(n+1)}{2} \tilde{T}\gamma$. Hence, the cost of policy $\mathcal{P}$ is

$$Z(\mathcal{P}) = \frac{nK}{\tilde{T}} + \frac{n+1}{2} \tilde{T}\gamma.$$

Minimizing with respect to $\tilde{T}$ gives

$$Z(\mathcal{P}) = \sqrt{2n(n+1)K\gamma}.$$

Consequently,

$$\frac{Z^H}{Z^{LB}} \geq \frac{Z^H}{Z(\mathcal{P})} = \frac{2n\sqrt{K\gamma}}{\sqrt{2n(n+1)K\gamma}}.$$

The limit of this last quantity is $\sqrt{2}$ (as n goes to infinity); hence, along with Theorem 7.2.4, we see that an example can be constructed where the worst-case ratio is arbitrarily close to $\sqrt{2}$.

The Tactical Model

For the tactical version of the EWLSP, a simple heuristic denoted HW first proposed by Hadley and Whitin (1963) is to solve

$$\text{Problem } P^{HW} : C^{HW} = Min \sum_i \left(h_i T_i + \frac{K_i}{T_i} \right)$$

s.t.

$$\sum_i \gamma_i T_i \leq v,$$

$$T \geq 0.$$

We show that the HW heuristic has a worst-case performance bound of 2 with respect to all feasible policies. We do so by proving that the solution to the following nonlinear program provides a lower bound on the cost of any feasible policy.

$$\text{Problem } P^{LB} : C^{LB} = Min \sum_i \left(h_i T_i + \frac{K_i}{T_i} \right)$$

s.t.

$$\frac{1}{2} \sum_i \gamma_i T_i \leq v, \tag{7.4}$$

$$T \geq 0.$$

Lemma 7.2.5 C^{LB} *is a lower bound on the cost of any feasible inventory policy.*

Proof. Consider any feasible policy $\mathcal{P}$ over the interval $[0, t]$ that places m_i orders for item i in $[0, t]$. From Lemma 7.2.2, we have $\forall t > 0$,

$$v \geq V(\mathcal{P}, t) \geq \frac{1}{2} \sum_i \frac{\gamma_i}{m_i} t.$$

The average inventory holding and carrying cost incurred over the interval $[0, t]$ is

$$C(\mathcal{P}, t) = \frac{1}{t} \sum_i \left[m_i K_i + 2 h_i \int_{\tau=0}^t I_i(\tau) d\tau \right]$$

$$\geq \sum_i \left[K_i \left(\frac{m_i}{t} \right) + h_i \left(\frac{t}{m_i} \right) \right]. \tag{7.5}$$

Again, the last inequality follows from Property 7.1.1.

Thus, by replacing $\frac{t}{m_i}$ with T_i for all $i \geq 1$, we see that minimizing (7.5) subject to $\frac{1}{2} \sum_i \gamma_i t/m_i \leq v$ provides a lower bound on $C(\mathcal{P}, t)$. ∎

We now prove the worst-case bound.

Theorem 7.2.6

$$\frac{C^{HW}}{C^{LB}} \leq 2.$$

Proof. Let $T^{LB} = \{T_1^{LB}, T_2^{LB}, \ldots, T_n^{LB}\}$ be the optimal solution to P^{LB}. Obviously, $T_i' = \frac{1}{2} T_i^{LB}$ is feasible for P^{HW}. Hence,

$$C^{HW} \leq \sum_i \left(h_i T_i' + \frac{K_i}{T_i'} \right)$$

$$= \frac{1}{2} \sum_i h_i T_i^{LB} + 2 \sum_i \frac{K_i}{T_i^{LB}}$$

$$\leq 2 C^{LB}.$$

∎

As in the strategic version, the worst-case bound provided by the above theorem can be shown to be tight. To do so, consider the case where all items are identical with $K_i = K$, $h_i = 0$ and $\gamma_i = \gamma = \frac{1}{n}$ for all $i \in N$. The solution to problem P^{HW} is clearly $T_i = v$ for all $i \in N$, so $C^{HW} = \frac{nK}{v}$. Consider now a feasible policy $\mathcal{P}$ with identical reorder intervals denoted by $\tilde{T}$ such that an order for item i is placed at times $\tilde{T}[\frac{(i-1)}{n} + k]$ for $k \geq 0$. The maximum volume occupied by policy $\mathcal{P}$ is $\frac{(n+1)}{2} \tilde{T} \gamma$. So $\tilde{T} = \frac{2v}{(n+1)\gamma}$ is feasible and $C(\mathcal{P}) = \frac{K(n+1)}{2v}$. Hence,

$$\lim_{n \to \infty} \frac{C^{HW}}{C(\mathcal{P})} = \lim_{n \to \infty} \frac{nK/v}{K(n+1)/2v} = 2.$$

By performing a similar analysis, one can obtain worst-case bounds on the performance of heuristics for other versions of the EWLSP. For instance, for the *joint replenishment* version of the strategic model, where an additional setup cost K_0 is incurred whenever an order for one or more items is placed, the worst-case bound of a heuristic, similar to the one described for the EWLSP, can be shown to be $\sqrt{3}$. The worst-case bound on the tactical version of the joint replenishment model can be shown to be $2\sqrt{2}$.

7.3 A Single-Warehouse Multiretailer Model

7.3.1 Introduction

Many distribution systems involve replenishing the inventories of geographically dispersed retailers. Consider a distribution system in which a single warehouse supplies a set of retailers with a single product. Each retailer faces a constant retailer-specific demand that must be met without shortage or backlogging. The warehouse faces orders for the product from the different retailers and in turn places orders to an outside supplier. A fixed, facility-dependent, setup cost is charged each time the warehouse or the retailers receive an order, and an inventory carrying cost is accrued at each facility at a constant facility-dependent rate. The objective is to determine simultaneously the timing and sizes of retailer deliveries to the warehouse as well as replenishment strategies at the warehouse so as to minimize the long-run average inventory purchasing and carrying costs.

In the absence of a fixed setup cost charged when the warehouse places an order, the problem can be decomposed into an economic lot size model for each retailer. That is, the existence of this cost ties together the different retailers, requiring the warehouse to coordinate its orders and deliveries to the different retailers. It is well known that optimal policies can be very complex, and thus the problem has attracted a considerable amount of attention in recent years (see Graves and Schwarz 1977, Roundy 1985). The latter paper presents the best approach currently available for this model; it suggests a set of power-of-two reorder intervals for each facility and shows that the cost of this solution is within 6 % of a lower bound on the optimal cost. In this section, we present this method along with the worst-case bound.

7.3.2   Model and Analysis

Consider a single warehouse (indexed by 0) that supplies n retailers, indexed $1, 2, \ldots, n$. We will use the term *facility* to designate either the warehouse or a retailer. We make the following assumptions.

- Each retailer faces a constant demand rate of D_i units, for $i = 1, 2, \ldots, n$.

- The setup cost for an order at a facility is K_i, for $i = 0, 1, \ldots, n$.

- The holding cost is h'_0 at the warehouse and h'_i at retailer i, with $h'_i \geq h'_0$ for each $i = 1, 2, \ldots, n$.

- No shortages are allowed.

As demonstrated by several researchers, policies for this problem may be quite complex, and thus it is of interest to restrict our attention to a subset of all feasible policies. A popular subset of policies is the set of *nested* and *stationary* policies. A nested policy is characterized by having each retailer place an order whenever the warehouse does. As in the previous section, stationarity implies that reorder intervals are constant for each facility. It is easy to show that any policy should satisfy the zero-inventory-ordering property. Roundy (1985) showed that, although appealing from a coordination point of view, nested policies may perform arbitrarily badly in one-warehouse, multiretailer systems. We therefore will not restrict ourselves to nested policies. We concentrate on policies where each retailer's reorder intervals are a power-of-two multiple of a base planning period T_B. Below, we assume the base planning period is fixed. The worst-case bound reduces to 1.02 if it can be chosen optimally although we omit this extension.

Let's first determine the cost of an arbitrary power-of-two policy $T = \{T_0, T_1, \ldots, T_n\}$ that satisfies the zero-inventory-ordering property. If we consider the inventory at the warehouse, then it does not have the saw-toothed pattern. To overcome this difficulty, it is convenient to introduce the notion of *system* inventory as well as *echelon* holding cost rates. Retailer i's system inventory is defined as the inventory at retailer i plus the inventory at the warehouse that is destined for retailer i. If we consider the *system* inventory of retailer i, then it has the saw-toothed pattern. Echelon holding cost rates are defined as $h_0 = h'_0$ and $h_i = h'_i - h'_0$. For simplicity, define $g_i = \frac{1}{2} h_i D_i$ and $g^i = \frac{1}{2} h_0 D_i$ for each $i = 1, 2, \ldots, n$. To compute the cost of such a policy, we separate each item in the warehouse's inventory into categories depending on the retailer for which the item is destined. Let $\mathcal{H}_i(T_0, T_i)$ be the average cost of holding inventory for retailer i at the warehouse and at retailer i. We claim

$$\mathcal{H}_i(T_0, T_i) = g_i T_i + g^i \max\{T_0, T_i\}.$$

To prove this, consider the two cases:

Case 1: $T_i \geq T_0$. Since T is a power-of-two policy, $T_i \geq T_0$ implies that the warehouse places an order every time the retailer does. Therefore, the warehouse never holds inventory for retailer i, and the average holding cost is

$$\frac{1}{2} h'_i T_i D_i = \frac{1}{2} (h_i + h_0) T_i D_i = (g_i + g^i) T_i.$$

Case 2: $T_i < T_0$. Consider the portion of the warehouse inventory that is destined for retailer i. Using the echelon holding cost rates, that is, inventory at retailer i is charged at a rate of h_i and system inventory is charged at a rate of h_0, we have

$$\mathcal{H}_i(T_0, T_i) = \frac{1}{2} h_i D_i T_i + \frac{1}{2} h_0 D_i T_0 = g_i T_i + g^i T_0.$$

Therefore, the average cost of a power-of-two policy T is given by

$$\sum_{i \geq 0} \frac{K_i}{T_i} + \sum_{i \geq 1} \mathcal{H}_i(T_0, T_i). \tag{7.6}$$

Our objective then is to find the power-of-two policy T that minimizes (7.6).

Our approach to solving this problem is to first minimize the average cost over all vectors $T \geq 0$. That is, we solve this problem when the restriction to power-of-two vectors is relaxed. We then round the solution T to a vector whose elements are the power-of-two multiple of T_B.

For a fixed value of T_0, we consider the following problem:

$$b_i(T_0) = \inf_{T_i > 0} \left\{ \frac{K_i}{T_i} + \mathcal{H}_i(T_0, T_i) \right\}. \tag{7.7}$$

To solve this problem, let $\tau_i' \doteq \sqrt{\frac{K_i}{g_i+g^i}}$, let $\tau_i \doteq \sqrt{\frac{K_i}{g_i}}$, and note that $\tau_i' \leq \tau_i$ for all $i \geq 1$. Then one can show that

$$b_i(T_0) = \begin{cases} 2\sqrt{K_i(g_i + g^i)} & \text{if } T_0 < \tau_i', \\ \frac{K_i}{T_0} + (g_i + g^i)T_0 & \text{if } \tau_i' \leq T_0 \leq \tau_i, \\ 2\sqrt{K_i g_i} + g^i T_0 & \text{if } \tau_i < T_0. \end{cases}$$

That is, if $T_0 < \tau_i'$, it is best to choose $T_i^* = \tau_i'$. If $\tau_i' \leq T_0 \leq \tau_i$, then choose $T_i^* = T_0$. If $T_0 > \tau_i$, it is best to choose $T_i^* = \tau_i$.

We now consider minimizing

$$B(T_0) \doteq \frac{K_0}{T_0} + \sum_{i=1}^n b_i(T_0)$$

over all $T_0 > 0$. The function B is of the form

$$\frac{K(T_0)}{T_0} + M(T_0) + H(T_0)T_0$$

over any interval where $K()$, $M()$, and $H()$ are constant. For any T_0, define the sets $G(T_0) \doteq \{i : T_0 < \tau_i'\}$, $E(T_0) \doteq \{i : \tau_i' \leq T_0 \leq \tau_i'\}$, and $L(T_0) \doteq \{i : \tau_i < T_0\}$. Then $K()$, $M()$, and $H()$ are constant on those intervals where $G()$, $E()$, and $L()$ do not change. To find the minimum of B, consider the intervals induced by the $2n$ values τ_i' and τ_i for $i = 1, 2, \ldots, n$. Say T_0 falls in some specific interval; then we set

$$T_i^* = \begin{cases} \tau_i' & \text{if } i \in G(T_0), \\ T_0 & \text{if } i \in E(T_0), \\ \tau_i & \text{if } i \in L(T_0). \end{cases}$$

The sets G, E, and L change only when T_0 crosses a *breakpoint* τ_i' or τ_i for some $i \geq 1$. Specifically, if T_0 moves from right to left across τ_i, retailer i moves from L to E. If T_0 moves from right to left across τ_i', retailer i moves from E to G. This suggests a simple algorithm to minimize $B(T_0)$. Start with T_0 larger than the largest breakpoint, and let $L = \{1, 2, \ldots, n\}$ and $G = E = \emptyset$. We then successively decrease T_0, moving from interval to interval. On each interval we need only check that $\sqrt{\frac{K(T_0)}{H(T_0)}}$ falls in the same subinterval as T_0. In this case, we set $T_0^* = \sqrt{\frac{K(T_0)}{H(T_0)}}$ since $B(T_0)$ is strictly convex in T_0. Let $B^* \doteq B(T_0^*) = \inf_{T_0 \geq 0}\{B(T_0)\}$; then this value is clearly a lower bound on the cost of any power-of-two policy.

We now want to prove that this value is a lower bound on the cost of any policy. For notational convenience, we abbreviate $G^* = G(T_0^*)$, $E^* = E(T_0^*)$, and $L^* = L(T_0^*)$. Let $K = K_0 + \sum_{i \in E^*} K_i$, $G = \sum_{i \in E^*}(g_i + g^i) + \sum_{i \in L^*} g^i$, and $M = 2\sqrt{KG}$. We also define for each $i \geq 0$

$$G_i = \begin{cases} g_i + g^i, & \text{if } i \in G^*, \\ g_i, & \text{if } i \in L^*, \\ \frac{K_i}{(T_0^*)^2}, & \text{if } i \in E^* \cup \{0\}, \end{cases}$$

$G^i = g^i + g_i - G_i$, and $M_i = 2\sqrt{K_i G_i}$. In this way, we can write B^* as

$$B^* = M + \sum_{i \in L^* \cup G^*} M_i. \qquad (7.8)$$

We now prove that B^* is a lower bound on any policy. We first show that, in fact, $B^* = \sum_{i \geq 0} M_i$. From (7.8), we need only show that $M = \sum_{i \in E^* \cup \{0\}} M_i$,

$$M = 2\sqrt{KG} = 2\frac{K}{T_0^*}$$

$$= 2 \sum_{i \in E^* \cup \{0\}} \frac{K_i}{T_0^*}$$

$$= 2 \sum_{i \in E^* \cup \{0\}} \frac{K_i}{\sqrt{K_i/G_i}}$$

$$= 2 \sum_{i \in E^* \cup \{0\}} \sqrt{K_i G_i}$$

$$= \sum_{i \in E^* \cup \{0\}} M_i.$$

Consider any policy over an interval $[0, t']$ for $t' > 0$. We show that the total cost associated with this policy over $[0, t']$ is at least $B^* t'$. Let m_i be the number of orders placed by facility $i \geq 0$ in the interval $[0, t']$. Let $I_i(t)$ be the inventory at facility $i \geq 1$ at time t, and let $S_i(t)$ be the system inventory of facility $i \geq 1$ at time t. Clearly, the total inventory holding cost is

$$\sum_{i \geq 1} \int_0^{t'} \left(h_i I_i(t) + h_0 S_i(t) \right) dt.$$

We will show that this is no smaller than

$$\sum_{i \geq 1} \int_0^{t'} \left(H_i I_i(t) + H^i S_i(t) \right) dt,$$

where $H_i = \frac{2G_i}{D_i}$ and $H^i = \frac{2G^i}{D_i}$ for each $i = 0, 1, \ldots, n$. For this purpose, consider the quantity $H_i I_i(t) + H^i S_i(t)$ for each $i \geq 1$. There are three cases to consider.
Case 1: $i \in G^*$. Then $G_i = g_i + g^i$ and $G^i = g_i + g^i - G_i = 0$, and since $S_i(t) \geq I_i(t)$ for all $t > 0$, we have

$$h_i I_i(t) + h_0 S_i(t) \geq H_i I_i(t) + H^i S_i(t).$$

Recall that $h_i = \frac{2g_i}{D_i}, h_0 = \frac{2g^i}{D_i}, H_i = \frac{2G_i}{D_i}$, and $H^i = \frac{2G^i}{D_i}$.
Case 2: $i \in L^*$. Then $G_i = g_i$ and $G^i = g_i + g^i - G_i = g^i$; hence,

$$h_i I_i(t) + h_0 S_i(t) = H_i I_i(t) + H^i S_i(t).$$

Case 3: $i \in E^*$. Then $G_i = \frac{K_i}{(T_0^*)^2}$ and $G^i = g_i + g^i - G_i$. Observe that, by definition, if $i \in E$, then $\tau_i' \leq T_0^* \leq \tau_i$, which implies $g_i \leq G_i \leq g_i + g^i$. Since $S_i(t) \geq I_i(t)$ for all $t \geq 0$, then

$$h_i I_i(t) + h_0 S_i(t) = H_i I_i(t) + H^i S_i(t) + (H_i - h_i)(S_i(t) - I_i(t))$$
$$\geq H_i I_i(t) + H^i S_i(t). \tag{7.9}$$

Therefore, our lower bound on the inventory holding cost can be written as

$$\sum_{i \geq 1} \int_0^{t'} \left(H_i I_i(t) + H^i S_i(t) \right) dt = \sum_{i \geq 0} \int_0^{t'} H_i I_i(t) dt,$$

where we have defined $I_0(t) = \frac{1}{H_0} \sum_{i \geq 1} H^i S_i(t)$.

Hence, the total cost per unit of time under this policy is at least

$$\frac{1}{t'} \sum_{i \geq 0} \left(K_i m_i + \int_0^{t'} H_i I_i(t) dt \right) \geq \sum_{i \geq 0} \left(K_i \frac{m_i}{t'} + G_i \frac{t'}{m_i} \right)$$

$$\geq 2 \sum_{i \in L^* \cup G^*} \sqrt{K_i G_i} + 2 \sum_{i \in E^* \cup \{0\}} \sqrt{K_i G_i}$$

$$= \sum_{i \geq 0} M_i = B^*,$$

where the first inequality follows from Property 7.1.1 and the fact that $G_0 = \sum_{i \geq 1} G^i$ (see Exercise 7.7). We have thus established that B^* is a lower bound on the total cost per unit time of any policy.

Finally, for each $i \in G^* \cup L^*$, select a power-of-two policy (a value of k) such that

$$\frac{1}{\sqrt{2}} T_i^* \leq T_B 2^k \leq \sqrt{2} T_i^*.$$

For each $i \in E^* \cup \{0\}$, select a power-of-two policy (a value of k) such that

$$\frac{1}{\sqrt{2}} T_0^* \leq T_B 2^k \leq \sqrt{2} T_0^*.$$

It is a simple exercise (Exercise 7.4) to show that the policy constructed in this manner has cost at most 1.06 times the cost of the lower bound.

7.4 Exercises

Exercise 7.1. Consider the economic lot size model, and let K be the setup cost, h the holding cost per item per unit of time, and D the demand rate. Shortage is not allowed and the objective is to find an order quantity so as to minimize the long-run average cost. That is, the objective is to minimize

$$C(Q) = \frac{KD}{Q} + \frac{hQ}{2},$$

where Q is the order quantity. Suppose the warehouse can order only an integer multiple of q units. That is, the warehouse can order q, or $2q$, or $3q$, and so on.

(a) Prove that the optimal order quantity Q^* has the following property. There exists an integer m such that $Q^* = mq$ and

$$\sqrt{\frac{m-1}{m}} \le \frac{Q^e}{Q^*} \le \sqrt{\frac{m+1}{m}},$$

where Q^e, the economic order quantity, is

$$Q^e = \sqrt{\frac{2KD}{h}}.$$

(b) Suppose now that $m \ge 2$. Show that $C(Q^*) \le 1.06C(Q^e)$.

Exercise 7.2. (Zavi 1976) Consider the economic lot size model with infinite horizon and deterministic demand D items per unit of time. When the inventory level is zero, production of Q items starts at a rate of P items per unit of time, $P \ge D$. The setup cost is $K\$$ and the holding cost is $h\$/item/time$. Every time production starts at a level of P items/time, we incur a cost of αP, $\alpha > 0$.

(a) What is the optimal production rate?

(b) Suppose that due to technological constraints, P must satisfy $2D \le P \le 3D$. What are the optimal production rate and the optimal order quantity?

Exercise 7.3. Consider the economic lot size model over the infinite horizon. Assume that when an order of size Q is placed, the items are delivered by trucks of capacity q, and thus the number of trucks used to deliver Q is $\lceil \frac{Q}{q} \rceil$, where $\lceil m \rceil$ is the smallest integer greater than or equal to m. The setup cost is a linear function of the number of trucks used: It is $K_0 + \lceil \frac{Q}{q} \rceil K$. The holding cost is $h\ \$/item/time$, and shortage is not allowed. What is the optimal reorder quantity?

Exercise 7.4. Prove that the heuristic for the single-warehouse, multiretailer model described in Sect. 7.3 provides a solution within 1.06 of the lower bound.

Exercise 7.5. Consider the power-of-two policies described in the single-product model of Sect. 7.1.3. Describe how you could generate a power-of-three policy (a policy where each $T_i = 3^k T_B$ for some integer $k \ge 0$). What is the effectiveness (in terms of worst-case performance) of the best power-of-three policy?

Exercise 7.6. (Porteus 1985) The Japanese concept of JIT (just-in-time) advocates reducing setup cost as much as possible. To analyze this concept, consider the economic lot size model with constant demand of D items per year, holding cost h \$ per item per year, and **current setup cost** K_0. Suppose **you can** lease a new technology that allows you to reduce the setup cost from K_0 to K at an annual leasing cost of $A - Bln(K)$ dollars. That is, reducing the setup cost from the current setup cost, K_0, to K will annually cost $A - Bln(K)$ dollars. Of course, we assume that $A - B\ln(K_0) = 0$, which implies that using the current setup cost requires no leasing cost. What is the optimal setup cost? What is the optimal order quantity in this case?

Exercise 7.7. Show that in the proof of the lower bound, B^*, for the single-warehouse, multiretailer model, we have $G_0 = \sum_{i \geq 1} G^i$.

Exercise 7.8. Prove (7.9).

8

Economic Lot Size Models with Varying Demands

Our analysis of inventory models so far has focused on situations where demand was both known in advance and constant over time. We now relax this latter assumption and turn our attention to systems where demand is known in advance yet varies with time. This is possible, for example, if orders have been placed in advance, or contracts have been signed specifying deliveries for the next few months. In this case, a *planning horizon* is defined as those periods where demand is known. Our objective is to identify optimal inventory policies for single-item models as well as heuristics for the multi-item case. We also present extensions to single-item models with price-dependent demand.

8.1 The Wagner–Whitin Model

Assume we must plan a sequence of orders, or production batches, over a T-period planning horizon. In each period, a single decision must be made: the size of the order or production batch.

We make the following assumptions:

- Demand during period t is known and denoted by $d_t > 0$.

- The per-unit order cost is c and a fixed order cost K is incurred every time an order is placed; that is, if y units are ordered, the order cost is $cy + K\delta(y)$ [where $\delta(y) = 1$ if $y > 0$, and 0 otherwise].

- There is a holding cost $h > 0$ per unit per period.

D. Simchi-Levi et al., *The Logic of Logistics: Theory, Algorithms, and Applications for Logistics Management*, Springer Series in Operations Research and Financial Engineering, DOI 10.1007/978-1-4614-9149-1_8, © Springer Science+Business Media New York 2014

- Initial inventory is zero.

- Lead times are zero; that is, an order arrives as soon as it is placed.

- All ordering and demand occur at the start of the period. Inventory holding cost is charged on the amount on hand at the end of the period.

The problem is to decide how much to order in each period so that demands are met without backlogging and the total cost, including the cost of ordering and holding inventory, is minimized. This basic model was first analyzed by Wagner and Whitin (1958b) and has now been called the Wagner–Whitin model.

Let y_t be the amount ordered in period t, and let I_t be the amount of product in inventory at the end of period t. Using these variables, we can formulate the problem as follows:

$$\text{Problem } WW: \quad \text{Min} \sum_{t=1}^{T} \left[K\delta(y_t) + cy_t + hI_t \right]$$

$$\text{s.t.} \quad I_t = I_{t-1} + y_t - d_t, \quad t = 1, 2, \ldots, T, \tag{8.1}$$

$$I_0 = 0, \tag{8.2}$$

$$I_t, y_t \geq 0, \quad t = 1, 2, \ldots, T. \tag{8.3}$$

Here constraints (8.1) are called the *inventory-balance* constraints, while (8.2) simply specifies the initial inventory. Note that the inventory can also be rewritten as $I_t = \sum_{i=1}^{t}(y_i - d_i)$, and therefore the I_t variables can be eliminated from the formulation.

In the above model, it is clear that the total variable order cost incurred will be fixed and independent of the schedule of orders, and thus we ignore this cost in our analysis until we talk about models with price-dependent demand.

Wagner and Whitin make the following important observation.

Theorem 8.1.1 *Any optimal policy is a zero-inventory-ordering policy, that is, a policy in which*

$$y_t I_{t-1} = 0, \text{ for } t = 1, 2, \ldots, T.$$

Proof. The proof is quite simple. By contradiction, assume there is an optimal policy in which an order is placed in period t even though the inventory level at the beginning of this period $[I_{t-1}]$ is positive. We will demonstrate the existence of another policy with a lower total cost. Evidently, the I_{t-1} items of inventory were ordered in various periods prior to t. Thus, if we instead order these items in period t, we save all the holding cost incurred from the time they were each ordered. ∎

Thus, ordering only occurs when inventory is zero. A simple corollary is that in an optimal policy, *an order is of size equal to satisfy demands for an integer number of subsequent periods.*

Using the above property, Wagner and Whitin developed a dynamic programming algorithm to determine those periods when ordering takes place. By constructing a simple acyclic network with nodes $V = \{1, 2, \ldots, T+1\}$, we can view the problem of determining a policy as a shortest-path problem. Formally, let ℓ_{ij}, the length of arc (i, j) in this network, be the cost of ordering in period i to satisfy the demands in periods $i, i+1, \ldots, j-1$, for all $1 \leq i < j \leq T+1$. That is,

$$\ell_{ij} = K + h \sum_{k=i}^{j-1} (k-i) d_k.$$

All other arcs have $\ell_{ij} = +\infty$. The length of the shortest path from node 1 to node $T+1$ in this acyclic network is the minimal cost of satisfying the demands for periods 1 through T. The optimal policy, that is, a specification of the periods in which an order is placed, can be easily reconstructed from the shortest path itself. This procedure is clearly $O(T^2)$.

Most of the assumptions made above can be relaxed without changing the basic solution methodology. For example, one can consider problem data that are period-dependent (e.g., c_t, h_t, or K_t). The assumption of zero lead times can be relaxed if one assumes the lead times are known in advance and deterministic. In that case, if an order is required in period t, then it is ordered in period $t - L$, where L is the lead time.

Researchers have also considered order costs that are general concave functions of the amount ordered, that is, $c_t(y)$. The problem can be formulated as a network flow problem with concave arc costs. This was the approach of Zangwill (1966), who also extended the model to handle backlogging although the solution method is only computationally attractive for small problems.

The Wagner–Whitin model can also be useful if demands during periods well into the future are not known. This idea is embodied in the following theorem.

Theorem 8.1.2 *Let t be the last period a setup occurs in the optimal order policy associated with a T-period problem. Then for any problem of length $T^* > T$, it is necessary to consider only periods $\{j : t \leq j \leq T^*\}$ as candidates for the last setup. Furthermore, if $t = T$, the optimal solution to a T^*-period problem has $y_t > 0$.*

This result is useful since it shows that if an order is placed in period t, the optimal policy for periods $1, 2, \ldots, t-1$ does not depend on demands beyond period t.

Surprisingly, even though the Wagner–Whitin solution procedure is extremely efficient, often simple approximate yet intuitive heuristics may be more appealing to managers. For example, this may be the reason for the popularity of the Silver and Meal (1973) heuristic or the part-period balancing heuristic of Dematteis (1968). One important reason is the sensitivity of the *optimal* strategy to changes in forecasted demands d_t, $t = 1, 2, \ldots, T$. Indeed, in practice, these forecasted demands are typically modified "on the fly." These changes typically imply changes in the optimal strategy. Some of the previously mentioned heuristics are not as sensitive to these changes while producing optimal or near-optimal strategies. For another approach, see Federgruen and Tzur (1991).

Researchers have shown that it is possible to take advantage of the special cost structure in the Wagner–Whitin model and use it to develop faster exact algorithms [i.e., $O(T)$]. This includes the work of Aggarwal and Park (1993) and Park, Federgruen and Tzur (1991) and Wagelmans et al. (1992).

We sketch here the $O(T)$ algorithm of Wagelmans et al., which is the most intuitive of those proposed. It is a backward dynamic programming approach. Define $d_{ij} = \sum_{t=i}^{j} d_t$ for $i, j = 1, 2, \ldots, T$, that is, the demand from period i to period j. To describe the algorithm, we will slightly change the way we account for the holding cost. If an item is ordered in period i, then we are charged $H_i \doteq (T - i + 1)h$ per unit. That is, we incur the holding cost until the end of the time horizon. As long as we remember to subtract the constant $h \sum_{i=1}^{T} d_{1i}$ from our final cost, then we are charged exactly the right amount. With this in mind, define $G(i)$ to be cost of an optimal solution with a planning horizon from period i to period T, for $i = 1, 2, \ldots, T$. For convenience, define $G(T + 1) = 0$. Then

$$G(i) = \min_{i < t \leq T+1} \{K + H_i d_{i,t-1} + G(t)\}$$

$$= K + \min_{i < t \leq T+1} \{H_i d_{i,t-1} + G(t)\}. \tag{8.4}$$

The final cost is then $G(1) - h \sum_{i=1}^{T} d_{1i}$. Using this recursion, which is just a reformulation of the shortest-path recursion discussed earlier, we clearly find that the complexity is $O(T^2)$. Wagelmans et al.'s $O(T)$ algorithm is based on the crucial observation that with careful implementation, the total amount of time spent finding the period that minimizes (8.4) over the entire running of the algorithm is $O(T)$.

Consider the calculation of $G(i)$. It is useful to plot the points $(d_{jT}, G(j))$ for $j = i+1, i+2, \ldots, T+1$, where the point $(d_{T+1,T}, G(T+1))$ is simply the origin. Let $\mathcal{E}$ be the lower convex envelope of these points; then define the function $g(x) = y$ if and only if $(x, y) \in \mathcal{E}$. It is clear that g is a piecewise linear convex function on $[0, d_{i+1,T}]$, with $g(d_{i+1,T}) = G(i + 1)$ and $g(0) = 0$. See Fig. 8.1.

Define the breakpoints of g to be all the points x where g changes slope in addition to the points $x = 0$ and $x = d_{i+1,T}$. If x is a breakpoint, then $x = d_{jT}$ for some period $j \in \{i + 1, i + 2, \ldots, T + 1\}$. Let there be r breakpoints and let $i + 1 = t(1) < t(2) < \ldots < t(r) = T + 1$ denote the corresponding periods. These periods are called *efficient* because of the following.

Theorem 8.1.3

$$\min_{i < t \leq T+1} \{H_i d_{i,t-1} + G(t)\} = \min_{1 \leq p \leq r} \{H_i d_{i,t(p)-1} + G(t(p))\}.$$

Proof. Suppose that j (with $i + 1 < j < T + 1$) is not an efficient period, and let k and ℓ (with $k < j < \ell$) be the two consecutive efficient periods straddling j. The slope of g on $[d_{\ell T}, d_{kT}]$ is equal to $[G(k) - G(\ell)]/d_{k,\ell-1}$; hence,

$$g(d_{jT}) = G(\ell) + \frac{G(k) - G(\ell)}{d_{k,\ell-1}} d_{j,\ell-1}.$$

Furthermore, $G(j) \geq g(d_{jT})$.

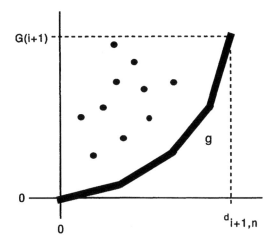

FIGURE 8.1. The plotted points and the function g

There are two cases to consider.

Case 1: $H_i \geq \frac{G(k)-G(\ell)}{d_{k,\ell-1}}$. Then

$$H_i d_{i,j-1} + G(j) \geq H_i d_{i,k-1} + H_i d_{k,j-1} + g(d_{jT})$$

$$\geq H_i d_{i,k-1} + \frac{G(k)-G(\ell)}{d_{k,\ell-1}} d_{k,j-1} + G(\ell) + \frac{G(k)-G(\ell)}{d_{k,\ell-1}} d_{j,\ell-1}$$

$$= H_i d_{i,k-1} + G(k).$$

Case 2: $H_i < \frac{G(k)-G(\ell)}{d_{k,\ell-1}}$. Then

$$H_i d_{i,j-1} + G(j) \geq H_i d_{i,\ell-1} - H_i d_{j,\ell-1} + g(d_{jT})$$

$$\geq H_i d_{i,\ell-1} - \frac{G(k)-G(\ell)}{d_{k,\ell-1}} d_{j,\ell-1} + G(\ell) + \frac{G(k)-G(\ell)}{d_{k,\ell-1}} d_{j,\ell-1}$$

$$= H_i d_{i,\ell-1} + G(\ell).$$

In both cases, the minimum occurs at an efficient period.   ∎

Being able to quickly find the efficient period p that achieves the minimum is therefore crucial to the complexity of the algorithm. This step is aided by the following result.

Lemma 8.1.4 *Let k and ℓ, $k < \ell$ be two consecutive efficient periods. If*

$$\frac{G(k)-G(\ell)}{d_{k,\ell-1}} < H_i,$$

then

$$H_i d_{i,k-1} + G(k) < H_i d_{i,\ell-1} + G(\ell);$$

otherwise,

$$H_i d_{i,k-1} + G(k) \geq H_i d_{i,\ell-1} + G(\ell).$$

Proof. Suppose that $\frac{G(k)-G(\ell)}{d_{k,\ell-1}} < H_i$; then $G(k) < H_i d_{k,\ell-1} + G(\ell)$. Adding $H_i d_{i,k-1}$ to both sides results in $H_i d_{i,k-1} + G(k) < H_i d_{i,\ell-1} + G(\ell)$. The other case can be shown in a similar fashion. ∎

We now describe specifically how to find the efficient period achieving the minimum in (8.4). This is done by keeping an up-to-date list L of the current efficient periods. Let $\ell(p)$ be the index of the efficient period immediately following efficient period p; that is, $p < \ell(p)$. From Lemma 8.1.4 and the convexity of g, it follows that the value of j that achieves the minimum of

$$\min_{i<j\leq T+1} \{H_i d_{i,j-1} + G(j)\}$$

corresponds to the period $q(i)$ defined by

$$q(i) \doteq \min\left[T+1, \min\left\{p \in L \mid p < T+1 \text{ and } \frac{G(p) - G(\ell(p))}{d_{p,\ell(p)-1}} < H_i\right\}\right],$$

because then

$$H_i d_{i,p-1} + G(p) \geq H_i d_{i,\ell(p)-1} + G(\ell(p)), \quad \text{for } p \in L \text{ and } p < q(i),$$

and

$$H_i d_{i,p-1} + G(p) < H_i d_{i,\ell(p)-1} + G(\ell(p)), \quad \text{for } p \in L \text{ and } p \geq q(i).$$

In fact, it is easy to determine $q(i)$ from $q(i+1)$. Note that $q(i+1) \in L$ and as long as $q(i+1)$ is efficient, it has the same successor $\ell(i+1)$ in L. Using the definition of $q(i+1)$, we obtain

$$\frac{G(q(i+1)) - G(\ell(q(i+1)))}{d_{q(i+1),\ell(q(i+1))-1}} < H_{i+1} \leq H_i.$$

Hence, it follows that $q(i) \leq q(i+1)$; that is, *the values of $q(i)$ are decreasing in i.* Therefore, starting at $q(i+1)$, we successively decrement by one until we find $q(i)$. The total amount of time spent searching for $q(i)$ in the entire algorithm is therefore $O(T)$.

To complete the complexity result, we must be able to quickly update the list of efficient periods, that is, update the lower convex envelope. After calculating $G(i)$ and plotting the point $(d_{iT}, G(i))$, we search for the smallest efficient period $t(s)$ such that the slope of the line segment connecting $(d_{iT}, G(i))$ to $(d_{t(s),T}, G(t(s)))$ is greater than the slope of the line segment connecting $(d_{t(s+1),T}, G(t(s+1)))$ to $(d_{t(s),T}, G(t(s)))$ (thus maintaining convexity). Then the new efficient periods are i and the periods from $t(s)$ to $t(r) \equiv T+1$; the efficient periods between $i+1$ and $t(s)-1$ become inefficient. Since a period can become inefficient at most once, one can verify that the total amount of work spent updating the list L over the entire algorithm is $O(T)$.

8.2   Models with Capacity Constraints

An important generalization of the Wagner–Whitin model is the inclusion of upper bounds on the amount that can be ordered or produced in a given period. This corresponds to adding the following constraints to Problem WW:

$$y_t \leq C_t, \quad t = 1, 2, \ldots, T. \tag{8.5}$$

The values $C_t \geq 0$ correspond to the maximum amount that can be ordered (or produced) in period t due to, for example, limited production capacities.

In this case, the problem is not as simple as before; Florian et al. (1980) show that, in general, the problem is $\mathcal{NP}$-Complete . Florian and Klein (1971) propose a dynamic programming approach that involves solving a sequence of acyclic shortest-path problems for the special case where $C_t = C$ for all t. Love (1973) devises an algorithm based on characterizing the extreme points of the solution space for the general problem. The branch-and-bound algorithm of Baker et al. (1978) seems to be the most computationally effective although it is not polynomial.

We sketch here the approach of Florian and Klein. For now, assume unequal capacities; most of the structural results proved by Florian and Klein hold in this more general case. Clearly, a feasible solution exists if and only if

$$\sum_{j=1}^{i} C_j \geq \sum_{j=1}^{i} d_j, \quad \text{for } i = 1, 2, \ldots, T.$$

We therefore assume this is satisfied. Let

$$\mathcal{P} = \{y \in \mathbb{R}^T : y \text{ satisfies } (8.1), (8.2), (8.3) \text{ and } (8.5)\},$$

and let D be the set of extreme points of $\mathcal{P}$. Since the objective function is concave (why?), we know an optimal solution will exist in D.

Florian and Klein prove the following *inventory decomposition property.*

Theorem 8.2.1 *Suppose that the constraint*

$$I_k = 0, \text{ for some } k \in [1, \ldots, T-1],$$

is added to Problem WW and

$$\sum_{j=k+1}^{i} C_j \geq \sum_{j=k+1}^{i} d_j, \quad \text{for } i = k+1, \ldots, T$$

holds. Then an optimal solution to the original problem can be found by independently finding solutions to the problems for the first k periods and for the last $T-k$ periods.

This is clearly a generalization of Theorem 8.1.2. Following this idea, call a period t a *regeneration point* if $I_t = 0$. Define a *production sequence* S_{ij}, where $0 \leq i < j \leq T$, to be

$$S_{ij} = \{(y_{i+1}, y_{i+2}, \ldots, y_j) \mid I_i = I_j = 0, I_k > 0 \text{ for } i < k < j\}.$$

Clearly, any production plan can be decomposed into a set of production sequences. Define a production sequence S_{ij} to be *capacity-constrained* if the production level in at most one period k ($i+1 \leq k \leq j$) satisfies $0 < y_k < C_k$ and all other production levels are either zero or at their capacities.

The authors then characterize the extreme points of $\mathcal{P}$ in the following way.

Theorem 8.2.2

$y \in D \iff y$ *consists of capacity-constrained production sequences only.*

This characterization is done in several steps. First:

Lemma 8.2.3 *If $y \in D$, then y consists only of capacity-constrained production sequences.*

Proof. Suppose $y \in D$ and S_{ij} is a production sequence of y that is not capacity-constrained. This means there are at least two periods, say k and ℓ ($i+1 \leq k < \ell \leq j$), in which $0 < y_k < C_k$ and $0 < y_\ell < C_\ell$. Without loss of generality, we can assume there are only two periods of this type.

Let

$$\delta = \frac{1}{2}\min\{y_k, C_k - y_k, y_\ell, C_\ell - y_\ell, \min_{i+1 \leq t < j} I_t\},$$

and let e_n be the $(j-i)$-component vector with a 1 in the nth position and 0s everywhere else. Define two production sequences

$$S'_{ij} = S_{ij} - \delta e_{k-i} + \delta e_{\ell-i}$$

and

$$S''_{ij} = S_{ij} + \delta e_{k-i} - \delta e_{\ell-i}.$$

Note that production sequence S'_{ij} simply represents a shifting of production from period k to period ℓ, while sequence S''_{ij} represents the opposite shift. They are clearly feasible, and since $\delta > 0$, they are distinct. However, $S_{ij} = \frac{1}{2}(S'_{ij} + S''_{ij})$, a contradiction. ∎

Lemma 8.2.4 *If y' and y'' are distinct feasible production plans and $y = \frac{1}{2}(y' + y'')$, then y' and y'' share all the regeneration points of y.*

Proof. Let period k be a regeneration point of y. Then

$$0 = \sum_{t=1}^{k}(y_t - d_t) = \frac{1}{2}\left[\sum_{t=1}^{k}(y'_t - d_t) + \sum_{t=1}^{k}(y''_t - d_t)\right] = \frac{1}{2}(I'_k + I''_k).$$

Since $I'_k, I''_k \geq 0$, both I'_k and I''_k must be zero. ∎

Lemma 8.2.5 *If a feasible plan y consists only of capacity-constrained production sequences, then $y \in D$.*

Proof. Assume to the contrary that $y \notin D$. Then there exist feasible plans y' and y'' such that $y = \frac{1}{2}(y' + y'')$.

From Lemma 8.2.4, y' and y'' share the regeneration points of y. Let i and j be two such successive regeneration points, and let S_{ij}, S'_{ij}, and S''_{ij} be the associated distinct production sequences of y, y', and y'', respectively. Evidently,

$$S_{ij} = \frac{1}{2}(S'_{ij} + S''_{ij}).$$

We show that the only possibility is $S_{ij} = S'_{ij} = S''_{ij}$. For this purpose, consider any period k, $i + 1 \le k \le j$, and observe that y_k can take only three possible values. Either $y_k = 0$, in which case $y'_k = y''_k = 0$, or $y_k = C_k$, in which case $y'_k = y''_k = C_k$ or $0 < y_k < C_k$. Since S_{ij} is a capacity-constrained sequence, at most one period, say period ℓ, $i + 1 \le \ell \le j$, has $0 < y_\ell < C_\ell$. But the total production between period $i + 1$ and period j must be equal to total demands over the same periods, and hence $y_\ell = y'_\ell = y''_\ell$. Consequently, $S_{ij} = S'_{ij} = S''_{ij}$. ∎

This completes the proof of Theorem 8.2.2.

It is now clear that an optimal solution must be made up of a sequence of optimal capacity-constrained production sequences. However, determining these sequences can be quite tedious and computationally expensive. To make the problem tractable, Florian and Klein consider the case where the capacity constraints are identical and equal to C. The demand between any two periods, say periods i and j, can then be written as $mC + p$, where m is an integer and $p < C$. Then

Corollary 8.2.6 *If $C_t = C$ for all t, an optimal production sequence has a number of periods in which the production levels are equal to C, at most one period where the production level is $0 < p < C$, and the remaining periods have zero production levels.*

This simplifies the problem considerably; for example, consider determining the optimal production sequence between regeneration points i and j. From Corollary 8.2.6, in each period $k \in \{i + 1, i + 2, \ldots, j\}$, the production is 0, C, or p for some $p \in (0, C)$. Let $Y_k = \sum_{\ell=i+1}^{k} y_k$, for $i < k \le j$, that is, the amount produced between periods $i + 1$ and k in this production sequence. Then Y_k can only take on values in $\{0, p, C, C + p, 2C, \ldots, mC, mC + p\}$.

Thus, we can construct a network where the vertices correspond to the possible values of Y_k for each $i < k \le j$ with directed edges (Y_k, Y_{k+1}) defined by the following:

- If $Y_k = \ell C$, $\ell = 0, 1, \ldots, m$, then there are three edges emanating from this vertex: one to $Y_{k+1} = \ell C$ (corresponding to no production in period k); one to $Y_{k+1} = \ell C + p$ (corresponding to production of p in period k); and one to $Y_{k+1} = (\ell + 1)C$ (corresponding to production of C in period k).

- If $Y_k = \ell C + p$, $\ell = 0, 1, \ldots, m$, then there are two edges emanating from this vertex: one to $Y_{k+1} = \ell C + p$ (corresponding to no production in period k) and one to $Y_{k+1} = (\ell+1)C+p$ (corresponding to production of C in period k).

After creating an artificial initial vertex Y_0, we see that every path from Y_0 to Y_j represents a feasible capacity-constrained production sequence. If we assign arc costs equal to the cost of producing and storing the corresponding product amounts, it is clear that finding the optimal production sequence from i to j is no harder than solving the shortest-path problem on this network. The complexity of this procedure is clearly proportional to $(j-i)^2$, thus determining that the optimal production sequence between all pairs of periods is $O(T^4)$.

To determine the optimal production plan over the entire planning horizon, Florian and Klein solve another shortest-path problem on a network similar to the one formulated in Sect. 8.1. That is, the length of an arc (i, j) in this network is the total cost of the optimal production sequence from i to j. After solving the shortest-path problem, we can find the optimal set of regeneration points by checking the shortest path. This step is $O(T^2)$.

8.3   Multi-Item Inventory Models

In many practical situations, the coordination of inventory and ordering policies involves a variety of different products, and this complicates the problem considerably. Consider the uncapacitated case once again, and assume there are n products. Each product faces a known demand during the next T periods. In addition, a fixed order cost of K_i is incurred every time product i is ordered.

For each product i, define the following:

- Let y_{it} be the amount of product i ordered in period t, for $t = 1, 2, \ldots, T$.

- Let h_i be the inventory holding cost for product i.

- Let I_{it} be the amount of product i in inventory at the start of period t, for $t = 1, 2, \ldots, T$.

- Let d_{it} be the demand in period t for product i, for $t = 1, 2, \ldots, T$.

If we make the same assumptions as in the Wagner–Whitin model, the problem is then

$$\text{Problem } P: \text{ Min} \sum_{t=1}^{T} \sum_{i=1}^{n} \left[K_i \delta(y_{it}) + h_i I_{it} \right]$$

$$\text{s.t. } I_{it} = I_{i,t-1} + y_{it} - d_{it}, \quad i = 1, 2, \ldots, n, \ t = 1, 2, \ldots, T, \qquad (8.6)$$

$$I_{i0} = 0, \quad i = 1, 2, \ldots, n, \qquad (8.7)$$

$$I_{it}, y_{it} \geq 0, \quad i = 1, 2, \ldots, n, \ t = 1, 2, \ldots, T. \qquad (8.8)$$

Here (8.6) are inventory-balance constraints for each product, while (8.7) specify the starting inventory for each product.

It is easy to see that P decomposes into m single-product problems. Each of these single-product problems can be solved using the algorithms for the Wagner–Whitin model.

A more realistic version of this problem is when a *joint setup cost* K_0 is present. This cost is incurred whenever *any* product is ordered. The problem then becomes

$$\text{Problem } P': \quad Min \sum_{t=1}^{T} \left[K_0 \delta(\sum_{i=1}^{n} y_{it}) + \sum_{i=1}^{m} \left(K_i \delta(y_{it}) + h_i I_{it} \right) \right]$$

$$s.t. \ (8.6), (8.7), \text{ and } (8.8).$$

Unfortunately, this problem is considerably more difficult to solve than the simple Wagner–Whitin model. In fact, Arkin et al. (1989) prove that it is $\mathcal{NP}$-Complete. Several researchers have proposed heuristics for this problem, including Silver (1976), Atkins and Iyogun (1988), and Joneja (1990). We present here Joneja's approach.

Joneja's *cost-covering* heuristic proceeds period by period in a forward direction. Specifically, at period t, the ordering policy of periods $1, 2, \ldots, t-1$ has been determined and the decision is which items to order, if any, in period t. Let t_i be the last period in which item i was ordered. Let H_{it} denote the total inventory holding cost incurred by item i since period t_i assuming no order for item i is placed in period t. That is,

$$H_{it} = h_i \sum_{j=t_i+1}^{t} (j - t_i) d_{ij}.$$

Intuitively, if we forget for the moment the joint order cost and $H_{it} > K_i$, then it is worth ordering item i in period t, since it costs more to keep an item in inventory from period t_i (the last time item i was ordered) to t than to order it in period t. The quantity $\max\{H_{it} - K_i, 0\}$ can be seen as the *savings* that are accrued by ordering item i in period t. This approach is basically the Silver–Meal heuristic adapted to the multiple-item case. With the joint order cost present, an order should be placed only if the total savings accrued by ordering a set of items in period t exceeds the joint order cost. Therefore, Joneja proposes the following ordering rule.

Rule 1. In period t, order those items i such that $H_{it} \geq K_i$ if $\sum_{i=1}^{n} \max\{H_{it} - K_i, 0\} \geq K_0$.

Joneja shows that this single rule is not quite strong enough to ensure that the schedule of orders is cost-efficient. For instance, consider the following example with two products. The holding costs are equal ($h_1 = h_2 = 1$). Pick an integer m and set the demands to

$$d_{1t} = 0, \text{ for } t = 1, 2, \ldots, m - 1,$$

$$d_{1m} = \frac{K_0 + K_1}{m - 1},$$

$$d_{2t} = 0, \text{ for } t = 1, 2, \ldots, m,$$

$$d_{2,m+1} = \frac{K_0 + K_2}{m}.$$

With Rule 1, item 1 will be ordered at time m, but not item 2. Item 2 will be ordered at time $m + 1$. If both items were ordered at time m, then we pay $h_2 d_{2,m+1} = \frac{K_0 + K_2}{m}$ in extra holding costs but save K_0 in ordering costs. Therefore, for large m, we see that we can be far from optimal.

To counteract this behavior, Joneja proposes the following additional feature. Let t_0 be the time at which the last joint order was placed, and assume item i was not included in this order (since $H_{it_0} < K_i$). It may, in some cases, be advantageous to order item i at time t_0 even though Rule 1 would specify the opposite. Define

$$S_{it} = h_i(t_0 - t_i) \sum_{j=t_0}^{t} d_{ij}.$$

Then S_{it} is the savings in inventory holding cost accrued by ordering item i at time t_0. Since a joint order is already placed in period t_0, the following rule was proposed.

Rule 2. In period t, if the last joint order was in period t_0, item i was not ordered in period t_0, and $S_{it} \geq K_i$, then order item i in period t_0.

Computational experiments with this heuristic, whose complexity is $O(nT)$, show that it produces solutions fairly close to optimal.

8.4   Single-Item Models with Pricing

The previous models focusing solely on inventory replenishment can be naturally extended to settings in which demand is endogenously determined by pricing decisions. For simplicity, we only consider the extension of Problem WW. Specifically, assume that at the beginning of period t ($t = 1, 2, \ldots, T$), we can set a selling price p_t in addition to the ordering quantity y_t. The demand in period t is assumed to be a continuous function of the current period selling price p_t, denoted as $d_t(p_t)$. We are now facing an integrated inventory and pricing model. The objective is to find a sequence of order quantities x_t and prices p_t so as to maximize the total profit over the planning horizon.

Upon denoting a pricing plan $(p_1, p_2, \ldots, p_T)$ and its corresponding demand sequence $(d_1(p_1), d_2(p_2), \ldots, d_T(p_T))$, we find that a mathematical model for the integrated inventory and pricing problem is

$$
\begin{aligned}
Max \quad & \sum_{t=1}^{T} p_t d_t(p_t) - C(d_1(p_1), d_2(p_2), \ldots, d_T(p_T)) \\
s.t. \quad & p_t \in [\underline{p}_t, \bar{p}_t], t = 1, 2, \ldots, T,
\end{aligned}
\tag{8.9}
$$

where a lower bound $\underline{p}_t$ and an upper bound $\bar{p}_t$ on the selling price p_t are imposed to prevent a low profit margin and an unreasonable high price, respectively. In the objective function of the above problem, the first term is the total revenue, and the second term $C(d_1(p_1), d_2(p_2), \ldots, d_T(p_T))$ is the minimum ordering and inventory holding cost over the planning horizon for a given pricing plan, which is the optimal objective value of Problem WW when $(d_1, d_2, \ldots, d_T)$ is replaced by $(d_1(p_1), d_2(p_2), \ldots, d_T(p_T))$. Notice that unlike the models in the previous sections, the variable order cost cannot be ignored here.

To develop an efficient algorithm, observe that Theorem 8.1.1 still holds; that is, any optimal replenishment policy is a zero-inventory-ordering policy. To see this, simply note that given any fixed price sequence $(p_1, p_2, \ldots, p_T)$, Problem (8.9) is equivalent to Problem WW. It is clear that for any ordering plan with the zero-inventory-ordering property, it suffices to specify the ordering periods. Specifically, if periods i and j $(i < j)$ are two consecutive ordering periods in such an ordering plan, the zero-inventory-ordering property implies that the demand at period t $(i \leq t < j)$ is filled by the order placed at period i only, and thus the marginal cost of satisfying period t's demand is given by $c + (t - i)h$. The associated optimal price for period t with $i \leq t < j$ can then be derived, independent of other periods' prices, by solving the following optimization problem:

$$
\begin{aligned}
v_{it} = \quad & \max \quad p_t d_t(p_t) - (c + (t - i)h)d_t(p_t) \\
& s.t. \quad p_t \in [\underline{p}_t, \bar{p}_t],
\end{aligned}
\tag{8.10}
$$

where the objective function is the profit of period t, when taking into account the marginal cost of satisfying the demand of that period. The dynamic programming algorithm of Wagner and Whitin can be easily adopted here. We can construct the same acyclic network. The only difference is that when we define the acyclic network, the length of arc (i, j) with $1 \leq i < j \leq T + 1$ is given by

$$
\ell_{ij} = K - \sum_{t=i}^{j-1} v_{it},
$$

which is the negative of the maximum total profit obtained from satisfying demand from period i to period $j - 1$ with a single order at period i. With this modification, the length of a shortest path from node 1 to node $T + 1$ in the acyclic network gives the negative of the maximum total profit over the planning horizon. The optimal ordering periods can be easily reconstructed from the shortest path, and the optimal prices can be derived through Problem (8.10) once the corresponding ordering periods are known. The algorithm involves solving $O(T^2)$ single-variable optimization problems of the form (8.10) and finding a shortest path in the acyclic network that can be done in $O(T^2)$.

The extension presented here first appeared in Wagner and Whitin (1958a) around the same time they developed Problem WW. The algorithm in Sect. 8.2 can also be easily modified to deal with integrated inventory and pricing models with capacity constraints. We refer to Deng and Yano (2006), Geunes et al. (2006), and Chen and Simchi-Levi (2012) for details. For integrated inventory and pricing models with stochastic demand, see Chap. 10.

8.5 Exercises

Exercise 8.1. Assume order costs are general concave and time-dependent functions of the number of items produced. Also, assume holding costs are general concave and time-dependent functions of the number of items held in inventory. Prove that the zero-inventory-ordering property holds in this general setting as well.

Exercise 8.2. The Silver–Meal heuristic works as follows. Let $d_1, d_2, \ldots, d_n$ be the demands in the n-period planning horizon. Define $C(T)$ to be the per-period average holding and setup cost under the condition that the current order covers demand in the next T periods. Then $C(1) = K$, $C(2) = \frac{1}{2}(K + hd_2)$, and so on. In the Silver–Meal heuristic, we calculate these until $C(i) > C(i-1)$. In this case, we stop and produce in period 1 to meet the demand of the first $i-1$ periods. We then start over with the ith period.

Construct an example where the Silver–Meal heuristic provides a nonoptimal solution.

Exercise 8.3. Consider the integrated inventory and pricing model in Sect. 8.4 with one additional requirement that the prices are the same throughout the planning horizon. Develop an efficient algorithm to solve it. Can you extend your algorithm to cases in which the parameters are time-dependent?

9

Stochastic Inventory Models

9.1 Introduction

The inventory models considered so far are all deterministic in nature; demand is assumed to be known and either constant over the infinite horizon or varying over a finite horizon. In many logistics systems, however, such assumptions are not appropriate. Typically, demand is a random variable whose distribution may be known.

Stochastic inventory models have attracted considerable attention in the last three decades. The pioneering work of Arrow, Harris and Marschak (1951), Scarf (1960), Iglehart (1963a and b), and Veinott and Wagner (1965) for a single warehouse, Clark and Scarf (1960) for multi-echelon systems, Eppen and Schrage (1981) and Federgruen and Zipkin (1984a–c) for distribution systems, and Rosling (1989) for assembly systems all represent milestones in our understanding of complex stochastic logistics systems. More recently, the works of Zheng (1991), Zheng and Federgruen (1991), Chen and Zheng (1994), and Zipkin (2008) reveal new insights and provide more efficient algorithms for these problems. For recent reviews, we refer the reader to Lee and Nahmias (1993), Porteus (1990), and Zipkin (2000).

In this chapter, we review some of the main results in stochastic inventory models. We start with the analysis of a single-warehouse model. To build our intuition, Sect. 9.2 considers a single-period model. In Sects. 9.3 and 9.4, we show that the insight obtained in the previous section can be used to analyze a multiperiod model. Section 9.5 extends the analysis further to the infinite-horizon

D. Simchi-Levi et al., *The Logic of Logistics: Theory, Algorithms, and Applications*   151
for Logistics Management, Springer Series in Operations Research and Financial Engineering,
DOI 10.1007/978-1-4614-9149-1_9, © Springer Science+Business Media New York 2014

model. Models with positive lead times are addressed in Sect. 9.6. Finally, Sect. 9.7 describes the development of interesting bounds on the optimal cost for multi-echelon systems.

9.2 Single-Period Models

9.2.1 The Model

Consider a risk-neutral company that designs, produces, and sells winter fashion items such as ski jackets and coats. About six months before the winter season, the company must commit itself to specific production quantities for all its products. Since there is no clear indication as to how the market will respond to the new designs, these decisions are typically based on realized sales from the last few years, current economic conditions, and professional judgment.

To assist management in selecting production quantities, the marketing department assumes that demand D for each new product is randomly distributed, generated from a product-specific distribution with continuous cdf $F(\cdot)$. Additional information available to the decision makers includes the variable production cost per unit c, the selling price per unit r, and the salvage value per unit v. Clearly, these variables should satisfy $r > c > v$; otherwise, the problem can trivially be solved.

Since demand is a random variable, the decision concerning how many units to produce is based on the *expected cost* $z(y)$, which is a function of the amount produced, y. This expected cost is

$$z(y) = cy - rE[\min(y, D)] - vE[\max(0, y - D)] \quad \text{for } y \geq 0,$$

where $E(\cdot)$ denotes the expectation. Note that

$$E[\min(y, D)] = \int_0^y D dF(D) + y \int_y^\infty dF(D).$$

Adding and subtracting the quantity $r \int_y^\infty D dF(D)$ to $z(y)$, we get

$$z(y) = cy - rE[D] - r \int_y^\infty (y - D)dF(D) - v \int_0^y (y - D)dF(D). \tag{9.1}$$

The objective is, of course, to choose y so as to minimize the expected cost $z(y)$. This is known as the *newsboy problem* or *newsvendor problem*.

Taking the derivative of $z(y)$ with respect to y and using the Leibnitz rule, we get the first-order optimality condition:

$$c - r(1 - \Pr\{D \leq y\}) - v \Pr\{D \leq y\} = 0,$$

which implies that the optimal production quantity S should satisfy

$$\Pr\{D \leq S\} = \frac{r - c}{r - v}.$$

Since, by assumption, $r - c < r - v$ and $F(D)$ is continuous, a finite value S, $S > 0$, always exists. In addition, it can easily be verified that the expected cost $z(y)$ is *convex* for $y \in (0, \infty)$ and that the value of $z(y)$ tends to infinity as $y \to \infty$. Hence, the quantity S is a minimizer of $z(y)$.

Observe that, implicitly, three assumptions have been made in the above analysis. First, there is no initial inventory. Second, there is no fixed setup cost for starting production. Third, the excess demand is lost; that is, if the demand D happens to be greater than the produced quantity y, then the additional revenue $r(D - y)$ is lost.

The tools developed so far allow us to extend the above results to models with initial inventory y_0 and setup cost K. We now relax the first two assumptions. Observe that the expected cost of producing $(y - y_0)$ units is

$$K - cy_0 + z(y).$$

Hence, S clearly minimizes this expected cost if we decide to produce. Consequently, there are two cases to consider.

1. If $y_0 \geq S$, we should not produce anything.

2. If $y_0 < S$, the best we can do is to raise the inventory to level S. However, this is optimal only if $-cy_0 + z(y_0)$, the cost associated with not producing anything, is larger than or equals $K - cy_0 + z(S)$, the cost associated with producing $S - y_0$. That is, if $y_0 < S$, it is optimal to produce $S - y_0$ only if $z(y_0) \geq K + z(S)$.

Let s be a number such that

$$z(s) = K + z(S).$$

The discussion above implies that the optimal policy is of the (s, S) type.

Definition 9.2.1 *An (s, S) policy is a policy in which we order $S - y_0$ if the initial inventory level y_0 is at or below s, and do not order otherwise.*

The quantity S is called the *order-up-to level*, while s is referred to as the *reorder point*. In the special case with zero fixed ordering cost, we have $s = S$ and the policy reduces to a base stock policy: When the initial inventory level is no more than S, make an order to raise the inventory level to S; otherwise, no order is placed.

9.3   Finite-Horizon Models

9.3.1   Model Description

We are now ready to consider the finite-horizon (multiperiod) inventory problem. This problem can be described as follows. At the beginning of each period, for example, each week or every month, the inventory of a certain item at the warehouse is reviewed and the inventory level is noted. Then an order may be placed to raise the inventory level up to a certain level. Replenishment orders arrive instantly. The cases with the nonzero lead times will be discussed in Sect. 9.6.

We assume that demands for successive periods are independently and identically distributed (iid). If the demand exceeds the inventory on hand, then the additional demand is backlogged and is filled when additional inventory becomes available. Thus, the backlogged units are viewed as negative inventory. The inventory left over at the end of the final period has a value of c per unit, and all unfilled demand at this time can be backlogged at the same cost c. As we shall see, these assumptions ensure that the expected (gross) revenue in each period is a constant, and therefore we will not include the revenue term in our formulation. Lost sales models are addressed in 9.6.

Costs include ordering, holding, and backorder costs. Ordering cost consists of a setup cost, K, charged every time the warehouse places a replenishment order, and a proportional purchase cost c. There are a holding cost of h^+ for each unit of the inventory on hand at the end of a period and a backorder cost of h^- per unit whenever demand exceeds the inventory on hand. To avoid triviality, we assume $h^-, h^+ > 0$ (why?). The objective is to determine an inventory policy that minimizes the expected cost over T periods. In what follows, we show that an (s_t, S_t) policy is optimal. Of course, an (s_t, S_t) policy is similar to the (s, S) policy described earlier except that the parameters s and S may vary from period to period.

To characterize the optimal policy for the finite-horizon model, we first develop a dynamic programming formulation of the problem. Let x_t be the inventory level at the beginning of period t (before possible ordering).

If the inventory level immediately after ordering is y, then the expected one-period inventory holding and backorder cost for that period is

$$G(y) = h^+ \int_D \max(y - D, 0) dF(D) + h^- \int_D \max(D - y, 0) dF(D), \qquad (9.2)$$

which is called the one-period *loss function*. Since the maximum of convex functions is convex and convexity is preserved under integration, we see that $G(y)$ is convex.

Given a policy $Y = (y_1, y_2, \cdots, y_T)$, where y_t is the order-up-to level (random variable) of period t and may be contingent upon other variables, the sum of the total expected proportional purchasing cost and salvage value P_Σ is given by

$$P_\Sigma = E\left[\sum_{t=1}^{T} c(y_t - x_t) - c(y_T - D_T)\right],$$

where D_t is the realized demand in period t. Noting that $x_{t+1} = y_t - D_t$, we have

$$P_\Sigma = cE[y_1 - x_1 + y_2 - (y_1 - D_1) + \cdots + y_T - (y_{T-1} - D_{T-1}) + D_T - y_T]$$
$$= cTE[D].$$

Thus, P_Σ is independent of the ordering policy, and we can drop off the linear ordering cost component from the formulation. This observation is quite intuitive, since all backlogged demand is filled at the end of the last period, while all remaining inventory left at this period is salvaged, both at the same price c. We also remark that whenever possible, we will suppress the subscript t from D_t (because demands are iid) and y_t.

Recall that x_t is the inventory level, prior to ordering, at the beginning of period t. To formulate the dynamic program, define the following two expected cost functions. Let $G_t(x_t)$ be the *expected cost* for the remaining $T - t + 1$ periods if *we do not order* in period t, and act *optimally in the remaining $T - t$ periods*. Let $z_t(x_t)$ be the *minimal expected cost* incurred through the remaining $T - t + 1$ periods if we *act optimally in period t and all the remaining $T - t$ periods*. It follows that for $t = 1, 2, \ldots, T$,

$$G_t(y) = G(y) + \int_D z_{t+1}(y - D)dF(D) \tag{9.3}$$

and

$$z_t(x) = Min_{y \geq x} \ \{K\delta(y - x) + G_t(y)\}, \tag{9.4}$$

where $z_{T+1}(x) = 0$ for any x, and $\delta(u)$ is 1 if $u > 0$ and it is 0 otherwise.

Note that if we order up to the level $y > x_t$ in period t, the cost for the final $T - t + 1$ periods is $K + G_t(y)$.

Notice that the functions $G_t(y)$ and $z_t(y)$ are not convex and may even have many local minima. In order to show that an (s, S) policy is optimal for this model, we employ the concept of K-convexity, introduced by Scarf (1960), which provides us with a powerful tool to analyze stochastic inventory models with fixed ordering cost.

9.3.2 K-Convex Functions

Definition 9.3.1 *A real-valued function f is called K-convex for $K \geq 0$ if, for any $x_0 \leq x_1$ and $\lambda \in [0, 1]$,*

$$f((1 - \lambda)x_0 + \lambda x_1) \leq (1 - \lambda)f(x_0) + \lambda f(x_1) + \lambda K. \tag{9.5}$$

Below we summarize properties of K-convex functions.

Lemma 9.3.2   (a) *A real-valued convex function is also 0-convex and hence K-convex for all $K \geq 0$. A K_1-convex function is also a K_2-convex function for $K_1 \leq K_2$.*

(b) *If $f_1(y)$ and $f_2(y)$ are K_1-convex and K_2-convex, respectively, then for $\alpha, \beta \geq 0$, $\alpha f_1(y) + \beta f_2(y)$ is $(\alpha K_1 + \beta K_2)$-convex.*

(c) *If $f(y)$ is K-convex and ζ is a random variable, then $E_\zeta[f(y - \zeta)]$ is also K-convex, provided $E[|f(y - \zeta)|] < \infty$ for all y.*

(d) *Assume that* f *is a continuous* K-*convex function and* $f(y) \to \infty$ *as* $|y| \to \infty$. *Let* S *be a minimum point of* f *and* s *be any element of the set*

$$\{x | x \leq S, f(x) = f(S) + K\}.$$

Then the following results hold.

(i) $f(S) + K = f(s) \leq f(y)$, *for all* $y \leq s$.

(ii) $f(y)$ *is a nonincreasing function on* $(-\infty, s)$.

(iii) $f(y) \leq f(z) + K$ *for all* y, z *with* $s \leq y \leq z$.

Proof. Parts (a), (b), and (c) are straightforward and are left as an exercise. Hence, we focus on part (d).

Let S be a minimum point of function f and let s be any element of the set

$$\{x | x \leq S, f(x) = f(S) + K\}.$$

The existence of s and S is guaranteed since f is continuous and $f(y) \to \infty$ as $|y| \to \infty$.

Consider any y and y' with $y \leq y' \leq s$; there exists a $\lambda \in [0, 1]$ such that $y' = (1 - \lambda)y + \lambda S$. The K-convexity of the function $f(x)$ implies that

$$f(y') \leq (1 - \lambda)f(y) + \lambda(f(S) + K) = (1 - \lambda)f(y) + \lambda f(s). \tag{9.6}$$

Part (d) (i) follows from (9.6) upon letting $y' = s$, which immediately implies part (d) (ii).

Finally, consider any y and z with $s \leq y \leq z$. If $y \leq S$, there exists a $\lambda \in [0, 1]$ such that $y = (1 - \lambda)s + \lambda S$. Since $f(x)$ is K-convex and S is a global minimizer of the function f, we have

$$f(y) \leq (1 - \lambda)(f(s) - K) + \lambda f(S) + K = f(S) + K \leq f(z) + K.$$

If $y \geq S$, there exists a $\lambda \in [0, 1]$ such that $y = (1 - \lambda)S + \lambda z$. Again, the K-convexity of the function f and the definition of S imply that

$$f(y) \leq (1 - \lambda)f(S) + \lambda f(z) + \lambda K \leq f(z) + K.$$

∎

Figure 9.1 gives an illustration of the properties of K-convex functions in Lemma 9.3.2 part (d).

Proposition 9.3.3 *If* $f(x)$ *is a* K-*convex function, then function*

$$\phi(x) = Min_{y \geq x} \quad Q\delta(y - x) + f(y)$$

is max$\{K, Q\}$-*convex.*

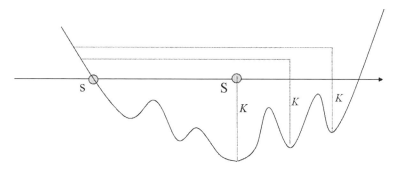

FIGURE 9.1. Illustration of the properties of K-convex functions

Proof. We only need to discuss the case $K \geq Q$. In fact, when $K \leq Q$, the K-convexity of $f(x)$ implies the Q-convexity of $f(x)$, and the Q-convexity of the function $\phi(x)$ follows from the case for $K \geq Q$. Hence, we assume that $K \geq Q$.

Let $E = \{x \mid \phi(x) = f(x)\}$ and $O = \{x \mid \phi(x) < f(x)\}$. We show that for any x_0, x_1 and $\lambda \in [0, 1]$ with $x_0 \leq x_1$,

$$\phi(x_\lambda) \leq (1 - \lambda)\phi(x_0) + \lambda\phi(x_1) + \lambda K,$$

where $x_\lambda = (1 - \lambda)x_0 + \lambda x_1$. We consider four different cases.

Case 1: $x_0, x_1 \in E$. In this case,

$$
\begin{aligned}
\phi(x_\lambda) &\leq f(x_\lambda) \\
&\leq (1 - \lambda)f(x_0) + \lambda f(x_1) + \lambda K \\
&= (1 - \lambda)\phi(x_0) + \lambda\phi(x_1) + \lambda K,
\end{aligned}
$$

where the second inequality follows from the K-convexity of the function $f(x)$.

Case 2: $x_0, x_1 \in O$. In this case, let $\phi(x_i) = Q + f(y_i)$ for $i = 0, 1$ with $y_i \geq x_i$ and let $y_\lambda = (1 - \lambda)y_0 + \lambda y_1$. It is clear that $y_0 \leq y_1$ and $y_\lambda \geq x_\lambda$. Furthermore,

$$
\begin{aligned}
\phi(x_\lambda) &\leq Q + f(y_\lambda) \\
&\leq (1 - \lambda)(Q + f(y_0)) + \lambda(Q + f(y_1)) + \lambda K \\
&= (1 - \lambda)\phi(x_0) + \lambda\phi(x_1) + \lambda K,
\end{aligned}
$$

where the second inequality follows from the K-convexity of the function $f(x)$.

Case 3: $x_0 \in E, x_1 \in O$. Let $\phi(x_1) = Q + f(y_1)$ with $y_1 \geq x_1$. Let $x_\lambda = (1 - \mu)x_0 + \mu y_1$ with $\mu \leq \lambda$. Then

$$
\begin{aligned}
\phi(x_\lambda) &\leq f(x_\lambda) \\
&\leq (1 - \mu)f(x_0) + \mu f(y_1) + \mu K \\
&= (1 - \lambda)\phi(x_0) + \lambda\phi(x_1) + \mu K \\
&\quad + (\lambda - \mu)(f(x_0) - f(y_1)) - \lambda Q \\
&\leq (1 - \lambda)\phi(x_0) + \lambda\phi(x_1) + \mu K - \lambda Q + (\lambda - \mu)Q \\
&\leq (1 - \lambda)\phi(x_0) + \lambda\phi(x_1) + \lambda K,
\end{aligned}
$$

where the second inequality follows from the K-convexity of the function $f(x)$ and the third inequality holds since $f(x_0) \leq Q + f(y_1)$.

Case 4: $x_0 \in O, x_1 \in E$. Let $\phi(x_0) = Q + f(y_0)$ for $y_0 \geq x_0$. We distinguish between two different cases.

Subcase 1: $x_\lambda \leq y_0$. In this case,

$$
\begin{aligned}
\phi(x_\lambda) &\leq Q + f(y_0) \\
&= (1 - \lambda)(Q + f(y_0)) + \lambda f(x_1) + \lambda(Q + f(y_0) - f(x_1)) \\
&\leq (1 - \lambda)\phi(x_0) + \lambda\phi(x_1) + \lambda Q,
\end{aligned}
$$

where the last inequality holds since $f(y_0) \leq f(x_1)$.

Subcase 2: $x_\lambda \geq y_0$. Let $x_\lambda = (1 - \mu)y_0 + \mu x_1$ with $\mu \leq \lambda$. Then

$$
\begin{aligned}
\phi(x_\lambda) &\leq f(x_\lambda) \\
&\leq (1 - \mu)f(y_0) + \mu f(x_1) + \mu K \\
&= (1 - \lambda)\phi(x_0) + \lambda\phi(x_1) + \mu K \\
&\quad + (\lambda - \mu)(f(y_0) - f(x_1)) - (1 - \lambda)Q \\
&\leq (1 - \lambda)\phi(x_0) + \lambda\phi(x_1) + \lambda K,
\end{aligned}
$$

where the second inequality follows from the K-convexity of the function $f(x)$ and the last inequality holds since $f(y_0) \leq f(x_1)$. ∎

9.3.3 Main Results

It remains to show that an (s_t, S_t) policy is optimal for every t, $t = 1, 2, \ldots, T$. For this purpose, it is sufficient to prove that the function $G_t(y)$ is K-convex, and $G_t(y) \to \infty$ as $|y| \to \infty$, for each period t, $t = 1, 2, \ldots, T$.

Theorem 9.3.4   (a) For any $t = 1, 2, \ldots, T$, $G_t(y)$ and $z_t(y)$ are continuous and $\lim_{|y| \to \infty} G_t(y) = \infty$.

(b) For any $t = 1, 2, \ldots, T$, $G_t(y)$ and $z_t(y)$ are K-convex.

(c) For any $t = 1, 2, \ldots, T$, there exist two parameters s_t and S_t such that it is optimal to make an order to raise the inventory level to S_t when the initial inventory level is no more than s_t and to order nothing otherwise.

Proof. We prove by induction. For $t = T$, $G_T(y) = G(y)$ for all y. Hence, $G_T(y)$ is continuous and K-convex (in fact, convex), and $\lim_{|y| \to \infty} G_T(y) = \infty$.

Assume that $G_t(y)$ is continuous and K-convex, and $\lim_{|y| \to \infty} G_t(y) = \infty$. Then Lemma 9.3.2 part (d) allows us to show that there exist two parameters s_t and S_t with $s_t \leq S_t$ such that S_t minimizes $G_t(y)$ and $G_t(s_t) = G_t(S_t) + K$. Furthermore,

$$
z_t(y) = \begin{cases} K + G_t(S_t), & \text{if } y \leq s_t, \\ G_t(y), & \text{otherwise.} \end{cases}
$$

Since $G_t(s_t) = G_t(S_t) + K$, $z_t(y)$ is continuous and Proposition 9.3.3 implies that $z_t(y)$ is K-convex.

Finally, $G_{t-1}(y) = G(y) + E[z_t(y - D)]$. Therefore, $G_{t-1}(y)$ is continuous, and from Lemma 9.3.2 part (c), $G_{t-1}(y)$ is K-convex. Moreover, $\lim_{|y| \to \infty} G_{t-1}(y) = \infty$, since $z_t(y) \geq G_t(S_t)$ for any y. ∎

So far we assume that demands are identically distributed and the cost parameters, c, h^+ and h^-, are time-independent. These assumptions can be easily relaxed and an (s, S) policy is still optimal. Indeed, in Chap. 10, we analyze the finite-horizon inventory and pricing model, including the inventory model analyzed in this section as a special case, under more general assumptions.

9.4   Quasiconvex Loss Functions

The above proof on the optimality of (s_t, S_t) policies relies on the fact that the one-period loss function $G(y)$ is convex. In many practical situations, this assumption is not appropriate. For instance, consider the previous model, but assume that whenever a shortage occurs, an emergency shipment is requested. Suppose further that this emergency shipment incurs a fixed cost plus a linear cost proportional to the shortage level. It can be easily shown that the new loss function $G(y)$ is, in general, not convex.

To overcome this difficulty, Veinott (1966) offers a different yet elegant proof for the optimality of (s_t, S_t) policies under the assumption that $G(y)$ is quasiconvex. Here we provide a slightly simplified proof suggested by Chen (1996) for the model considered here. Recall the concept of quasiconvexity. A function f is quasiconvex on a convex set X if for any x and $y \in X$ and $0 \leq \lambda \leq 1$,

$$f(\lambda x + (1 - \lambda)y) \leq \max\{f(x), f(y)\}.$$

As we already pointed out in Chap. 2, a convex function is also quasiconvex, and f is quasiconvex if

$$-f(x) \text{ is unimodal.}$$

Consider the dynamic program (9.3)–(9.4). In the analysis below, we use the following assumptions on $G(y)$.

(i) $G(y)$ is continuous and quasiconvex.

(ii) $G(y) > \inf_x G(x) + K$ as $|y| \to \infty$.

Other assumptions on ordering costs and demands are the same as in the previous section.

If (i) and (ii) hold, there is a number y^* that minimizes $G(y)$. In addition, there are two numbers $\underline{s}(\leq y^*)$ and $\overline{S}(\geq y^*)$ such that

$$G(\overline{S}) = G(y^*) + K, \tag{9.7}$$
$$G(\underline{s}) = G(y^*) + K. \tag{9.8}$$

It is also worth mentioning that $G(y)$ is nonincreasing in y on $(-\infty, y^*]$ and non-decreasing in y on (y^*, ∞).

To prove the optimality of an (s_t, S_t) policy for all t, we need the next two lemmas.

Lemma 9.4.1 *For $t = 1, \ldots, T$, and $y \leq y'$,*

$$z_t(y) \leq z_t(y') + K \quad and \tag{9.9}$$

$$G_t(y') - G_t(y) \geq G(y') - G(y) - K. \tag{9.10}$$

Proof. It follows that

$$
\begin{aligned}
z_t(y) &= \min\{G_t(y), K + \min_{x \geq y} G_t(x)\} \\
&\leq K + \min_{x \geq y} G_t(x) \\
&\leq K + \min_{x \geq y'} G_t(x) \\
&\leq K + z_t(y').
\end{aligned}
$$

We also provide an alternative proof here. The result obviously holds for $y' = y$. Now assume that $y' > y$. Suppose that at the beginning of the period, the inventory level prior to any ordering is y. Consider the following strategy: We first raise the inventory level up to y' and then act optimally as if we started with the inventory level y' (prior to any ordering). Such a strategy incurs cost equal to $K + z_t(y')$. Because this strategy is not necessarily optimal, it follows that

$$z_t(y) \leq K + z_t(y'),$$

which also proves (9.9).

Inequalities in (9.9) imply that

$$
\begin{aligned}
G_t(y') - G_t(y) &= G(y') - G(y) + E[z_{t+1}(y' - D)] - E[z_{t+1}(y - D)] \\
&\geq G(y') - G(y) - K,
\end{aligned}
$$

which completes the proof. ∎

A function $f : \Re \to \Re$ is called non-K-decreasing if for any x and x' with $x \leq x'$, $f(x) \leq f(x') + K$. The above lemma thus implies that $z_t(y)$ and $G_t(y) - G(y)$ are non-K-decreasing. The following lemma, on the other hand, illustrates that $z_t(y)$ and $G_t(y)$ are nonincreasing for $y \leq y^*$.

Lemma 9.4.2 *For $t = 1, \ldots, T$ and $y \leq y' \leq y^*$,*

$$G_t(y') - G_t(y) \leq G(y') - G(y) \leq 0 \quad and \tag{9.11}$$

$$z_t(y') \leq z_t(y). \tag{9.12}$$

Proof. The proof is by induction. Note that $G(y)$ is decreasing in y for $y \leq y^*$.

For $t = T$, $G_T(y') - G_T(y) = G(y') - G(y) \leq 0$, which implies that $\min_{x \geq y'} G_T(x)$ $= \min_{x \geq y} G_T(x)$. Then

$$
\begin{aligned}
z_T(y') &= \min\{G_T(y'), K + \min_{x \geq y'} G_T(x)\} \\
&\leq \min\{G_T(y), K + \min_{x \geq y'} G_T(x)\} \\
&= \min\{G_T(y), K + \min_{x \geq y} G_T(x)\} = z_T(y).
\end{aligned}
$$

Assume that for $t + 1 \geq 0$ and $y \leq y' \leq y^*$,

$$
\begin{aligned}
G_{t+1}(y') - G_{t+1}(y) &\leq G(y') - G(y) \leq 0 \quad \text{and} \\
z_{t+1}(y') &\leq z_{t+1}(y).
\end{aligned}
$$

Now it follows immediately that

$$
\begin{aligned}
G_t(y') - G_t(y) &= G(y') - G(y) + E[z_{t+1}(y' - D)] - E[z_{t+1}(y - D)] \\
&= G(y') - G(y) + E[z_{t+1}(y' - D) - z_{t+1}(y - D)] \\
&\leq G(y') - G(y) \leq 0 \tag{9.13}
\end{aligned}
$$

and

$$
\begin{aligned}
z_t(y') &= \min\{G_t(y'), K + \min_{x \geq y'} G_t(x)\} \\
&\leq \min\{G_t(y), K + \min_{x \geq y} G_t(x)\} = z_t(y).
\end{aligned}
$$

This completes the proof. ∎

We are now ready to show the optimality result.

Theorem 9.4.3 (Veinott 1966) *If* (i) *and* (ii) *hold, an* (s_t, S_t) *policy is optimal for the model (9.4). Moreover,* $\underline{s} \leq s_t \leq y^*$ *and* $y^* \leq S_t \leq \overline{S}$.

Proof. The proof proceeds in several steps. We start with the assumption that $G_t(y)$ is continuous in y. This assumption will be confirmed at the end.
(1) S_t is a global minimizer of $G_t(y)$. For this purpose, we first show that $G_t(y)$ is decreasing for $y \leq y^*$, which follows directly from (9.11). Because $G_t(y)$ is continuous, there exists a number S_t that minimizes $G_t(y)$ over $[y^*, \overline{S}]$. Now it is clear that S_t minimizes $G_t(y)$ on $(-\infty, \overline{S})$. By the definition of $\overline{S}$ and Lemma 9.4.1, it follows that for $y > \overline{S}(\geq y^*)$,

$$
\begin{aligned}
G_t(y) - G_t(y^*) &\geq G(y) - G(y^*) - K \\
&\geq G(\overline{S}) - G(y^*) - K = 0,
\end{aligned}
$$

where $G(y) \geq G(\overline{S})$ due to the quasiconvexity of $G(y)$. Hence, S_t is indeed a global minimizer of $G_t(y)$ and $y^* \leq S_t \leq \overline{S}$.

(2) There exists a number s_t such that

$$G_t(S_t) + K = G_t(s_t) \text{ and } \underline{s} \leq s_t \leq y^*.$$

The definitions of S_t, $\underline{s}$, and y^* imply that

$$G_t(S_t) + K - G_t(\underline{s}) \leq G_t(y^*) + K - G_t(\underline{s})$$
$$\leq G(y^*) + K - G(\underline{s}) = 0,$$

where the first inequality follows from the definition that S_t is the minimizer of $G_t(y)$, while the second inequality holds due to Lemma 9.4.2. From the definition of y^* and Lemma 9.4.1, we see

$$G_t(S_t) + K - G_t(y^*) \geq G(S_t) - G(y^*) - K + K \geq 0.$$

Together with the continuity assumption of $G_t(y)$ and the fact that $G_t(y)$ is decreasing on $(-\infty, y^*]$, the above two inequalities imply that there exists a number s_t such that

$$G_t(S_t) + K = G_t(s_t) \text{ and } \underline{s} \leq s_t \leq y^*.$$

(3) For $y^* < y < y'$,

$$[K + G_t(y')] - G_t(y) \geq 0.$$

This follows directly from Lemma 9.4.1 and the fact that $G(y') \geq G(y)$:

$$G_t(y') - G_t(y) \geq G(y') - G(y) - K \geq -K.$$

Note that this observation implies that placing an order does not reduce the expected cost when $y > y^*$.

(4) We conclude, therefore, that an (s_t, S_t) policy is optimal.

(5) It remains to prove that $G_t(y)$ is continuous in y.

Again, we proceed by induction. It is true for $t = T$ because $G_T(y) = G(y)$ [by assumption (i)]. Suppose now that $G_{t+1}(y)$ is continuous for $t < T$. From (4),

$$z_{t+1}(y) = \begin{cases} K + G_{t+1}(S_t) & \text{if } y \leq s_t, \\ G_{t+1}(y) & \text{if } y \geq s_t. \end{cases}$$

Hence, z_{t+1} is continuous. Finally, the continuity of $E[z_{t+1}(y-D)]$ follows from the continuity of function z_{t+1} and the uniform continuity theorem, which basically says that a continuous function is uniformly continuous over a compact set. ∎

The above proof for the optimality of (s_t, S_t) policies is based on the assumption that demands are independent and identically distributed. If demands are not independent and identically distributed, Lemma 9.4.2 will generally fail to hold for the following reason. In the proof of Lemma 9.4.2, we require that $z_{t+1}(y' - D) - z_{t+1}(y - D) \leq 0$ for all D in (9.13), which holds only if $y - D \leq y' - D \leq y^*$. When demands are not independent and identically distributed, the minimizer of $G(y)$ may vary from period to period, and the requirement that $z_{t+1}(y' - D) - z_{t+1}(y - D) \leq 0$ may not be met. In the proof based on K-convexity, however, no requirement is imposed upon demands. Thus, while the result in this section is more general than the results of Sect. 9.3 when demands are independent and identically distributed, it is not a generalization of the first.

9.5   Infinite-Horizon Models

In this section, we consider a *discrete-time* infinite-horizon model in which an order may be placed by the warehouse at the beginning of any period. To simplify the analysis, we focus on discrete inventory levels and assume a discrete distribution of the one-period demand D. Let $p_j = \Pr\{D = j\}$ for $j = 0, 1, 2, \ldots$. The objective is to minimize the long-run expected cost per period. All other assumptions and notations are identical to those in the previous section.

This problem has attracted considerable attention in the last three decades. The intuition developed in the previous section (for the finite-horizon models) suggests and is proved by Iglehart (1963b) and Veinott and Wagner (1965), that an (s, S) policy is optimal for the infinite-horizon case. A simple proof is proposed by Zheng (1991). Various algorithms have been suggested by Veinott and Wagner (1965), Bell (1970), and Archibald and Silver (1978) as well as others; see, for instance, Porteus (1990) or Zheng and Federgruen (1991). This section describes a simple proof for the optimality of a stationary (s, S) policy given by Zheng (1991) and sketches an algorithm developed by Zheng and Federgruen (1991) for finding the optimal (s, S) policy. We follow those papers, as well as the insight provided in Denardo (1996).

Let $c(s, S)$ be the long-run average cost associated with the (s, S) policy. Given a period and an initial inventory y, recall that the loss function $G(y)$ is the expected holding and shortage cost minus revenue at the end of the period. In what follows, the loss function $G(y)$ is assumed to be quasiconvex and $G(y) \to \infty$ as $|y| \to \infty$.

Let $M(j)$ be the expected number of periods that elapse until the next order is placed when starting with $s + j$ units of inventory. That is, $M(j)$ is the expected number of periods until the total demand is no less than j units. It is obvious that for all j, we have

$$M(j) = \sum_{k=0}^{j} p_k[1 + M(j-k)] + \sum_{k=j+1}^{\infty} p_k \qquad (9.14)$$

$$= \sum_{k=0}^{\infty} p_k M(j-k) + 1,$$

with $M(j) = 0$ for $j \leq 0$.

Let $\mathcal{F}(s, y)$ be the expected total cost in all periods until placing the next order, when we start with y units of inventory.

Observe that since orders are received immediately, each time an order is placed, the inventory level increases to S. Hence, replenishment times can be viewed as *regeneration points*; see Ross (1970). The theory of regeneration processes tells us that

$$c(s, S) = \frac{\mathcal{F}(s, S)}{M(S - s)}. \qquad (9.15)$$

That is, $c(s, S)$, the long-run average cost, is the ratio of the expected cost between successive regeneration points and the expected time between successive regeneration points.

To calculate $M(S - s)$, one need only solve the recursive equation (9.14). In addition,

$$\mathcal{F}(s, S) = K + H(s, S),$$

where $H(s, y)$ is the expected holding and shortage cost until placing the next order, when starting with y units of inventory. How can we calculate the quantity $H(s, S)$? For this purpose, observe that $M(j + 1) \geq M(j)$ and let

$$m(j) \doteq M(j + 1) - M(j),$$

for $j = 1, 2, 3, \ldots$. To interpret $m(j)$, observe that for any j, $j < S - s$, $M(j + 1)$ is the expected time until demand exceeds j units. Thus, the definition of $M(j)$ implies that $m(j)$ is the expected number of periods, prior to placing the next order, for which the inventory level is exactly $S - j$. Hence,

$$H(s, S) = \sum_{j=0}^{S-s-1} m(j)G(S - j). \tag{9.16}$$

An alternative way of computing $H(s, y)$ is as follows:

$$H(s, y) = G(y) + \sum_{j=0}^{\infty} p_j H(s, y - j), \text{ for } y > s, \tag{9.17}$$

and $H(s, y) = 0$ for $y \leq s$. To summarize, for a (s, S) policy, we have

$$c(s, S) = \frac{K + \sum_{j=0}^{S-s-1} m(j)G(S - j)}{M(S - s)}.$$

Let y^* be any minimizer of the loss function G. Zheng and Federgruen's algorithm as well as Zheng's proof are essentially based on the following results, which characterize the properties of the optimal (s, S) policies.

Lemma 9.5.1 *For any given (s, S) policy, there exists another (s', S') policy with $s' < y^* \leq S'$ such that $c(s', S') \leq c(s, S)$.*

Proof. Observe that $G(y)$ is a quasiconvex function of y and therefore $G(y)$ is nonincreasing for $y \leq y^*$ and nondecreasing for $y \geq y^*$. Consider now $s \geq y^*$. Equation (9.16) together with the quasiconvexity of $G(y)$ implies that $H(s-1, S-1) \leq H(s, S)$. Hence, $c(s, S) \geq c(s-1, S-1)$. Suppose now that $y^* > S$. A similar argument shows that $H(s+1, S+1) \leq H(s, S)$, and hence $c(s, S) \geq c(s+1, S+1)$, which completes the proof.∎

The following result is useful for our analysis.

Lemma 9.5.2 *Assume $s^0 < y^* < S$. For a given ρ, we have that*

(a) If $\rho \leq G(s^0)$, then for any $s < s^0$, there exists $0 < \beta \leq 1$ such that

$$c(s, S) \geq \beta c(s^0, S) + (1 - \beta)\rho.$$

(b) If $\rho \geq G(s^0 + 1)$, then for any $s^0 < s < y^*$, there exists $0 < \beta \leq 1$ such that

$$c(s^0, S) \leq \beta c(s, S) + (1 - \beta)\rho.$$

Proof. For part (a), let $\beta = M(S - s^0)/M(S - s)$ and observe that $0 < \beta \leq 1$. From the definition of $c(s, S)$, we have

$$c(s, S) = \frac{K + \sum_{j=0}^{S-s^0-1} m(j)G(S - j) + \sum_{j=S-s^0}^{S-s-1} m(j)G(S - j)}{M(S - s)}$$

$$= \frac{c(s^0, S)M(S - s^0) + \sum_{j=S-s^0}^{S-s-1} m(j)G(S - j)}{M(S - s)}$$

$$\geq \frac{c(s^0, S)M(S - s^0) + \sum_{j=S-s^0}^{S-s-1} m(j)\rho}{M(S - s)}$$

$$= \beta c(s^0, S) + (1 - \beta)\rho,$$

where the inequality holds since the loss function G is quasiconvex. Finally, the proof of part (b) follows from a similar argument and is left as an exercise.   ∎

We are ready to provide a useful characterization of the optimal reorder levels for a given order-up-to level.

Lemma 9.5.3 *For a given order-up-to level S, a reorder level $s^0 < y^*$ is optimal [i.e., $c(s^0, S) = \min_{s \leq S} c(s, S)$] if*

$$G(s^0) \geq c(s^0, S) \geq G(s^0 + 1). \tag{9.18}$$

Similarly, for any order-up-to level S, there exists an optimal reorder level s^0 such that $s^0 < y^*$ and (9.18) holds.

Proof. The optimality of s^0 for s^0 satisfying (9.18) follows from Lemma 9.5.2 upon letting $\rho = c(s^0, S)$.

We now prove the second part of the result. For any $s < y^*$, there exists an $s^0 < y^*$ such that $G(s^0) \geq c(s, S) \geq G(s^0 + 1)$ since $G(y) \to \infty$ for $y \to \infty$ and $c(s, S) > \min_x G(x)$. Upon letting $\rho = c(s, S)$, Lemma 9.5.2 implies that $c(s^0, S) \leq c(s, S)$. If s^0 satisfies (9.18), then we are done; otherwise, there exist $s^1 < y^*$ such that $s^1 > s^0$ and $G(s^1) \geq c(s^0, S) \geq G(s^1 + 1)$. Again from Lemma 9.5.2, we have $c(s^1, S) \leq c(s^0, S)$. If s^1 satisfies (9.18), we are done; otherwise, repeat this process. This process has to be finite since y^* is an upper bound, and thus we end up with a reorder point satisfying (9.18), which is optimal from the first part of the result.   ∎

An immediate byproduct of the lemma is an algorithm for finding an optimal reorder point s^0 for any given S.

Corollary 9.5.4 *For any value of S, $s^0 = \max\{y < y^* | c(y, S) \leq G(y)\}$ is the optimal reorder level associated with S.*

Proof. Let

$$\alpha = \frac{M(S - s - 1)}{M(S - s)}$$

and observe that (9.15) and (9.16) imply that

$$c(s, S) = \alpha c(s + 1, S) + (1 - \alpha)G(s + 1). \tag{9.19}$$

The definition of s^0 implies that

$$G(s^0) \geq c(s^0, S) \text{ and } G(s^0 + 1) < c(s^0 + 1, S).$$

In addition, using (9.19), we have $c(s^0, S) > G(s^0 + 1)$. Hence, (9.18) holds and by Lemma 9.5.3, s^0 is an optimal reorder point associated with S. ∎

Lemma 9.5.5 *For two order-up-to levels $S^0, S \geq y^*$, let s^0 and s with $s^0, s < y^*$ be the corresponding optimal reorder points, respectively. Moreover, assume that*

$$G(s^0) \geq c(s^0, S^0) \geq G(s^0 + 1).$$

The (s, S) policy improves on (has smaller cost than) (s^0, S^0) if and only if

$$c(s^0, S) < c(s^0, S^0).$$

Proof. We only need to show that if $c(s, S) < c(s^0, S^0)$, then $c(s^0, S) < c(s^0, S^0)$. By contradiction, assume $c(s^0, S) \geq c(s^0, S^0)$. Upon letting $\rho = c(s^0, S) \geq c(s^0, S^0) \geq G(s^0 + 1)$, we have from Lemma 9.5.2 part (b) that

$$c(s, S) \geq c(s^0, S) \geq c(s^0, S^0),$$

which is a contradiction. ∎

Finally, we provide a characterization of the optimal order-up-to level for a given reorder level s. For this purpose, define

$$\phi(i, s, S) = \begin{cases} 0, & \text{if } i \leq s, \\ G(i) - c(s, S) + \sum_{j=0}^{\infty} p_j \phi(i - j, s, S), & \text{otherwise} . \end{cases} \tag{9.20}$$

From the recursive forms (9.14) and (9.17), we have that for $i > s$,

$$\phi(i, s, S) = H(s, i) - c(s, S)M(i - s)$$

and $\phi(S, s, S) = -K$.

Lemma 9.5.6 *For a given reorder level s, if an order-up-to level S^0 is optimal $(c(s, S^0) = \inf_{s \leq S} c(s, S))$, then $c(s, S^0) \geq G(S^0)$.*

Proof. Assume that $c(s,S) < G(S)$ for some $S \geq s$. Then there exists an inventory level i with $s < i < S$ such that $-K = \phi(S,s,S) > \phi(i,s,S)$. This implies that

$$c(s,S) > \frac{K + H(s,i)}{M(i-s)} = c(s,i).$$

Thus, S cannot be optimal. ∎

From the above proof, we can also see that if an order-up-to level S is optimal for a given reorder level s, then $\phi(i,s,S) \geq -K = \phi(S,s,S)$ for any i. The following characterization of the properties of the best (s,S) policy is an immediate consequence of Lemma 9.5.1, Lemma 9.5.3, Lemma 9.5.6, and the above observation.

Lemma 9.5.7 *There exists an (s^*, S^*) policy such that the following hold:*

(a) $c^* \equiv c(s^*, S^*) = \inf_{s \leq S} c(s,S)$;

(b) $s^* < y^* \leq S^*$;

(c) $G(s^*) \geq c^* \geq G(s^* + 1)$;

(d) $G(S^*) \leq c^*$;

(e) $\phi(i) \geq -K = \phi(S^*)$ *for any i, where $\phi(i) \equiv \phi(i, s^*, S^*)$.*

Furthermore, these results suggest the following simple algorithm. Start with $S^0 = y^*$ and find the best reorder point s^0 applying Corollary 9.5.4. Now increase S by increments of 1 each time comparing $c(s^0, S^0)$ to $c(s^0, S)$. If $c(s^0, S) < c(s^0, S^0)$, set $S^0 = S$ and find the corresponding reorder point. Continue until you've identified (s^0, S^0) such that no $S, S > S^0$ has $c(s^0, S) < c(s^0, S^0)$ and $G(S) > c(s^0, S^0)$.

So far we've characterized the properties of the best (s,S) policy, the (s^*, S^*) policy, and we've described how to find such a policy. We are now ready to prove that this stationary (s^*, S^*) policy is optimal for the infinite-horizon model. Of course, as is common for the general infinite-horizon dynamic program, one might attempt to prove that there exists a function h such that the following optimality equation holds:

$$h(x) + c^* = Min_{y \geq x} \quad \{K\delta(y-x) + G(y) + \sum_{j=0}^{\infty} p_j h(y-j)\}. \qquad (9.21)$$

In fact, one can prove that the function ϕ defined in Lemma 9.5.7 satisfies the above optimality equation (9.21). Unfortunately, since the function h is unbounded, there is no result in dynamic programming that allows us to claim the optimality of the stationary (s^*, S^*) policy without further justification. Hence, we follow a different approach. In particular, we focus on a relaxed model where negative order is allowed and whenever a negative order is placed, a fixed cost K is charged.

We construct a bounded function h satisfying the optimality equation for the relaxed model

$$h(x) + c^* = Min_y \quad \{K\delta(|y - x|) + G(y) + \sum_{j=0}^{\infty} p_j h(y - j)\}. \tag{9.22}$$

The construction of function h is as follows:

$$h(i) = \begin{cases} 0, & \text{if } i \leq s^*, \\ O(i), & \text{for } s^* < i \leq S^*, \\ \min\{0, O(i)\}, & \text{otherwise,} \end{cases} \tag{9.23}$$

where $O(i) = G(i) - c^* + \sum_{j=0}^{\infty} p_j h(i - j)$.

We now prove that $-K \leq h(i) \leq 0$ for any i. First notice that $O(i) = \phi(i)$ for $i \leq S^*$; hence, from Lemma 9.5.7 part (e), we have that $h(i) \geq -K$ for $i \leq S^*$. Moreover, using Lemma 9.5.7 part (c), we can show that $h(i) \leq 0$ for any $s^* < i \leq S^*$ and consequently, $h(i) \leq 0$ for any i. Thus, it suffices to prove $O(i) \geq -K$ for $i > S^*$. Assume to the contrary that there exists an i' such that $O(i') < -K$, and without loss of generality let i' be the smallest one. Then $h(i) \geq -K$ for any $i < i'$. In addition, there must exist an i'' such that $S^* < i'' < i'$ and $O(i'') > 0$; otherwise, for any i with $s^* < i \leq i'$, $O(i) \leq 0$, and therefore $h(i) = O(i) = \phi(i) \geq -K$ from Lemma 9.5.7 part (e). This implies that $G(i'') > c^*$. However, since G is quasiconvex and $i'' > S^* \geq y^*$, we can prove by induction that $h(i) \geq -K$ for any i. This is a contradiction since $h(i') = O(i') < -K$. Hence, $-K \leq h(i) \leq 0$ for any i.

In summary, $-K \leq h(i) \leq 0$ for any i, $h(S^*) = -K$ and $O(i) \geq -K$ for any $i > s^*$. It is straightforward to verify that $h(i)$ satisfies the optimality equation of the relaxed model (9.22), and a modified (s^*, S^*) policy attains the minimization in the optimality equation. In the modified policy, make an order to raise the inventory level to S^* whenever the initial inventory level is no more than s^*; do not make any order when the initial inventory lies between $s^* + 1$ and S^*; for an inventory level above S^*, make a negative order to reduce the inventory level to S^* or do nothing depending on which choice is more cost-effective.

We claim that the modified (s^*, S^*) policy is optimal for the relaxed model and its associated long-run average cost c^* is optimal. Indeed, this claim follows from well-known results for infinite-horizon dynamic programming under an average cost criterion since as we just proved, the function h is bounded; for details, one may refer to any standard dynamic programming textbook, for instance, Theorem 2.1, p. 93, in Ross (1983). Also observe that the modified (s^*, S^*) policy is different from the (s^*, S^*) policy in at most one period: When the initial inventory level is too high, we may make a negative order to reduce the inventory level to S^* and after that the inventory level will never exceed S^*. Because the outcome of a finite number of periods will not affect the long-run average cost, it is safe to claim that the stationary (s^*, S^*) policy is optimal for the relaxed model and its associated cost c^* is the optimal average cost. Finally, notice that the stationary (s^*, S^*) policy is feasible for the original model, and the optimal average cost of the original model is no less than the optimal average cost of the modified model. Thus, this stationary (s^*, S^*) policy is optimal for the original infinite-horizon model and its associated cost c^* is the optimal long-run average cost.

9.6   Models with Positive Lead Times

So far we have assumed zero lead times. Under this assumption, if demand that cannot be filled right away is lost instead of being backlogged, the analysis in the previous sections can be modified and our main results are still valid (you are asked to show that the analysis in Sect. 9.3 can be carried over to lost sales models in the exercise). However, if a fixed delivery lead time has to be incorporated, there is a significant difference between lost sales models and backlogging models.

For the case with backlogging, the models discussed in the previous sections but allowing for positive lead times can be transformed into corresponding ones with zero lead time using a fairly simple cost accounting scheme, proposed by Scarf (1960) for the finite-horizon models. In this scheme, the cost allocated to period t is the ordering cost plus the expected inventory holding and backorder cost of period $t+l$ instead of period t, where l denotes the lead time. The rationale behind this is that the inventory cost incurred by the ordering decision at period t will only take effect after the order arrives l periods later.

To calculate the inventory holding and backorder cost of period $t + l$, simply observe that it only depends on the *inventory position* at the warehouse, defined as the inventory at that warehouse plus inventory in transit to the warehouse, at period t. Indeed, the on-hand inventory level at period $t+l$ is the difference of the inventory position at period t immediately after placing the order, referred to as the inventory order-up-to position, and the cumulative demand from period t to period $t + l - 1$. Thus, given the inventory order-up-to position y at period t, we can write the expected inventory holding and backorder cost of period $t + l$ as

$$\hat{G}(y) = h^+ \int_{\hat{D}} \max(y - \hat{D}, 0)d\hat{F}(\hat{D}) + h^- \int_{\hat{D}} \max(\hat{D} - y, 0)d\hat{F}(\hat{D}), \qquad (9.24)$$

where $\hat{F}$ is the cdf of the total demand during the lead time plus one period.

A backlogging model with a positive lead time l can then be transformed into a new one with zero lead time in which $\hat{G}(y)$ is treated as the one-period loss function. With this understanding, the dynamic program (9.3)–(9.4) in Sect. 9.3 can be modified by replacing the loss function $G(\cdot)$ with $\hat{G}(\cdot)$ so that for $t = 1, 2, \ldots, T - l$,

$$\hat{G}_t(y) = \hat{G}(y) + \int_D \hat{z}_{t+1}(y - D)dF(D)$$

and

$$\hat{z}_t(x) = Min_{y \geq x} \ \{K\delta(y - x) + \hat{G}_t(y)\}.$$

It is not hard to show that the optimal inventory order-up-to position can be determined by solving the above dynamic program, and all the analysis and structural properties in Sects. 9.3 and 9.4 can be carried over to this new dynamic program. For the infinite-horizon counterpart, it suffices to replace the loss function $G(\cdot)$ by $\hat{G}(\cdot)$ without changing the analysis in Sect. 9.5.

Unfortunately, this observation no longer works for lost sales models. In fact, though the inventory cost incurred by the ordering decision at period t is still a function of the on-hand inventory level at period $t + l$, its dependence on the inventory at the warehouse and in transit to the warehouse at period t is more complicated, and the inventory position at period t alone is not sufficient. Consequently, the structure of the optimal policy is much more complex. In the remainder of this section, we focus on a stochastic inventory model with lost sales, positive lead time, and zero setup cost. Our analysis is based on Zipkin (2008), in which the concept of $L^{\natural}$-convexity introduced in Chap. 2 plays a critical role.

To completely describe the inventory system, at the beginning of period t after receiving the order placed l periods ago but before placing an order at the current period, let s_i $(i = 0, \ldots, l-1)$ be the inventory level at the warehouse plus the amount of inventory that will arrive within i periods. The state of the inventory system can be described by $s = [s_0, s_1, \ldots, s_{l-1}]$. Note that s_{l-1} is the inventory position before placing an order. Let s_l be the inventory order-up-to position. Given the realized demand D at period t, the state of the next period, $\tilde{s} = [\tilde{s}_0, \tilde{s}_1, \ldots, \tilde{s}_{l-1}]$, is given by

$$\tilde{s}_i = s_{i+1} - s_0 \wedge D \;\forall\; i = 0, 1, \ldots, l-1,$$

and the expected cost for the remaining $T - t + 1$ periods immediately after an order is placed to raise the inventory position to s_l if we act optimally in the remaining $T - t$ periods can be represented as

$$\check{G}_t(s, s_l, D) = h^+ \max(s_0 - D, 0) + h^- \max(D - s_0, 0) + \check{z}_{t+1}(\tilde{s}).$$

The expected cost incurred through the remaining $T - t + 1$ periods if we act optimally in period t and the remaining $T - t$ periods, $\check{z}_t(s)$, can then be derived by the following dynamic program:

$$\check{z}_t(s) = Min_{s_l \geq s_{l-1}} \;\{c(s_l - s_{l-1}) + E[\check{G}_t(s, s_l, D)]\},$$

where the feasible set of the states is

$$\mathcal{S} = \{(s_0, s_1, \ldots, s_{l-1}) : 0 \leq s_0 \leq s_1 \leq \ldots \leq s_{l-1}\}.$$

Note that unlike the backlogging models, we need to keep the linear ordering cost component in the formulation.

Lemma 9.6.1 *For any $s \in \mathcal{S}$ and $s_l \geq s_{l-1}$, $\check{z}_t(s)$ and $\check{G}_t(s, s_l, D)$ are $L^{\natural}$-convex.*

Proof. Clearly, $\mathcal{S}$ is $L^{\natural}$-convex from Proposition 2.3.3 part (e), and so does the set $\{(s, s_l) : s \in \mathcal{S}, s_{l-1} \leq s_l\}$. From Proposition 2.3.4 parts (c) and (e), it suffices to show the $L^\natural$-convexity of $\check{G}_t(s, s_l, D)$. For this purpose, we claim that $\check{G}_t(s, s_l, D)$ equals the optimal objective value of the following optimization problem:

$$
\begin{aligned}
&Min && h^+(s_0 - u) + h^-(D - u) + \check{z}_{t+1}(s_2 - u, \ldots, s_{l-1} - u, s_l - u) \\
&\text{s.t.} && 0 \leq u \leq D, \\
& && u \leq s_0.
\end{aligned}
$$

To see it, note that the optimization problem allows the firm to hold inventory even when there is unsatisfied demand. However, since we face a stationary system, it can never be optimal to hold inventory and reject demand at the same time since it is more cost-effective to fill one-unit current demand than one-unit future demand by avoiding any additional holding cost. Therefore, the optimal u is $\min(D, s_0)$ and our claim is correct. Finally, Proposition 2.3.4 part (d) implies that the above objective function is $L^\natural$-convex. Thus, Proposition 2.3.4 part (e) is applicable and $\check{G}_t(s, s_l, D)$ is $L^\natural$-convex. ∎

Having established the $L^\natural$-convexity of $\check{z}_t(s)$ and $\check{G}_t(s, s_l, D)$, we can derive the monotonicity of the optimal inventory order-up-to position $s_l^*(s)$ and the optimal ordering quantity $s^*(s) - s_{l-1}$. Let x_0 be the on-hand inventory and x_i the inventory in transit that is to arrive in i period $(i > 0)$ and $x = [x_0, x_1, \ldots, x_{l-1}]$.

Theorem 9.6.2 *The optimal inventory order-up-to position $s_l^*(s)$ is increasing in s. However, $s_l^*(s + \xi e) \leq s_l^*(s) + \xi$ for any $\xi \geq 0$. Thus, the optimal ordering quantity $x_l^*(x)$ as a function of x satisfies the following inequalities:*

$$
x_l^*(x) \geq x_l^*(x + \xi e_1) \geq x_l^*(x + \xi e_2) \geq \ldots \geq x_l^*(x + \xi e_{l-1}) \geq x_l^*(x) - \xi \,\forall\, \xi \geq 0,
$$

where e_i is the unit vector with 1 at the ith element.

Proof. The claim on the optimal inventory order-up-to position $s_l^*(s)$ follows directly from Lemma 2.3.5. To prove the claim on the optimal ordering quantity $x_l^*(x)$, note that $x_l^*(x) = s_l^*(s) - s_{l-1}$. Thus, for any $i \leq l - 2$ and $\xi \geq 0$,

$$
\begin{aligned}
x_l^*(x + \xi e_{i+1}) &= s_l^*(s_0, \ldots, s_i, s_{i+1} + \xi, \ldots, s_{l-1} + \xi) - s_{l-1} - \xi \\
&\leq s_l^*(s_0, \ldots, s_{i-1}, s_i + \xi, \ldots, s_{l-1} + \xi) - s_{l-1} - \xi \\
&= x_l^*(x + \xi e_i) \\
&\leq s_l^*(s + \xi e) - s_{l-1} - \xi \\
&\leq s_l^*(s) - s_{l-1} \\
&= x_l^*(x),
\end{aligned}
$$

where e is the all-1s vector. Here the first two inequalities hold since $s_l^*(s)$ is increasing in s, and the last inequality holds since $s_l^*(s + \xi e) \leq s_l^*(s) + \xi$ for $\xi \geq 0$. To prove that $x_l^*(x + \xi e_{l-1}) \geq x_l^*(x) - \xi$, note that

$$
x_l^*(x + \xi e_{l-1}) = s_l^*(s + \xi e_{l-1}) - s_{l-1} - \xi \geq s_l^*(s) - s_{l-1} - \xi = x_l^*(x) - \xi,
$$

where the inequality follows from the monotonicity of $s_l^*(s)$ in s. ∎

If $x_l^*(x)$ is differentiable, the above theorem implies that

$$0 \geq \frac{\partial x_l^*(x)}{\partial x_1} \geq \frac{\partial x_l^*(x)}{\partial x_2} \geq \ldots \geq \frac{\partial x_l^*(x)}{\partial x_{l-1}} \geq -1;$$

that is, the optimal ordering quantity has bounded monotone sensitivities. Specifically, it decreases in the on-hand inventory and the inventory in transit. In addition, it is more sensitive to newer outstanding orders than older outstanding orders and the on-hand inventory with bounded sensitivities.

9.7   Multi-Echelon Systems

Consider a distribution system with a single warehouse, denoted by the index 0, and n retailers, indexed from 1 to n. Incoming orders from an outside vendor with unlimited stock are received by the warehouse that replenishes the retailers. We refer to the warehouse or the retailers as *facilities*. The transportation lead time to facility $i = 0, 1, 2, \ldots, n$, is a constant L_i.

As in the previous section, we analyze a discrete-time model in which customer demands are independent and identically distributed and are faced only by the retailers. Every time a facility places an order, it incurs a setup cost K_i, $i = 0, 1, 2, \ldots, n$. The echelon inventory holding cost (see Chap. 7) is h_i^+ at facility i, $i = 0, 1, 2, \ldots, n$. Finally, demand is backlogged at a penalty cost of h_i^-, $i = 1, 2, \ldots, n$, per unit per period. The objective is to find a centralized strategy, that is, a strategy that uses systemwide inventory information, so as to minimize the long-run average system cost.

As the reader no doubt understands, the analysis of stochastic distribution models is quite difficult and finding an optimal strategy is close to impossible; consider the difficulty involved in finding an approximate solution for its deterministic, constant-demand counterpart; see Chap. 7. As a result, limited literature is available. The rare exceptions are the approximate strategy suggested by Eppen and Schrage (1981) and the lower bounds developed by Federgruen and Zipkin (1984a–c) and Chen and Zheng (1994). We briefly describe these two bounds here.

For this purpose, let the *echelon inventory position* at a facility be defined as the echelon inventory at that facility plus inventory in transit to that facility.

Consider the following approach suggested by Federgruen and Zipkin (1984a–c). Given an inventory position y_i at retailer i, let the loss function $G_i(y_i)$ be

$$G_i(y_i) = h_i^+ \max\{0, y_i - D\} + (h_i^- + h_0^+) \max\{0, D - y_i\},$$

where D is the total demand faced by retailer i during $L_i + 1$ periods (see the end of the previous section for a discussion).

Consider now any inventory policy with echelon inventory of y units at the warehouse and inventory position y_i at retailer i. The expected one-period holding and shortage cost in the system is

$$G(y) = h_0^+ (y - \mu) + \sum_{i=1}^{n} G_i(y_i),$$

where μ is the expected single-period systemwide demand. Since, by definition, $y \geq \sum_{i=1}^{n} y_i$, a lower bound on $G(y)$ is obtained by finding

$$G_0(y) \doteq Min_{y_1,\ldots,y_n} \left\{ h_0^+(y - \mu) + \sum_{i=1}^{n} G_i(y_i) \mid \sum_{i=1}^{n} y_i \leq y \right\}. \qquad (9.25)$$

Thus, a lower bound on the long-run average system cost C^{FZ} is obtained by solving a single-facility inventory problem with loss function G_0 and setup cost K_0. Notice that this bound does not take into account the retailer-specific setup costs. This is incorporated in the next-lower bound of Chen and Zheng (1994).

To describe their lower bound, consider the following *assembly-distribution system* associated with the original distribution system. In the assembly-distribution system, each retailer sells a product consisting of two components. A basic component, denoted by a_0, and a retailer-specific component, denoted by a_i. Each retailer receives component a_0 from the warehouse, which receives it from the outside supplier. On the other hand, component a_i is supplied directly from the vendor to retailer i. The arrival of a basic component at retailer i is coordinated with the arrival of component a_i. That is, at the time the warehouse delivers basic components to retailer i, the same number of a_i components is shipped to the retailer from the supplier. These two shipments arrive at the same time and the final product is assembled, each containing one basic component and one a_i component.

To ensure that the original distribution system and the assembly-distribution system are, in some sense, equivalent, we allocate cost in the new system as follows. Associated with retailer i is a single-facility inventory model with setup cost K_i, holding cost h_i^+, and shortage cost $h_0^+ + h_i^-$. The delivery lead time to the facility is L_i and the demand is distributed according to the demand faced by retailer i. This is, of course, a standard inventory model for which an (s_i, S_i) policy is optimal. Let C_i be the long-run average cost associated with this optimal policy. Given an inventory position y, let $G_i(y)$ be the associated loss function. Finally, let

$$G_i^i(y) = \begin{cases} C_i & \text{if } y \leq s_i, \\ G_i(y) & \text{if } y > s_i, \end{cases}$$

and $G_i^0(y) = G_i(y) - G_i^i(y)$.

In the assembly-distribution system, costs are charged as follows. A setup cost K_0 is allocated to the basic component and a setup cost K_i to each component a_i, and an expected holding and penalty cost, that is, loss function, of G_i^0 to the basic component and G_i^i to component a_i. Notice that since shipments are coordinated, there is no difference between the long-run average cost in the original system and in the assembly-distribution system.

To find a lower bound on the long-run average cost of the original system, we consider a relaxation of the assembly-distribution system in which the basic components can be sold independently of the other components. Thus, C_i, $i = 1, 2, \ldots, n$, is exactly the long-run average cost associated with the distribution of component a_i. Let C_0 be a lower bound on the long-run average cost of the basic component. Consequently, $\sum_{i=0}^{n} C_i$ is a lower bound on the long-run average cost of the original distribution system.

It remains to find C_0. This is obtained following the approach suggested by Federgruen and Zipkin and described above. For this purpose, we replace G_i by G_i^0 in (9.25) and take C^{FZ} as C_0.

9.8   Exercises

Exercise 9.1. In (9.1), we assume that $F(D)$ is continuous. Now suppose that $F(D)$ is not necessarily continuous. Does there exist an S such that $z(y)$ is minimized at $y = S$? If there exists such an S, how can you determine it?

Exercise 9.2. Prove (9.19).

Exercise 9.3. It is now June and your company has to make a decision regarding how many ski jackets to produce for the coming winter season. It costs c dollars to produce one ski jacket, which can be sold for r dollars. Ski jackets not sold during the winter season are lost. Suppose your marketing department estimates that demand during the season can take one of the values $D_1, D_2, \ldots, D_k$, $k \geq 3$. Since this is a new product, they do not know what probabilities to attach to each possible demand D_i; that is, they do not have estimates of p_i, the probability that demand during the winter season will be D_i, $i = 1, 2, \ldots, k$. They have, however, a good estimate of average demand μ and the variance of the demand σ^2. Your objective is to find the production quantity y that will protect you against the worst probability distribution possible while maximizing profit. For this purpose, you would like to consider the following optimization model:

$$\text{MAXIMIZE}_y \ \text{MINIMIZE}_{p_1 \ldots, p_k \in \mathcal{P}} \ \text{Average Profit}, \qquad (9.26)$$

where $\mathcal{P}$ is the set of all possible discrete distribution functions with mean μ and variance σ^2.

(a) Write an expression for the average profit as a function of the production quantity y and the unknown probabilities $p_1, p_2, \ldots, p_k$.

(b) Suppose we have already determined the production quantity y. Write a linear program that identifies the worst possible distribution, that is, the one that minimizes average profit.

(b) Given a value of y, characterize the worst possible distribution; that is, identify the number of demand points that have positive probabilities in the probability distribution found in the previous question.

(c) Can you formulate a linear program that finds the optimal production quantity; that is, can you write a linear program that solves equation (9.26)?

Exercise 9.4. Consider the following discrete version of the newsboy problem. Demand for product can take the values $D_1, D_2, \ldots, D_n$, $n \geq 3$, with probabilities $p_1, p_2, \ldots, p_n$, where $\sum_{i=1}^{n} p_i = 1$. Let r be a known selling price per unit and c be a known cost per unit. Our objective is to find an order quantity y that maximizes expected profit. Prove that the optimal order quantity that maximizes the expected profit must be one of the demand points, $D_1, D_2, \ldots, D_n$.

Exercise 9.5. Prove Lemma 9.3.2, parts (a), (b), and (c).

Exercise 9.6. Consider the newsboy problem with demand D being a random variable whose density, $f(D)$, is known. Let r be a known selling price per unit and c be a known cost per unit. Assume no initial inventory and no salvage value. The objective is to find an order quantity y that maximizes expected profit.

(a) Let a **service level** be defined as the probability that demand is no more than the order quantity, y. Our objective is to find the order quantity, y, that maximizes expected profit subject to the requirement that the service level is **at least** α. What is the optimal order quantity as a function of α, c, r, and $f(D)$?

(b) Suppose there is no service-level requirement; however, there is a capacity constraint, C, on the amount we can order. That is, the order quantity, y, cannot be more than C. What is the optimal order quantity, y, that maximizes expected profit subject to the capacity constraint, C?

(c) Suppose there is a service-level requirement, α, and a capacity constraint, C. What is the optimal order quantity, y, that maximizes expected profit subject to the constraints that service level is at least α and the capacity constraint, C?

Exercise 9.7. Prove that a function $f : \Re \to \Re$ is K-convex if and only if for any $z \geq 0$, $b > 0$, and any y, we have

$$K + f(y + z) \geq f(y) + \frac{z}{b}(f(y) - f(y - b)).$$

Exercise 9.8. Prove Lemma 9.5.2, part (b).

Exercise 9.9. (Pang 2011) If a function $f : \Re \to \Re$ is K-convex and non-K-decreasing, then for $\lambda \geq \gamma \geq 0$, the function $f(\lambda x^+ - \gamma(-x)^+)$ is K-convex. Use this result to show that the analysis in Sect. (9.3) can be carried over to lost sales models with zero lead time.

10

Integration of Inventory and Pricing

10.1 Introduction

In the previous chapters, we analyzed the traditional inventory models, which focus on effective replenishment strategies and typically assume that a commodity's price is exogenously determined. In recent years, however, a number of industries have used innovative pricing strategies to manage their inventory effectively. For example, techniques such as *revenue management* have been applied in the airlines, hotels, and rental car agencies—integrating price, inventory control, and quality of service; see Kimes (1989). In the retail industry, to name another example, dynamically pricing commodities can provide significant improvements in profitability, as shown by Gallego and van Ryzin (1994).

These developments call for models that integrate inventory control and pricing strategies. Such models are clearly important not only in the retail industry, where price-dependent demand plays an important role, but also in manufacturing environments in which production/distribution decisions can be complemented with pricing strategies to improve the firm's bottom line.

The coordination of replenishment strategies and pricing policies has been the focus of many papers, starting with the work of Whitin (1955), who analyzed the celebrated newsvendor problem with price-dependent demand. For a review, the reader is referred to Eliashberg and Steinberg (1991), Petruzzi and Dada (1999), Federgruen and Heching (1999), Yano and Gilbert (2002), Elmaghraby and Keskinocak (2003), Chan, Shen, Simchi-Levi and Swann (2004), or Chen and Simchi-Levi (2012).

D. Simchi-Levi et al., *The Logic of Logistics: Theory, Algorithms, and Applications for Logistics Management*, Springer Series in Operations Research and Financial Engineering, DOI 10.1007/978-1-4614-9149-1_10, © Springer Science+Business Media New York 2014

In this chapter, we review some of the main progress on stochastic models. We first present regularity conditions on demand modeling and some commonly used demand models in Sect. 10.2. We then analyze the single-period models in Sect. 10.3. The finite-horizon models based on Chen and Simchi-Levi (2004a) are analyzed in Sect. 10.4, followed in Sect. 10.5 by an alternative approach due to Huh and Janakiraman (2008). A brief description of extensions and challenges is presented in Sect. 10.6. Finally, in Sect. 10.7, we focus on risk-averse inventory (and pricing) models based on Chen, Sim, Simchi-Levi and Sun (2007).

10.2   Demand Models

To make optimal pricing decisions, it is pivotal that we know the volume of a product that customers are willing to purchase at a specific price. The relationship between the volume and price gives rise to a demand model. The demand of a product can depend on many variables other than price, such as quality, brand name, and competitor's prices; these variables change in each scenario. Here, however, we restrict our discussion to demand models of a single product in which price is the only variable.

Economic theory provides us with basic demand models derived from the classical rational choice theory of consumer behavior (we refer to van Ryzin 2012 and Chap. 7 in Talluri and van Ryzin (2004) , as well as the references therein for more details). Built upon this theory, several regularity conditions are imposed on deterministic demand functions of a single product. Let $[\underline{p}, \bar{p}]$ be the feasible domain of price.

Assumption 10.2.1 *For a given selling price $p \in [\underline{p}, \bar{p}]$, the demand function, $D(p)$, satisfies the following conditions:*

(a) $D(p)$ is continuous in p.

(b) $D(p)$ is strictly decreasing in p and thus has an inverse $D^{-1}(d)$.

(c) $D(p) \in [0, +\infty)$.

(d) The revenue, $D^{-1}(d)d$, is concave in d.

These regularity conditions are quite intuitive and not restrictive in most cases. They are imposed to avoid unnecessary technical complications. Some commonly used deterministic demand functions include

- the linear demand $d(p) = b - ap$ for $p \in [0, b/a]$ ($a > 0$ and $b \geq 0$),

- the exponential demand $d(p) = e^{b-ap}$ ($a > 0$ and $b > 0$),

- the iso-elasticity demand $d(p) = bp^{-a}$ ($a > 1$ and $b > 0$) [note that the price elasticity of demand, $e(p)$, is the relative change in demand in response to a relative change in price, i.e., $e(p) = -\frac{pd'(p)}{d(p)}$],

- the Logit demand $d(p) = \Omega \frac{e^{-ap}}{1+e^{-ap}}$, which is the product of the market size Ω and the probability that a customer with a coefficient of price sensitivity a buys at price p.

It is straightforward to check that the above demand functions satisfy the regularity conditions in Assumption 10.2.1.

In stochastic settings, the demand of a product, denoted by $D(p, \epsilon)$, is often represented as a function of the price p and a random noise ϵ independent of p. Sometimes it is important to specify the format that the random noise ϵ enters the demand function. For our purpose, we focus on stochastic demand models under the following assumptions.

Assumption 10.2.2 *For a given price p, the demand is given by*

$$D(p, \epsilon) = \alpha D(p) + \beta, \qquad (10.1)$$

where $D(p)$ satisfies Assumption 10.2.1, $\epsilon = (\alpha, \beta)$, and α is a nonnegative random variable with $E[\alpha] = 1$ and $E[\beta] = 0$.

An implicit assumption here is that the realized demand $D(p, \epsilon)$ is always nonnegative, which imposes some conditions on the selling price and the two random variables α and β. Observe that, by scaling and shifting, the assumptions $E[\alpha] = 1$ and $E[\beta] = 0$ can be made without loss of generality.

A special case of this demand function is the **additive** demand function. In this case, the demand function is of the form $D(p, \epsilon) = D(p) + \beta$. Another special case of the demand function (10.1) is the model with **multiplicative** demand. In this case, the demand function is of the form $D(p, \epsilon) = \alpha D(p)$, where α is a random variable.

Observe that for additive demand in which $\alpha = 1$, the demand variance is independent of the price, while the coefficient of variation (the ratio of standard deviation and mean) is dependent on the price. In contrast, for multiplicative demand in which $\beta = 0$, the coefficient of variation does not depend on the price while the variance does. In single-product settings with decreasing expected demand $d(p)$, a higher price leads to a higher uncertainty for additive demand but a lower uncertainty for multiplicative demand.

10.3   Single-Period Stochastic Models

We start by analyzing a single-period problem in which a risk-neutral retailer has to decide on its stock level and the selling price of a single product. In this problem, demand is stochastic and depends on the selling price. In particular, we assume that the demand follows Assumption 10.2.2. An ordering and pricing decision is made before the realization of the demand uncertainty. The unit ordering cost is c and unsatisfied demand is filled with an emergency order. Let $h(x)$ be the inventory holding/disposal cost or the emergency ordering cost when the inventory level after satisfying the demand is x. A common form of $h(x)$ is as follows:

$$h(x) = h^+ \max(0, x) + h^- \max(0, -x), \tag{10.2}$$

where h^+ is the unit inventory holding/disposal cost if h^+ is nonnegative or the unit salvage value if it is negative, and h^- is the unit cost for the emergency order. We assume that $h(x)$ is convex and 0 is a minimizer of the function $cx + h(x)$. For $h(x)$ having the particular form (10.2), the above assumptions imply that

$$h^- \geq c \geq \max\{0, -h^+\}.$$

That is, the salvage value is no more than the normal unit ordering cost, which in turn is no more than the unit cost for the emergency order.

For a given stock level y and a selling price p, the expected profit of the retailer is calculated as follows:

$$v(y, p) = E[pD(p, \epsilon)] - cy - E[h(y - D(p, \epsilon))].$$

Assumption 10.2.2 implies that there is a one-to-one correspondence between the selling price p and the expected demand d. Thus, we have an equivalent representation for the retailer's expected profit:

$$\phi(y, d) = R(d) - cy - E[h(y - \alpha d - \beta)].$$

The objective of the retailer is to find a stock level and a selling price, correspondingly an associated expected demand, so as to maximize the retailer's expected profit, namely,

$$\max_{y \geq 0, d \in [\underline{d}, \bar{d}]} \phi(y, d), \tag{10.3}$$

where $\underline{d}$ and $\bar{d}$ are the lower and upper bounds of the expected demand corresponding to the upper and lower bounds of the selling price; that is,

$$\underline{d} = D^{-1}(\bar{p}), \text{ and } \bar{d} = D^{-1}(\underline{p}).$$

Notice that $\phi(y, d)$ is jointly concave in y and d, and hence the above optimization can be solved efficiently.

Our intention here is to compare the selling prices under deterministic and stochastic demands. In particular, we show that there is a significant difference between the additive demand case and the multiplicative demand case. Before we proceed to our main result of this section, we need the following lemma.

Lemma 10.3.1 *Let f be a convex function over $\Re$. Then for any $x, d, \eta \geq 0$,*

$$E[f(x - \alpha d)] \leq E[f(x + \eta - \alpha(d + \eta))],$$

where α is a nonnegative random variable with $E[\alpha] = 1$.

Proof. Notice that a convex function has nondecreasing differences. Hence, we have that for any $x, d, \eta, \alpha \geq 0$,

$$f(x - \alpha d) - f(x - \alpha d - (\alpha - 1)\eta) \leq f(x) - f(x - (\alpha - 1)\eta).$$

Taking expectation on both sides of the above inequality and using Jensen's inequality give us the result. ∎

Now we are ready to present one of our main results of this section. For simplicity, we assume that the expected revenue is strictly concave in the expected demand so that the optimization problem (10.3) has a unique optimal d.

Theorem 10.3.2 *Under the assumption that the expected revenue is strictly concave in the expected demand, the optimal selling price for the additive demand case equals the optimal selling price for the deterministic demand case, which, on the other hand, is no more than the optimal selling price for the multiplicative demand case.*

Proof. It suffices to prove that there exist $(y_d^*, d_d^*), (y_a^*, d_a^*)$, and (y_m^*, d_m^*) optimal for problem (10.3) with deterministic, additive, and multiplicative demand, respectively, such that $d_a^* = d_d^* \geq d_m^*$.

First, notice that

$$\phi(y, d) = R(d) - cd - E[c(y - \alpha d - \beta) + h(y - \alpha d - \beta)].$$

For the deterministic demand case with $\alpha = 1$ and $\beta = 0$, since 0 is a minimizer for the function $cx + h(x)$, it is optimal to set a selling price such that the realized demand is d_d^*, which solves

$$\max_{d \in [\underline{d}, \bar{d}]} R(d) - cd,$$

and to order the demand exactly, that is, $y_d^* = d_d^*$.

Now we prove that there exists an optimal solution (y_a^*, d_a^*) for problem (10.3) with additive demand such that $d_a^* = d_d^*$. If $d_a^* < d_d^*$, then $(y_a^* + \eta, d_a^* + \eta)$ gives an objective value no less than that given by (y_a^*, d_a^*) for a sufficiently small positive η. If $d_a^* > d_d^*$, we distinguish between two cases. First, $y_a^* > 0$. In this case, $(y_a^* - \eta, d_a^* - \eta)$ gives an objective value no less than that given by (y_a^*, d_a^*) for a sufficiently small positive η. Second, $y_a^* = 0$. In this case, $(0, d_a^* - \eta)$ gives an objective value no less than that given by (y_a^*, d_a^*) for a sufficiently small positive η, since 0 is a minimizer of the function $cx + h(x)$. Therefore, there exists an optimal solution (y_a^*, d_a^*) for problem (10.3) with additive demand such that $d_a^* = d_d^*$.

Finally, we argue that there exists an optimal solution (y_m^*, d_m^*) for problem (10.3) with multiplicative demand such that $d_m^* \leq d_d^*$. Assume that $d_m^* > d_d^*$. Again, we distinguish between two cases. First, $y_m^* > 0$. In this case, Lemma 10.3.1 implies that $(y_m^* - \eta, d_m^* - \eta)$ gives an objective value no less than that given by (y_m^*, d_m^*) for a sufficiently small positive η. Second, $y_m^* = 0$. Similar to the argument for the additive demand case, $(0, d_m^* - \eta)$ gives an objective value no less than that given by (y_m^*, d_m^*) for a sufficiently small positive η, since 0 is a minimizer of the function $cx+h(x)$. Therefore, there exists an optimal solution (y_m^*, d_m^*) for problem (10.3) with multiplicative demand such that $d_m^* \le d^*$. ∎

The above theorem thus implies that there is a significant difference between the additive demand case and the multiplicative demand case. To understand this difference, notice that the variance of the additive demand is independent of the selling price, while the variance of the multiplicative demand is a decreasing function of the selling price. Thus, for the multiplicative demand case, the retailer tends to choose a higher selling price so as to decrease the variability of the demand.

In the above discussion, we assume a zero initial inventory level and zero fixed ordering cost. Now let x be the initial inventory level, let y be the target stock level, and also assume that the fixed ordering cost is K. In this case, we face the following problem:

$$\max_{y \ge x, d \in [\underline{d}, \bar{d}]} -K\delta(y-x) + \phi(y, d) + cx, \tag{10.4}$$

where $\delta(u) = 1$ for $u > 0$ and $\delta(0) = 0$.

In the following, we will show that a simple policy, referred to as the $(s, S, \mathbf{p})$ policy, is optimal for problem (10.4). In such a policy, the inventory is managed based on an (s, S) policy and the optimal price $\mathbf{p}(x)$ is a function of the initial inventory level x. Moreover, for the special case with zero fixed ordering cost, a base stock list price policy is optimal: The inventory is managed based on a base stock policy, and the optimal price is a nonincreasing function of the initial inventory level.

Theorem 10.3.3 *For problem (10.4), an $(s, S, \mathbf{p})$ policy is optimal. Furthermore, for the special case with zero fixed ordering cost, a base stock list price policy is optimal.*

Proof. First, notice that Theorem 2.2.6 implies that the function $\phi(x, d)$ is supermodular. Thus, from Theorem 2.2.8, there exists a nondecreasing function $d(x)$ such that $d(x)$ maximizes $\phi(x, d)$ for any given inventory level x.

Now let S be a maximizer of the function $\phi(x, d(x)) + cx$ and let s satisfy

$$\phi(s, d(s)) + cs = \phi(S, d(S)) + cS - K.$$

Since $\phi(x, d)$ is jointly concave in (x, d), $\phi(x, d(x))$ is a concave function. This allows one to show that the optimal inventory level is managed based on the (s, S) policy. Moreover, the optimal price is a function of the initial inventory level: If x is no more than s, the optimal price is $D^{-1}(d(S))$; if x is greater than s, the optimal price is the $D^{-1}(d(x))$.

Finally, for the special case with zero ordering cost, we have $s = S$ and the optimal selling price is $D^{-1}(d(\max(S, x)))$. Hence, a base stock list price policy is optimal. ∎

10.4   Finite-Horizon Models

10.4.1   Model Description

In this section, we focus on a finite-horizon model. Unlike the single-period models, the structure of the optimal policies are significantly different between the additive demand case and the multiplicative demand case, as we will demonstrate in this section.

Consider a firm that has to make replenishment and pricing decisions over a finite time horizon with T periods.

Demands in different periods are independent of each other. For each period t, $t = 1, 2 \ldots, T$, let d_t be the demand and p_t be the selling price in period t. We assume that $d_t = \alpha_t D_t(p_t) + \beta_t$, which is time-dependent and satisfies Assumption 10.2.2. Notice that in this section, the random perturbation ϵ, the demand function $D(p, \epsilon)$, and the expected revenue function $R(d)$ are indexed by t to denote time dependence. The selling price p_t is restricted in an interval. In particular, let $\underline{p}_t$ and $\bar{p}_t$ be the lower and upper bounds of the selling price p_t, respectively.

Let x_t be the inventory level at the beginning of period t, just before placing an order. Similarly, y_t is the inventory level at the beginning of period t after placing an order. The ordering cost function includes both a fixed cost and a variable cost and is calculated for every t, $t = 1, 2, \ldots$, as

$$K\delta(y_t - x_t) + c_t(y_t - x_t).$$

Lead time is assumed to be zero, and hence an order placed at the beginning of period t arrives immediately before demand for the period is realized.

Unsatisfied demand is backlogged. Let x be the inventory level carried over from period t to the next period. Since we allow backlogging, x may be positive or negative. A cost $h_t(x)$ is incurred at the end of period t, which represents the inventory holding cost when $x > 0$ and the shortage cost if $x < 0$.

Given a discount factor γ with $0 < \gamma \le 1$, an initial inventory level, $x_1 = x$, and a pricing and replenishment policy, let

$$V_T^\gamma(x) = \sum_{t=1}^{T} \gamma^{t-1}(-K\delta(y_t - x_t) - c_t(y_t - x_t) - h_t(x_{t+1}) + p_t D_t(p_t, \epsilon_t)) \quad (10.5)$$

be the T-period total discounted profit for a realization of the random perturbations ϵ_t, where $x_{t+1} = y_t - D_t(p_t, \epsilon_t)$.

The objective is to decide on ordering and pricing policies to maximize the total expected discounted profit over the entire planning horizon. That is, the objective is to maximize

$$E[V_T^\gamma(x)] \quad (10.6)$$

for any initial inventory level x and any $0 < \gamma \le 1$.

To find the optimal strategy that maximizes (10.6), let $v_t(x)$ be the maximum total expected discounted profit when $T - t$ periods remain in the planning horizon and the inventory level at the beginning of period t is x. A natural dynamic program that can be applied to find the policy maximizing (10.6) is as follows. For $t = 1, 2, \ldots, T$,

$$v_t(x) = c_t x + \max_{y \geq x, \bar{p}_t \geq p \geq \underline{p}_t} -K\delta(y - x) + \hat{g}_t(y, p) \tag{10.7}$$

with $v_{T+1}(x) = 0$ for any x, where

$$\hat{g}_t(y, p) := -c_t y + E[pD_t(p, \epsilon_t) - h_t(y - D_t(p, \epsilon_t)) + \gamma v_{t+1}(y - D_t(p, \epsilon_t))].$$

Observe that the single-period profit function

$$-c_t(y - x) + E[pD_t(p, \epsilon_t) - h_t(y - D_t(p, \epsilon_t))]$$

is not necessarily a concave function of the selling price p, since $D_t(p, \epsilon_t)$ may be a nonlinear function of p. Fortunately, for the general demand functions (10.1), we can represent the formulation (10.7) only with respect to the expected demand rather than with respect to the price, which allows us to show that the single-period profit function is jointly concave in terms of the inventory level and expected demand. Let

$$\underline{d}_t = D_t(\bar{p}_t) \text{ and } \bar{d}_t = D_t(\underline{p}_t).$$

Assumption 10.2.1 implies that there is a one-to-one correspondence between the selling price $p_t \in [\underline{p}_t, \bar{p}_t]$ and the expected demand $D_t(p_t) \in [\underline{d}_t, \bar{d}_t]$. Denote the expected demand at period t by $d = D_t(p)$. Also, let

$$\phi_t(x) = v_t(x) - c_t x, \ h_t^{\gamma}(y) = h_t(y) + (c_t - \gamma c_{t+1})y, \text{ and } \hat{R}_t(d) = R_t(d) - c_t d, \tag{10.8}$$

where $c_{T+1} = 0$ and R_t is the expected revenue function with

$$R_t(d) = dD_t^{-1}(d),$$

which, by Assumption 10.2.1, is a concave function of expected demand d. These functions, $\phi_t(x), h_t^{\gamma}(y),$ and $\hat{R}_t(d)$, allow us to transform the original problem into a problem with zero variable ordering cost.

Specifically, the dynamic program (10.7) can be written as

$$\phi_t(x) = \max_{y \geq x} -K\delta(y - x) + f_t(y) \tag{10.9}$$

with $\phi_{T+1}(x) = 0$ for any x, where

$$f_t(y) = \max_{\bar{d}_t \geq d \geq \underline{d}_t} g_t(y, d), \tag{10.10}$$

with

$$g_t(y, d) = H_t^{\gamma}(y, d) + \gamma E[\phi_{t+1}(y - \alpha_t d - \beta_t)] \tag{10.11}$$

and

$$H_t^{\gamma}(y, d) := -E[h_t^{\gamma}(y - \alpha_t d - \beta_t)] + \hat{R}_t(d).$$

Thus, most of our focus is on the transformed problem (10.9), which has a similar structure to problem (10.7). In this transformed problem, one can think of h_t^γ as being the holding and shortage cost function, $\tilde{R}_t$ as being the revenue function, and the variable ordering cost is equal to zero.

For technical reasons, we need the following assumption on the revenue functions and on the holding and shortage cost functions.

Assumption 10.4.1 For $t = 1, 2, \ldots$, h_t is convex and $H_t^\gamma(y, d)$ is well defined for any y and $d \in [\underline{d}_t, \bar{d}_t]$. Therefore, $H_t^\gamma(y, d)$ is jointly concave in y and d and consequently,

$$Q_t^\gamma(x) := \max_{\bar{d}_t \geq d \geq \underline{d}_t} H_t^\gamma(x, d) \tag{10.12}$$

is concave. Furthermore, we assume that for any t,

$$\lim_{|x| \to \infty} Q_t^\gamma(x) = -\infty.$$

Notice that one can think of $H_t^\gamma(y, d)$ as being the expected single-period profit excluding the ordering cost for a given inventory level y and a selling price associated with a given expected demand, and $Q_t^\gamma(x)$ as being the maximum expected single-period profit excluding the ordering cost for a given inventory level x by choosing the best selling price.

10.4.2 Symmetric K-Convex Functions

To motivate the technique used for characterizing the optimal policies for the integrated inventory and pricing models, it is useful to relate our problem to the celebrated stochastic inventory control problem discussed in Chap. 9. In that problem, demand is assumed to be exogenously determined, while here demand depends on price. Other assumptions regarding the framework of the model are similar to those made in Chap. 9. In order to prove that an (s, S) policy is optimal for the stochastic inventory models, we employed the concept of K-convexity. It is clear from Definition 9.3.1 that one significant difference between K-convexity and traditional convexity is that (9.5) is not symmetric with respect to x_0 and x_1, and thus it cannot be trivially extended to multidimensional space.

It turns out that this asymmetry is the main barrier when trying to identify the optimal policy to the integrated inventory and pricing problem with nonadditive demand functions. Indeed, there exist counterexamples that show that the function ϕ_t is *not* necessarily K-concave and an (s, S) inventory policy is not necessarily optimal for the finite-horizon model with multiplicative demand functions. This motivates the development of a new concept, the *symmetric K-concave function*, which allows us to characterize the optimal policy in the general demand case.

Definition 10.4.2 A function $f : \Re^n \to \Re$ is called symmetric K-convex for $K \geq 0$ if, for any $x_0, x_1 \in \Re^n$ and $\lambda \in [0, 1]$,

$$f((1 - \lambda)x_0 + \lambda x_1) \leq (1 - \lambda)f(x_0) + \lambda f(x_1) + \max\{\lambda, 1 - \lambda\}K. \qquad (10.13)$$

A function f is called symmetric K-concave if $-f$ is symmetric K-convex.

Observe that similar to the concept of convexity, the symmetric K-convexity is defined in a multidimensional space while the K-convexity is only defined in one-dimensional space. Moreover, a K-convex function is a symmetric K-convex function. The following results describe properties of symmetric K-convex functions, properties that are parallel to those summarized in Lemma 9.3.2 and Proposition 9.3.3.

Lemma 10.4.3 *(a) A real-valued convex function is also symmetric 0-convex and hence symmetric K-convex for all $K \geq 0$. A symmetric K_1-convex function is also a symmetric K_2-convex function for $K_1 \leq K_2$.*

(b) If $f_1(y)$ and $f_2(y)$ are symmetric K_1-convex and symmetric K_2-convex, respectively, then for $\alpha, \beta \geq 0$, $\alpha f_1(y) + \beta f_2(y)$ is symmetric $(\alpha K_1 + \beta K_2)$-convex.

(c) If $f(y)$ is symmetric K-convex and ζ is a random variable, then $E[f(y - \zeta)]$ is also symmetric K-convex, provided $E[|f(y - \zeta)|] < \infty$ for all y.

(d) Assume that $f : \Re \to \Re$ is a continuous symmetric K-convex function and $f(y) \to \infty$ as $|y| \to \infty$. Let S be a global minimizer of f and s be any element from the set

$$X := \{x | x \leq S, f(x) = f(S) + K \text{ and } f(x') \geq f(x) \text{ for any } x' \leq x\}.$$

Then we have the following results:

(i) $f(s) = f(S) + K$ and $f(y) \geq f(s)$ for all $y \leq s$.
(ii) $f(y) \leq f(z) + K$ for all y, z with $(s + S)/2 \leq y \leq z$.

Proof. Parts (a), (b), and (c) follow directly from the definition of symmetric K-convexity. Hence, we focus on part (d). Since f is continuous and $f(y) \to \infty$ as $|y| \to \infty$, X is not empty. Part (d)(i) is a direct consequence of the fact that $s \in X$.

To prove part (d)(ii), we consider two cases. First, for any y, z with $S \leq y \leq z$, there exists $\lambda \in [0, 1]$ such that $y = (1 - \lambda)S + \lambda z$, and we have from the definition of symmetric K-convexity that

$$f(y) \leq (1 - \lambda)f(S) + \lambda f(z) + \max\{\lambda, 1 - \lambda\}K \leq f(z) + K,$$

where the second inequality follows from the fact that S minimizes $f(x)$.

In the second case, consider y such that $S \geq y \geq (s + S)/2$. In this case, there exists $1 \geq \lambda \geq 1/2$ such that $y = (1 - \lambda)s + \lambda S$, and from the definition of symmetric K-convexity, we have that

$$f(y) \leq (1 - \lambda)f(s) + \lambda f(S) + \lambda K = f(S) + K \leq f(z) + K$$

since $f(s) = f(S) + K$. Hence, (i) and (ii) hold. ∎

Figure 10.1 provides an illustration of the property of a symmetric K-convex function in Lemma 10.4.3 part (d). Notice that there might exist a set $A \subseteq (s, (s + S)/2)$ such that $f(x) > f(S) + K$ for $x \in A$.

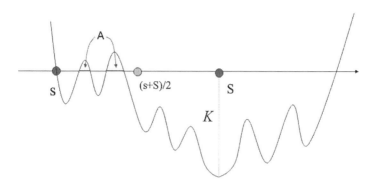

FIGURE 10.1. Illustration of the properties of a symmetric K-convex function

We now present another important property of symmetric K-convex functions, which allows us to prove the symmetric K-concavity of the functions $\phi_t(x)$ and $g_t(y, d)$ by dynamic programming backward induction.

Proposition 10.4.4 *If $f : \Re \to \Re$ is a symmetric K-convex function, then the function*

$$\phi(x) = \min_{y \leq x} Q\delta(x - y) + f(y)$$

is symmetric $\max\{K, Q\}$-convex. Similarly, the function

$$\psi(x) = \min_{y \geq x} Q\delta(x - y) + f(y)$$

is also symmetric $\max\{K, Q\}$-convex.

Proof. We only need to prove the symmetric $\max\{K, Q\}$-convexity of function $\phi(x)$. The second part of the result follows from the symmetric property of the symmetric K-convexity.

If $Q \geq K$, we know that $f(x)$ is also a symmetric Q-convex function by Lemma 10.4.3 part (a). Hence, it suffices to prove that in the case $K \geq Q$, the symmetric K-convexity of the function $f(x)$ implies the symmetric K-convexity of the function $\phi(x)$. Thus, in the remaining part of the proof, we assume that $K \geq Q$.

Observe that $\phi(x) \leq f(x)$ for any x and $\phi(x) \leq Q + f(y)$ for any $y \leq x$. Let $E = \{x \mid \phi(x) = f(x)\}$ and $R = \{x \mid \phi(x) < f(x)\}$. We want to show that for any x_0, x_1 and $\lambda \in [0, 1]$ with $x_0 \leq x_1$,

$$\phi(x_\lambda) \leq (1 - \lambda)\phi(x_0) + \lambda\phi(x_1) + \max\{\lambda, 1 - \lambda\}K, \qquad (10.14)$$

where $x_\lambda = (1 - \lambda)x_0 + \lambda x_1$. We will consider four different cases.

Case 1: $x_0, x_1 \in E$. In this case,

$$
\begin{aligned}
\phi(x_\lambda) &\leq f(x_\lambda) \\
&\leq (1 - \lambda)f(x_0) + \lambda f(x_1) + \max\{\lambda, 1 - \lambda\}K \\
&= (1 - \lambda)\phi(x_0) + \lambda\phi(x_1) + \max\{\lambda, 1 - \lambda\}K,
\end{aligned}
$$

where the second inequality follows from the symmetric K-convexity of the function $f(x)$.

Case 2: $x_0, x_1 \in R$. In this case, let $\phi(x_i) = Q + f(y_i)$ for $i = 0, 1$, with $y_i \leq x_i$, and let $y_\lambda = (1 - \lambda)y_0 + \lambda y_1$. It is clear that $y_0 \leq y_1$ and $y_\lambda \leq x_\lambda$. Furthermore,

$$
\begin{aligned}
\phi(x_\lambda) &\leq Q + f(y_\lambda) \\
&< (1 - \lambda)(Q + f(y_0)) + \lambda(Q + f(y_1)) + \max\{\lambda, 1 - \lambda\}K \\
&= (1 - \lambda)\phi(x_0) + \lambda\phi(x_1) + \max\{\lambda, 1 - \lambda\}K,
\end{aligned}
$$

where the second inequality follows from the symmetric K-convexity of the function $f(x)$.

Case 3: $x_0 \in R, x_1 \in E$. Let $\phi(x_0) = Q + f(y_0)$ with $y_0 \leq x_0$. We will distinguish between two cases.

Subcase 1: $f(y_0) - f(x_1) \leq K - Q$. In this case,

$$
\begin{aligned}
\phi(x_\lambda) &\leq Q + f(y_0) \\
&= (1 - \lambda)(Q + f(y_0)) + \lambda f(x_1) + \lambda(Q + f(y_0) - f(x_1)) \\
&\leq (1 - \lambda)\phi(x_0) + \lambda\phi(x_1) + \lambda K.
\end{aligned}
$$

Subcase 2: $f(y_0) - f(x_1) \geq K - Q$. Let $x_\lambda = (1 - \mu)y_0 + \mu x_1$ with $\lambda \leq \mu$. Then

$$
\begin{aligned}
\phi(x_\lambda) &\leq f(x_\lambda) \\
&\leq (1 - \mu)f(y_0) + \mu f(x_1) + \max\{\mu, 1 - \mu\}K \\
&= (1 - \lambda)\phi(x_0) + \lambda\phi(x_1) + \max\{\mu, 1 - \mu\}K \\
&\quad + (\mu - \lambda)(f(x_1) - f(y_0)) - (1 - \lambda)Q \\
&\leq (1 - \lambda)\phi(x_0) + \lambda\phi(x_1) + \max\{\mu, 1 - \mu\}K - (1 - \mu)Q - (\mu - \lambda)K \\
&\leq (1 - \lambda)\phi(x_0) + \lambda\phi(x_1) + \max\{\lambda, 1 - \lambda\}K,
\end{aligned}
$$

where the second inequality follows from the symmetric K-convexity of the function $f(x)$, and the third inequality follows from the assumption that $f(y_0) - f(x_1) \geq K - Q$.

Case 4: $x_0 \in E, x_1 \in R$. Let $\phi(x_1) = Q + f(y_1)$ for $y_1 \leq x_1$. Again, we distinguish between two different cases.

Subcase 1: $y_1 \leq x_\lambda$. In this case,

$$
\begin{aligned}
\phi(x_\lambda) &\leq Q + f(y_1) \\
&= (1-\lambda)f(x_0) + \lambda(Q + f(y_1)) + (1-\lambda)(Q + f(y_1) - f(x_0)) \\
&\leq (1-\lambda)\phi(x_0) + \lambda\phi(x_1) + (1-\lambda)Q,
\end{aligned}
$$

where the last inequality holds since $f(y_1) \leq f(x_0)$.

Subcase 2: $y_1 \geq x_\lambda$. Let $x_\lambda = (1-\mu)x_0 + \mu y_1$ with $\lambda \leq \mu$. Then

$$
\begin{aligned}
\phi(x_\lambda) &\leq f(x_\lambda) \\
&\leq (1-\mu)f(x_0) + \mu f(y_1) + \max\{\mu, 1-\mu\}K \\
&= (1-\lambda)\phi(x_0) + \lambda\phi(x_1) + \max\{\mu, 1-\mu\}K \\
&+ (\mu-\lambda)(f(y_1) - f(x_0)) - \lambda Q,
\end{aligned}
\tag{10.15}
$$

where the second inequality follows from the symmetric K-convexity of the function $f(x)$. On the other hand, since $x_0 \leq x_\lambda$,

$$
\begin{aligned}
\phi(x_\lambda) &\leq Q + f(x_0) \\
&= (1-\lambda)\phi(x_0) + \lambda\phi(x_1) \\
&+ \lambda(f(x_0) - f(y_1)) + (1-\lambda)Q.
\end{aligned}
\tag{10.16}
$$

If $\mu \leq \frac{1}{2}$, then inequality (10.15) implies inequality (10.14) since $\max\{\mu, 1-\mu\} = 1 - \mu \leq 1 - \lambda$ and $f(y_1) \leq f(x_0)$.

Now assume that $\mu \geq \frac{1}{2}$. Multiplying (10.15) by λ/μ and (10.16) by $(\mu-\lambda)/\mu$ and adding them together, we have

$$
\phi(x_\lambda) \leq (1-\lambda)\phi(x_0) + \lambda\phi(x_1) + \lambda K - (\frac{\lambda}{\mu} - (1-\lambda))Q.
\tag{10.17}
$$

If $\lambda \geq \frac{1}{2}$, then $\frac{\lambda}{\mu} - (1-\lambda) \geq 0$, which, together with inequality (10.17), implies (10.14). On the other hand, if $\lambda \leq \frac{1}{2}$, we have that

$$
\lambda K - (\frac{\lambda}{\mu} - (1-\lambda))Q = (1-\lambda)K + (2\lambda - 1)(K - Q) + \lambda Q(1 - \frac{1}{\mu}) \leq (1-\lambda)K,
$$

which, together with inequality (10.17), implies (10.14). ∎

In the following, we show that, like convex functions, the symmetric K-convexity can be preserved under optimization operations.

Lemma 10.4.5 *Let $f(\cdot, \cdot) : \Re^n \times \Re^m \to \Re$ be symmetric K-convex. Assume that for a given $x \in \Re^n$, there is an associated set $C(x) \subset \Re^m$ and*

$$
C := \{(x, y) \mid y \in C(x), x \in \Re^n\}
$$

is convex. Furthermore, assume that

$$
\phi(x) = \min_{y \in C(x)} f(x, y)
$$

is well defined and the minimization is attainable for any x. Then f is symmetric K-convex.

Proof. For any $x_0, x_1 \in \Re^n$ and $\lambda \in [0, 1]$, let $y_0 \in C(x_0)$ and $y_1 \in C(x_1)$ such that $\phi(x_0) = f(x_0, y_0)$ and $\phi(x_1) = f(x_1, y_1)$. Then

$$(1 - \lambda)y_0 + \lambda y_1 \in C((1 - \lambda)x_0 + \lambda x_1)$$

and

$$
\begin{aligned}
\phi((1 - \lambda)x_0 + \lambda x_1) &\leq f((1 - \lambda)x_0 + \lambda x_1, (1 - \lambda)y_0 + \lambda y_1) \\
&\leq (1 - \lambda)f(x_0, y_0) + \lambda f(x_1, y_1) + \max\{\lambda, 1 - \lambda\}K \\
&= (1 - \lambda)\phi(x_0) + \lambda\phi(x_1) + \max\{\lambda, 1 - \lambda\}K.
\end{aligned}
$$

Therefore, f is symmetric K-convex. ∎

In the following, we focus on characterizing the optimal solution for the finite-horizon model. Specifically, our objective is to identify pricing and replenishment policies that solve (10.7) or its equivalent (10.9).

However, under the additive demand model, this concept is not needed. Indeed, we prove in Sect. 10.4.3 that, for additive demand functions, the function ϕ_t is K-concave and hence the optimal policy for problem (10.9) is an $(s, S, \mathbf{p})$ policy. Formally, in this policy, every period, t, the inventory policy is characterized by two parameters, the reorder point, s_t, and the order-up-to level, S_t. An order of size $S_t - x_t$ is made at the beginning of period t if the initial inventory level at the beginning of the period, x_t, is smaller than s_t. Otherwise, no order is placed. The selling price in period t, p_t, is a function of the inventory level after an order was made.

It turns out that for the additive demand model, the optimal policy and the analysis are significantly different from the optimal policy for the general demand case. In fact, in this case, the symmetric K-convexity is not needed. Specifically, we show, in Sect. 10.4.3, that when the demand function is additive, the function ϕ_t is K-concave for any t, and hence an $(s, S, \mathbf{p})$ policy is optimal for problem (10.9). Formally, in this policy, every period, t, the inventory policy is characterized by two parameters, the reorder point, s_t, and the order-up-to level, S_t. An order of size $S_t - x_t$ is made at the beginning of period t if the initial inventory level at the beginning of the period, x_t, is smaller than s_t. Otherwise, no order is placed. The selling price in period t, p_t, is a function of the inventory level after an order was made.

For more general demand functions, that is, multiplicative plus additive functions, the function ϕ_t is not necessarily K-concave and an $(s, S, \mathbf{p})$ policy is not necessarily optimal. Indeed, in this case, we show, in Sect. 10.4.4, that ϕ_t is symmetric K-concave, which allows us to characterize the optimal policy for the general demand model. Finally, in Sect. 10.4.5, we show that our results imply that in the special case with zero fixed cost and general demand functions, a base stock list price policy is optimal.

10.4.3   Additive Demand Functions

In the additive demand model, the demand function is assumed to be of the form

$$d_t = D_t(p_t) + \beta_t,$$

where β_t is a random variable.

Observe that a special case of this demand function is the *additive linear demand function* in which $d_t = b_t - a_t p_t + \beta_t$, with $b_t, a_t > 0$ for $t = 1, 2, \ldots, T$.

In the following, we show, by induction, that $f_t(y)$ is a K-concave function of y and $\phi_t(x)$ is a K-concave function of x. Therefore, the optimality of an $(s, S, \mathbf{p})$ policy follows directly from Lemma 9.3.2.

To prove that $f_t(y)$ is a K-concave function of y, we need the following lemma.

Lemma 10.4.6 *Assume that r is concave on a bounded interval $[\underline{d}, \bar{d}]$ and $w : \Re \to \Re$ is continuous. Then there exists an optimal solution $d(x)$ of the optimization problem*

$$f(x) = \max_{d \in [\underline{d}, \bar{d}]} r(d) + w(x - d) \qquad (10.18)$$

such that $x - d(x)$ is nondecreasing in x. If w is K-concave, then f is also K-concave.

Proof. Define a new variable $\tilde{d} = x - d$. We have that

$$f(x) = \max_{x - \tilde{d} \in [\underline{d}, \bar{d}]} r(x - \tilde{d}) + w(\tilde{d}).$$

Our assumption, together with Theorem 2.2.6, implies that the objective function of the above optimization problem is supermodular in $(x, \tilde{d})$. It is also easy to verify that the constraint set is a lattice. Thus, from Theorem 2.2.8, there exists an optimal solution $\tilde{d}(x)$ that is nondecreasing in x. The first part of the lemma follows since $d(x) = x - \tilde{d}(x)$ is optimal to problem (10.18).

To prove the second part of the lemma, take any x, x' with $x \leq x'$ and $\lambda \in [0, 1]$. Since f is K-concave and $\tilde{d}(x)$ is nondecreasing, we have that

$$
\begin{aligned}
f((1 - \lambda)x + \lambda x') &\geq r((1 - \lambda)d(x) + \lambda d(x')) + w((1 - \lambda)\tilde{d}(x) + \lambda \tilde{d}(x')) \\
&\geq (1 - \lambda)r(d(x)) + \lambda r(d(x')) + (1 - \lambda)w(\tilde{d}(x)) \\
&\quad + \lambda w(\tilde{d}(x')) - \lambda k \\
&= (1 - \lambda)f(x) + \lambda f(x') - \lambda k,
\end{aligned}
$$

where the first inequality and the equality follow from the definition of $\tilde{d}(\cdot)$ and $d(\cdot)$, and the second inequality from the concavity of $r(d)$ and K-concavity of $w(x)$ as well as the monotonicity of $\tilde{d}(x)$. Thus, f is K-concave. ∎

We are now ready to prove our main results for the additive demand model.

Theorem 10.4.7 *(a) For any $t = 1, 2, \ldots, T$, $g_t(y, d)$ is jointly continuous in (y, d), and hence for any fixed y, $g_t(y, d)$ has a finite maximizer $d_t(y)$ such that $y - d_t(y)$ is nondecreasing in y. Furthermore,*

$$\lim_{|y| \to \infty} g_t(y, d) = -\infty \ \text{for any} \ d \in [\underline{d}_t, \bar{d}_t] \ \text{uniformly} .$$

(b) For any $t = 1, 2, \ldots, T$, $f_t(y)$ and $\phi_t(x)$ are K-concave.

(c) For any $t = 1, 2, \ldots, T$, there exist s_t and S_t with $s_t \leq S_t$ such that it is optimal to order $S_t - x_t$ and set the selling price $p_t(x_t) = D_t^{-1}(d_t(S_t))$ when the initial inventory level $x_t < s_t$, and not to order anything and set $p_t(x_t) = D_t^{-1}(d_t(x_t))$ when $x_t \geq s_t$.

Proof. We prove by induction. For period T, part (a) follows directly from Assumption 10.4.1. Parts (b) and (c) hold since $f_T(y)$ is concave.

Assume that parts (a), (b), and (c) hold for $t + 1$. Since $g_{t+1}(y, d)$ is continuous, $f_{t+1}(y) = \max_{d \in [\underline{d}_t, \bar{d}_t]} g_{t+1}(y, d)$, and hence, $\phi_{t+1}(x) = \max\{f_t(x), K + \min_{y \geq x} f_t(y)\}$ are also continuous, which implies that $g_t(y, d)$ is continuous in (y, d) as well. Note that $g_t(y, d)$ can be expressed as the sum of a concave function of d and a K-concave function of $y - d$ since $\phi_{t+1}(x)$ is K-concave. Therefore, from Lemma 10.4.6, $f_t(y)$ is K-concave, and from Proposition 9.3.3, ϕ_t is K-concave as well. Thus, part (b) holds for period t.

In addition, we have again by Lemma 10.4.6 that for any fixed y, $g_t(y, d)$ has a finite maximizer $d_t(y)$ such that $y - d_t(y)$ is nondecreasing in y. The optimality of the (s_{t+1}, S_{t+1}) policy implies that

$$E[\phi_{t+1}(y - d - \beta_t)] \leq \phi_{t+1}(S_{t+1})$$

for any (y, d), and hence, $\lim_{|y| \to \infty} g_t(y, d) = -\infty$ for any $d \in [\underline{d}_t, \bar{d}_t]$ uniformly by Assumption 10.4.1. Therefore, part (a) holds for period t.

We now prove part (c). Since $f_t(y)$ is K-concave, Lemma 9.3.2 part (d) implies that there exist s_t and S_t such that S_t maximizes $f_t(y)$ and s_t is the smallest value of y such that $f_t(S_t) = f_t(y) + K$, and

$$\phi_t(x) = \begin{cases} -K + f_t(S_t) & \text{if } x \leq s_t, \\ f_t(x) & \text{if } x \geq s_t. \end{cases}$$

Hence, part (c) holds.

Finally, the optimality of the price function $p_t(x_t)$ follows from the definition of $d_t(y)$. ∎

Thus, Theorem 10.4.7 implies that an $(s, S, \mathbf{p})$ policy is optimal when the demand is additive. In addition, there exists an optimal solution $d_t(y)$ maximizing (10.10) such that $y - d_t(y)$ is a nondecreasing function of y. That is, the higher the initial inventory level at the beginning of time period t, y_t, the higher the expected inventory level at the end of period t, $y_t - d_t(y_t)$.

An interesting question is whether a list price policy is optimal, as is the case for the single-period model with no fixed cost. Unfortunately, this property does not hold for the finite-horizon model, as illustrated by Chen and Simchi-Levi (2004a). Indeed, although there is incentive to lower the selling price to reduce inventory, there is also incentive to raise the price in order to slow the depletion of inventory and delay the incurring of fixed ordering cost.

10.4.4 General Demand Functions

In this section, we focus on the model with general demand functions (10.1). Observe that the additive demand function analyzed in the previous section is a special case of the general demand function (10.1). More importantly, multiplicative demand functions of the form $d_t = \alpha_t D_t(p)$, where $D_t(p) = a_t p^{-b_t}$ $(a_t > 0, b_t > 1)$, or demand functions of the form $d_t = \beta_t + \alpha_t(b_t - a_t p)$ $(a_t > 0, b_t > 0)$ are also special cases.

To characterize the optimal policy for the model with the demand functions (10.1), one might consider using the same approach applied in Sect. 10.4.3. Unfortunately, in this case, the function $y - \alpha_t d_t(y)$ is not necessarily a nondecreasing function of y for all possible α_t, as is the case for additive demand functions. Hence, the approach employed in Sect. 10.4.3 does not work in this case. In fact, as demonstrated in Chen and Simchi-Levi (2004a), the function $g_t(y, d_t(y))$ and $\phi_t(x)$ are in general not K-concave, and an $(s, S, \mathbf{p})$ policy is not necessarily optimal.

To overcome these difficulties, we apply the concept of symmetric K-convexity introduced in Sect. 10.4.2. Specifically, in the following, we show, by induction, that $g_t(y, d)$ is a symmetric K-concave function of (y, d) and $\phi_t(x)$ is a symmetric K-concave function of x. Hence, a characterization of the optimal pricing and ordering policies follows from Lemma 10.4.3.

Theorem 10.4.8 (a) For any t, $g_t(y, d)$ is continuous in (y, d), and hence for any fixed y, $g_t(y, d)$ has a finite maximizer $d_t(y)$. Furthermore,

$$\lim_{|y| \to \infty} g_t(y, d) = -\infty \text{ for any } d \in [\underline{d}_t, \bar{d}_t] \text{ uniformly .}$$

(b) For any $t = 1, 2, \ldots, T$, $g_t(y, d)$ and $\phi_t(x)$ are symmetric K-concave.

(c) For any $t = 1, 2, \ldots, T$, there exist s_t and S_t with $s_t \leq S_t$ and a set $A_t \subseteq [s_t, (s_t + t)/2]$ such that it is optimal to order $S_t - x_t$ and set the selling price $p_t = D_t^{-1}(d_t(S_t))$ when the initial inventory level $x_t < s_t$ or $x_t \in A_t$, and not to order anything and set $p_t = D_t^{-1}(d_t(x_t))$ otherwise.

Proof. The proof of part (a) is similar to that of part (a) in Theorem 10.4.7. We now focus on part (b).

By induction, $\phi_{T+1}(x) = 0$ is symmetric 0-concave. From the symmetric K-concavity of $\phi_{t+1}(x)$, we have that $E[\phi_{t+1}(y - \alpha_t d - \beta_t)]$ is symmetric K-concave. Also, we have that $H_t^\gamma(y, d)$ is concave by Assumption 10.4.1. Hence, $g_t(y, d)$ is symmetric γK-concave, and hence by Lemma 10.4.5, the function $f_t(y)$ is symmetric γK-concave. Finally, $f_t(y)$ is symmetric K-concave by Lemma 10.4.3 part (a), and the symmetric K-concavity of $\phi_t(x)$ follows from Proposition 10.4.4. Thus, part (b) holds.

We now prove part (c). From Lemma 10.4.3 part (d), we have

$$\phi_t(x) = \begin{cases} -K + f_t(S_t) & \text{if } x \in I_t, \\ f_t(x) & \text{if } x \notin I_t, \end{cases}$$

where S_t is the maximizer of $f_t(y)$ and

$$I_t = \{y \le S_t \mid f_t(y) \le f_t(S_t) - K\}.$$

Furthermore, $\phi_t(x) \ge f_t(x)$ for any x and $\phi_t(x) \ge -K + f_t(S_t)$ for any $x \le S_t$.

Let s_t be defined as the smallest value of y such that $f_t(S_t) = f_t(y) + K$. Note that from Lemma 10.4.3 part (d), $(-\infty, s_t] \subset I_t$ and $[(s_t + S_t)/2, \infty) \subset (I_t)^c$, the complement of I_t. Part (c) follows from Lemma 10.4.3 and part (b) by defining

$$A_t = I_t \cap [s_t, (s_t + S_t)/2].$$

Again, the optimality of the price function $p_t(x_t)$ follows from the definition of $d_t(y)$. ∎

Theorem 10.4.8 thus implies that the optimal policy for problem (10.7) is an $(s, S, A, \mathbf{p})$ policy. Such a policy is characterized by two parameters, s_t and S_t, and a set $A_t \subseteq [s_t, (s_t + S_t)/2]$, possibly empty. When the inventory level x_t at the beginning of the period t is less than s_t or x_t is in the set A_t, an order of size $S_t - x_t$ is made. Otherwise, no order is placed. Thus, it is possible that an order will be placed when the inventory level $x_t \in [s_t, (s_t + S_t)/2]$, depending on the problem instance. In any case, if an order is placed, it is always to raise the inventory level to S_t.

10.4.5 Special Case: Zero Fixed Ordering Cost

We now apply our results to the zero-fixed-cost case.

Corollary 10.4.9 *Consider our model with zero fixed ordering cost and general demand functions (10.1). In this case, a base stock list price policy is optimal.*

Proof. By Theorem 10.4.8, the functions $\phi_t(x)$ and $f_t(y)$, $t = 1, 2, \ldots, T$, are symmetric 0-concave and hence, from Definition 10.4.2, concave. The optimality of the base stock inventory policy follows directly from the concavity of $f_t(y)$ for $t = 1, 2, \ldots, T$.

We now show that $d_t(y)$ is nondecreasing, and therefore the optimal price $p_t(y)$ is nonincreasing. In fact, in the zero fixed ordering cost case, $g_t(y, d)$ can be expressed as $r(d) + E[w(y - \alpha d)]$ for some concave function w, and thus Theorem 2.2.6 implies that $g_t(y, d)$ is supermodular. From Theorem 2.2.8, there exists $d_t(y)$, which is nondecreasing. ∎

10.5 Alternative Approach to the Optimality of $(s, S, \mathbf{p})$ Policies

The analysis in the previous section builds upon the concept of K-convexity and its extension, symmetric K-convexity. In this section, we provide an alternative approach developed by Huh and Janakiraman (2008) to prove the optimality of $(s, S, \mathbf{p})$ policies for the model in Sect. 10.4. The approach is essentially an extension of the one in Sect. 9.4 for stochastic inventory models to integrated inventory and pricing models. Similar to Sect. 9.4, we focus on stationary systems. That is, the inventory holding and backorder cost function h_t and parameters c_t, $\underline{d}_t$, and $\bar{d}_t$ are all time-independent, and the random variables (α_t, β_t) $(t = 1, 2 \ldots, T)$ are iid across time. Thus, we drop the subscript t of $\underline{d}_t$, $\bar{d}_t$, and H_t^γ in the dynamic program (10.9)–(10.10) in this section.

We assume that $H^\gamma(y, d)$ is continuous in (y, d) and $\lim_{|x| \to \infty} Q^\gamma(x) = -\infty$, where $Q^\gamma(x) = \max_{d \in [\underline{d}, \bar{d}]} H^\gamma(x, d)$. With these assumptions, the functions ϕ_t and f_t are continuous, and $\lim_{|x| \to \infty} \phi_t(x) = \lim_{|x| \to \infty} f_t(x) = -\infty$. Thus, the related optimal solutions exist in the dynamic program (10.9)–(10.10). Let y^0 be a global maximizer of $Q^\gamma(x)$. We make the following assumption.

Assumption 10.5.1   (a) *For any y^1 and y^2 with $y^2 \leq y^1 \leq y^0$ and $d^2 \in [\underline{d}, \bar{d}]$, there exists a $d^1 \in [\underline{d}, \bar{d}]$ such that*

$$H^\gamma(y^1, d^1) \geq H^\gamma(y^2, d^2) \text{ and } y^1 - \alpha d^1 - \beta \geq y^2 - \alpha d^2 - \beta$$

for any realization of (α, β).

(b) *For any y^1 and y^2 with $y^0 \leq y^1 \leq y^2$ and $d^2 \in [\underline{d}, \bar{d}]$, there exists a $d^1 \in [\underline{d}, \bar{d}]$ such that*

$$H^\gamma(y^1, d^1) \geq H^\gamma(y^2, d^2) \text{ and } y^1 - \alpha d^1 - \beta \leq \max\{y^2 - \alpha d^2 - \beta, y^0\}$$

for any realization of (α, β).

Observe that given (y, d) and the realization of (α, β) at a period, $y - \alpha d - \beta$, represents the initial inventory level of the next period. Assumption 10.5.1 indicates that for any (y^2, d^2), if y^1 is closer to y^0—the inventory level that attains the highest single-period expected profit—than y^2, we can always find an expected demand level d^1 (correspondingly a selling price) such that (y^1, d^1) incurs a higher single-period expected profit than (y^2, d^2). This implies that Q^γ is quasiconcave. In addition, the initial inventory level of the next period resulting from (y^1, d^1) is closer to y^0 than that from (y^2, d^2), or one can raise the inventory level $y^1 - \alpha d^1 - \beta$ to y^0 by placing an order in the case with $y^0 \leq y^1 \leq y^2$. The approach in this section is applicable for demand models more general than the linear one presented here, for which we refer to Huh and Janakiraman (2008).

We now present conditions under which Assumption 10.5.1 is valid. Recall the definitions of h_t^γ and $\hat{R}$ in (10.8).

Proposition 10.5.2 *Assume additive demand, namely, $\alpha = 1$. If $E[h_t^\gamma(x - \beta)]$ is quasiconvex and $\hat{R}(d)$ is quasiconcave, then Assumption 10.5.1 holds.*

Proof.   Let d^0 be a global maximizer of $\hat{R}$ over $[\underline{d}, \bar{d}]$ and let x^0 be a global minimizer of $E[h_t^\gamma(x - \beta)]$. Since $H^\gamma(y, d) = \hat{R}(d) - E[h_t^\gamma(y - d - \beta)]$, $y^0 = d^0 + x^0$ is a global maximizer of Q^γ.

First, consider the case with $y^2 \leq y^1 \leq y^0$. For any fixed $d^2 \in [\underline{d}, \bar{d}]$, define

$$d^1 = \min\{(y^1 - y^2) + d^2, d^0\}.$$

If $d^1 = (y^1 - y^2) + d^2$, we have that

$$y^1 - d^1 - \beta = y^2 - d^2 - \beta, \ d^2 \leq d^1 \leq d^0$$

and

$$\begin{aligned} H^\gamma(y^1, d^1) = \hat{R}(d^1) - E[h_t^\gamma(y^1 - d^1 - \beta)] \\ \geq \hat{R}(d^2) - E[h_t^\gamma(y^2 - d^2 - \beta)] = H^\gamma(y^2, d^2), \end{aligned}$$

where the inequality follows from the quasiconcavity of $\hat{R}$.
   If $d^1 = d^0$, we have that $d^0 \leq (y^1 - y^2) + d^2$. Hence,

$$y^0 - d^0 - \beta \geq y^1 - d^1 - \beta \geq y^2 - d^2 - \beta,$$

which implies that

$$\begin{aligned} H^\gamma(y^1, d^1) = \hat{R}(d^1) - E[h_t^\gamma(y^1 - d^1 - \beta)] \\ \geq \hat{R}(d^2) - E[h_t^\gamma(y^2 - d^2 - \beta)] = H^\gamma(y^2, d^2), \end{aligned}$$

where the inequality follows from the definition of d^0 and the quasiconvexity of $E[h_t^\gamma(x - \beta)]$. Thus, Assumption 10.5.1 part (a) holds.
   For the case with $y^2 \leq y^1 \leq y^0$, define

$$d^1 = \max\{(y^1 - y^2) + d^2, d^0\}.$$

Assumption 10.5.1 part (b) follows from a similar argument.   ∎

Lemma 10.5.3 *Under Assumption 10.5.1, for any $y_t^2 \leq y_t^1 \leq y^0$,*

$$f_t(y_t^1) \geq f_t(y_t^2);$$

for $y^0 \leq y_t^1 \leq y_t^2$,

$$f_t(y_t^1) \geq f_t(y_t^2) - \gamma K.$$

Proof. At period t, we compare two systems with initial inventory levels y_t^1 and y_t^2, referred to as systems y^1 and y^2, respectively. Assume that system y^2 follows an optimal strategy that attains $f_t(y_t^2)$, the expected total discounted profit, for the remaining $T - t + 1$ periods if we act optimally in the remaining $T - t + 1$ periods except that no order is placed at period t. For a given sample path of the system, namely, a realization of system uncertainties, denote (x_l^2, y_l^2, d_l^2) as the inventory level before placing the order, the inventory level after receiving the order, and the expected demand level of period l of system y^2 following the optimal strategy, respectively. We will construct a feasible strategy for system y^1 and compare its expected total discounted profit with $f_t(y_t^2)$. Let (x_l^1, y_l^1, d_l^1) be the inventory level before placing the order, the inventory level after receiving the order, and the expected demand level of period l of system y^1 along this sample path under this strategy, respectively. Note that $x^i_{l+1} = y^i_l - \alpha_l d^i_l - \beta_l$ for a realization (α_l, β_l) and $i = 1, 2$.

We now describe the process of constructing a strategy for system y^1. For the case with $y^2_t \leq y^1_t \leq y^0$, according to Assumption 10.5.1 part (a), we can pick a $d^1_t \in [\underline{d}, \bar{d}]$ such that

$$H^\gamma(y^1_t, d^1_t) \geq H^\gamma(y^2_t, d^2_t) \text{ and } x^2_{t+1} \leq x^1_{t+1} \leq y^0. \tag{10.19}$$

For the case with $y^0 \leq y^1_t \leq y^2_t$, according to Assumption 10.5.1 part (b), we can select a $d^1_t \in [\underline{d}, \bar{d}]$ such that

$$H^\gamma(y^1_t, d^1_t) \geq H^\gamma(y^2_t, d^2_t) \text{ and } x^1_{t+1} \leq \max\{x^2_{t+1}, y^0\}.$$

In either case, we end up with two possibilities:

$$(1) \ x^1_{t+1} < y^2_{t+1} \text{ or } (2) \ y^2_{t+1} \leq x^1_{t+1} \leq y^0.$$

In the first case, for system y^1, place an order to raise the inventory level from x^1_{t+1} to y^2_{t+1} at period $t + 1$, use d^2_{t+1}, and follow the strategy of system y^2 thereafter. In the second case, for system y^1, order nothing at period $t+1$. Thus, $y^1_{t+1} = x^1_{t+1}$ and $y^2_{t+1} \leq y^1_{t+1} \leq y^0$. We then repeat the process at period $t+1$ and later periods if necessary until systems y^1 and y^2 end up with the same inventory level or we reach the end of the planning horizon.

From the description of the process, for $y^2_t \leq y^1_t \leq y^0$, it is clear that at any period l with $l \geq t$, the realized profit of system y^1 is always no less than that of system y^2. Note that when $x^1_{t+1} < y^2_{t+1}$, system y^2 must have placed an order at period $t+1$. Since this is true along any sample path, we have that $f_t(y^1_t) \geq f_t(y^2_t)$.

The case with $y^0 \leq y^1_t \leq y^2_t$ is similar except that when $x^1_{t+1} < y^2_{t+1}$, system y^1 needs to place an order at period $t + 1$, while system y^2 may not. Therefore, $f_t(y^1_t) \geq f_t(y^2_t) - \gamma K$. ∎

The above lemma allows us to show the optimality of $(s, S, \mathbf{p})$ policy under Assumption 10.5.1.

Theorem 10.5.4 *Consider the finite-horizon model described in Sect. 10.4. If the system is stationary and Assumption 10.5.1 holds, then an $(s, S, \mathbf{p})$ policy is optimal.*

Proof. Let S_t be a global maximizer of f_t with $S_t \geq y^0$. Its existence is guaranteed by our assumptions on H^γ and Lemma 10.5.3. Let $s_t = \min\{x | f_t(x) = f_t(S_t) - K\}$. Since f_t is continuous and $\lim_{|x| \to \infty} f_t(x) = -\infty$, s_t is well defined. In addition, $s_t \leq y^0$ since from Lemma 10.5.3, $f_t(y) \geq f_t(S_t) - \gamma K$ for any y with $S_t \geq y \geq y^0$.

We show that the (s_t, S_t) inventory policy is optimal. To see this, note that for any $y^1_t \leq s_t$, $f_t(y^1_t) \leq f_t(s_t) = f_t(S_t) - K$. Thus, it is optimal to place an order to raise the inventory level to S_t. On the other hand, for any inventory level y_t^1 above s_t, it is optimal not to place an order. In fact, if $y_t^1 \in [s_t, y^0]$,

$$f_t(y_t^1) \geq f_t(s_t) = f_t(S_t) - K,$$

and if $y_t^1 \geq y^0$, from Lemma 10.5.3,

$$f_t(y_t^1) \geq f_t(y_t^2) - \gamma K \geq f_t(y_t^2) - K \ \forall \ y^0 \leq y_t^1 \leq y_t^2,$$

which implies that it is better off not to place an order.   ∎

Interestingly, using a similar approach but under a less restrictive assumption, we can show that a stationary $(s, S, \mathbf{p})$ policy is optimal for a corresponding infinite-horizon model with stationary parameters and general demand functions under the discounted profit criterion.

Assumption 10.5.5   *(a) For any y^1 and y^2 with $y^2 \leq y^1 \leq y^0$ and $d^2 \in [\underline{d}, \bar{d}]$, there exists $d^1 \in [\underline{d}, \bar{d}]$ such that*

$$H^\gamma(y^1, d^1) \geq H^\gamma(y^2, d^2).$$

(b) The same as Assumption 10.5.1 part (b).

Again, the above assumption implies that Q^γ is quasiconcave.

Proposition 10.5.6 *If $H^\gamma(y, d)$ is quasiconcave in (y, d), then Assumption 10.5.5 holds.*

Proof. Let d^0 be the global maximizer of $H^\gamma(y^0, d)$ for $d \in [\underline{d}, \bar{d}]$. For any y^1 and y^2 with $y^2 \leq y^1 \leq y^0$ or $y^0 \leq y^1 \leq y^2$ and d^2, let $\lambda \in [0, 1]$ such that $y^1 = (1 - \lambda)y^0 + \lambda y^2$. Define

$$d^1 = (1 - \lambda)d^0 + \lambda d^2.$$

From the quasiconcavity of H^γ, we have that

$$H^\gamma(y^1, d^1) \geq \min\{H^\gamma(y^0, d^0), H^\gamma(y^2, d^2)\} = H^\gamma(y^2, d^2).$$

For $y^0 \leq y^1 \leq y^2$ and any realization of (α, β),

$$\begin{aligned}
y^1 - \alpha d^1 - \beta &= (1 - \lambda)(y^0 - \alpha d^0 - \beta) + \lambda(y^2 - \alpha d^2 - \beta) \\
&\leq \max\{y^0 - \alpha d^0 - \beta, y^2 - \alpha d^2 - \beta\} \\
&\leq \max\{y^0, y^2 - \alpha d^2 - \beta\}.
\end{aligned}$$

Thus, Assumption 10.5.5 holds.   ∎

When $\hat{R}$ is concave and h^γ is convex, H^γ is concave and thus quasiconcave. It would be interesting to identify other conditions under which H^γ is quasiconcave or Assumption 10.5.5 holds.

Lemma 10.5.7 *Under Assumption 10.5.5, for any $y_t^2 \leq y_t^1 \leq y^0$ or $y^0 \leq y_t^1 \leq y_t^2$,*

$$f_t(y_t^1) \geq f_t(y_t^2) - \gamma K.$$

Proof. The proof is similar to that for Lemma 10.5.3. The only exception is that for the case with $y_t^2 \leq y_t^1 \leq y^0$, we may no longer claim $x_{t+1}^2 \leq x_{t+1}^1$ in (10.19) under Assumption 10.5.5. In this case, we cannot exclude the possibility that system y^1 places an order at period $t+1$ to raise the inventory level from x_{t+1}^1 to y_{t+1}^2 while system y^2 does not order at period $t+1$. Thus, for $y_t^2 \leq y_t^1 \leq y^0$, $f_t(y_t^1) \geq f_t(y_t^2) - \gamma K$ instead of $f_t(y_t^1) \geq f_t(y_t^2)$.   ∎

Theorem 10.5.8 *Consider the infinite-horizon counterpart of the model described in Sect. 10.4 with stationary parameters and $\gamma \in (0, 1)$. If Assumption 10.5.5 holds, then a stationary $(s, S, \mathbf{p})$ policy is optimal.*

Proof. For the infinite-horizon counterpart of the model described in Sect. 10.4 with stationary parameters and $\gamma \in (0, 1)$, we can show that f_t is well defined and continuous and $\lim_{|x|\to\infty} f_t(x) = -\infty$. In addition, it is independent of t, and thus we drop the subscript of f_t in the proof.

Let S be a global maximizer of f and $s = \max\{x | f(x) = f(S) - K, x \leq \min\{S, y^0\}\}$. We show that it is optimal to follow the (s, S) policy. The definition of s implies that for any $x \in (s, \min\{S, y^0\}]$, it is optimal not to order. From Lemma 10.5.7, for any given x with $x \geq \min\{S, y^0\}$, if $S \leq x \leq y^0$,

$$f(x) \geq f(S) - \lambda K \geq f(S) - K;$$

and if $x \geq y^0$, then for any $y \geq x$,

$$f(x) \geq f(y) - \lambda K \geq f(y) - K.$$

Thus, for any $x \geq s$, it is optimal not to order.

It remains to prove that for any $x \leq s$, it is optimal to place an order to raise the inventory level to S. First, observe that at the inventory level s, the expected profit generated by the strategy of not ordering now but ordering up to S at the next period is given by $Q^\gamma(s) + \gamma(f(S) - K)$, which by definition is no more than $f(s) = f(S) - K$. Therefore,

$$Q^\gamma(s) \leq (1 - \gamma)(f(S) - K).$$

Assume that it is not optimal to order for some $x \leq s$. Start with the initial inventory level x at any period, without loss of generality, at the first period. Given an optimal strategy, let τ be the first time that an order is placed for a realization of the uncertainties. Clearly, τ is a stopping time and we have that

$$x = x_1 \geq x_2 \ldots \geq x_\tau,$$

where x_t denotes the inventory level at the beginning of period t. At period $t < \tau$, no order is placed and the profit is no more than $Q^\gamma(x_t)$. Thus, for the realization of the uncertainties, the total discounted profit from periods 1 to $\tau - 1$ is no more than

$$\sum_{t < \tau} \gamma^{t-1} Q^{\gamma}(x_t).$$

Since Q^{γ} is quasiconcave, $Q^{\gamma}(x_t) \le Q^{\gamma}(s)$ for $t < \tau$ and the above profit is no more than

$$\sum_{t < \tau} \gamma^{t-1} Q^{\gamma}(s) \le (f(S) - K)(1 - \gamma^{\tau-1}).$$

At period τ, it is optimal to raise the inventory level to S, since S is a global maximizer of f and $x \le S$. Conditioning on τ, the net present value of the total expected discounted profit from periods τ to $T = \infty$ is $(1 - \gamma)^{\tau-1}(f(S) - K)$. Therefore, starting with x, the total expected discounted profit over the infinite planning horizon is no more than $f(S) - K$. However, this profit can be obtained by placing an order at period 1 to raise the inventory level to S. Thus, for any $x \le s$, it is optimal to place an order to raise the inventory level to S. ∎

10.6   Extensions and Challenges

In Sect. 10.4, we show by employing the classic concept of K-convexity that an $(s, S, \mathbf{p})$ policy is optimal for the additive demand case. By using a weaker concept of symmetric K-convexity, we show that an $(s, S, A, \mathbf{p})$ policy is optimal for the general demand case. Theorem 10.5.8 in Sect. 10.5 shows that a stationary $(s, S, \mathbf{p})$ policy is optimal for the infinite-horizon counterpart with stationary parameters and general demand functions under the discounted profit criterion. Based on the concept of symmetric K-concavity, Chen and Simchi-Levi (2004b) provide a unified proof for the optimality of a stationary $(s, S, \mathbf{p})$ policy for the infinite-horizon model under either the discounted profit or the average profit criterion. Table 10.1 is a summary of structural results of the inventory (and pricing) models.

TABLE 10.1. Summary of results for the inventory (and pricing) problems

	Inventory model	Joint inventory and pricing model		
No fixed cost	Base stock policy	Base stock list price policy		
Fixed ordering cost	(s,S) Policy	Finite-horizon case		Infinite-horizon case
		Additive demand	General demand	
		$(s,S,\mathbf{p})$ Policy	$(s,S,A,\mathbf{p})$ Policy	$(s,S,\mathbf{p})$ Policy

Of course, it is appropriate to point out that many of our results in this chapter may not hold for problems with discrete prices (see Chen 2003). Indeed, if price is restricted to take values from a discrete set, even the single-period profit function may not be concave, and our analysis no longer works. This fact imposes a significant challenge for solving the integrated inventory and pricing models, since in order to solve these models, one usually discretizes inventory levels and discrete prices. Thus, a natural question is whether one can design efficient algorithms by employing the structural results of optimal policy identified in previous subsections.

Another challenge for the integrated inventory and pricing models analyzed in this section is the zero-lead-time assumption. This is not the case for the standard inventory control problems with backlogging. In fact, for the standard stochastic inventory models, the structural results of the optimal policies can be generally extended to models with deterministic lead time, as we pointed out at the beginning of Sect. 9.6. The idea is to transfer a model with a positive lead time to one with a similar structure while with zero lead time. However, this technique is not valid here, since for our models with a positive lead time, the two decisions—the ordering decision and the pricing decision—will take effect at different times. Yet, as demonstrated by Pang et al. (2012) , the concept of $L^{\natural}$-convexity can still be useful, and the results in Sect. 9.6 can be extended to the integrated inventory and pricing models with positive lead times and backlogging. Specifically, they show that the optimal ordering quantity decreases in the on-hand inventory and the inventory in transit and is more sensitive to newer outstanding orders than older outstanding orders and the on-hand inventory with bounded sensitivities. The optimal price decreases in the on-hand inventory and the inventory in transit as well. However, it is more sensitive to older outstanding orders and the on-hand inventory than newer outstanding orders. More recently, Chen et al. (2012a) extend the models and results to perishable products with finite lifetimes.

In this chapter, we restrict our effort to backlogging models. Lost sales models are much more complicated to deal with. In this case, even in the single-period model, the expected revenue $pE[\min(x, D(p, \epsilon))]$ as a function of p and x may not be well behaved. Many papers in the literature focus on the existence and uniqueness of the optimal solutions, the concavity or quasiconcavity of the expected profit functions, and comparative statics analysis. We refer to Chen and Simchi-Levi (2012) for references.

Finally, a few recent papers analyze integrated inventory and pricing models with price adjustment cost (Chen, Zhou and Chen 2011 and Chen and Hu 2012) and models in which demand depends on not only the current price but also previous prices (Chen et al. 2012c). Again, we refer to Chen and Simchi-Levi (2012) for a survey of these new developments.

10.7 Risk-Averse Inventory Models

All the inventory (and pricing) models discussed so far focus on risk-neutral decision makers, that is, inventory managers who are insensitive to profit variations. Evidently, not all inventory managers are risk-neutral; many planners are willing to trade off lower expected profit for downside protection against possible losses. Indeed, experimental evidence suggests that for some products, the so-called high-profit products, decision makers are risk-averse; see Schweitzer and Chachon (2000) for more details. Unfortunately, traditional inventory control models fall short of meeting the needs of risk-averse planners. For instance, traditional inventory models do not suggest mechanisms to reduce the chance of unfavorable profit levels. Thus, it is important to incorporate the notions of risk aversion in a broad class of inventory models.

The literature on risk-averse inventory models is quite limited and mainly focuses on single-period problems or is based on mean–variance tradeoffs. For instance, Lau (1980) analyzes the classical newsvendor model, in which he maximizes the decision maker's expected utility of total profit or the probability of achieving a certain level of profit. Eeckhoudt, Gollier and Schlesinger (1995) focus on the impact of risk and risk aversion in the newsvendor model when risk is measured by expected utility functions.

Chen and Federgruen (2000) analyze the mean–variance tradeoffs in newsvendor models as well as some standard infinite-horizon inventory models. Specifically, in the infinite-horizon models, Chen and Federgruen focus on the mean–variance tradeoff of customer waiting time as well as the mean–variance tradeoffs of inventory levels. Martínez-de-Albéniz and Simchi-Levi (2006) study the mean–variance tradeoffs faced by a manufacturer signing a portfolio of option contracts with its suppliers and having access to a spot market.

Assuming a linear ordering cost, Bourakiz and Sobel (1992) minimize the expected exponential utility of the present value of costs over a finite planning horizon or an infinite horizon. In particular, they show that a base stock policy is optimal.

So far, all the papers referenced above assume that demand is exogenous. A rare exception is Agrawal and Seshadri (2000) who consider a risk-averse retailer that has to decide on its ordering quantity and selling price for a single period. They demonstrate that different assumptions on the demand–price function may lead to different properties of the selling price.

In this section, we discuss a general framework for incorporating risk aversion in multiperiod inventory (and pricing) models, in which risk is measured based on increasing and concave utility functions. Our analysis is based on Chen, Sim, Simchi-Levi and Sun (2007).

The assumptions made in the risk-averse models are similar to those in the joint inventory and pricing models analyzed in Sect. 10.4. One exception is that demand is a *linear* function of the selling price; that is, $D_t(p)$ is a linear function of p. More importantly, the objective of the risk-averse decision maker is to maximize the expected utility of the total discounted profit over the planning horizon. That is, the objective is to maximize

$$E[u(V_T^\gamma(x))] \tag{10.20}$$

for any initial inventory level x and any given $0 < \gamma \le 1$, where $u(\cdot)$ is a utility function and $V_T^\gamma(x))$ is defined in (10.5).

We require the utility function, $u(x)$, to be *increasing* so that more is always preferred over less. Of course, if $u(x)$ is a linear and increasing function, the model (10.20) yields the same optimal solution as the risk-neutral model of (10.6). We also assume that the utility function is *concave* so that the marginal satisfaction of gaining a dollar is never more than the marginal loss of satisfaction associated with losing the same amount of money. It is appropriate to point out that expected utility theory is widely used in microeconomics and finance literature.

In the next subsection, we discuss the risk-averse framework based on a general increasing and concave utility function. This is followed by a subsection on models based on an important special case, the exponential utility.

10.7.1   Expected Utility Risk-Averse Models

Unlike the risk-neutral models analyzed in Sect. 10.4, the objective function (10.20) in its current form appears not to be decomposable and is not amenable to the dynamic programming approach. To deal with this issue, we introduce a new variable w to denote the wealth accumulated from the beginning of the planning horizon up to the current period. Thus, the state of the problem at period t can now be modeled as the inventory level x_t and the accumulated wealth from period T to period t, w_t.

Consider the expected utility measure. Let $W_t(x, w)$ be the maximum utility achievable starting at the beginning of period t with an initial inventory level x and an accumulated wealth w. The dynamic program can be written as follows. Let

$$W_{T+1}(x, w) = u(w),$$

and for $t = 1, 2, \ldots, T$,

$$W_t(x, w) = \max_{y \geq x, \bar{p}_t \geq p \geq \underline{p}_t} E[W_{t+1}(x_+, \bar{w}_+)], \qquad (10.21)$$

where

$$x_+ = y - D_t(p, \epsilon_t)$$

and

$$\bar{w}_+ = w + \gamma^{t-1}(-K\delta(y - x) - c_t(y - x) + pD_t(p, \epsilon_t) - h_t(y - D_t(p, \epsilon_t)). \quad (10.22)$$

We would like to emphasize that in this section, $D_t(p, \epsilon_t)$ is linear in p. Also notice that here we assume, without loss of generality, that $W_{T+1}(x, w)$ is independent of x, which implies zero salvage value. Finally, we have

$$\max E[u(V_T^\gamma(x))] = W_1(x, 0).$$

Instead of working with the dynamic program (10.21), we find that it is more convenient to work with an equivalent formulation. Let

$$U_t(x, w) = W_t(x, w - \gamma^{t-1}c_t x).$$

The dynamic program (10.21) becomes

$$U_t(x, w) = \max_{y \geq x, \bar{p}_t \geq p \geq \underline{p}_t} E[U_{t+1}(x_+, w_+)], \tag{10.23}$$

where

$$w_+ = w + \gamma^{t-1}(-K\delta(y - x) + f_t(y, p, \epsilon_t))$$

and

$$f_t(y, p, \epsilon_t) = -(c_t - \gamma c_{t+1})y + (p - \gamma c_{t+1})D_t(p, \epsilon_t) - h_t(y - D_t(p, \epsilon_t)). \tag{10.24}$$

We have the following observation, which can be easily verified by induction.

Lemma 10.7.1 *For any period t and fixed x, $U_t(x, w)$ is increasing in w.*

Interestingly, this observation allows us to show that a wealth-dependent base stock inventory policy is optimal when there is zero fixed ordering cost.

Theorem 10.7.2 *Assume that $K = 0$. In this case, $U_t(x, w)$ is jointly concave in x and w for any period t. Furthermore, a wealth-dependent base stock inventory policy is optimal for the risk-averse inventory (and pricing) problem (10.20).*

Proof. We prove by induction. Obviously, $U_{T+1}(x, w)$ is jointly concave in x and w. Assume that $U_{t+1}(x, w)$ is jointly concave in x and w. We now prove that a wealth-dependent base stock inventory policy is optimal and $U_t(x, w)$ is jointly concave in x and w.

First, notice that for any realization of ϵ_t, f_t is jointly concave in (y, p), which implies that w_+ is jointly concave in (w, x, y, p).

Since x_+ is a linear function of (y, p) and w_+ is jointly concave in (w, x, y, p), Lemma 10.7.1 allows us to show that $U_{t+1}(x_+, w_+)$ is jointly concave in (w, x, y, p). This implies that $E[U_{t+1}(x_+, w_+)]$ is jointly concave in (w, x, y, p).

We now prove that a w-dependent base stock inventory policy is optimal. Let $y^*(w)$ be an optimal solution for the problem

$$\max_y \left\{ \max_{\bar{p}_t \geq p \geq \underline{p}_t} E[U_{t+1}(x_+, w_+)] \right\}.$$

Since $E[U_{t+1}(x_+, w_+)]$ is concave in y for any fixed w, it is optimal to order up to $y^*(w)$ when $x < y^*(w)$ and not to order otherwise. In other words, a state-dependent base stock inventory policy is optimal.

Finally, according to Proposition 2.1.15, $U_t(x, w)$ is jointly concave. ∎

Recall that in the case of a risk-neutral decision maker, a base stock list price policy is optimal. Theorem 10.7.2 thus implies that in the case of an increasing concave utility risk measure, the optimal policy is quite different. Indeed, in these cases, the base stock level depends on the total profit accumulated from the beginning of the planning horizon, and it is not clear whether a list price policy is optimal.

Stronger results exist for models based on the exponential utility risk measure, as is demonstrated in the next subsection.

10.7.2   Exponential Utility Risk-Averse Models

We now focus on exponential utility functions of the form $u(w) = b(1-\exp(-w/b))$ with parameter $b > 0$. The beauty of exponential utility functions is that we can essentially separate x and w as is illustrated in the next theorem.

Theorem 10.7.3 *For any time period t, there exists a function $G_t(x)$ such that*

$$U_t(x, w) = u(w + \gamma^{t-1} G_t(x)).$$

Proof. We prove by induction. For $t = T+1$, $G_{T+1}(x) = 0$ for any x. Assume that there exists a function $G_{t+1}(x)$ such that

$$U_{t+1}(x, w) = u(w + \gamma^t G_{t+1}(x)).$$

From the recursion (10.21), we have that

$$
\begin{aligned}
U_t(x, w) &= \max_{y \geq x, \bar{p}_t \geq p \geq \underline{p}_t} bE[1 - \exp(-(w_+ + \gamma^t G_{t+1}(y - D_t(p, \epsilon_t))/b)] \\
&= b - b\exp(-w/b)\min_{y \geq x, \bar{p}_t \geq p \geq \underline{p}_t} \\
&\quad \exp(\gamma^{t-1}/b(K\delta(y - x) - L_t(y, p)/b)) \\
&= u(w + \gamma^{t-1} G_t(x)),
\end{aligned}
$$

where

$$L_t(y, p) = -b/\gamma^{t-1} \ln\left(E[\exp(-\gamma^{t-1}(f_t(y, p, \epsilon_t) + \gamma G_{t+1}(y - D_t(p, \epsilon_t))/b)]\right)$$

and

$$G_t(x) = \max_{y \geq x, \bar{p}_t \geq p \geq \underline{p}_t} -K\delta(y - x) + L_t(y, p). \tag{10.25}$$

Thus, the result is true. ∎

The theorem thus implies that the optimal policy is independent of the accumulated wealth when exponential utility functions are used, which significantly simplifies the problem. In fact, the optimal policy can be found by solving problem (10.25). Furthermore, this theorem, together with Theorem 10.7.2, implies that when there is zero fixed ordering cost, a base stock inventory policy is optimal under the exponential utility risk criterion independent of whether or not price is a decision variable.

Before we present our main result for the problem with $K > 0$, recall the famous Hölder inequality.

Theorem 10.7.4 *Assume $p, q > 0$ with $1/p + 1/q = 1$. If f and g are continuous functions on $\Re$ with $\int_{\Re} |f(x)|^p dx < \infty$ and $\int_{\Re} |g(x)|^q d(x) < \infty$, then*

$$\int_{\Re} |f(x)g(x)|dx \leq \left(\int_{\Re} |f(x)|^p dx\right)^{1/p} \left(\int_{\Re} |f(x)|^q dx\right)^{1/q}.$$

An important corollary of the Hölder inequality is as follows.

Theorem 10.7.5 *If a function f is convex, K-convex, or symmetric K-convex, then the function*

$$g(x) = \ln(E[\exp(f(x - \xi))])$$

is also convex, K-convex, or symmetric K-convex, respectively.

Proof. We only prove the case with K-convexity; the other two cases can be proven by following similar steps.

Define $M(x) = E[\exp(f(x - \xi))]$. It suffices to prove that for any x_0, x_1 with $x_0 \le x_1$ and any $\lambda \in [0, 1]$,

$$M(x_\lambda) \le M(x_0)^{1-\lambda} M(x_1)^\lambda \exp(\lambda K),$$

where $x_\lambda = (1 - \lambda)x_0 + \lambda x_1$. Notice that

$$
\begin{aligned}
M(x_\lambda) &\le E[\exp((1 - \lambda)f(x_0 - \xi) + \lambda f(x_1 - \xi) + \lambda K)] \\
&= \exp(\lambda K)E[\exp((1 - \lambda)f(x_0 - \xi)) \exp(\lambda f(x_1 - \xi))] \\
&\le \exp(\lambda K)E[\exp(f(x_0 - \xi))]^{1-\lambda}E[\exp(f(x_1 - \xi))]^\lambda \\
&= M(x_0)^{1-\lambda} M(x_1)^\lambda \exp(\lambda K),
\end{aligned}
$$

where the first inequality holds since f is K-convex and the second inequality follows from the Hölder inequality with $1/p = 1 - \lambda$ and $1/q = \lambda$. ∎

We can now present the optimal policy for the risk-averse multiperiod inventory (and pricing) problem with exponential utility function.

Theorem 10.7.6   (a) *If price is not a decision variable (i.e., $\underline{p}_t = \bar{p}_t$ for each t), $G_t(x)$ and $L_t(y, p)$ are K-concave and an (s, S) inventory policy is optimal.*

(b) *If price is a decision variable, $G_t(x)$ and $L_t(y, p)$ are symmetric K-concave and an $(s, S, A, \mathbf{p})$ policy is optimal.*

Proof. We only provide a sketch of the proof; the complete proof is left as an exercise. The main idea of the proof is as follows: If $G_{t+1}(x)$ is K-concave when price is not a decision variable (or symmetric K-concave when price is a decision variable), then, by Theorem 10.7.5, $L_t(y, p)$ is K-concave (or symmetric K-concave). The remaining parts follow directly from Lemma 9.3.2 and Proposition 9.3.3 for K-concavity (or Lemma 10.4.3 and Proposition 10.4.4 for symmetric K-concavity). ∎

We observe the similarities and differences between the optimal policy under the exponential utility measure and the one under the risk-neutral case. Indeed, when demand is exogenous, that is, price is not a decision variable, an (s, S) inventory policy is optimal for the risk-neutral case; see Theorem 9.3.4. Theorem 10.7.6 implies that this is also true under the exponential utility measure. Similarly, for the more general inventory and pricing problem, Theorem 10.4.8 implies that an

TABLE 10.2. Summary of results for finite-horizon risk-neutral and risk-averse models

	Price not a decision		Price is a decision	
	$K = 0$	$K > 0$	$K = 0$	$K > 0$
Risk-neutral model	*Base stock*	(s, S)	*Base stock List price*	$(s, S, A, \mathbf{p})$
Exponential utility	*Base stock*	(s, S)	*Base stock*	$(s, S, A, \mathbf{p})$
Increasing & concave utility	*Wealth-dependent Base stock*	?	*Wealth-dependent Base stock*	?

$(s, S, A, \mathbf{p})$ policy is optimal for the risk-neutral case. Interestingly, this policy is also optimal for the exponential utility case.

Of course, the results for the risk-neutral case are a bit stronger. Indeed, if demand is additive, Theorem 10.4.7 suggests that an $(s, S, \mathbf{p})$ policy is optimal. Unfortunately, it is not clear whether this result still holds for the risk-averse inventory and pricing problem under exponential risk measure.

The structural results of the optimal policies for the finite-horizon risk-averse models as well as risk-neutral models are summarized in Table 10.2. For infinite-horizon models with exponential utility and fixed ordering cost, Chen and Sun (2012) prove that like the infinite-horizon risk-neutral model, a stationary $(s, S, \mathbf{p})$ is optimal.

10.8 Exercises

Exercise 10.1. Prove Theorem 10.7.5 by Exercise 2.4.

Exercise 10.2. Complete the proof of Theorem 10.7.6.

Exercise 10.3. Recall the single-period model analyzed in Sect. 10.3. We modify the model as follows. Instead of placing an emergency order to satisfy shortages, we assume that unsatisfied demand is lost. In this case, $h(x)$ is the penalty cost for lost sales if $x < 0$. Show that the optimal selling for the additive demand case is no more than that for the deterministic demand case, which in turn is no more than that for the multiplicative demand case.

Exercise 10.4. Building on the concept of symmetric K-convexity, Ye and Duenyas (2007) introduce the concept of (K, Q)-convexity. A real-valued function f is called (K, Q)-convex for $K, Q \geq 0$ if, for any x_0, x_1 with $x_0 \leq x_1$ and $\lambda \in [0, 1]$,

$$f((1-\lambda)x_0 + \lambda x_1) \leq (1-\lambda)f(x_0) + \lambda f(x_1) + \lambda K + (1-\lambda)Q - \min\{\lambda, 1-\lambda\}\min\{K, Q\}.$$

It is easy to see that $(K, 0)$-convexity is exactly the K-convexity and the (K, K)-convexity is the symmetric K-convexity. Prove the following.

(a) A (K, Q)-convex function is also (K', Q')-convex for $K \leq K'$ and $Q \leq Q'$. A real-valued convex function is $(0, 0)$-convex and hence (K, Q)-convex for all $K, Q \geq 0$.

(b) If $g_1(y)$ and $g_2(y)$ are (K_1, Q_1)-convex and (K_2, Q_2)-convex, respectively, and $(K_1 - Q_1)(K_2 - Q_2) \geq 0$, then for $\alpha, \beta \geq 0$, $\alpha g_1(y) + \beta g_2(y)$ is $(\alpha K_1 + \beta K_2, \alpha Q_1 + \beta Q_2)$-convex.

(c) If $g(y)$ is (K, Q)-convex and w is a random variable, then $E\{g(y - w)\}$ is also (K, Q)-convex, provided $E\{|g(y - w)|\} < \infty$ for all y.

(d) Assume that g is a continuous (K, Q)-convex function with $K \geq Q$ and $g(y) \to \infty$ as $|y| \to \infty$. Define

$$S = \min\{ x \mid g(x) \leq g(y), \text{ for any } y\},$$

$$s = \min\{ x \mid g(x) = g(S) + K\},$$

$$s' = \sup\{x \mid x \leq S, g(x') \geq g(S) + (K - Q) \text{ for any } x' \leq x\},$$

and

$$u = \inf\{x \mid x \geq S, g(x') \geq g(S) + Q \text{ for all } x' \geq x\}.$$

Then $s \leq s' \leq S \leq u$, and we have the following results.

(i) $g(s) = g(S) + K$ and $g(y) \geq g(s)$ for all $y \leq s$.

(ii) $g(u) = g(S) + Q$ and $g(y) \geq g(u)$ for all $y \geq u$.

(iii) $g(y) \leq g(z) + Q$ for all y, z with $z \leq y \leq s'$.

(iv) $g(y) \leq g(z) + K$ for all y, z with $s' \leq y \leq z$.

(v) $g(y) \leq g(z) + K$ for all y, z with $(s + S)/2 \leq y \leq z$.

Exercise 10.5. (Chen and Simchi-Levi 2009) Given a (K, Q)-convex function f, prove that the function

$$g(x) = \min_{y \leq x} Q\delta(y - x) + f(y)$$

is also (K, Q)-convex, where $\delta(x) = 1$ for $x > 0$ and $\delta(x) = 0$ otherwise. Similarly,

$$h(x) = \min_{y \geq x} K\delta(y - x) + f(y)$$

is also (K, Q)-convex.

Exercise 10.6. (Chen and Simchi-Levi 2009) Assume that $f : \Re \to \Re$ is (K, Q)-convex. Prove that there exists a convex function $\underline{f}(x)$ such that

$$\underline{f}(x) \leq f(x) \leq \underline{f}(x) + \max\{K, Q\}, \text{ for any } x.$$

Exercise 10.7. (Chen, Zhang and Zhou 2010) A function f is quasi-K-concave with changeover a if it is increasing on $(-\infty, a]$ and non-K-increasing on $[a, \infty)$. Prove the following statement: If $r(\cdot)$ is a differentiable concave function and $w(\cdot)$ is a continuously differentiable quasi-K-concave function with some finite changeover ξ^0, then the function $f(\cdot)$ defined by $f(x) = \max_{d \in [\underline{d}, \bar{d}]} r(d) + w(x - d)$ is quasi-K-concave with a finite changeover no less than ξ^0. Is the differentiability assumption dispensable?

Part III

Competition, Coordination and Design Models

11

Supply Chain Competition and Collaboration Models

In this chapter, we analyze decentralized supply chain systems with independent retailers, each of which—facing uncertain demand—needs to decide its stock level and selling price in a single period. In Sect. 11.1, the retailers compete on prices for which noncooperative game theory is appropriate. In Sect. 11.2, the retailers do not compete and have incentives to form coalitions that place joint orders and share inventory due to risk-pooling effects and economies of scale. The model and analysis in Sects. 11.1 and 11.2 are based on Bernstein and Federgruen (2004) and Chen (2009), respectively. Our intention is to provide a snapshot of the applications of game theory to supply chain management. For surveys on this topic, see Cachon and Netessine (2004) and Nagarajan and Sošić (2008).

11.1 Inventory and Pricing Competition

Consider a system with n independent retailers. Let $N = \{1, 2, \ldots, n\}$ denote the set of retailers. Each retailer faces a single-period problem similar to the one in Sect. 10.3. Specifically, retailer i ($i \in N$), facing demand uncertainty, has to decide on its stock level y_i and a selling price $p_i \in [\underline{p}_i, \bar{p}_i]$ of a single product before the realization of the demand uncertainty. Demand is filled as much as possible from the on-hand inventory, and unsatisfied demand is filled with an emergency order. For retailer i, let c_i be the unit ordering cost, h_i^- the unit emergency ordering cost, and h_i^+ the unit inventory holding/disposal cost if h_i^+ is nonnegative or the unit salvage value if it is negative. Assume that

D. Simchi-Levi et al., *The Logic of Logistics: Theory, Algorithms, and Applications for Logistics Management*, Springer Series in Operations Research and Financial Engineering, DOI 10.1007/978-1-4614-9149-1_11, © Springer Science+Business Media New York 2014

$$h_i^- \geq c_i \geq \max\{0, -h_i^+\}.$$

As we pointed out in Sect. 10.3, this implies that the salvage value is no more than the normal unit ordering cost, which in turn is no more than the unit cost for the emergency order. To avoid trivial cases, we also assume that $\underline{p}_i \geq h_i^-$.

The demand of retailer i ($i \in N$), $D_i(p, \alpha_i)$, is a deterministic function of the prices of all retailers, $p = (p_1, \ldots, p_n)$, times a nonnegative random noise α_i with a continuous cdf $F_i(\cdot)$. That is,

$$D_i(p, \alpha_i) = d_i(p)\alpha_i.$$

Without loss of generality, assume $E[\alpha_i] = 1$. The expected demand $d_i(p) = E[D_i(p, \alpha_i)]$ ($i \in N$) is assumed to be differentiable in p. We also assume

$$\frac{\partial d_i(p)}{\partial p_i} \leq 0, \quad \frac{\partial d_i(p)}{\partial p_j} \geq 0 \ \forall \ j \neq i.$$

The first inequality indicates that a higher selling price of a retailer leads to a lower demand of itself, which is reasonable under most circumstances. The second inequality implies that a higher selling price of a retailer leads to higher demands of other retailers, which implies that the products offered by the retailers are substitutable. One additional assumption is imposed on $d_i(p)$.

Assumption 11.1.1 *For $i \in N$, $d_i(p)$ is log-supermodular; that is, $\log d_i(p)$ is supermodular.*

Several plausible demand functions satisfy these assumptions:

- the linear demand

$$d_i(p) = b_i - \sum_{j \in N} a_{ij} p_j \ (b_i > 0, a_{ii} > 0, \text{ and } a_{ij} \leq 0 \ \forall \ i, j \in N, i \neq j);$$

- the exponential demand

$$d_i(p) = e^{b_i - \sum_{j \in N} a_{ij} p_j} \ (a_{ii} > 0 \text{ and } a_{ij} \leq 0 \ \forall i, j \in N, i \neq j);$$

- the Cobb–Douglas demand

$$d_i(p) = b_i \Pi_{j \in N} p_j^{-a_{ij}} \ (b_i > 0, a_{ii} > 1, \text{ and } a_{ij} \leq 0 \ \forall \ i, j \in N, i \neq j);$$

- the constant elasticity of substitution demand

$$d_i(p) = \Omega \frac{p_i^{-r-1}}{\sum_{j \in N} p_j^{-r}} \ (\Omega > 0, r > 0);$$

- the Logit demand

$$d(p) = \Omega \frac{e^{-a_i p_i}}{1 + \sum_{j \in N} e^{-a_j p_j}} \quad (\Omega > 0, a_i > 0 \ \forall i \in N).$$

Given the vector of stock levels $y = (y_1, \ldots, y_n)$ and the price vector p, retailer i's strategy set is $\hat{S}_i = [0, \infty) \times [\underline{p}_i, \bar{p}_i]$ and its expected profit is given by

$$v_i(y, p) = p_i d_i(p) - c_i y_i - h_i^+ E[\max(y_i - d_i(p)\alpha_i, 0) - h_i^- E[\max(d_i(p)\alpha_i - y_i, 0)]$$

$$= p_i d_i(p) - \ell_i \left(\frac{y_i}{d_i(p)} \right) d_i(p),$$

where

$$\ell_i(y) = h_i^- - (h_i^- - c_i)y + (h_i^- + h_i^+)E[\max(y - \alpha_i, 0)].$$

Since a retailer's expected profit depends on other retailers' decisions, the system can be modeled as a noncooperative game $(N, \{\hat{S}_i\}_{i \in N}, \{v_i\}_{i \in N})$. Assume that all the costs, demand functions, and the structure of the game are common knowledge. The concept of Nash equilibrium is a natural predication of the outcome of the system.

Since a retailer's stocking decision has no impact on other retailers, its best stock level in response to the price vector p can be easily derived as the optimal ordering quantity of a newsvendor problem. Specifically, retailer i's optimal stock level is given by

$$y_i(p) = d_i(p)F_i^{-1}(\rho_i),$$

where $F_i^{-1}(\rho_i)$ is a solution of $\ell_i'(y) = 0$ and $\rho_i = \frac{h_i^- - c_i}{h_i^- + h_i^+}$. With this stock level, retailer i's reduced expected profit is now a function of the price vector only:

$$\pi_i(p) = d_i(p)(p_i - \ell_i \left(F_i^{-1}(\rho_i) \right)).$$

We end up with a reduced game $(N, \{S_i\}_{i \in N}, \{\pi_i\}_{i \in N})$, where $S_i = [\underline{p}_i, \bar{p}_i]$. Since $\pi_i(p)$ can be easily shown to be log-supermodular by Assumption 11.1.1, the game $(N, \{S_i\}_{i \in N}, \{\pi_i\}_{i \in N})$ belongs to the class of log-supermodular games and thus shares the properties of supermodular games in Theorem 3.1.4.

Theorem 11.1.2 *Under Assumption 11.1.1, the set of Nash equilibria of the reduced game $(N, \{S_i\}_{i \in N}, \{\pi_i\}_{i \in N})$ is nonempty and has a largest and a smallest element. In addition, the largest and smallest Nash equilibria are increasing in c_i, h_i^+, and h_i^- ($i \in N$).*

Proof. It remains to prove the second part. From Theorem 3.1.4, it suffices to show that $\pi_i(p)$ has increasing differences in (p_i, w_i) for any fixed p_{-i}, where w_i is c_i, h_i^+, or h_i^-. Given the formulation of $\pi(p)$, we only need to prove that $\frac{\partial \ell_i \left(F_i^{-1}(\rho_i) \right)}{\partial c_i}$, $\frac{\partial \ell_i \left(F_i^{-1}(\rho_i) \right)}{\partial h_i^+}$, and $\frac{\partial \ell_i \left(F_i^{-1}(\rho_i) \right)}{\partial h_i^-}$ are nonnegative.

Since $F_i^{-1}(\rho_i)$ is a solution of $\ell_i'(y) = 0$, we have that

$$\frac{\partial \ell_i\left(F_i^{-1}(\rho_i)\right)}{\partial h_i^-} = (1 - F_i^{-1}(\rho_i)) + E[\max(F_i^{-1}(\rho_i) - \alpha_i, 0)]$$

$$+\ell_i'(y)|_{y=F_i^{-1}(\rho_i))}\frac{\partial F_i^{-1}(\rho_i)}{\partial h_i^-}$$

$$= (1 - F_i^{-1}(\rho_i)) + E[\max(F_i^{-1}(\rho_i) - \alpha_i, 0)]$$

$$\geq 0.$$

Similarly,

$$\frac{\partial \ell_i\left(F_i^{-1}(\rho_i)\right)}{\partial c_i} = F_i^{-1}(\rho_i) \geq 0$$

and

$$\frac{\partial \ell_i\left(F_i^{-1}(\rho_i)\right)}{\partial h_i^+} = E[\max(F_i^{-1}(\rho_i) - \alpha_i, 0)] \geq 0.$$

Thus, the largest and smallest Nash equilibria are increasing in c_i, h_i^+, and h_i^-. ∎

Since the expected profit of a retailer is nondecreasing in the other retailers' prices, the largest Nash equilibrium is preferable by all retailers. Indeed, let p^* be a Nash equilibrium and $\bar{p}^*$ be the largest Nash equilibrium. We have that for $i \in N$,

$$\pi_i(\bar{p}^*) \geq \pi_i(p_i^*, \bar{p}_{-i}^*) \geq \pi_i(p^*),$$

where the first inequality follows from the definition of Nash equilibrium and the second one holds since $\pi_i(p)$ is nondecreasing in p_j $(j \neq i)$.

For any equilibrium, p^*, of the reduced game, it is clear that $(y(p^*), p^*)$ is an equilibrium of the game $(N, \{\hat{S}_i\}_{i\in N}, \{v_i\}_{i\in N})$, where $y(p) = (y_1(p), \ldots, y_n(p))$. It would be interesting to see how $y(p^*)$ changes with c_i, h_i^+, h_i^- when p^* is either the largest or the smallest equilibrium.

Finally, from (3.1) in Chap. 3, a sufficient condition for the uniqueness of a Nash equilibrium is the diagonally dominant condition:

$$-\frac{\partial^2 \log \pi_i(p)}{\partial^2 p_i} > \sum_{j \neq i} \frac{\partial^2 \log \pi_i(p)}{\partial p_i \partial p_j}, i \in N.$$

In the exercise, you are asked to provide conditions on the demand functions listed earlier under which the diagonally dominant condition holds.

11.2   Inventory Centralization Games

In recent years, many companies have started exploring innovative collaboration strategies in an effort to improve their supply chain efficiency and ultimately the bottom line. Firms are employing strategies such as forming long-term alliances and building collaborative logistics to reduce their supply chain costs. There are numerous examples of collaboration in supply chains. For instance, Good Neighbor Pharmacy is a retailers' cooperative network of 2,700 independently owned and operated pharmacies, and Affiliated Foods Midwest supplies more than 850 independent food-retailer members in 12 midwestern U.S. states with a full line of grocery products.

Indeed, to compete with big box retailers, it is common for independent grocery stores, hardware stores, and pharmacies to form retailers' cooperative groups, business entities that employ economies of scale on behalf of retailer-members to get discounts from manufacturers and to pool marketing. To join a retailers' cooperative, a store would typically pay a membership fee and purchase certain stock in the cooperative in return for its voting share. In addition, a store is usually required to purchase a minimum amount of inventory from the cooperative. The operating profits of the cooperative are returned to the member stores in cash or stock rebate (see Stankevich 1996). Over the years, retail cooperative groups have developed a variety of popular groupwide programs, such as insurance, pension plans, inventory management, pricing assistance, logistics, warehousing, store design and layout, site selection, and employee training (see Ghosh 1994).

These innovative strategies raise a variety of important and challenging questions on managing supply chains. For instance, for a group of companies in a supply chain, how should they cooperate, what possible outcomes can be achieved, and how do the players share the costs and benefits? Indeed, getting all players to agree on how to share costs and benefits was identified as one of the major barriers to collaborative commerce according to a European Chemical Transport Association white paper.

In this section, we consider a distribution system with multiple retailers that may place joint orders and keep inventory at a central warehouse. The retailers are interested in this type of cooperation for two reasons. First, retailers can take advantage of the risk-pooling effect by delaying the allocation of inventory. Second, exploiting economies of scale allows retailers to reduce their costs or increase their profits.

The cost-allocation problem among the retailers can be modeled as a cooperative game, referred to as an *inventory centralization game*. We will show that under certain conditions, an inventory centralization game has a nonempty core, which implies that no group of retailers will be better off by deviating from the cooperation.

11.2.1 Model

Assume that in the distribution system, there are m warehouses and n retailers. The retailers order from the outside suppliers through the warehouses and sell a single type of goods in a single period. Let $W = \{1, 2, \cdots, m\}$ and $N = \{1, 2, \cdots, n\}$ be the sets of suppliers and retailers, respectively. The retailers are assumed to be noncompeting and allowed to make their selling price decisions. Each retailer's demand depends on its own selling price and a common random variable—the market signal, ω. To satisfy their demand, the retailers, taking advantage of risk-pooling effects, may form coalitions to place joint orders through the warehouses before observing the market signal, while the inventories are allocated to the retailers after the market signal is revealed. Let $Z_j \subseteq W$ be the set of warehouses that can be used to supply retailer j if she does not cooperate with other retailers. If retailer j together with some other retailers decides to form a group S, referred to as a coalition, by placing joint orders and sharing inventory, her demand can be served by the inventory at any warehouse in $\cup_{j \in S} Z_j$.

The sequence of events is as follows. Before observing the realization of the market signal, each warehouse places an order by paying an ordering cost of $c_i(y_i)$ for an order quantity y_i at warehouse i. After the market signal ω is revealed, the retailers decide their selling prices $p_j(\omega)$, which depend on the market signal ω, and all goods at the warehouses are allocated to the retailers; say, $x_{ij}(\omega)$ units of goods are shipped from warehouse i to retailer j. The transportation cost of sending one unit of goods from warehouse i to retailer j is s_{ij}. For each retailer j, if the total amount of goods received from the warehouses is more than the realized demand, a per-unit holding cost of h_j^+ for excess inventory is incurred. On the other hand, we make the following assumption regarding unsatisfied demand.

Assumption 11.2.1 *Unsatisfied demand at retailer j is filled by an emergency order, which incurs a per-unit emergency ordering cost of h_j^-.*

The demand of each retailer is random and depends on the realization of the market signal ω and its own selling price. Specifically, we focus on demand functions of the following forms:

Assumption 11.2.2 *For $j \in N$, the demand function of retailer j given its price p_j satisfies*

$$\tilde{d}_j = D_j(p_j, \omega) := \beta_j(\omega) - \alpha_j(\omega) p_j, \tag{11.1}$$

where α_j and β_j are two nonnegative random variables, represented as functions of the market signal ω.

To avoid technical complications, we assume that the sample space Ω of ω is finite. However, this assumption can be relaxed if necessary.

We further assume that $\underline{p}_j$ and $\bar{p}_j$ are the lower and upper bounds of $p_j(\omega)$, respectively. Thus, the feasible set of retailer j's price decision $p_j(\cdot)$ is given by

$$P_j = \{p_j(\cdot) : \underline{p}_j \leq p_j(\omega) \leq \bar{p}_j, \forall \, \omega \in \Omega\}.$$

The inventory centralization problem for a coalition of retailers $S \subseteq N$ can be formulated as a two-stage stochastic programming model with recourse. In this model, y_i, $i = 1, 2, \cdots, m$, is the first-stage decision variable. After the market signal ω is revealed, a recourse decision should be made, which is the amount of goods sent from i to j, namely, $x_{ij}(\omega)$, for all $i \in \cup_{j \in S} Z_j$ and $j \in S$, and the selling price $p_j(\omega)$. Let $v_j(\omega)$ be the total amount of goods received by retailer j. For the coalition S, the objective is to maximize the expected total profit of all retailers in S.

Denote the maximum expected profit of the coalition S by $v(S)$, which can be written as the optimal value of the following two-stage stochastic programming problem with recourse:

$$V(S) = \quad Max \quad \sum_{j \in S} \{ E[R_j(p_j(\omega), \omega) - f_j(v_j(\omega) - \beta_j(\omega) + \alpha_j(\omega)p_j(\omega))] \}$$

$$- \sum_{i \in \cup_{j \in S} Z_j} (c_i(y_i) + \sum_{j \in S} s_{ij} E[x_{ij}(\omega)])$$

$$\text{s.t.} \quad y_i - \sum_{j \in S} x_{ij}(\omega) = 0, \quad i \in \cup_{j \in S} Z_j, \omega \in \Omega,$$

$$v_j(\omega) - \sum_{i \in \cup_{j \in S} Z_j} x_{ij}(\omega) = 0, \quad j \in S, \omega \in \Omega,$$

$$x_{ij}(\omega) \geq 0, \quad j \in S, i \in \cup_{j \in S} Z_j, \omega \in \Omega,$$

$$p_j(\cdot) \in P_j, \quad j \in S,$$

$$(11.2)$$

where the maximization is taken over $(y_i, x_{ij}(\cdot), p_j(\cdot), v_j(\cdot))$, R_j is the realized revenue function for a given selling price r and a realization of the marker signal ω:

$$R_j(r, \omega) = r(\beta_j(\omega) - \alpha_j(\omega)r),$$

and f_j represents the inventory holding cost or emergency ordering cost

$$f_j(\chi) = h_j^+ \chi^+ + h_j^- (-\chi)^+.$$

In the above model, the term in the first summation in the objective function is the expected revenue minus the expected inventory holding cost and emergency ordering cost. The term in the second summation in the objective function is the regular ordering cost and the transportation cost. The first constraint implies that no warehouse holds inventory. The second constraint specifies that the total amount of goods received by a retailer equals the total amount sent to the retailer from the warehouses.

Now the pair (N, V) with V given by (11.2) for each coalition $S \subseteq N$ defines a cooperative inventory centralization game.

11.2.2 Inventory Games with a Linear Ordering Cost

In this subsection, we assume that the ordering cost $c_i(y_i)$ is linear; by slightly abusing the notation, we also use c_i to denote the unit ordering cost. Since the realized revenue $R_j(p_j(\omega), \omega)$ is concave in $p_j(\omega)$ by Assumption 11.2.2 and $f_j(\chi)$ is convex, problem (11.2) is a concave maximization problem with linear constraints, which allows us to apply the elegant duality theory for convex minimization problems with linear constraints. For this purpose, define the Lagrangian function

$$L_S(y, p, v, x, \lambda, \mu, \pi) = \sum_{j \in S} \left(E[R_j(p_j(\omega), \omega) - f_j(v_j(\omega) - \beta_j(\omega) + \alpha_j(\omega)p_j(\omega))] \right)$$

$$- \sum_{i \in \cup_{j \in S} Z_j} \left(c_i y_i + \sum_{j \in S} s_{ij} E[x_{ij}(\omega)] \right)$$

$$+ \sum_{i \in \cup_{j \in S} Z_j} E[\lambda_i(\omega)(y_i - \sum_{j \in S} x_{ij}(\omega))]$$

$$+ \sum_{j \in S} E[\mu_j(\omega)(\sum_{i \in \cup_{j \in S}} x_{ij}(\omega) - v_j(\omega)))]$$

$$+ \sum_{j \in S, i \in \cup_{j \in S} Z_j} E[\pi_{ij}(\omega) x_{ij}(\omega)]$$

$$= \sum_{j \in S} \psi_j(p_j, v_j, \mu_j) + \sum_{i \in \cup_{j \in S} Z_j} y_i(E[\lambda_i(\omega)] - c_i)$$

$$+ \sum_{j \in S, i \in \cup_{j \in S} Z_j} E[x_{ij}(\omega)(\pi_{ij}(\omega) - s_{ij} - \lambda_i(\omega) + \mu_j(\omega))],$$

where $y = (y_i)_{i \in \cup_{j \in S} Z_j}, d = (d_j)_{j \in S}, v = (v_j)_{j \in S}, \lambda = (\lambda_i)_{i \in \cup_{j \in S} Z_j}, \mu = (\mu_j)_{j \in S}, \pi = (\pi_{ij})_{j \in S, i \in \cup_{j \in S} Z_j}$, and

$$\psi_j(p_j, v_j, \mu_j) = E[R_j(p_j(\omega), \omega) - f_j(v_j(\omega) - \beta_j(\omega) + \alpha_j(\omega)p_j(\omega)) - \mu_j(\omega)v_j(\omega)]. \tag{11.3}$$

Consider the dual function $\gamma_S(\lambda, \mu, \pi)$ defined by

$$\gamma_S(\lambda, \mu, \pi) = \begin{array}{ll} Sup & L_S(y, p, v, x, \lambda, \mu, \pi) \\ \text{s.t.} & p_j(\cdot) \in P_j, \quad j \in S. \end{array}$$

The duality theorem for convex minimization problems with linear constraints implies that $V(S)$ is equal to the optimal objective value of the dual problem (see, for instance, page 299 of Bertsekas 1995):

$$V(S) = \begin{array}{ll} Min & \gamma_S(\lambda, \mu, \pi) \\ \text{s.t.} & \pi_{ij}(\omega) \geq 0, \quad j \in S, i \in \cup_{j \in S} Z_j, \omega \in \Omega. \end{array} \tag{11.4}$$

Let $(\lambda^*, \mu^*, \pi^*)$ be optimal for the dual problem (11.4) with $S = N$. Then, again, the duality theorem implies that

$$V(N) = \begin{array}{ll} Max & L_N(y, p, v, x, \lambda^*, \mu^*, \pi^*) \\ \text{s.t.} & p_j(\cdot) \in P_j, \quad j \in N. \end{array} \tag{11.5}$$

Define for $j \in N$,

$$l_j = \begin{array}{ll} Max & \psi_j(p_j, v_j, \mu_j^*) \\ \text{s.t.} & p_j(\cdot) \in P_j. \end{array}$$

We claim that $(l_1, l_2, \ldots, l_n)$ is in the core of the cooperative game (N, V).

Theorem 11.2.3 *The vector $l = (l_1, l_2, \ldots, l_n)$ is in the core of the cooperative game (N, V).*

Proof. Notice that in the optimization problem (11.5), no constraint is imposed on the decision variables y_i and $x_{ij}(\omega)$. Thus, we must have

$$E[\lambda_i^*(\omega)] - c_i = 0, \quad i \in \cup_{j \in N} Z_j, \omega \in \Omega, \tag{11.6}$$

and

$$\pi_{ij}^*(\omega) - s_{ij} - \lambda_i^*(\omega) + \mu_j^*(\omega) = 0, \quad j \in N, i \in \cup_{j \in N} Z_j, \omega \in \Omega. \tag{11.7}$$

Therefore,

$$L_S(y, p, v, x, \lambda^*, \mu^*, \pi^*) = \sum_{j \in S} \psi_j(p_j, v_j, \mu^*).$$

This, together with (11.5), implies that

$$\sum_{j \in N} l_j = V(N).$$

In addition, since $(\lambda^*, \mu^*, \pi^*)$ is feasible for problem (11.4),

$$\sum_{j \in S} l_j = \gamma_S(\lambda^*, \mu^*, \pi^*) \geq V(S).$$

Thus, $l = (l_1, l_2, \ldots, l_n)$ is in the core of the cooperative game (N, V). ∎

From the above proof, we know that the optimal dual variables $(\lambda^*, \mu^*, \pi^*)$ must satisfy constraints (11.6) and (11.7). We now provide some intuition of the dual variables and the constraints. In the dual, we attempt to allocate the ordering cost and the transportation cost to each unit of goods received by the retailers. Specifically, let the dual variable $\mu_j^*(\omega)$ be the charge for each unit of goods received by retailer j to compensate for its ordering cost and transportation cost and $\lambda_i^*(\omega)$ be a charge for each unit of goods sent out by warehouse i to compensate its ordering cost if the market signal turns out to be ω. The constraint (11.6) implies that the average unit charge by warehouse i should be enough to cover its ordering cost c_i. Since $\pi_{ij}^*(\omega) \geq 0$, the dual constraint (11.7) implies that this unit charge $\mu_j^*(\omega)$ at retailer j should be no more than the unit price, $\lambda_i^*(\omega)$, charged by warehouse i plus the transportation cost s_{ij}. On the other hand, if there is a shipment from warehouse i to retailer j, then $\pi_{ij}^*(\omega) = 0$ by the complementarity slackness condition, and the dual constraint (11.7) implies that this unit charge $\mu_j^*(\omega)$ is enough to compensate for the unit price, $\lambda_i^*(\omega)$, charged by warehouse i plus the transportation cost s_{ij}.

11.2.3 Inventory Games with Quantity Discounts

In this subsection, we assume that the supplier provides quantity discounts to encourage large orders, or a third-party carrier provides volume discounts to encourage larger shipments. Specifically, we make the following assumption.

Assumption 11.2.4 *We assume that $c_i(y)/y$ is nonincreasing. That is, the larger the ordering quantity, the lower the average unit ordering cost.*

For technical reasons, we assume that $c_i(y)$ is lower semicontinuous. That is, $\underline{\lim}_{y \to x} c_i(y) \geq c_i(x)$ for any x. Further, we assume $c_i(y) \to \infty$ as $y \to \infty$. Under these assumptions, problem (11.2) has an optimal solution for any $S \subseteq N$.

Our assumption on the ordering cost is quite general. Indeed, we don't require $c_i(x)$ to be continuous, monotone, convex, or concave. Moreover, it includes several commonly used discounts: incremental discounts and all-units discounts. The concave ordering cost analyzed in Chen and Zhang (2009) and the less-than-truckload (LTL) volume discount function (see Muriel and Simchi-Levi 2003) are also important special cases.

Given this general ordering cost structure, unfortunately, the corresponding cooperative game may have an empty core. Indeed, in a special case of the inventory centralization games in which price is not a decision variable, Chen and Zhang (2009) show that for a distribution system with multiple warehouses, the core of the corresponding cooperative game may be empty even if the ordering costs involve only fixed costs and demand is deterministic.

Thus, in this subsection, we focus on inventory centralization games with a single warehouse (N, V). Since we analyze inventory games with a single warehouse, in the following analysis, we drop the index associated with the warehouses. In this case, the value of a coalition S can be defined as

$$V(S) = \begin{array}{cc} Max & -c(y) + g(y, S) \\ \text{s.t.} & y \geq 0, \end{array} \tag{11.8}$$

where

$$g(y, S) = E[g_S(y, \omega)], \tag{11.9}$$

with

$$\begin{array}{cc} g_S(y, \omega) = & Max \quad \sum_{j \in S} g_j(p_j, x_j, \omega) \\ & \text{s.t.} \quad y - \sum_{j \in S} x_j = 0, \\ & \qquad x_j \geq 0, \quad j \in S, \\ & \qquad \underline{p}_j \leq p_j \leq \bar{p}_j, \quad j \in S, \end{array}$$

and

$$g_j(p_j, x_j, \omega) = R_j(p_j, \omega) - f_j(x_j - \beta_j(\omega) + \alpha_j(\omega)p_j) - s_j x_j.$$

It is clear that given the general quantity discount function $c(y)$, the objective function of the above optimization problem is neither convex nor concave. Thus, analyzing it directly appears to be quite challenging. To get around this challenge, we construct another inventory centralization game $(N, \tilde{V})$ with a linear ordering cost, which is known to have a nonempty core, such that $\tilde{V}(S) \geq V(S)$ for any $S \subset N$ and $\tilde{V}(N) = V(N)$. If this could be done, then we could prove that any element in the core of the game $(N, \tilde{V})$ is in the core of the game (N, V). We show that this is true by first proving that for Problem (11.8), the bigger a coalition is, the larger the optimal ordering quantity should be. For this purpose, we define for any given scalar $\hat{c} \geq 0$ an inventory centralization game $(N, V_{\hat{c}})$ with the ordering cost being $\hat{c}x$. In this game, for any $S \subseteq N$,

$$V_{\hat{c}}(S) = \begin{array}{ll} Max & -\hat{c}y + g(y, S) \\ \text{s.t.} & y \geq 0. \end{array}$$

For any nonempty set $S \subseteq N$, let $y^*(S)$ be the smallest optimal solution of problem (11.8) that is guaranteed to exist when $c(y)$ is lower-semicontinuous and $\lim_{y \to +\infty} c(y) = +\infty$.

Lemma 11.2.5 *For any given $S \subseteq N$, let $y^*(S)$ be the smallest optimal ordering quantity for the postponed pricing model (11.8)–(11.9). We have $y^*(S_1) \leq y^*(S_2)$ for $S_1 \subseteq S_2$.*

Proof. We prove this result by contradiction. Assume that there exist $S_1, S_2 \subseteq N$ with $S_1 \subset S_2$ such that $y^*(S_1) > y^*(S_2)$. Let $(x_j^1(\omega), p_j^1(\omega))_{j \in S_1}$ be the optimal inventory allocation and pricing associated with the optimal ordering quantity $y^*(S_1)$ for problem (11.8)–(11.9) with $S = S_1$. Similarly, let $(x_j^2(\omega), p_j^2(\omega))_{j \in S_2}$ be the optimal inventory allocation and pricing associated with the optimal ordering quantity $y^*(S_2)$ for problem (11.8)–(11.9) with $S = S_2$.

The definition of $y^*(S_1)$ and $y^*(S_2)$ implies that

$$- c(y^*(S_1)) + g_{S_1}(p^1, x^1) > -c(y^*(S_2)) + g_{S_1}(p^3, x^3) \tag{11.10}$$

for any $p_j^3(\cdot) \in P_j$ and $x_j^3(\cdot)$ with

$$y^*(S_2) = \sum_{j \in S_1} x_j^3(\omega), \ \forall \, \omega \in \Omega, \tag{11.11}$$

where for any $S \subseteq N$, $(x_j(\omega), p_j(\omega))_{j \in S}$,

$$\hat{g}_S(p, x) = \sum_{j \in S} E[g_j(p_j(\omega), x_j(\omega), \omega)].$$

Similarly,

$$- c(y^*(S_2)) + g_{S_2}(p^2, x^2) \geq -c(y^*(S_1)) + g_{S_2}(p^4, x^4) \tag{11.12}$$

for any $p_j^4(\cdot) \in P_j^{(p)}$ and $x_j^4(\cdot)$ with

$$y^*(S_1) = \sum_{j \in S_2} x_j^4(\omega), \ \forall \, \omega \in \Omega.$$

Specifically, let

$$p_j^4(\omega) = p_j^2(\omega), x_j^4(\omega) = x_j^2(\omega), \forall\, j \in S_2 \setminus S_1, \omega \in \Omega.$$

This, together with inequality (11.12), implies that

$$- c(y^*(S_2)) + g_{S_1}(p^2, x^2) > -c(y^*(S_1)) + g_{S_1}(p^4, x^4) \tag{11.13}$$

for any $p_j^4(\cdot) \in P_j$ and $x_j^4(\cdot)$ ($j \in S_1$) with

$$y^*(S_1) - y^*(S_2) = \sum_{j \in S_1} (x_j^4(\omega) - x_j^2(\omega)), \ \forall\, \omega \in \Omega. \tag{11.14}$$

Adding the two inequalities (11.10) and (11.13) together gives us that

$$g_{S_1}(p^1, x^1) + g_{S_1}(p^2, x^2) > g_{S_1}(p^3, x^3) + g_{S_1}(p^4, x^4) \tag{11.15}$$

for any $p_j^3(\cdot), p_j^4(\cdot) \in P_j$ ($j \in S_1$) $(x_j^3(\cdot))_{j \in S_1}$ satisfying (11.11) and $(x_j^4(\cdot))_{j \in S_1}$ satisfying (11.14).

Define

$$\lambda(\omega) = \frac{y^*(S_1) - y^*(S_2)}{y^*(S_1) - \sum_{j \in S_1} x_j^2(\omega)}.$$

Since

$$\sum_{j \in S_1} x_j^2(\omega) \le \sum_{j \in S_2} x_j^2(\omega) = y^*(S_2) < y^*(S_1),$$

we have that $\lambda(\omega) \in [0, 1]$. For $j \in S_1$, let

$$x_j^3(\omega) = (1 - \lambda(\omega))x_j^1(\omega) + \lambda(\omega)x_j^2(\omega),$$

$$p_j^3(\omega) = (1 - \lambda(\omega))p_j^1(\omega) + \lambda(\omega)p_j^2(\omega),$$

and

$$x_j^4(\omega) = \lambda(\omega)x_j^1(\omega) + (1 - \lambda(\omega))x_j^2(\omega),$$

$$p_j^4(\omega) = \lambda(\omega)p_j^1(\omega) + (1 - \lambda(\omega))p_j^2(\omega).$$

It is clear that $(x_j^3(\cdot))_{j \in S_1}$ satisfies (11.11) and $(x_j^4(\cdot))_{j \in S_1}$ satisfies (11.14). In addition,

$$x_j^3(\omega) + x_j^4(\omega) = x_j^1(\omega) + x_j^2(\omega)$$

and

$$p_j^3(\omega) + p_j^4(\omega) = p_j^1(\omega) + p_j^2(\omega).$$

Thus, the concavity of the realized revenue function R_j implies that

$$R_j(p_j^3(\omega), \omega) + R_j(p_j^4(\omega), \omega) \ge R_j(p_j^1(\omega), \omega) + R_j(p_j^2(\omega), \omega),$$

and the convexity of f_j implies that

$$-f_j(x_j^3(\omega) - \beta_j(\omega) + \alpha_j(\omega)p_j^3(\omega)) - f_j(x_j^4(\omega) - \beta_j(\omega) + \alpha_j(\omega)p_j^4(\omega))$$
$$\geq -f_j(x_j^1(\omega) - \beta_j(\omega) + \alpha_j(\omega)p_j^1(\omega)) - f_j(x_j^2(\omega) - \beta_j(\omega) + \alpha_j(\omega)p_j^2(\omega)).$$

Adding the two inequalities together and taking expectation with respect to ω give us an inequality that contradicts the inequality (11.15). Thus, $y^*(S_1) \leq y^*(S_2)$. $\blacksquare$

It is also important to point out that Lemma 11.2.5 is independent of how the retailers' demands are correlated. In addition, this result is true for any ordering cost as long as the relevant quantities are well defined.

Lemma 11.2.6 *There exists a scalar $\hat{c}^*$ such that for any $S \subseteq N$, $V_{\hat{c}^*}(S) \geq V(S)$ and $V_{\hat{c}^*}(N) = V(N)$.*

Proof. We consider two cases. First, assume that $y^*(N) = 0$. Lemma 11.2.5 implies that $y^*(S) = 0$ for any nonempty set $S \subseteq N$. If we choose a sufficiently large $\hat{c}^*$, say $\hat{c}^* \geq Max_{j \in N} \max\{\bar{p}_j, q_j\}$, it is easy to see that 0 is also an optimal solution for problem $\max_{y \geq 0} -\hat{c}^* y + g(y, S)$. Thus, in this case, $V_{\hat{c}^*}(S) = V(S)$ for any $S \subseteq N$.

We now assume that $y^*(N) > 0$. For simplicity of notation, let $y^* = y^*(N)$. Upon denoting $c^* = c(y^*)/y^*$, we have that

$$\begin{aligned} V(N) &= -c(y^*) + g(y^*, N) \\ &= -c^* y^* + g(y^*, N) \\ &\leq \max_{y \geq 0} -c^* y + g(y, N) \\ &= V_{c^*}(N). \end{aligned}$$

Since $y^* > 0$, we have that

$$\begin{aligned} V(N) &= \max_{y \geq 0} -c(y) + g(y, N) \\ &> g(0, N) \\ &= \lim_{\hat{c} \to \infty} V_{\hat{c}}(N). \end{aligned}$$

The continuity of $V_{\hat{c}}(N)$ as a function of $\hat{c}$, together with the above two inequalities, implies that there exists a $\hat{c}^*$ such that $V(N) = V_{\hat{c}^*}(N)$.

Define $\hat{x} = \sup\{x \geq 0 : c(x)/x \geq \hat{c}^*\}$. Let $\hat{y}^*$ be the smallest optimal solution for the problem $\min_{y \geq 0} -\hat{c}^* y + g(y, N)$. We claim that $\hat{y}^* \leq \hat{x}$.

Assume to the contrary that $\hat{y}^* > \hat{x}$. The definition of $\hat{x}$ together with the monotonicity of $c(x)/x$ implies that $c(\hat{y}^*)/\hat{y}^* < \hat{c}^*$. Thus,

$$\begin{aligned} V_{\hat{c}^*}(N) &= -\hat{c}^* \hat{y}^* + g(\hat{y}^*, N) \\ &< -c(\hat{y}^*) + g(\hat{y}^*, N) \\ &\leq -c(y^*) + g(y^*, N) \\ &= V(N), \end{aligned}$$

which contradicts the fact that $V(N) = V_{\hat{c}^*}(N)$. Thus, $\hat{y}^* \leq \hat{x}$.

Define a new function $\tilde{c}(x)$ as follows:

$$\tilde{c}(x) = \begin{cases} \hat{c}^* x, & \text{for } 0 \le x < \hat{x}, \\ c(x), & \text{otherwise} . \end{cases}$$

The following properties of $\tilde{c}(x)$ will be useful for our analysis. First, $\tilde{c}(x) \le c(x)$ for any x. This follows directly from the monotonicity of $c(x)/x$. Second, $\tilde{c}(x)$ preserves the lower semicontinuity of $c(x)$. To show this, it suffices to prove that $\tilde{c}(x)$ is lower-semicontinuous at $x = \hat{x}$. Notice that

$$\lim_{y \to \hat{x}+} \tilde{c}(y) = \lim_{y \to \hat{x}+} c(y) \ge c(\hat{x}) = \tilde{c}(\hat{x}),$$

while

$$\lim_{y \to \hat{x}-} \tilde{c}(y) = \hat{c}^* \hat{x} \ge \lim_{y \to \hat{x}+} \hat{x} c(y)/y \ge c(\hat{x}) = \tilde{c}(\hat{x}), \tag{11.16}$$

where the first inequality and the second inequality follow from the definition of $\hat{x}$ and the lower semicontinuity of $c(x)$, respectively. Notice that (11.16) implies that

$$\tilde{c}(y) \le \hat{c}^* y, \quad \text{for any } 0 \le y \le \hat{x}. \tag{11.17}$$

Third, $\hat{y}^*$ is also optimal for the problem $\max_{y \ge 0} -\tilde{c}(y) + g(y, N)$. Indeed, the definition of $\hat{y}^*$ together with the fact $\hat{y}^* \le \hat{x}$ implies that for any $0 \le y < \hat{x}$,

$$-\tilde{c}(\hat{y}^*) + g(\hat{y}^*, N) \ge -\hat{c}^* \hat{y}^* + g(\hat{y}^*, N) \ge -\hat{c}^* y + g(y, N) = -\tilde{c}(y) + g(y, N),$$

where the first inequality follows from (11.17). For $y \ge \hat{x}$,

$$\begin{aligned} -\tilde{c}(\hat{y}^*) + g(\hat{y}^*, N) &\ge -\hat{c}^* \hat{y}^* + g(\hat{y}^*, N) \\ &= -c(y^*) + g(y^*, N) \\ &\ge -c(y) + g(y, N) \\ &= -\tilde{c}(y) + g(y, N), \end{aligned}$$

where the first equality follows from the definition of $\hat{y}^*$ and $\hat{c}^*$. Thus, $\hat{y}^*$ is also optimal for the problem $\max_{y \ge 0} -\tilde{c}(y) + g(y, N)$.

We are now ready to prove that for any $S \subset N$, $V_{\hat{c}^*}(S) \ge V(S)$. Let $\tilde{y}^*(S)$ be the smallest optimal solution for the problem $\max_{y \ge 0} -\tilde{c}(y) + g(y, S)$. Notice that $\tilde{c}(x)$ is lower-semicontinuous. Hence, $\tilde{y}^*(S)$ is well defined. Lemma 11.2.5 implies that for any $S \subset N$, $\tilde{y}^*(S) \le \hat{y}^* \le \hat{x}$.

We claim that

$$\hat{c}^* \tilde{y}^*(S) = \tilde{c}(\tilde{y}^*(S)). \tag{11.18}$$

Indeed, if $\tilde{y}^*(S) < \hat{x}$, we have from the definition of $\tilde{c}(\cdot)$ that $\hat{c}^* \tilde{y}^*(S) = \tilde{c}(\tilde{y}^*(S))$. On the other hand, if $\tilde{y}^*(S) = \hat{x}$ and $\hat{c}^* \tilde{y}^*(S) > \tilde{c}(\tilde{y}^*(S)) = c(\tilde{y}^*(S))$, we have that $\hat{y}^* = \tilde{y}^*(S)$ and

$$V_{c^*}(N) = -\hat{c}^* \hat{y}^* + g(\hat{y}^*, N) < -c(\hat{y}^*) + g(\hat{y}^*, N) \le V(N),$$

which is a contradiction. Thus, in this case, (11.18) follows from (11.17).

Finally, we have that

$$
\begin{aligned}
V_{\hat{c}^*}(S) &= \max_{y \geq 0} -\hat{c}^* y + g(y, S) \\
&\geq -\hat{c}^* \tilde{y}^*(S) + g(\tilde{y}^*(S), S) \\
&= -\tilde{c}(\tilde{y}^*(S)) + g(\tilde{y}^*(S), S) \\
&= \max_{y \geq 0} -\tilde{c}(y) + g(y, S) \\
&\geq \max_{y \geq 0} -c(y) + g(y, S) \\
&= V(S),
\end{aligned}
$$

where the second equality follows from (11.18) and the last inequality from the fact that $\tilde{c}(y) \leq c(y)$ for any $y \geq 0$. The proof is now complete. ∎

We can now prove that the core of (N, V) is nonempty.

Theorem 11.2.7 *Under Assumption 11.2.4, the inventory centralization game (N, V) with the characteristic value function defined by (11.8)–(11.9) has a non-empty core. Let $(N, V_{\hat{c}^*})$ be the inventory centralization game with marginal ordering cost $\hat{c}^*$, where $\hat{c}^*$ is defined in Lemma 2. Then any element in the core of $(N, V_{\hat{c}^*})$ is also in the core of (N, V).*

Proof. The proof is straightforward. Let $l = (l_j)_{j \in N}$ be an element in the core of $(N, V_{\hat{c}^*})$. We have that for any $S \subseteq N$,

$$
\sum_{j \in S} l_j \geq V_{\hat{c}^*}(S) \geq V(S).
$$

In addition,

$$
\sum_{j \in N} l_j = V_{\hat{c}^*}(N) = V(N).
$$

Hence, $l = (l_j)_{j \in N}$ is also in the core of (N, V). Since $(N, V_{\hat{c}^*})$ is an inventory centralization game with a linear ordering cost, Theorem 11.2.3 implies that it has a nonempty core. Thus, (N, V) has a nonempty core as well. ∎

Our approach to prove the nonemptiness of the core of a cooperative game (N, V) with quantity discount suggests a way to find an allocation in the core in three steps. First, solve

$$
V(N) = \max_{y \geq 0} -c(y) + g(y, N).
$$

Second, given $V(N)$, find a $\hat{c}^*$ such that

$$
V(N) = \max_{y \geq 0} -\hat{c}^* y + g(y, N).
$$

Third, find an allocation in the core of the inventory centralization game $(N, V_{\hat{c}^*})$ with a linear ordering cost by employing the duality approach in Sect. 11.2.2. Theorem 11.2.7 implies that this allocation is in the core of (N, V).

11.3 Exercises

Exercise 11.1. For the linear demand, the exponential demand, the Cobb–Douglas demand, the constant elasticity of substitution demand, and the Logit demand in Sect. 11.1, provide conditions under which the payoff functions satisfy the diagonally dominant conditions.

Exercise 11.2. Show that the inventory centralization games are not convex in general.

Exercise 11.3. The inventory centralization game analyzed in Sect. (11.2) assumes that pricing decisions are made after the market signal is revealed. Do we have similar results if pricing decisions are made before the realization of the market signal?

Exercise 11.4. Consider a special case of the inventory centralization game with a single warehouse, the newsvendor game, in which pricing decisions are fixed, the ordering cost is linear, and the retailers have identical transportation costs s_j, inventory holding costs h_j^+, and emergency ordering costs h_j^-. Simplify the dual derived in Sect. (11.2.2). Find an optimal solution of the dual in closed form and an associated (dual-based) core allocation. Does the dual-based core allocation satisfy any of the monotonicity properties in Sect. 3.2?

12
Procurement Contracts

12.1 Introduction

The inventory models discussed in Chap. 9 focus on characterizing the optimal replenishment policy for a single facility given some assumptions, such as lead time and yield, of its supplier. This of course emphasizes the need, in many cases, to develop direct relationships with suppliers. These relationships can take many forms, both formal and informal, but often, to ensure adequate supplies and timely deliveries, buyers and suppliers typically agree on supply contracts. These contracts address issues that arise between a buyer and a supplier, whether the buyer is a manufacturer purchasing raw materials from a supplier or a retailer purchasing manufactured goods from a manufacturer. In a supply contract, the buyer and supplier may agree on

- pricing and volume discounts,

- minimum and maximum purchase quantities,

- delivery lead times,

- product or material quality,

- product return policies.

As we will see, supply contracts are very powerful tools that can be used for far more than ensuring adequate supply and demand for goods.

D. Simchi-Levi et al., *The Logic of Logistics: Theory, Algorithms, and Applications for Logistics Management*, Springer Series in Operations Research and Financial Engineering, DOI 10.1007/978-1-4614-9149-1_12, © Springer Science+Business Media New York 2014

To illustrate the importance and impact of different types of supply contracts on supply chain performance, consider a typical two-stage supply chain consisting of a retailer and a supplier. In such a supply chain, the retailer places orders trying to maximize its own profit and the supplier reacts to the orders placed by the retailer. This process is referred to as a sequential supply chain since decisions are made sequentially. Thus, in a sequential supply chain, each party determines its own course of action independent of the impact of its decisions on other parties; clearly, this cannot be an effective strategy for supply chain partners.

It is natural to look for mechanisms that enable supply chain entities to move beyond this sequential process and toward global optimization. Of course, this may be quite difficult since, in a typical supply chain, different parties may have different, sometimes even conflicting, objectives. Thus, it is important to identify mechanisms that maximize the efficiency of the supply chain while allowing different parties to focus on their own objectives. One way to achieve this goal is to use contracts specifying the transactions between supply chain parties such that every party's objective is aligned with the objective of the entire supply chain. We will refer to such a contract as a contract that **coordinates** the supply chain.

To illustrate how supply contracts can be used to coordinate the supply chain, we investigate in this chapter a simplified supply chain consisting of two risk-neutral decision makers, a supplier and a retailer. The retailer faces uncertain demands and needs to procure a certain quantity of a single product from the supplier. The supplier then produces and delivers the order to the retailer before demand is realized. The two parties negotiate and form a contract regarding the terms of the transactions.

A simple example of such a contract is the wholesale contract that we have seen in the analysis of the newsvendor problem; see Chap. 9, Sect. 9.2, in which the supplier specifies a wholesale price, while the retailer places an order to the supplier and the payment is proportional to the quantity purchased by the retailer. Unfortunately, as we will see in the next section, this simple wholesale contract does not coordinate the supply chain in general.

Several supply contracts have been proposed to achieve system efficiencies. Among those contracts, the buy-back contracts and the revenue-sharing contracts are commonly used in some industries due to their effectiveness and simplicity. In fact, under the setting to be specified later on in this chapter, the two contracts coordinate the supply chain; that is, these contracts allow supply chain partners to achieve global optimization, in other words, maximize supply chain expected profit.

Furthermore, in these contracts, the retailer's optimal strategy, namely, the optimal ordering quantity, together with the supplier's optimal strategy, namely, the optimal cost parameters specified in the contracts, consists of a Nash equilibrium. Thus, neither the retailer nor the supplier could increase its profit by unilaterally deviating from its optimal strategies.

Interestingly, the buy-back contracts and the revenue-sharing contracts are shown to be equivalent under our model setting. The literature on supply contracts that coordinate the supply chain system is quite extensive and is still expanding. We refer the reader to the review paper by Cachon (2003) for more details.

Of course, effective supply contracts are important not only in the retail industry. In the electronics industry, there has been a marked increase in purchasing volume as a percentage of the firm's total sales. For instance, between 1998 and 2000, outsourcing in the electronic industry increased from 15% of all components to 40%. This increase in the level of outsourcing implies that the procurement function becomes critical for an original equipment manufacturer (OEM) to remain in control of its destiny. As a result, many OEMs focus on closely collaborating with the suppliers of their strategic components. In some cases, this is done using effective supply contracts that try to coordinate the supply chain.

A different approach has been applied by OEMs for nonstrategic components. In this case, products can be purchased from a variety of suppliers, and flexibility to market conditions is perceived as more important than a permanent relationship with the suppliers. Indeed, commodity products, for instance, electricity, computer memory, steel, oil, grain, or cotton, are typically available from a large number of suppliers and can be purchased in spot markets. Because these are highly standard products, switching from one supplier to another is not considered a major problem.

Thus, in this chapter, we also introduce and analyze portfolio contracts based on the recent work of Martínez-de-Albéniz and Simchi-Levi (2005). In these contracts, the buyer signs a portfolio of supply contracts, contracts that provide the buyer with the appropriate tradeoff between price and flexibility.

12.2 Wholesale Price Contracts

In a wholesale contract, the supplier specifies a wholesale price and in return, the retailer decides how much to order from the supplier. Specifically, when the retailer places an order, its payment to the supplier is proportional to the quantity it orders. Thus, in this case, the retailer is facing a newsvendor problem and chooses the optimal ordering quantity according to the newsvendor model we analyzed in Chap. 9 Sect. 9.2. Of course, the supplier anticipates the reaction of the retailer and takes it into account when deciding its wholesale price. This is the so-called Stackelberg game between the supplier and the retailer, in which the supplier is the leader and the retailer is the follower.

The setting in this model is as follows. The retailer places an order from the supplier before the realization of the uncertain demand and sells the product to its customers at a unit price r. Let F be the cumulative distribution function of the demand. The function F is assumed to be strictly increasing and differentiable. For simplicity, we assume that unsatisfied demand is lost and there is no penalty cost for lost sales. In addition, leftover inventory is salvaged with unit price v. Finally, we assume that the supplier has no production capacity limit, and its unit production cost is c with $v < c < r$.

Before proceeding to analyze the Stackelberg game between the supplier and the retailer, we first discuss the optimal production quantity of the entire system assuming that the supplier and the retailer belong to a centralized system. In this case, the objective is to maximize the system's expected profit. Given the production quantity q, the profit for the total supply chain is

$$
\begin{aligned}
\pi^0(q) &= -cq + rE_D[\min(q, D)] + vE_D[\max(q - D, 0)] \\
&= (r - c)q - (r - v)E_D[\max(q - D, 0)].
\end{aligned}
$$

This is exactly the classical newsvendor problem analyzed in Chap. 9, Sect. 9.2. Thus, the optimal production quantity for the total supply chain is

$$
q^0 = F^{-1}\left(\frac{r - c}{r - v}\right),
$$

where F^{-1} is the inverse function of the cumulative distribution function F.

We now analyze the Stackelberg game between the supplier and the retailer. Assume for now that the unit wholesale price of the supplier is w. As we already noticed, the retailer is facing a newsvendor model. Again from the analysis of the newsvendor model, we can determine the optimal ordering quantity for the retailer as follows:

$$
q(w) = F^{-1}\left(\frac{r - w}{r - v}\right). \tag{12.1}
$$

Of course, here we assume $r > w > v$ to avoid trivial cases. Notice that $q(w) \geq q^0$ only if $w \leq c$. However, this implies that the supplier makes a nonpositive profit. Thus, the supplier prefers a higher wholesale price, and in this case, the retailer always tends to order less than q^0, the quantity that is optimal for the entire supply chain. We refer to this behavior as **double marginalization**. Of course, this behavior has an intuitive explanation. Since the retailer bears all the risk for overstocking, it has no incentive to order more and thus tries to reduce its risk exposure by reducing inventory levels.

As we already pointed out, the supplier anticipates this behavior of the retailer when setting its wholesale price. From (12.1), there is a one-to-one correspondence between the optimal ordering quantity of the retailer and the wholesale price set by the supplier, since F is strictly increasing. Therefore, given the optimal ordering quantity of the retailer q, the wholesale price is

$$
w(q) = r - (r - v)F(q).
$$

The objective of the supplier is to maximize its own profit, which can be written as a function of the ordering quantity of the retailer:

$$
\pi_s(q) = (w(q) - c)q = ((r - c) - (r - v)F(q))q.
$$

Of course, if the cumulative distribution function F is too general, there is no guarantee that the supplier has a unique optimal wholesale price. Hence, we focus on demands with increasing generalized failure rate (IGFR) distributions, namely, distributions such that $qF'(q)/(1 - F(q))$ is increasing. Notice that several commonly used distributions, such as the normal distribution and the exponential distribution, are IGFR distributions.

We now show that for IGFR demand distributions, the optimal wholesale price of the supplier is unique. First, observe that the first-order optimality condition implies that the retailer's ordering quantity, associated with the supplier optimal wholesale price, satisfies

$$\pi_s'(q) = -(c - v) + (r - v)(1 - F(q) - F'(q)q) = 0,$$

or

$$1 - \frac{qF'(q)}{1 - F(q)} = \frac{c - v}{r - v}\frac{1}{1 - F(q)}. \tag{12.2}$$

Notice that the left-hand side of the above equation is decreasing in q while the right-hand side of the equation is increasing in q. Hence, there is a unique solution q^* for Equation (12.2), and therefore $w(q^*)$ is the unique optimal wholesale price of the supplier. Furthermore, it is easy to verify that $\pi_s'(q) < 0$ for $q < q^*$ and $\pi_s'(q) > 0$ for $q > q^*$. In other words, $\pi_s(q)$ is decreasing for $q < q^*$ and increasing for $q > q^*$. Thus, $\pi_s(q)$ is unimodal.

In summary, for the wholesale contract, there exists a unique Nash equilibrium for the Stackelberg game between the supplier and the retailer when the demand distribution is IGFR. In addition, in such a contract, the retailer always orders less than the quantity that would be optimal for the entire supply chain due to the fact that it bears all the risks of overstocking. Thus, the wholesale contract does not coordinate the supply chain.

12.3 Buy-Back Contracts

The previous discussion reveals that wholesale contracts do not coordinate the supply chain, since the retailer bears all the risks of overstocking and tends to order less than the amount that would be optimal for the entire system. Thus, one might expect that the retailer is willing to order more and hence improve the performance of the supply chain if the supplier would share some of its risks.

Buy-back contracts provide such a mechanism for the supplier to share the risks with the retailer. In such a contract, the supplier specifies a wholesale price w_b and a buy-back price b. This contract is similar to the wholesale price contract; that is, the retailer orders from the supplier according to a wholesale price w_b. However, one significant difference is that in addition to a unit salvage value v for unsold items, the retailer can get a refund from the supplier for a unit price b.

Given a wholesale price w_b, a buy-back price b, and an order quantity q, the retailer's expected profit is

$$\begin{aligned}
\pi_r^b(w_b, b, q) &= -w_b q + r E_D[\min(q, D)] + (b + v)E_D[\max(q - D, 0)] \\
&= (r - w_b)q - (r - b - v)E_D[\max(q - D, 0)].
\end{aligned}$$

Consider now a wholesale price w_b and a buy-back price b satisfying the following requirements:

$$r - w_b = \lambda(r - c) \text{ and } r - b - v = \lambda(r - v)$$

for some $\lambda \in [0, 1]$, or alternatively,

$$w_b = r - \lambda(r - c) \text{ and } b = (1 - \lambda)(r - v).$$

This implies that the expected profit of the retailer is given by

$$\pi_r^b(w_b, b, q) = \lambda \pi^0(q).$$

Hence, the optimal order quantity of the retailer equals q^0, the optimal production quantity of the entire supply chain. Similarly, the expected profit of the supplier is given by

$$\pi_s^b(w_b, b, q) = (1 - \lambda)\pi^0(q).$$

Thus, the supplier's optimal production quantity is also equal to q^0. Therefore, the system's expected profit is maximized and the buy-back contract coordinates the supply chain. Furthermore, in this case, the retailer receives λ of the system's expected profit and the supplier seizes $(1 - \lambda)$ of the system's expected profit.

12.4   Revenue-Sharing Contracts

A different contract that allows for risk sharing between suppliers and retailers is the so-called revenue-sharing contract. In a revenue-sharing contract, the retailer and the supplier agree on the wholesale price, typically a discounted wholesale price, and in return the supplier receives a given fraction of the revenue from each unit sold by the retailer. Of course, since the supplier receives some of the revenue, it has an incentive to reduce the wholesale price and hence increase the amount ordered by the retailer.

Assume that the wholesale price is w_r and the supplier receives a fraction $(1 - \phi)$ of the retailer's revenue. Thus, the retailer's profit is

$$
\begin{aligned}
\pi_r^r(w_r, \phi, q) &= -w_r q + \phi(r E_D[\min(q, D)] + v E_D[\max(q - D, 0)]) \\
&= (\phi r - w_r)q - \phi(r - v)E_D[\max(q - D, 0)].
\end{aligned}
$$

If we choose ϕ and w_r such that

$$\phi r - w_r = \lambda(r - c)$$

and

$$\phi = \lambda$$

for some $\lambda \in [0, 1]$, then

$$\pi_r^r(w_r, \phi, q) = \lambda \pi^0(q).$$

Similarly, the supplier's expected profit is given by

$$\pi_s^r(w_b, b, q) = (1 - \lambda)\pi^0(q).$$

Thus, both the retailer's optimal ordering quantity and the the supplier's optimal production quantity equal q^0, the optimal production quantity for the entire supply chain. Hence, the system's expected profit is achieved, and the revenue-sharing contract coordinates the supply chain.

Furthermore, if the wholesale price is $w_r(\lambda) = \lambda c$ and the retailer shares a fraction $\phi(\lambda) = \lambda$ of its expected revenue with the supplier, the retailer receives a fraction λ of the system's expected profit and the supplier seizes $(1 - \lambda)$ of the system's expected profit.

Notice that both the buy-back contract with parameters $(w_b(\lambda), b(\lambda))$ and the revenue-sharing contract with parameters $(w_r(\lambda), \phi(\lambda))$ coordinate the supply chain and have the same allocation of the system's expected profit to the supplier and the retailer.

In fact, a revenue-sharing contract with parameters $(w_r(\lambda), \phi(\lambda))$ is equivalent to the following contract: The wholesale price is $w_r(\lambda) + (1 - \lambda)r$, and the retailer receives r for each sold unit and gets a refund equal to $(1 - \lambda)r - (1 - \phi(\lambda))v$ from the supplier for each salvaged unit. It is easy to verify that this is exactly the buy-back contract with parameters $(w_b(\lambda), b(\lambda))$.

The following example illustrates the impact of supply contracts in practice.

Until 1998, video rental stores used to purchase copies of newly released movies from the movie studios for about \$65 and rent them to customers for \$3. Because of the high purchase price, rental stores did not buy enough copies to cover peak demand, which typically occurs during the first 10 weeks after a movie is released on video. The result was a low customer service level; in a 1998 survey, about 20% of customers could not get their first choice of movie. Then, in 1998, Blockbuster Video entered into a revenue-sharing contract with the movie studios in which the wholesale price was reduced from \$65 to \$8 per copy, and, in return, studios were paid about 30–45% of the rental price of every rental. This revenue-sharing contract had a huge impact on Blockbuster revenue and market share. Today, revenue sharing is used by most large video rental stores; see Cachon and Lariviere (2005).

12.5 Portfolio Contracts

A recent trend for many industrial manufacturers has been outsourcing; firms are considering outsourcing everything from production and manufacturing to the procurement function itself. Indeed, in the mid-1990s, there was a significant increase in purchasing volume as a percentage of the firm's total sales. Between 1998 and 2000, outsourcing in the electronics industry increased from 15% of all components to 40%.

Of course, the increase in the level of outsourcing implies that the procurement function becomes critical for a manufacturer to remain in control of its destiny. Thus, an effective procurement strategy has to focus on both driving costs down and reducing risks. These risks include both *inventory* and *financial* risks.

By inventory risks, we refer to inventory shortages, while financial risks refer to the purchasing price, which is uncertain if the procurement strategy depends on spot markets.

A traditional procurement strategy that eliminates financial risk is the use of *fixed commitment* contracts. These contracts specify a fixed amount of supply to be delivered at some point in the future; the supplier and the manufacturer agree on both the price and the quantity delivered to the manufacturer. Thus, in this case, the manufacturer bears no financial risk while taking huge inventory risks due to uncertainty in demand and the inability to adjust order quantities.

One way to reduce inventory risk is through *option* contracts, in which the buyer prepays a relatively small fraction of the product price up-front, in return for a commitment from the supplier to reserve capacity up to a certain level. The initial payment is typically referred to as a *reservation price* or *premium*. If the buyer does not *exercise* the option, the initial payment is lost. The buyer can purchase any amount of supply up to the option level, by paying an additional price, agreed to at the time the contract is signed, for each unit purchased. This additional price is referred to as the *execution price* or *exercise price*. Of course, the total price (reservation plus execution price) paid by the manufacturer for each purchased unit is typically higher than the unit price in a fixed commitment contract.

Evidently, option contracts provide the manufacturer with the flexibility to adjust order quantities depending on realized demand, and hence these contracts reduce inventory risk. Thus, these contracts shift risks from the manufacturer to the supplier since the supplier is now exposed to customer demand uncertainty. This is in contrast to fixed commitment contracts in which the manufacturer takes all the risk.

Thus, consider a single-period model in which the manufacturer can procure a single product from multiple sources. For example, consider automotive manufacturing companies purchasing steel or PC manufacturers procuring memory units.

The manufacturer faces stochastic demand D and sells the finished product at a unit selling price r. Unsold items have a unit salvage value v. Most importantly, we assume that there are a total of n suppliers and before the planning horizon, the retailer signs an option contract with each supplier. That is, the manufacturer reserves capacity x_i with the ith supplier for a reservation cost v_i per unit of capacity reserved and pays an execution fee of w_i for each unit ordered from the supplier, after demand is realized. Thus, the procurement strategy of the retailer is a portfolio contract consisting of n option contracts with parameters (v_i, w_i, x_i).

The class of portfolio contracts contains several widely used contracts. This includes, for instance, long-term contracts, buy-back contracts, and *flexibility* contracts. A long-term contract specifies a fixed amount of supply, x, to be delivered at a predetermined time in the future for a given price, $\hat{v}$. Thus, it is equivalent to a portfolio contract consisting of only one option contract with parameters $(\hat{v}, 0, x)$, that is, with positive reservation price and zero execution cost. In the long-term contract, the buyer bears all the risks of overstocking or understocking due to uncertain demand and its inability to adjust order quantity.

A flexibility contract specifies a fixed amount of supply, x, for a given price, $\hat{v}$. In addition, in this contract, the amount to be delivered and paid can differ from the specified quantity by no more than a given percentage, say α, determined upon signing the contract. That is, the order quantity is within the interval $[(1 - \alpha)x, (1 + \alpha)x]$. The flexibility contract is equivalent to a portfolio contract consisting of a long-term contract with parameters $(\hat{v}, 0, (1 - \alpha)x)$ and an option contract with parameters $(0, \hat{v}, 2\alpha x)$.

Given that the retailer has to procure q units from the suppliers, it can choose an appropriate combination of suppliers so that its cost is minimized. Let $R(q)$ be the optimal cost for procuring q units from the n suppliers. We have

$$R(q) = \sum_{i=1}^{n} v_i x_i + \quad \min \quad \sum_{i=1}^{n} w_i q_i$$
$$\text{s.t.} \quad \begin{cases} \sum_{i=1}^{n} q_i = q, \\ 0 \leq q_i \leq x_i, \text{ for all } i = 1, 2, \ldots, n. \end{cases} \tag{12.3}$$

It is easy to prove that $R(q)$ is a convex piecewise linear function of q.

Given an initial inventory level I and the order quantity q, the buyer's expected profit is

$$f(I, q) = G(I + q) - R(q),$$

where

$$G(q) = rE[\min(q, D)] + vE[\max(0, q - D)].$$

In the following, we characterize the optimal replenishment policy for the retailer. First, we present a result that illustrates how the optimal order quantity changes monotonically as a function of the initial inventory level when the retailer's ordering cost function is convex while its revenue function is concave.

Theorem 12.5.1 *Assume that the ordering cost function R is convex and the revenue function G is concave. Moreover, $f(I, q) \to \infty$ for $q \to \infty$ for any I. Then there exists a function $q^*(I)$ solving*

$$\max_{q \geq 0} \ f(I, q) \tag{12.4}$$

such that $q^*(I)$ is nonincreasing and $I + q^*(I)$ is nondecreasing.

Proof. First, observe that q is an optimal solution for the optimization problem (12.4) if and only if $q' = -q$ is optimal for the following problem:

$$\max_{q' \leq 0} \ g(I, q') := G(I - q') - R(-q'). \tag{12.5}$$

Let $q'(I) = \min\{ q' \leq 0 \mid q'$ solves (12.5)$\}$. Since G is concave, Theorem 2.2.6 implies that $g(I, q')$ is supermodular. Therefore, from Theorem 2.2.8, we have that $q'(I)$ is nondecreasing. Thus, $q^*(I) = -q'(I)$ solves (12.4) and is nonincreasing.

To prove the remaining part of the theorem, observe that q is an optimal solution for the optimization problem (12.4) if and only if $I' = I + q$ is optimal for the following problem:

$$\max_{I' \geq I} \ g(I, I') := G(I') - R(I' - I). \tag{12.6}$$

Since R is convex, Theorem 2.2.6 implies that $g(I, I')$ is supermodular. Therefore, from Theorem 2.2.8 and the definition of $q^*(I)$, we have that $I'(I) = I + q^*(I)$ is nondecreasing. ∎

The above result implies that the optimal ordering quantity is a nonincreasing function of the initial inventory level, while the end period inventory level, $I + q^*(I) - E[D]$, is a nondecreasing function of the initial inventory level.

Finally, we characterize the structure of the optimal ordering policy of the retailer when a portfolio contract is employed by the retailer. As we already pointed out, the order cost function $R(q)$ is convex piecewise linear. In fact, without loss of generality, assume that $w_1 \leq w_2 \leq \ldots \leq w_n$. Define $z_0 = 0$ and $z_i = \sum_{j=1}^{i} x_j$ for $i = 1, 2, \ldots, n$. Then

$$R(q) = \sum_{j=1}^{n} v_j x_j + \sum_{j=1}^{i-1} w_j x_j + w_i(q - z_{i-1}), \text{ for } q \in [z_{i-1}, z_i].$$

Hence, for $q \in (z_{i-1}, z_i)$, $\frac{dR(q)}{dq} = w_i$, and for $q = z_i$, $\partial R(q) = [w_i, w_{i+1}]$.

Theorem 12.5.2 *If the ordering cost function $R(q)$ is given by (12.3), then there exist inventory levels f_i $(i = 1, 2, \ldots, 2n + 1)$ with*

$$\infty = f_0 \geq f_1 \geq \ldots \geq f_{2n} \geq f_{2n+1} = 0$$

such that

(a) *For $I \in [f_{2i}, f_{2i-1})$, it is optimal to set $I' = I + q$ to a constant level such that $w_i \in \partial G(I + q)$.*

(b) *For $I \in [f_{2i+1}, f_{2i})$, it is optimal to set the ordering quantity q to the constant level z_i.*

Proof. For $i = 1, 2, \ldots, n$, let

$$q_i = \max\{q^* \geq 0 \mid q^* \text{ maximizes } G(q) - w_i q \text{ subject to } q \geq 0\}.$$

Theorem 2.2.4 and Theorem 2.2.8 imply that $q_i \leq q_{i-1}$ for $i = 2, 3, \ldots, n$. Let $f_0 = \infty$, $f_{2n+1} = 0$,

$$f_{2i-1} = \max(q_i - z_{i-1}, 0), \text{ and } f_{2i} = \max(q_i - z_i, 0), i = 1, 2, \ldots, n.$$

Then $\infty = f_0 \geq f_1 \geq \ldots \geq f_{2n} \geq f_{2n+1} = 0$. We claim that f_i $(i = 0, 1, \ldots, 2n+1)$ satisfies parts (a) and (b).

First, notice that for $I \in [f_{2i}, f_{2i-1}) \neq \emptyset$, we have $q_i > 0$ and $q = q_i - I \in (z_{i-1}, z_i]$. The first-order optimality condition implies that $w_i \in \partial G(q_i)$. Hence, $q^*(I) = q_i - I$ is optimal for problem (12.4), since we have $0 \in \partial(G(I + q) - R(q))|_{q=q^*(I)}$. Thus, part (a) is true.

On the other hand, for $I \in [f_{2i+1}, f_{2i})$ with $i \geq 1$, we claim that the optimal ordering quantity $q^*(I) = z_i$. In fact, observe that for $I = f_{2i}$, we have

$q^*(I) = q_i - I = z_i$ and for $I = f_{2i+1} > 0$, we have $q^*(I) = q_{i+1} - I = z_i$. Thus, Theorem 12.5.1 implies that for $I \in [f_{2i+1}, f_{2i})$, $q^*(I) = z_i$ is optimal. Finally, for $I \geq f_1$, it is clear that $q^*(I) = 0$ is optimal. Hence, part (b) holds. ∎

Figure 12.1 illustrates the structure of the optimal ordering policy identified in Theorem 12.5.2 for a case with $n = 2$.

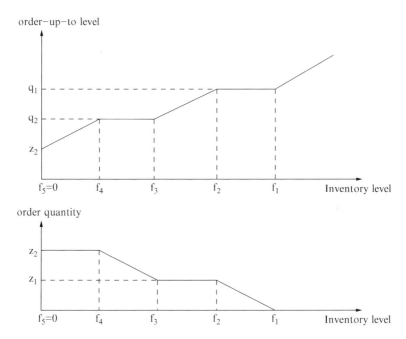

FIGURE 12.1. Illustration of the structure of the optimal ordering policy

12.6 Exercises

Exercise 12.1. Prove that the normal distribution and the exponential distribution have increasing generalized failure rate (IFGR).

Exercise 12.2. As we have shown in Sect. 12.2, wholesale contracts do not coordinate the supply chain in general. Now assume that the supplier is willing to provide an all-unit quantity discount. Design an all-unit quantity discount contract coordinating the supply chain; that is, find a per-unit wholesale price $w(q)$ as a decreasing function of the order quantity q such that the optimal ordering quantity of the retailer and the optimal production quantity of the supplier equal the optimal production quantity of the whole system.

Exercise 12.3. Show that a buy-back contract is a special case of portfolio contracts.

Exercise 12.4. Show that Theorem 12.5.2 implies the optimality of *a modified base stock policy*. In such a policy, there exist inventory target levels $b_i \geq 0$ ($i = 1, \cdots, n$) with $b_i \leq \max(0, b_{i+1} - x_{i+1})$ for $i = 1, \cdots, n-1$, such that it is optimal to order nothing if $I \geq b_i$, order $b_i - I$ if $I \in [\max(0, b_i - x_i), b_i]$, or order x_i otherwise.

Exercise 12.5. Consider a single manufacturer and a single supplier. Six months before demand is realized, the manufacturer has to sign a supply contract with the supplier. Let D be a random variable representing demand and $f(D)$ be the demand density function. Let p be the selling price, that is, the price at which the manufacturer sells products to consumers.

The sequence of events is as follows. Procurement contracts are signed in February and demand is realized during a short period of 10 weeks that starts in August. Components are delivered from the supplier to the manufacturer at the beginning of August, and the manufacturer produces items to customer orders. Thus, we can ignore any inventory holding cost. We will assume that unsold items at the end of the 10-week selling period have zero value. Finally, assume that the manufacturer can also purchase additional items in the spot market. Let s be a random variable representing the per-unit spot market price and $f(s)$ be its density function. The objective is to identify a procurement strategy so as to maximize expected profit.

Assume the supplier offers an option contract in which the per-unit reservation price is v and the per-unit execution price is w. Given the existence of the spot market, how much capacity should the manufacturer reserve with the supplier when the contract is signed in February?

13

Process Flexibility

13.1 Introduction

For many manufacturing firms, the ability to match demand and supply is key to their success. Failure to do so could lead to loss of revenue, reduced service levels, negative impact on reputation, and decline in the company's market share. Unfortunately, recent developments, such as intense market competition, product proliferation, and the increase in the number of products with a short life cycle, have created an environment where customer demand is volatile and unpredictable. In such an environment, traditional operations strategies such as building inventory, investing in capacity buffers, or increasing committed response time to consumers do not offer manufacturers a competitive advantage. Therefore, many manufacturers have started to adopt an operations strategy known as process flexibility to better respond to market changes without significantly increasing cost, inventory, or response time (see Simchi-Levi 2010).

Process flexibility is defined as the ability to "build different types of products in the same manufacturing plant or on the same production line at the same time" (Jordan and Graves 1995). For example, in "full" (process) flexibility, each plant is capable of producing all products. In this case, when the demand for one product is higher than expected while the demand for a different product is lower than expected, a flexible manufacturing system can quickly make adjustments by shifting production capacities appropriately. By contrast, in a "dedicated" strategy (sometimes called "no flexibility"), each plant is responsible for a single product and hence does not have the same ability to match supply with demand.

D. Simchi-Levi et al., *The Logic of Logistics: Theory, Algorithms, and Applications for Logistics Management*, Springer Series in Operations Research and Financial Engineering, DOI 10.1007/978-1-4614-9149-1_13, © Springer Science+Business Media New York 2014

Because of its effectiveness in responding to uncertainties, process flexibility has gained significant attention in several industries, in particular, in the automotive industry. Evidently, it is often too expensive to achieve a high degree of flexibility, for example, full flexibility, and as a result, sparse or partial flexibility is implemented instead. One set of sparse flexibility designs is the 2-flexibility designs. A flexibility design is a 2-flexibility design if each plant produces exactly two products and demand for each product can be satisfied from exactly two plants.

Of course, there are many ways to implement sparse designs, and the challenge is to identify an effective one. An important concept analyzed in the literature and applied in practice by various companies is the concept of the *long chain*. The first to observe the power of the long chain were Jordan and Graves (1995), who, through empirical analysis, showed that the long-chain design can provide almost as much benefit as full flexibility. In particular, Jordan and Graves (1995) found that for randomly generated demand, the expected amount of demand that can be satisfied by a long-chain design is very close to that of a full flexibility design.

Though the analysis and results can be extended to more general settings, for simplicity, we focus on *balanced* manufacturing systems, that is, manufacturing systems with an equal number of plants and products, and each plant has a equal capacity. Given a balanced manufacturing system, a flexibility design $\mathscr{A}$ is represented by the arc set of a *directed* bipartite graph, where an arc from plant node i to product node j implies that plant i is capable of producing product j. For example, if $\mathscr{A}$ is a dedicated design, then $\mathscr{A}$ has exactly n arcs such that each plant node is incident to one arc and each product node is incident to one arc. By contrast, if $\mathscr{A}$ is a full flexibility design, then $\mathscr{A}$ has arcs connecting every plant node to all product nodes.

Because $\mathscr{A}$ is represented by a bipartite graph, applying standard graph theory notation, we define an *undirected cycle* in $\mathscr{A}$ to be a set of arcs that forms a cycle when the arc directions are ignored. A flexibility design $\mathscr{A}$ is a long chain if its arcs form exactly one undirected cycle containing all plant and product nodes (see Fig. 13.1 for an example). A *closed chain* is defined as an induced subgraph in $\mathscr{A}$ that forms an undirected cycle, while an *open chain* is an induced subgraph in $\mathscr{A}$ that forms an undirected line (one arc less than an undirected cycle). Figure 13.1 presents an example of an open and a closed chain. It can be seen that any 2-flexibility design, where each product/plant node is incident to two arcs, is the union of a number of closed chains.

The results presented in this chapter are motivated by a few observations made in the literature regarding the effectiveness of the long-chain flexibility design. The first is an observation that has been made in (Graves 2008 and Hopp et al. 2004) regarding the performance of the long chain for a balanced system when product demands are independent and identically distributed (iid). The observation states that if one starts with a dedicated design and adds arcs to create the long chain, the *incremental benefits*, or the change in performance, associated with each added arc *is increasing*.

To illustrate this observation, consider an example with six plants and six products, where the demand for each product is equal to 0.8, 1, or 1.2 with equal

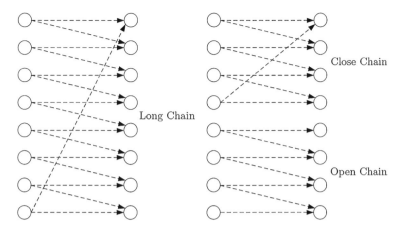

FIGURE 13.1. Configurations for flexibility designs

probabilities, and the capacity of each plant is 1. Then we start with a dedicated flexibility design (the dashed arcs in Fig. 13.2a) and add arcs $(1,2)$, $(2,3)$, ..., $(5,6)$, and $(6,1)$ one at a time, until we complete the long chain. Each time we add such an arc, we determine the expected sales associated with the resulting design at that time. Figure 13.2b displays the performance of the flexibility designs at different stages, as well as the incremental benefits when a new arc is added. The *incremental benefits increase as we add more arcs*. The biggest impact, surprisingly, occurs when we add the last arc and close the long chain.

This example also illustrates the second observation. The long chain is an effective flexibility design; in this case, it achieves the same performance as that of full flexibility. Indeed, numerous empirical papers reported that the long-chain flexibility design is more effective than other 2-flexibility designs, and its expected sales is almost the same as that of full flexibility.

To explain the effectiveness of the long chain, we start in Sect. 13.2 by establishing a supermodularity property and apply it to prove that the incremental benefits

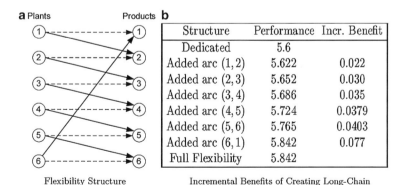

Structure	Performance	Incr. Benefit
Dedicated	5.6	
Added arc $(1,2)$	5.622	0.022
Added arc $(2,3)$	5.652	0.030
Added arc $(3,4)$	5.686	0.035
Added arc $(4,5)$	5.724	0.0379
Added arc $(5,6)$	5.765	0.0403
Added arc $(6,1)$	5.842	0.077
Full Flexibility	5.842	

Flexibility Structure   Incremental Benefits of Creating Long-Chain

FIGURE 13.2. The increase in incremental benefit

are increasing as the long chain is constructed. Section 13.3 illustrates that the performance of a long chain can be characterized by the difference in the performances of two open chains. This result allows us to show that the long-chain design maximizes expected sales among all 2-flexibility design strategies. In Sect. 13.4, we then apply the previous results to compare the performance of the long chain to that of full flexibility. Finally, we present some extensions beyond long chains and balanced systems in Sect. 13.5.

13.2   Supermodularity and Incremental Benefits of Long Chains

Consider a *balanced* manufacturing system of size n facing random demand. In such a system, there are n plants, each of which has unit capacity, and n different products. We will use the vector D to denote the random demand distribution, and d a particular random instance. Since in practice demand is never negative, D is assumed to be nonnegative.

For a balanced system of size n, we say that its demand, D, is *exchangeable* if $[D_1, \ldots, D_n]$ equals $[D_{\pi(1)}, .., D_{\pi(n)}]$ in distribution for any π that is a permutation of $\{1, 2, \ldots, n\}$. We note that any identically and independently distributed (iid) demand is exchangeable, but not all exchangeable demands are iid.

Next, we define several classes of flexibility designs for balanced manufacturing systems. For any integer $n \geq 2$, the dedicated design for a balanced system of size n, $\mathscr{D}_n$, is defined as $\mathscr{D}_n = \{(i,i) | i = 1, 2, \ldots, n\}$; a long-chain flexibility design for a balanced system of size n, $\mathscr{C}_n$, is defined as $\mathscr{C}_n = \mathscr{D}_n \cup \{(i, i+1) | i = 1, 2, \ldots, n-1\} \cup \{(n, 1)\}$; and the full flexibility design for a balanced system of size n, $\mathscr{F}_n$, is defined as $\mathscr{F}_n = \{(i,j) | i, j = 1, 2, \ldots, n\}$. In flexibility designs, we refer to an arc (i,i) as a *dedicated* arc and arc $(i,j), i \neq j$ as a *flexible* arc. We also define *open chain* $\mathscr{L}_k$ as $\mathscr{L}_k = \mathscr{D}_k \cup \{(i, i+1) | i = 1, \ldots, k-1\}$, for any integer $k \geq 0$. One can think of $\mathscr{L}_k$ as the open chain that connects plant 1 to product k. Note that $\mathscr{L}_k$ is simply $\mathscr{C}_k \setminus \{(k, 1)\}$.

Given a random instance of the demand, d, the maximum sales that can be achieved by a flexibility design $\mathscr{A}$ with arc capacities u, denoted by $P(d, \mathscr{A}, u)$, is defined as

$$
\begin{aligned}
P(d, \mathscr{A}, u) \quad = \quad &\text{Min} \quad &&\sum_{1 \leq i,j \leq n} f_{ij} \\
&\text{s.t.} &&\sum_{i=1}^{n} f_{ij} \leq d_j, \forall 1 \leq j \leq n, \\
& &&\sum_{j=1}^{n} f_{ij} \leq 1, \forall 1 \leq i \leq n, \\
& &&f_{ij} \leq u_{ij}, \forall 1 \leq i, j \leq n, \\
& &&f_{ij} \geq 0, \forall (i,j) \in \mathscr{A}, \\
& &&f_{ij} = 0, \forall (i,j) \notin \mathscr{A}.
\end{aligned}
$$

It is not difficult to see that this optimization problem is a max-flow problem, and as a result, we refer to f_{ij} as the flow on arc (i,j). Note that when the arc capacities are 1, the capacity constraints are redundant. In this case, the above optimization problem is referred to as the optimization problem associated with $P(d, \mathscr{A})$, or simply, $P(d, \mathscr{A})$, when there is no ambiguity.

Under random demands D, we define the *performance*, also referred to as expected sales, of $\mathscr{A}$ to be $E[P(D, \mathscr{A})]$, where $E[\cdot]$ is the expectation of a random variable with respect to D. For succinctness, we also use $[\mathscr{A}]$ to denote this quantity when the random vector D is given.

13.2.1 Supermodularity in Arc Capacities

In this subsection, we show that $P(d, \mathscr{A}, u)$ is supermodular in the capacities of flexible arcs in $\mathscr{A}$. For this purpose, we first present a classical result by Gale and Politof (1981) on supermodularity of the maximum-weight circulation problem. We then note that $P(d, \mathscr{A}, u)$ can be computed by solving an equivalent maximum-weight circulation problem, and the result by Gale and Politof (1981) applies.

In a maximum-weight circulation problem, we are given a directed graph G and for each arc γ, a weight w_γ, a lower bound l_γ, and an upper bound u_γ on the arc flow. A flow f is called a circulation if it satisfies the flow-balance constraints; that is, at each node, the inflow equals the outflow. The maximum-weight circulation problem is to find a feasible circulation f satisfying the lower and upper bound constraints such that the total weight $\sum_\gamma w_\gamma f_\gamma$ is maximized. For a set of arcs S and any vector ξ indexed by the arcs, let $\xi_S = (\xi_\gamma)_{\gamma \in S}$, a vector consisting of components of ξ with indices in S. The following theorem from Gale and Politof (1981) illustrates that the optimal objective value of the problem, denoted as $F(w, l, u)$, is supermodular in the capacities of arcs that are in series (pairwise). Note that two arcs α, β are said to be **in series** if, for any (undirected) cycle C containing both α and β, α and β have the same direction.

Theorem 13.2.1 *If S contains only arcs that are in series (pairwise), $F(w, l, u)$ is supermodular in u_S.*

Proof. In view of Theorem 2.2.2, we only need to show that for any two arcs α and β in S, $F(w, l, u)$ has increasing differences in (u_α, u_β) when other components of u are fixed. Given two capacity vectors u and u' with $u_{ij} = u'_{ij}$ for all (i, j) except α and β, assume without loss of generality that $u_\alpha > u'_\alpha$ and $u_\beta < u'_\beta$. Let f and f' be the optimal circulations in the graph given in the maximum-weight circulation problem with capacities u and u', respectively. It suffices to construct two circulations g and g' such that g and g' are feasible for the maximum-weight circulation problems with capacities $u \wedge u'$ and $u \vee u'$, respectively, and

$$g + g' = f + f'.$$

If $f \leq u \wedge u'$, or $f' \leq u \wedge u'$, simply let $g = f$ and $g' = f'$, or $g = f'$ and $g' = f$. Assume that $u'_\alpha < f_\alpha \leq u_\alpha$ and $u_\beta < f'_\beta \leq u'_\beta$. Let $\xi = f - f'$. Clearly, ξ is a circulation with $\xi_\alpha > 0$ and $\xi_\beta < 0$. We now show that ξ can be decomposed as the summation of several conformal circuits. That is, there exist simple cycles C_l and scalars $t_l > 0$ for $l = 1, \dots, \tau$ for some integer τ such that

$$\xi = \sum_{l=1}^{\tau} t_l f^l, \tag{13.1}$$

where $f_\gamma^l = sign(\xi_\gamma)$ if $\gamma \in C_l$ (regardless of direction) and zero otherwise. To see this, start from any node, say node i_0, incident to an arc (i_0, i_1) with $\xi_{i_0 i_1} \neq 0$. Without loss of generality, assume $\xi_{i_0 i_1} > 0$; otherwise, consider $-\xi$. Since ξ is a circulation, node i_1 must be incident to an arc (i_1, i_2) or (i_2, i_1) for some node i_2 such that either $\xi_{i_1 i_2} > 0$ or $\xi_{i_2 i_1} < 0$. Similarly, node i_2 must be incident to an arc (i_2, i_3) or (i_3, i_2) for some node i_3 such that either $\xi_{i_2 i_3} > 0$ or $\xi_{i_3 i_2} < 0$. Continue the process until the first time we meet a node that has appeared before, say node i_ν. Let i_κ be the node visited immediately before the second time we visit i_ν. In this case, we end up with a simple cycle $i_\nu i_{\nu+1} \cdots i_\kappa i_\nu$, denoted by C_1. Let C_1^+ and C_1^- be the sets of arcs on C_1 in the same direction as C_1 and in the opposite direction as C_1, respectively. Let

$$t_1 = \min_{\gamma \in C_1} |\xi_\gamma|,$$

and $f_\gamma^1 = 1$ for $\gamma \in C_1^+$ and -1 for $\gamma \in C_1^-$. Clearly, f^1 is a circulation with $f^1 - sign(\xi_\gamma)$ for $\gamma \in C_1$ and zero otherwise. Our construction implies that the circulation $\xi - t_1 f^1$ has at least one more arc with zero flow than the circulation ξ. By repeating the above construction, we can show that (13.1) holds.

Note that for any $l = 1, \dots, \tau$, f_α^l takes values 0 or 1 because $f_\alpha > f_\alpha'$ and f_β^l takes values 0 or -1 because $f_\beta < f_\beta'$. In addition, since α and β are in series, it is impossible to have $f_\alpha^l = 1$ and $f_\beta^l = -1$ simultaneously for any cycle f^l.

Define new circulations

$$g = f - \sum_{l=1:\tau, f_\alpha^l=1} t_l f^l, \quad g' = f' + \sum_{l=1:\tau, f_\alpha^l=1} t_l f^l.$$

We have that $g + g' = f + f'$ and

$$g = f' + \sum_{l=1:\tau, f_\alpha^l=0} t_l f^l, \quad g' = f - \sum_{l=1:\tau, f_\alpha^l=0} t_l f^l.$$

We claim that g and g' are feasible for the maximum-weight circulation problems with capacities $u \wedge u'$ and $u \vee u'$, respectively. Notice that for any arc γ, since f_γ^l has the same sign as $f_\gamma - f_\gamma'$ if $\gamma \in C_l$, we have that

$$g_\gamma = f_\gamma - \sum_{l=1:\tau, f_\alpha^l=1} t_l f_\gamma^l \in [\min(f_\gamma, f_\gamma'), \max(f_\gamma, f_\gamma')].$$

Similarly,

$$g_\gamma' \in [\min(f_\gamma, f_\gamma'), \max(f_\gamma, f_\gamma')].$$

Thus, for any arc γ different from α and β,

$$g_\gamma, g'_\gamma \in [l_\gamma, u_\gamma].$$

For the arc α, since $f^l_\alpha = 0$ for $\alpha \notin C_l$,

$$g_\alpha = f'_\alpha \in [l_\alpha, \min(u_\alpha, u'_\alpha)], \ g'_\alpha = f_\alpha \in [l_\alpha, \max(u_\alpha, u'_\alpha)].$$

For the arc β, since $f^l_\alpha = 1$ implies that $f^l_\beta = 0$, we have that

$$g_\beta = f_\beta \in [l_\beta, \min(u_\beta, u'_\beta)], \ g'_\beta = f'_\beta \in [l_\beta, \max(u_\beta, u'_\beta)].$$

Thus, g and g' are feasible for the maximum-weight circulation problems with capacities $u \wedge u'$ and $u \vee u'$, respectively. ∎

To apply the above theorem, we convert the optimization problem of computing $P(d, \mathscr{A}, u)$ to an equivalent maximum-weight circulation problem. Specifically, let $G(\mathscr{A})$ be the underlying graph, which contains $\mathscr{A}$, an additional node s, an arc from s to each of the plant nodes, and an arc from each of the product nodes to s. Set the weight of each plant to product arc (that is, the arcs in $\mathscr{A}$) to 1 and the weight of every other arc to zero. The upper bound (capacity) on the flow on an arc from s to plant i is set to be 1 for all $i = 1, 2, \ldots, n$; the upper bound for the flow on an arc connecting product j to s is set to be d_j for all $j = 1, 2, \ldots, n$; and the upper bound for the flow on every arc $(i, j) \in \mathscr{A}$ is set to be u_{ij}. Finally, we set the lower bound for the flow on every arc in $G(\mathscr{A})$ to be 0. It is straightforward to show that $P(d, \mathscr{A}, u)$ can be computed by identifying a circulation satisfying the lower and upper bounds on flows with a maximum weight.

The underlying graph of the maximum-weight circulation problem, $G(\mathscr{A})$, is illustrated in Fig. 13.3 for $\mathscr{A} = \mathscr{C}_5$, of long chain for a balanced system of size 5. Recall that flexible arcs in a long chain of size n are arcs from the set $\{(i, i+1) : i = 1, 2, \ldots, n-1\} \cup \{(n, 1)\}$.

Theorem 13.2.2 *Let $\mathscr{A}$ be a flexibility design for a balanced system of size n, and $\mathscr{A} \subseteq \mathscr{C}_n$. Let S be the set of all flexible arcs in $\mathscr{A}$. We have that $P(d, \mathscr{A}, u)$ is supermodular in u_S. Hence, for any subsets X, Y of S,*

$$P(d, \mathscr{A} \setminus (X \cap Y)) + P(d, \mathscr{A} \setminus (X \cup Y)) \geq P(d, \mathscr{A} \setminus X) + P(d, \mathscr{A} \setminus Y).$$

Proof. We first show that flexible arcs are in series (pairwise). Consider any two flexible arcs α and β and let C be an arbitrary (undirected) simple cycle in $G(\mathscr{C}_n)$ containing both α and β. If C does not contain node s, then C must be the undirected cycle that contains every plant to product arcs in $\mathscr{C}_n$. In that case, it is easy to verify that α and β have the same direction in C. Otherwise, suppose C contains s. In such a case, C can be decomposed into four pieces, X_1, X_2, α, and β, where X_1, X_2 are the two paths between α and β. Without loss of generality, we assume X_1 contains s. Since α and β cannot be incident to the same node, both X_1 and X_2 are nonempty. As X_2 does not contain s, all arcs in X_2 are plant-to-product

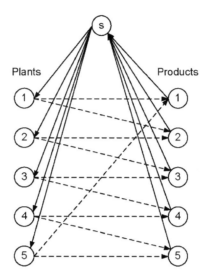

FIGURE 13.3. $G(\mathscr{C}_5)$ for the max-weight circulation associated with $P(d, \mathscr{C}_n, u)$

arcs (i.e., $X_2 \subseteq \mathscr{C}_n$). Because of the structure of $\mathscr{C}_n$, X_2 contains an odd number of arcs. Moreover, the path in $X_2 \cup \{\alpha\} \cup \{\beta\}$ has alternating directions for every two consecutive arcs and therefore, α and β have the same direction in C. This is illustrated by Fig. 13.4. Since this is true for any arbitrary undirected cycle C, α and β are in series in $G(\mathscr{C}_n)$. From Theorem 13.2.1, $P(d, \mathscr{A}, u)$ is supermodular in u_S.

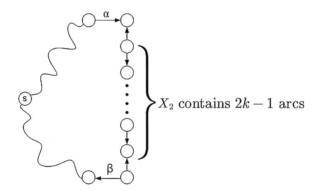

FIGURE 13.4. Illustration for the proof of Theorem 13.2.2

To complete the proof, define u and u' as follows:

$$u_\gamma = \begin{cases} 1, & \gamma \in \mathscr{A} \setminus X, \\ 0, & \gamma \in X, \end{cases} \quad u'_\gamma = \begin{cases} 1, & \gamma \in \mathscr{A} \setminus Y, \\ 0, & \gamma \in Y. \end{cases}$$

Clearly,

$$P(d, \mathscr{A} \setminus X) = P(d, \mathscr{A}, u),$$
$$P(d, \mathscr{A} \setminus Y) = P(d, \mathscr{A}, u'),$$
$$P(d, \mathscr{A} \setminus (X \cap Y)) = P(d, \mathscr{A}, u \vee u'),$$

and

$$P(d, \mathscr{A} \setminus (X \cup Y)) = P(d, \mathscr{A}, u \wedge u').$$

Thus, the inequality in the theorem holds since $P(d, \mathscr{A}, u)$ is supermodular in u_S and $X, Y \subseteq S$. ∎

Since the theorem holds for any realization of demand, it must also be true in expectation.

Corollary 13.2.3 *For any flexible arcs α, β in $\mathscr{A} \subseteq \mathscr{C}_n$, $E[P(D, \mathscr{A}, u)]$ is supermodular in u_α and u_β for any random distributions D.*

The corollary thus suggests that any two flexible arcs in the long chain complement each other. That is, the existence of one flexible arc increases the marginal benefit that can be gained when the other flexible arc is added.

13.2.2 Incremental Benefits in Long Chains

Corollary 13.2.3 is useful to prove that the incremental benefits associated with adding arcs to the long chain is increasing. Consider the following sequence of flexibility designs: $\mathscr{L}_1^n, \mathscr{L}_2^n, \mathscr{L}_3^n, \ldots, \mathscr{L}_n^n, \mathscr{C}_n$, where we define $\mathscr{L}_1^n = \mathscr{D}_n$ and $\mathscr{L}_k^n = \mathscr{L}_k \cup \{(i, i) | i = k + 1, \ldots, n\}$. In words, $\mathscr{L}_k^n$ is simply the open chain from plant 1 to product k plus the dedicated arcs connecting plants i to products i for all $k < i \leq n$. One can think of the sequence $\mathscr{L}_2^n, \ldots, \mathscr{L}_n^n, \mathscr{C}_n$ as different stages of constructing $\mathscr{C}_n$, by starting at $\mathscr{L}_1^n = \mathscr{D}_n$ and adding flexible arcs $(1, 2), (2, 3), \ldots, (n - 1, n), (n, 1)$ sequentially. Finally, recall that $\mathscr{C}_n$ is the long chain of size n. In the following, we show that the incremental benefit, $[\mathscr{L}_k^n] - [\mathscr{L}_{k-1}^n]$, is nondecreasing with k.

Theorem 13.2.4 *For any balanced system of size n with exchangeable demand, we have*

$$[\mathscr{L}_2^n] - [\mathscr{L}_1^n] \leq [\mathscr{L}_3^n] - [\mathscr{L}_2^n] \leq \ldots \leq [\mathscr{L}_n^n] - [\mathscr{L}_{n-1}^n] \leq [\mathscr{C}_n] - [\mathscr{L}_n^n].$$

Proof. Fix any $1 \leq k \leq n - 1$. Let $\alpha = (1, 2)$, $\beta = (k, k+1)$. Note that by definition,

$$E[P(D, \mathscr{L}_{k+1}^n)] = [\mathscr{L}_{k+1}^n]$$

and

$$E[P(D, \mathscr{L}_{k+1}^n \setminus \{\beta\})] = E[P(D, \mathscr{L}_k^n)] = [\mathscr{L}_k^n].$$

Let $D_\sigma = [D_2, D_3, \ldots, D_n, D_1]$. Observe that

$$E[P(D, \mathscr{L}_{k+1}^n \setminus \{\alpha, \beta\})] = E[P(D_\sigma, \mathscr{L}_{k+1}^n \setminus \{\alpha, \beta\})] = E[P(D, \mathscr{L}_{k-1}^n)] = [\mathscr{L}_{k-1}^n]$$

and

$$E[P(D, \mathscr{L}_{k+1}^n \setminus \{\alpha\})] = E[P(D_\sigma, \mathscr{L}_k^n)] = E[P(D, \mathscr{L}_k^n)] = [\mathscr{L}_k^n],$$

where the above equalities hold since the random vector D is exchangeable.

By Corollary 13.2.3, we have

$$E[P(D, \mathscr{L}_{k+1}^n)] + E[P(D, \mathscr{L}_{k+1}^n \setminus \{\alpha, \beta\})] \geq E[P(D, \mathscr{L}_{k+1}^n \setminus \{\alpha\})]$$
$$+ E[P(D, \mathscr{L}_{k+1}^n \setminus \{\beta\})].$$

Thus, $[\mathscr{L}_{k+1}^n] - [\mathscr{L}_k^n] \geq [\mathscr{L}_k^n] - [\mathscr{L}_{k-1}^n]$, for $k = 2, \ldots, n-1$.

To show $[\mathscr{L}_n^n] - [\mathscr{L}_{n-1}^n] \leq [\mathscr{C}_n] - [\mathscr{L}_n^n]$, let $\alpha = (1, 2)$, $\beta = (n, 1)$ and let $D_\sigma = [D_2, D_3, \ldots, D_n, D_1]$. Then

$$E[P(D, \mathscr{C}_n)] = [\mathscr{C}_n],$$
$$E[P(D, \mathscr{C}_n \setminus \{\beta\})] = [\mathscr{L}_n^n],$$
$$E[P(D, \mathscr{C}_n \setminus \{\alpha, \beta\})] = E[P(D_\sigma, \mathscr{L}_{n-1}^n)] = [\mathscr{L}_{n-1}^n],$$

and

$$E[P(D, \mathscr{C}_n \setminus \{\alpha\})] = E[P(D_\sigma, \mathscr{L}_n^n)] = [\mathscr{L}_n^n].$$

Again by Corollary 13.2.3,

$$E[P(D, \mathscr{C}_n)] + E[P(D, \mathscr{C}_n \setminus \{\alpha, \beta\})] \geq E[P(D, \mathscr{C}_n \setminus \{\alpha\})] + E[P(D, \mathscr{C}_n \setminus \{\beta\})],$$

which implies that $[\mathscr{C}_n] - [\mathscr{L}_n^n] \geq [\mathscr{L}_n^n] - [\mathscr{L}_{n-1}^n]$. This completes the proof. ∎

Observe that the proof of Theorem 13.2.4 requires the application of the supermodularity result (Theorem 13.2.2), which holds deterministically for any fixed-demand instance. By contrast, Theorem 13.2.4 holds only stochastically under exchangeable demand but does not hold for any fixed-demand instance.

13.3   Characterizing the Performance of Long Chains

In this section, we show that in a balanced system of size n with exchangeable demand, the performance of the long chain can be characterized by the difference between the performances of two open chains, which allows us to show that the long chain is optimal among a class of flexibility designs and develop an efficient algorithm to compute its expected performance. Like the previous section, we start by developing several properties of the long chain when the demand is deterministic.

13.3.1   Decomposition of a Long Chain

In this subsection, we fix an arbitrary demand instance d. Throughout the subsection, when some integer k appears in a statement, we are, in fact, referring to some $i \in \{1, ..., n\}$ *congruent* to k *modulo* n. For example, if plant $n+3$ appears in a statement, then we are referring to plant 3; and if $f_{n+1,n+2}$, the flow from plant $n+1$ to product $n+2$ appears in a statement, then we are referring to $f_{1,2}$, the flow from plant 1 to product 2.

First, we start with the following lemma.

Lemma 13.3.1 *Suppose* $P(d, \mathscr{C}_n \setminus \{\alpha\}) = P(d, \mathscr{C}_n)$, *where* α *is a flexible arc in* $\mathscr{C}_n$. *Then, for any set* $X \subseteq S$, *where* S *is the set of all flexible arcs in* $\mathscr{C}_n$, *we have that*

$$P(d, \mathscr{C}_n \setminus (X \cup \{\alpha\})) = P(d, \mathscr{C}_n \setminus X).$$

Proof. If $\alpha \in X$, the result is trivial as $X \cup \{\alpha\} = X$. Otherwise, by Theorem 13.2.2,

$$P(d, \mathscr{C}_n \setminus (X \cup \{\alpha\})) + P(d, \mathscr{C}_n) \geq P(d, \mathscr{C}_n \setminus X) + P(d, \mathscr{C}_n \setminus \{\alpha\}),$$

which, together with the assumption that $P(d, \mathscr{C}_n) = P(d, \mathscr{C}_n \setminus \{\alpha\})$, implies that

$$P(d, \mathscr{C}_n \setminus (X \cup \{\alpha\})) \geq P(d, \mathscr{C}_n \setminus X).$$

Since by definition, $P(d, \mathscr{C}_n \setminus (X \cup \{\alpha\})) \leq P(d, \mathscr{C}_n \setminus X)$, the lemma holds. ∎

Next, we show that the sales associated with $\mathscr{C}_n$ can be expressed as a sum of n quantities, where each quantity is the difference of the sales associated with two open chains in $\mathscr{C}_n$.

Theorem 13.3.2 *For any fixed-demand instance* d *on balanced system of size* n, *we have*

$$P(d, \mathscr{C}_n) = \sum_{i=1}^{n} (P(d, \mathscr{C}_n \setminus \{(i, i+1)\}) - P(d, \mathscr{C}_n \setminus \{(i-1, i), (i, i), (i, i+1)\})).$$

Proof. Since the demand instance d is fixed, for the sake of succinctness, we use $P(\mathscr{A})$ to denote $P(d, \mathscr{A})$ in the proof. We also define $\alpha_i = (i, i+1)$ and $\beta_i = (i, i)$ for $i = 1, 2, \ldots, n$ [note that $\alpha_n = (n, 1)$ as $n+1$ is congruent with 1 modular n]. We first claim that there is some i^* such that $P(\mathscr{C}_n) = P(\mathscr{C}_n \setminus \{\alpha_{i^*}\})$. Indeed, given an optimal solution of the maximum-flow problem defining $P(\mathscr{C}_n)$, f^*, if $f^*_{\alpha_i} > 0$ for any $i = 1, \ldots, n$, define a new flow $\hat{f}$ such that

$$\hat{f}_{\alpha_i} = f^*_{\alpha_i} - \delta, \hat{f}_{\beta_i} = f^*_{\beta_i} + \delta \ \forall \ i = 1, \ldots, n,$$

where $\delta = \min_{i=1:n} f^*_{\alpha_i}$. Let $f^*_{\alpha_{i^*}} = \delta$. Clearly, $\hat{f}$ is feasible for the design $\mathscr{C}_n \setminus \{\alpha_{i^*}\}$ and generates exactly the same amount of total flow as f^* for the design $\mathscr{C}_n$.

Thus, $\mathscr{C}_n$ is optimal for the maximum-flow problem defining $P(\mathscr{C}_n \setminus \{\alpha_{i*}\})$ and $P(\mathscr{C}_n) = P(\mathscr{C}_n \setminus \{\alpha_{i*}\})$. Without loss of generality, we assume that $i^* = n$, as we can always relabel each plant (and product) i by $i - i^*$.

For each $1 \leq k_1 \leq k_2 \leq n$, define $\mathscr{L}_{k_1 \to k_2} = \{(i, i) | i = k_1, k_1 + 1, \ldots, k_2\} \cup \{(i, i+1) | i = k_1, k_1 + 1, \ldots, k_2 - 1\}$, and for each $1 \leq k_2 < k_1 \leq n$, define $\mathscr{L}_{k_1 \to k_2} = \{(i, i) | i = k_1, k_1 + 1 \ldots, n, 1, 2, \ldots, k_2\} \cup \{(i, i+1) | i = k_1, \ldots, n, 1, \ldots, k_2 - 1\}$. One can think of $\mathscr{L}_{k_1 \to k_2}$ as the open chain connecting plant k_1 to product k_2 in the balanced system of size n.

By the definition of α and β, we have that for $i = 1, \ldots, n - 1$,

$$
\begin{aligned}
&P(\mathscr{C}_n \setminus \{\alpha_i\}) - P(\mathscr{C}_n \setminus \{\alpha_{i-1}, \alpha_i, \beta_i\}) \\
&= P(\mathscr{C}_n \setminus \{\alpha_i\}) - P(\mathscr{C}_n \setminus \{\alpha_{i-1}, \alpha_i\}) + \min\{1, d_i\} \\
&= P(\mathscr{C}_n \setminus \{\alpha_i, \alpha_n\}) - P(\mathscr{C}_n \setminus \{\alpha_{i-1}, \alpha_i, \alpha_n\}) + \min\{1, d_i\} \\
&= P(\mathscr{L}_{1 \to i}) + P(\mathscr{L}_{(i+1) \to n}) + \min\{1, d_i\} \\
&\quad - \left(P(\mathscr{L}_{1 \to (i-1)}) + P(\mathscr{L}_{(i+1) \to n}) + \min\{1, d_i\} \right) \\
&= P(\mathscr{L}_{1 \to i}) - P(\mathscr{L}_{1 \to (i-1)}),
\end{aligned}
\tag{13.2}
$$

where the first equality holds since $\mathscr{C}_n \setminus \{\alpha_{i-1}, \alpha_i\}$ is the union of two disjoint components $\{\beta_i\}$ and $\mathscr{C}_n \setminus \{\alpha_{i-1}, \alpha_i, \beta_i\}$, the second equality follows from Lemma 13.3.1, and the third equality holds since $\mathscr{C}_n \setminus \{\alpha_i, \alpha_n\}$ is the disjoint union of components $\mathscr{L}_{1 \to i}$ and $\mathscr{L}_{(i+1) \to n}$, and $\mathscr{C}_n \setminus \{\alpha_{i-1}, \alpha_i, \alpha_n\}$ is the disjoint union of $\mathscr{L}_{1 \to (i-1)}$, $\mathscr{L}_{(i+1) \to n}$ and $\{\beta_i\}$. Similarly,

$$
\begin{aligned}
&P(\mathscr{C}_n \setminus \{\alpha_1\}) - P(\mathscr{C}_n \setminus \{\alpha_n, \alpha_1, \beta_1\}) \\
&= P(\mathscr{C}_n \setminus \{\alpha_1\}) - P(\mathscr{C}_n \setminus \{\alpha_n, \alpha_1\}) + \min\{1, d_1\} \\
&= P(\mathscr{C}_n \setminus \{\alpha_1, \alpha_n\}) - P(\mathscr{C}_n \setminus \{\alpha_1, \alpha_n\}) + \min\{1, d_1\} \\
&= \min\{1, d_1\},
\end{aligned}
\tag{13.3}
$$

where the first equality holds since $\mathscr{C}_n \setminus \{\alpha_n, \alpha_1\}$ is the union of two disjoint components $\{\beta_1\}$ and $\mathscr{C}_n \setminus \{\alpha_n, \alpha_1, \beta_1\}$ and the second inequality follows from Lemma 13.3.1.

Finally,

$$
P(\mathscr{C}_n \setminus \{\alpha_n\}) - P(\mathscr{C}_n \setminus \{\alpha_{n-1}, \alpha_n, \beta_n\}) = P(\mathscr{L}_{1 \to n}) - P(\mathscr{L}_{1 \to (n-1)}).
\tag{13.4}
$$

Now, applying Equations (13.2)–(13.4), we obtain that

$$
\begin{aligned}
&\textstyle\sum_{i=1}^n (P(\mathscr{C}_n \setminus \{\alpha_i\}) - P(\mathscr{C}_n \setminus \{\alpha_{i-1}, \alpha_i, \beta_i\})) \\
&= \min\{1, d_1\} + \textstyle\sum_{i=2}^n (P(\mathscr{L}_{1 \to i}) - P(\mathscr{L}_{1 \to (i-1)})) \\
&= \min\{1, d_1\} + P(\mathscr{L}_{1 \to n}) - P(\mathscr{L}_{1 \to 1}) \\
&= P(\mathscr{L}_{1 \to n}) \\
&= P(\mathscr{C}_n \setminus \{\alpha_n\}) \\
&= P(\mathscr{C}_n).
\end{aligned}
$$

This completes the proof.   ∎

We note that $\mathscr{C}_n \setminus \{(i, i+1)\}$ is an open chain connecting plant $i+1$ to product i, while $\mathscr{C}_n \setminus \{(i-1, i), (i, i), (i, i+1)\}$ is an open chain connecting plant $i+1$ to product $i-1$.

13.3.2 Characterization and Optimality

With Theorem 13.3.2, we can now characterize the performance of the long chain using the performances of open chains.

Theorem 13.3.3 *For any balanced system of size n with exchangeable demand D, we have*

$$[\mathscr{C}_n] = n([\mathscr{L}_n] - [\mathscr{L}_{n-1}]).$$

Proof. Theorem 13.3.2 states that for any d that is an instance of D,

$$P(d, \mathscr{C}_n) = \sum_{i=1}^{n} (P(d, \mathscr{C}_n \setminus \{(i, i+1)\}) - P(d, \mathscr{C}_n \setminus \{(i-1, i), (i, i), (i, i+1)\})). \quad (13.5)$$

Since D is exchangeable, for any $1 \leq i \leq n$,

$$E[P(D, \mathscr{C}_n \setminus \{(i, i+1)\})] = [\mathscr{L}_n],$$
$$E[P(D, \mathscr{C}_n \setminus \{(i-1, i), (i, i), (i, i+1)\})] = [\mathscr{L}_{n-1}].$$

Thus, the theorem follows by integrating over all random instances of D on Equation (13.5). ∎

Theorem 13.3.3 provides insights on the performance of long chains. Indeed, it relates the expected performance of a long chain, $[\mathscr{C}_n]$, with the difference in the expected performances of two open chains, $[\mathscr{L}_n]$ and $[\mathscr{L}_{n-1}]$, which are much easier to compute and analyze.

An immediate corollary of Theorem 13.3.3 is that the long chain is optimal among all 2-flexibility designs.

Corollary 13.3.4 *Consider a balanced system of size n with exchangeable demand. Let $\mathbb{F}_2$ be the set of all 2-flexibility designs of the system. That is, $\mathbb{F}_2$ is the set of all flexibility designs where each plant node and each product node are incident to exactly two arcs. Then we have*

$$[\mathscr{C}_n] = \arg\max_{\mathscr{A} \in \mathbb{F}_2} [\mathscr{A}].$$

In words, the long chain maximizes expected sales among all 2-flexibility designs in the system.

Proof. Consider a 2-flexibility design $\mathscr{A} \in \mathbb{F}_2$. $\mathscr{A}$ must consist of several closed chains (i.e., induced subgraphs in $\mathscr{A}$ that form undirected cycles) denoted by

$SC_1, SC_2, ..., SC_k$. Let n_i be the number of products and plants in the closed chain SC_i. Since the system size is n, $\sum_{i=1}^{k} n_i = n$. Now, by Theorem 13.3.3, we have

$$
\begin{aligned}
[\mathscr{A}] &= \sum_{i=1}^{k} n_i([\mathscr{L}_{n_i}] - [\mathscr{L}_{n_i-1}]) \\
&= \sum_{i=1}^{k} n_i([\mathscr{L}_{n_i}^n] - [\mathscr{L}_{n_i-1}^n] + E[\min\{1, D_1\}]) \\
&\leq \sum_{i=1}^{k} n_i([\mathscr{L}_n^n] - [\mathscr{L}_{n-1}^n] + E[\min\{1, D_1\}]) \\
&= \sum_{i=1}^{k} n_i([\mathscr{L}_n] - [\mathscr{L}_{n-1}]) \\
&= n([\mathscr{L}_n] - [\mathscr{L}_{n-1}]) \\
&= [\mathscr{C}_n],
\end{aligned}
$$

where the second and third equalities follow from the definition of $\mathscr{L}_k^n$, and the inequality from Theorem 13.2.4. ∎

13.3.3   Computing the Performance of a Long Chain

In this subsection, we present a method to compute $[\mathscr{C}_n]$, the performance of the long chain, based on Theorem 13.3.2. We focus on balanced systems with iid demand. Because we consider systems of arbitrary sizes, we let D be an infinite random vector with iid entries, where D_i is the random demand for product i generated by a given distribution D for all $i \geq 1$.

Since $\frac{[\mathscr{C}_n]}{n}$ is the difference of $[\mathscr{L}_n]$ and $[\mathscr{L}_{n-1}]$ by Theorem 13.3.3, we first introduce a greedy algorithm that finds the optimal solution of the linear program associated with $P(d, \mathscr{L}_n)$, where d is an instance of D.

Finding the Optimal Solution f^* for $P(d, \mathscr{L}_n)$

Step 1: Set $f_{1,1}^* = \min(1, d_1)$.

Step 2: For $k = 2$ to n,
set $f_{k-1,k}^* = \min\{1 - f_{k-1,k-1}^*, d_k\}$ and $f_{k,k}^* = \min\{1, d_k - f_{k-1,k}^*\}$.

Showing that f^* is optimal for the open chain $\mathscr{L}_n$ is straightforward and is left as an exercise.

Given a random demand vector D, let F_{ij} be the random flow on arc (i, j) returned by the above algorithm, for $1 \leq i, j \leq n$. For each integer $1 \leq k \leq n-1$, define $W_k = 1 - F_{kk}$ and $W_0 = 0$. W_k can be thought of as the remaining capacity in plant k after the production of product k at plant k is determined.

To develop a method to compute the performance of the long chain, assume that the support of D lies in $\{\frac{i}{N} | i = 0, 1, 2, \ldots\}$ for some $N \geq 1$. Under this assumption, we let $p_i = \Pr(D = \frac{i}{N})$, for any $i = 0, 1, \ldots, 2N - 1$, and $p_{2N} = \Pr(D \geq 2)$, where $\Pr(\cdot)$ denotes the probability mass function of D.

Since the support of D lies in $\{\frac{i}{N} | i = 0, 1, 2, \ldots\}$, it is easy to see that the support of F_{kk} lies in $\{\frac{i}{N} | i = 0, 1, 2, \ldots, N\}$. Since $W_k = 1 - F_{kk}$, the support set of W_k is also $\{\frac{i}{N} | i = 0, 1, 2, \ldots, N\}$. As a result, the distribution of W_k can be described by a row vector q^k with $N + 1$ elements, where its ith component q_i^k equals $\Pr(W_k = \frac{i}{N})$, for $i = 0, 1, \ldots, N$. The following lemma illustrates how q^k can be computed.

Lemma 13.3.5 $q^{k+1} = q^k A = q^0 A^{k+1}$ for $0 \le k \le n - 1$, where $q^0 = [1\,0\,0\,\ldots\,0]$

$$
A = \begin{bmatrix}
\sum_{i=N}^{2N} p_i & p_{N-1} & p_{N-2} & \cdots & p_1 & p_0 \\
\sum_{i=N+1}^{2N} p_i & p_N & p_{N-1} & \cdots & p_2 & p_0 + p_1 \\
\vdots & \vdots & \vdots & \vdots & \vdots & \vdots \\
p_{2N-1} + p_{2N} & p_{2N-2} & p_{2N-3} & \cdots & p_N & \sum_{i=0}^{N-1} p_i \\
p_{2N} & p_{2N-1} & p_{2N-2} & \cdots & p_{N+1} & \sum_{i=0}^{N} p_i
\end{bmatrix}.
$$

Proof. Since W_0 is 0 by definition, $q^0 = [1\,0\,0\,\ldots\,0]$. Because the demand is independent and W_k only depends on $D_1, \ldots, D_k$, W_k is independent of D_{k+1}. Hence, we have for $1 \le i \le N - 1$,

$$
q_i^{k+1} = \Pr\left(W_{k+1} = \frac{i}{N}\right) = \sum_{j=0}^{N} \Pr(W_k = j)\Pr\left(D_{k+1} = 1 - \frac{i}{N} + \frac{j}{N}\right) = \sum_{j=0}^{N} q_j^k p_{N-i+j}.
$$

In addition,

$$
q_0^{k+1} = \Pr(W_{k+1} = 0) = \sum_{j=0}^{N} \Pr(W_k = j)\Pr\left(D_{k+1} \ge 1 + \frac{j}{N}\right) = \sum_{j=0}^{N} q_j^k \sum_{l=N+j}^{2N} p_{N+l}
$$

and

$$
q_N^{k+1} = \Pr(W_{k+1} = 1) = \sum_{j=0}^{N} \Pr(W_k = j)\Pr\left(D_{k+1} \le \frac{j}{N}\right) = \sum_{j=0}^{N} q_j^k \sum_{l=0}^{j} p_l.
$$

Thus, $q^{k+1} = q^k A$. ∎

A direct consequence of Lemma 13.3.5 is that the following matrix multiplications can be used to determine the performance of the long chain when demands are iid and the support of a product demand is a subset of $\{\frac{i}{N} | i = 0, 1, 2, \ldots\}$.

Theorem 13.3.6 $\frac{[\mathscr{C}_n]}{n} = [\mathscr{L}_n] - [\mathscr{L}_{n-1}] = q^{n-1}\pi = q^0 A^{n-1}\pi$, where π is a vector of size $N + 1$, with

$$
\pi_i = \sum_{j=1}^{N+i} j p_j + (N + i) \sum_{j=N+i+1}^{2N} p_j, \qquad \forall\, 0 \le i \le N.
$$

Proof. The algorithm described above implies that $[\mathscr{L}_n] - [\mathscr{L}_{n-1}]$ can be written as the expectation of $F_{n-1n} + F_{nn}$, which is equal to $E[\min\{1 + W_{n-1}, D_n\}]$. Moreover,

$$E[\min\{1 + W_{n-1}, D_n\}] = \sum_{i=0}^{N} \Pr\left(W_{n-1} = i\right) E[\min\{D_n, 1 + \tfrac{i}{N}\}]$$
$$= \sum_{i=0}^{N} q_i^{n-1}(\sum_{j=1}^{N+i} jp_j + (N+i)\sum_{j=N+i+1}^{2N} p_j).$$

Hence, we have that $[\mathscr{L}_n] - [\mathscr{L}_{n-1}] = q^{n-1}\pi$. Apply Theorem 13.3.3, and we are done. ∎

The matrix multiplication method developed here to compute the performance of the long chain is polynomial in N and n. Indeed, computing $q^0 A^{n-1}\pi$ requires $O(nN^2)$ operations if one sequentially evaluates $q^0 A^i$ for $i = 1, \ldots, k$, or $O(N^{2.807} \log n)$ operations if one starts by determining A^{n-1} using the classical algorithm from Strassen (1969).

The matrix multiplication method can be used to compute the per-product performance of the long chain for an infinite-size system. Observe that the matrix A is the transition matrix of a Markov chain with states $\frac{i}{N}$ for each $i = 0, 1, \ldots, N$. Since we focus on a balanced system with $E[D] = 1$, we can show that the states in the Markov chain can be partitioned into two sets, a set containing states, including state 0, that communicate with each other and the other set containing inessential states. It is well known that $\lim_{n\to\infty} q^0 A^{n-1} = q^*$, where $q^* A = q^*$, $q_0^* > 0$, and $q_j^* = 0$ for any inessential states j. Thus, to compute $\lim_{n\to\infty} \frac{[\mathscr{C}_n]}{n}$, one can solve for q^* by finding the eigenvectors of A and then compute $q^*\pi$, which gives $\lim_{n\to\infty} \frac{[\mathscr{C}_n]}{n}$.

We can apply the matrix multiplication method for general iid demands as an approximation algorithm to compute the performance of long chains. In this case, one can approximate the performance of the long chain by discretizing the demand distribution on the set of $\{\frac{i}{N}|i = 0, 1, 2, \ldots\}$ for some integer N. Clearly, as N increases, the error of the approximation decreases while the running time grows. Specifically, it is straightforward to show that the error of the approximation is bounded by $\frac{n}{2N}$. Interestingly, computational experience suggests that the error is much smaller than this bound.

Moreover, the matrix multiplication method is fairly fast even for large N. For example, when $N = 1000$ and $n = 100$, $q^0 A^{n-1}\pi$ can be computed within 2 s using Matlab on a standard 2.1 GHz laptop. Hence, even for general iid demands, the matrix multiplication method can quickly approximate the performance of a large long chain very accurately.

Figure 13.5 presents computational results obtained using the matrix multiplication method for three different iid demand distributions:

- **Normal**: Demand for a product is a discretized normal random variable with mean 1 and standard deviation of 0.33 on the support set of $\{\frac{i}{14}|i = 0, 1, \ldots, 28\}$;

- **Uniform**: Demand for a product is uniformly distributed on the set $\{\frac{i}{10}|i = 0, 1, 2, \ldots, 9, 11, 12, \ldots, 20\}$;

- **Asymmetric**: Demand for a product is equal to $\frac{4}{5}$ with probability 0.4, 1 with probability 0.5, and 2 with probability 0.1.

For each distribution, Fig. 13.5 depicts $\frac{\lfloor\mathscr{F}_n\rfloor}{n}$ (the per-product performance of full flexibility), $\frac{\lfloor\mathscr{C}_n\rfloor}{n}$ (the per-product performance of the long chain), and $\frac{\lfloor\mathscr{C}_n\rfloor}{\lfloor\mathscr{F}_n\rfloor}$ (the ratio between the performance of the long chain and the performance of full flexibility design) for $n = 1, \ldots, 30$.

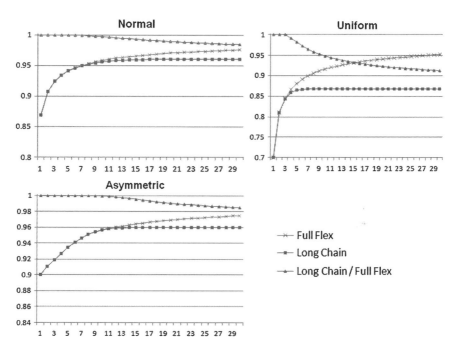

FIGURE 13.5. The performance of long chains vs. the performance of full flexibility

Figure 13.5 reveals several interesting observations. First, $\frac{\lfloor\mathscr{F}_n\rfloor}{n} - \frac{\lfloor\mathscr{C}_n\rfloor}{n}$, that is, the gap between the fill rates of full flexibility and the long chain, is increasing, while the ratio $\frac{\lfloor\mathscr{C}_n\rfloor}{\lfloor\mathscr{F}_n\rfloor}$ is decreasing. In addition, Fig. 13.5 suggests that the quantity $\frac{\lfloor\mathscr{C}_n\rfloor}{n}$, the fill rate of the long chain, is increasing but converges to a constant very quickly.

13.4 Performance of Long Chains

In this section, we present several analytical results that provide justifications to the observations in Fig. 13.5. The following theorem illustrates that the quantity $\frac{\lfloor\mathscr{F}_n\rfloor}{n} - \frac{\lfloor\mathscr{C}_n\rfloor}{n}$ is indeed nondecreasing with n. Its proof is a bit involved and hence omitted.

Theorem 13.4.1 *For any integer $n \geq 2$ and iid demand,*

$$\frac{[\mathscr{F}_n]}{n} - \frac{[\mathscr{C}_n]}{n} \leq \frac{[\mathscr{F}_{n+1}]}{n+1} - \frac{[\mathscr{C}_{n+1}]}{n+1} \leq \min\{1, E[D]\} - \gamma,$$

where $\gamma = \lim_{k \to \infty} \frac{[\mathscr{C}_k]}{k}$.

Note that the fill rate of $\mathscr{C}_n$ (and $\mathscr{F}_n$) is equal to $\frac{[\mathscr{C}_n]}{nE[D]}$ (and $\frac{[\mathscr{F}_n]}{nE[D]}$). Thus, Theorem 13.4.1 implies that the smaller the system size, the smaller the gap between the fill rate of full flexibility and that of the long chain. This suggests that the long chain is more effective relative to full flexibility for smaller systems.

Moreover, Theorem 13.4.1 can be used to bound the gap between the fill rate of full flexibility and that of the long chain for systems of *any size*. Since for many iid demand with $E[D] = 1$, γ is close to 1, as shown in Chou et al. (2010), Theorem 13.4.1 implies that for any size system, the performance of the long chain is close to that of full flexibility. For example, when D is normal with mean 1 and standard deviation 0.33, $\gamma = 0.96$. Therefore, for this demand distribution, we have that the gap between the fill rate of full flexibility and that of the long chain for systems of any size is at most 4 %.

Though the ratio of the performance of long chain to that of full flexibility, $\frac{[\mathscr{C}_n]}{[\mathscr{F}_n]} \geq \frac{[\mathscr{C}_{n+1}]}{[\mathscr{F}_{n+1}]}$, is observed to be nonincreasing empirically in Chou et al. (2008), whether it can be proven analytically remains an open question. Of course, if $\frac{[\mathscr{C}_n]}{[\mathscr{F}_n]} \geq \frac{[\mathscr{C}_{n+1}]}{[\mathscr{F}_{n+1}]}$ indeed holds, it follows that

$$\frac{[\mathscr{C}_n]}{[\mathscr{F}_n]} \geq \lim_{k \to \infty} \frac{[\mathscr{C}_k]/k}{[\mathscr{F}_k]/k} = \lim_{k \to \infty} \frac{\gamma}{\min\{E[D], 1\}}, \quad \forall n \geq 2, \tag{13.6}$$

where, as before, $\gamma = \lim_{k \to \infty} \frac{[\mathscr{C}_k]}{k}$ and $\lim_{k \to \infty} \frac{[\mathscr{F}_k]}{k} = \min\{E[D], 1\}$ by the weak law of large numbers. This would provide a lower bound on the ratio of the performance of the long chain to that of full flexibility for any system size.

A slightly weaker lower bound of the ratio when $E[D] = 1$ is provided by Simchi-Levi et al. (2012).

Corollary 13.4.2 *Suppose demand is iid and $E[D] = 1$; then*

$$\frac{[\mathscr{C}_n]}{[\mathscr{F}_n]} \geq 1 - \frac{(1 - \gamma)n}{[\mathscr{F}_n]},$$

where $\gamma = \lim_{k \to \infty} \frac{[\mathscr{C}_k]}{k}$.

Proof. By Theorem 13.4.1,

$$\frac{[\mathscr{F}_i]}{i} - \frac{[\mathscr{C}_i]}{i} \leq \frac{[\mathscr{F}_{i+1}]}{i+1} - \frac{[\mathscr{C}_{i+1}]}{i+1}, \quad \forall i \geq 1.$$

Thus,

$$\frac{[\mathscr{F}_i]}{i} - \frac{[\mathscr{F}_{i+1}]}{i+1} \leq \frac{[\mathscr{C}_i]}{i} - \frac{[\mathscr{C}_{i+1}]}{i+1}, \quad \forall i \geq 1. \tag{13.7}$$

Now, adding inequality (13.7) for all $i \geq n$, we have that

$$\frac{[\mathscr{F}_n]}{n} - \lim_{k \to \infty} \frac{[\mathscr{F}_k]}{k} \leq \frac{[\mathscr{C}_n]}{n} - \lim_{k \to \infty} \frac{[\mathscr{C}_k]}{k}. \tag{13.8}$$

But

$$\lim_{k \to \infty} \frac{[\mathscr{F}_k]}{k} = 1, \quad \lim_{k \to \infty} \frac{[\mathscr{C}_k]}{k} = \gamma,$$

and substituting those into inequality (13.8), we have

$$\frac{[\mathscr{F}_n]}{n} - 1 \leq \frac{[\mathscr{C}_n]}{n} - \gamma,$$

which leads to the inequality in the statement of this corollary. ∎

To explain the power of the lower bound in Corollary 13.4.2, let $\delta_n = \frac{n}{[\mathscr{F}_n]} - 1$, which implies that $1 - \frac{(1-\gamma)n}{[\mathscr{F}_n]} = \gamma - \delta_n(1 - \gamma)$. Since $\frac{[\mathscr{F}_n]}{n}$ is nondecreasing with n (readers are asked to prove this claim in the exercises), δ_n is nonincreasing. Thus, if $\delta_k \approx 0$ for some small integer k, then Corollary 13.4.2 provides a lower bound for $\frac{[\mathscr{C}_n]}{[\mathscr{F}_n]}$ that is close to γ for all $n \geq k$. Indeed, for many distributions with $E[D] = 1$, $\delta_k \approx 0$ for small k. For example, suppose the distribution of D is normal with mean 1 and standard deviation 0.33; then $\frac{3}{[\mathscr{F}_3]} = 1.08$, which implies $\delta_3 = 0.08$. Since $\gamma = 0.96$, by applying Corollary 13.4.2, we have that

$$\frac{[\mathscr{C}_n]}{[\mathscr{F}_n]} \geq \gamma - \delta_3(1 - \gamma) = 0.96 - 0.04 \times 0.08 = 0.9568, \quad \forall n \geq 3.$$

That is, when demand is normal with mean 1 and standard deviation 0.33, the long chain of any size greater than 2 achieves at least 95.68 % of the performance of full flexibility.

Finally, we focus on the per-product performance (and fill rate, which is linearly proportional to the per-product performance) of the long chain as a function of system size. We start by showing that $\frac{[\mathscr{C}_n]}{n}$ is nondecreasing with n under iid demand.

Theorem 13.4.3 *Under iid demand D, we have* $\frac{[\mathscr{C}_n]}{n} \leq \frac{[\mathscr{C}_{n+1}]}{n+1}$, *for any integer* $n \geq 2$.

Proof. Since D is iid, the first n (and $n + 1$) entries in D are exchangeable for a balanced system of size n (and $n + 1$). Thus, by Theorem 13.2.4, we have that

$$[\mathscr{L}_n^n] - [\mathscr{L}_{n-1}^n] \leq [\mathscr{L}_{n+1}^{n+1}] - [\mathscr{L}_n^{n+1}],$$

which is equivalent to

$$[\mathscr{L}_n] - [\mathscr{L}_{n-1}] - E[\min\{D, 1\}] \leq [\mathscr{L}_{n+1}] - [\mathscr{L}_n] - E[\min\{D, 1\}]$$

and hence implies that

$$[\mathscr{L}_n] - [\mathscr{L}_{n-1}] \leq [\mathscr{L}_{n+1}] - [\mathscr{L}_n].$$

Applying Theorem 13.3.3 completes the proof.   ∎

The theorem thus states that $\frac{[\mathscr{C}_n]}{n}$, as well as the fill rate associated with a long chain, increases with the number of products, n. This phenomenon is analogous to the classical "risk-pooling" effect associated with demand aggregation, except that here we aggregate across capacities.

Interestingly, Fig. 13.5 suggests that the fill rate of the long chain quickly converges to a constant. This is shown in the next theorem, where we prove that the convergence rate is exponential for arbitrary iid, nondegenerate demands.

Theorem 13.4.4 *When demands are iid and nondegenerate, there exist constants $c > 0$ and $K > 0$ such that*

$$\frac{[\mathscr{C}_{n+1}]}{n+1} - \frac{[\mathscr{C}_n]}{n} \leq K e^{-cn},$$

for any $n \geq 2$.

Proof. From the definition of W_{n-1}, we have that $\frac{[\mathscr{C}_n]}{n} = E[\min\{1 + W_{n-1}(D), D_n\}]$. Let $D^2 = [D_2, D_3, \ldots]$. We have

$$
\begin{aligned}
\frac{[\mathscr{C}_{n+1}]}{n+1} - \frac{[\mathscr{C}_n]}{n} &= E[\min\{1 + W_n(D), D_{n+1}\}] - E[\min\{1 + W_{n-1}(D^2), D_{n+1}\}] \\
&= E[\min\{1 + W_n(D), D_{n+1}\} - \min\{1 + W_{n-1}(D^2), D_{n+1}\}] \\
&\leq \Pr(W_n(D) \neq W_{n-1}(D^2)),
\end{aligned}
$$

where the last inequality is true because

$$\min\{1 + W_n(D), D_{n+1}\} - \min\{1 + W_{n-1}(D^2), D_{n+1}\}$$

never exceeds 1. Note that for any particular random instance d, $W_n(d) = W_{n-1}(d^2)$ if $W_i(d) = 0$ for some $1 \leq i \leq n$ or $W_i(d^2) = 1$ for some $1 \leq i \leq n - 1$. Thus,

$$\Pr(W_n(D) \neq W_{n-1}(D^2)) \leq \Pr(W_i(D) > 0, W_i(D^2) < 1, \forall 1 \leq i \leq n).$$

Therefore, we have

$$\frac{[\mathscr{C}_{n+1}]}{n+1} - \frac{[\mathscr{C}_n]}{n} \leq \Pr(W_i(D) > 0, W_i(D^2) < 1, \forall 1 \leq i \leq n).$$

Now, since D is nondegenerate and iid, there exists some t such that

$$p = \Pr\left(\sum_{j=1}^{t}(D_j - 1) \geq 1\right) > 0.$$

If some instance d satisfies the condition

$$W_i(d) > 0, W_i(d^2) < 1, \forall 1 \leq i \leq n,$$

then we must have that

$$\sum_{j=2+(k-1)t}^{kt+1} (d_j - 1) < 1$$

for any $1 \leq k \leq \lfloor \frac{n-1}{t} \rfloor$. Hence,

$$
\begin{aligned}
\frac{[\mathscr{C}_{n+1}]}{n+1} - \frac{[\mathscr{C}_n]}{n} &\leq \Pr(W_i(D) > 0, W_i(D^2) < 1, \forall 1 \leq i \leq n) \\
&\leq \Pr\left(\sum_{j=2+(k-1)t}^{kt+1}(D_j - 1) < 1, \forall 1 \leq k \leq \lfloor \frac{n-1}{t} \rfloor\right) \\
&= (1-p)^{\lfloor \frac{n-1}{t} \rfloor} \\
&\leq Ke^{-cn}
\end{aligned}
$$

for some constants $K > 0$ and $c > 0$. ∎

Figure 13.5 and Theorem 13.4.4 show that $\frac{[\mathscr{C}_n]}{n} \approx \frac{[\mathscr{C}_{n+t}]}{n+t}$ for any t provided that n is large. Hence, it implies that in a system with a large number of plants and products, it is not necessary to have a long-chain design that connects all the plants and products. A collection of several chains, each of which has a large number of plants and products, can be as effective.

13.5 Extensions

In the previous section, we focused on long chains for balanced systems and were mainly concerned with the average performance. In addition, we assumed that production quantities are decided only after demand is realized. Various extensions can be found in Chou et al. (2010, 2011, 2012). In Chou et al. (2010), in addition to analyzing the asymptotic average performance of long chains, the authors analyze a system with general demand and supply. They show using random sampling that there exists a sparse flexibility structure that achieves a performance nearly as well as the full flexibility structure on average. Specifically, in a manufacturing system with n plants and m products, there exists a flexible design $\mathscr{A}$ with $O\left(\frac{(n+m)U}{L\epsilon}\right)$ links such that $E[P(D, \mathscr{A})] \geq (1 - \epsilon)E[P(D, \mathscr{F})]$, where $U = \max \frac{D}{E[D]}$ and $L = \min \frac{D}{E[D]}$.

Since the random sampling approach reveals very few insights regarding the flexibility structure, Chou et al. (2011) show that the so-called graph expander structure, a sparse but highly connected graph, often used in communication networks, can achieve a performance nearly as good as the full flexibility structure in the worst case. For a balanced system, for any $\epsilon \in (0, 1)$, there exists a graph expander $\mathscr{A}$ with no more than Δ arcs incident to each plant node such that $P(d, \mathscr{A}) \geq (1 - \epsilon)P(d, \mathscr{F})$ for any instance d, as long as

$$\Delta \geq -\frac{1 + \log_2 U + (U + 1) \log_2 e}{-\log_2(1 - \epsilon)} + U + 1.$$

The authors extend the concept of graph expander to derive a similar result for general systems.

Finally, Chou et al. (2012) analyze a setting in which only a portion of the production can be postponed until the realization of demand. They show that in this case, though the performance of long chains deteriorates when the postponement level is moderate, a sparse structure with a small amount of additional flexibility can performnearly as well as the full flexibility structure.

13.6 Exercises

Exercise 13.1. (Murota and Shioura 2005) Consider the maximum-weight circulation problem described in Sect. 13.2. Let S be an arc set consisting of arcs that are in series (pairwise). Show that $F(w, l, u)$ is $M^{\natural}$-convex in w_S, and $L^{\natural}$ concave in l_S and in u_S.

Exercise 13.2. (Murota and Shioura 2005) In a directed graph, two arcs α, β are said to be **in parallel** if any (undirected) cycle C containing both α and β orients them in the opposite direction. Let P be an arc set consisting of arcs that are in parallel (pairwise). Show that $F(w, l, u)$ is $L^{\natural}$-convex in w_P, $M^{\natural}$-concave in l_P and in u_P.

Exercise 13.3. Show that f^* computed in the algorithm in Sect. 13.3.3 is indeed optimal for $P(d, \mathscr{L}_n)$.

Exercise 13.4. Show that $\frac{[\mathscr{F}_n]}{n}$ is nondecreasing in n. Note that the average sales of the full flexibility design, $[\mathscr{F}_n]$, is given by $E[\min\{n, \sum_{i=1}^{n} D_i\}]$.

14

Supply Chain Planning Models

14.1 Introduction

In the last decade, many companies have recognized that important cost savings and improved service levels can be achieved by effectively integrating production plans, inventory control, and transportation policies throughout their supply chains. The focus of this chapter is on planning models that integrate decisions across the supply chain for companies that rely on third-party carriers. These models are motivated in part by the great development and growth of many competing transportation modes, mainly as a consequence of deregulation of the transportation industry. This has led to a significant decrease in transportation costs charged by third-party distributors and, therefore, to an ever-growing number of companies that rely on third-party carriers for the transportation of their goods.

One important mode of transportation used in the retail, grocery, and electronic industries is the less-than-truckload (LTL) mode, which is attractive when shipment sizes are considerably less than truck capacity. Typically, LTL carriers offer volume, or quantity, discounts to their clients to encourage demand for larger, more profitable shipments. In this chapter, we model these discounts as a piecewise linear concave function of the quantity shipped.

Similarly, production costs can often be approximated by piecewise linear and concave functions of the quantity produced, that is, setup plus linear manufacturing costs. These economies of scale motivate the *shipper* to coordinate the production, routing, and timing of shipments over the transportation network to minimize systemwide costs. In what follows, we refer to this problem as the *shipper problem*.

D. Simchi-Levi et al., *The Logic of Logistics: Theory, Algorithms, and Applications for Logistics Management*, Springer Series in Operations Research and Financial Engineering, DOI 10.1007/978-1-4614-9149-1_14, © Springer Science+Business Media New York 2014

This planning model, while quite general, is based on several assumptions that are consistent with the view of modern logistics networks. Indeed, the model deals with situations in which all facilities are part of the same logistics network, and information is available to a central decision maker whose objective is to optimize the entire system. Thus, distribution problems in the retail and grocery industries are special cases of our model where the logistics network does not include manufacturing facilities.

The model also applies to situations in which suppliers and retailers are engaged in *strategic partnering*. For instance, in a vendor-managed inventory (VMI) partnership, point-of-sales data are transmitted to the supplier, which is responsible for the coordination of production and distribution, including managing retail inventory and shipment schedules. Hence, in this case, the model includes manufacturing facilities, warehouses, and retail outlets.

This deterministic tactical model is motivated, in part, by our experience with a number of companies that apply similar models on a rolling-horizon basis. That is, they consider forecast demand for the next 52 weeks and allow the model to generate a production, transportation, and inventory schedule for the entire planning horizon. The use of a rolling horizon implies that these companies employ the plan generated by the model only for a few time periods, say for the first three or four weeks. As time goes on, they update the demand forecasts and run the model again.

While this model is deterministic, in practice, safety stocks are determined exogenously and incorporated into the minimum inventory level that should be maintained at the beginning of each period. Of course, an important question when managing inventory in a complex supply chain is where to keep safety stock? The answer to this question clearly depends on the desired level of service, the logistic network, the demand forecast and forecast error, as well as lead times and lead-time variability. Thus, in Sect. 14.3, we discuss models for positioning and optimizing safety stock in the supply chain. We start in the next section with our modeling approach and results for the shipper problem.

14.2 The Shipper Problem

In this section, we focus on the shipper problem under piecewise linear and concave production and transportation costs, and use properties resulting from the concavity of the cost function to devise an efficient algorithm.

The objective of the shipper is to find a production plan, an inventory policy, and a routing strategy to minimize the total cost and satisfy all the demands. Backlogging of demands may be allowed, incurring a known penalty cost, which is a function of the length of the shortage period and the level of shortage. In this case, four different costs must be balanced to obtain an overall optimal policy: production costs; LTL shipping charges; holding costs incurred when carrying inventory at some facility; and penalty costs for delayed deliveries.

To formulate this tactical problem, we first incorporate the time dimension into the model by constructing the so-called expanded network. This expanded network is used to formulate the shipper problem as a set-partitioning problem. The formulation is found to have surprising properties, which are used to develop an efficient algorithm and to show that the linear programming relaxation of the set-partitioning formulation is tight in certain special cases (Sect. 14.2.4). Computational results, demonstrating the performance of the algorithm on a set of test problems, are reported in Sect. 14.2.5.

14.2.1 The Shipper Model

Consider a generic transportation network, $G = (N, A)$, with a set of nodes N representing the suppliers, warehouses, and customers. Customer demands for the next T periods are assumed to be deterministic, and each of them is considered a separate **commodity**, characterized by its origin, destination, size, and the time period when it is demanded. Our problem is to plan production and route shipments over time to satisfy these demands while minimizing the total production, shipping, inventory, and penalty costs.

A standard technique to efficiently incorporate the time dimension into the model is to construct the following **expanded network**. Let $\tau_1, \tau_2, \ldots, \tau_T$ be an enumeration of the model's relevant time periods. In the original network, G, each node i is replaced by a set of nodes $i_1, i_2, \ldots, i_T$. We connect node i_u with node j_v if and only if $\tau_v - \tau_u$ is exactly the time it takes to travel from i to j. Thus, arc $i_u \rightarrow j_v$ represents freight being carried from i to j starting at time τ_u and ending at time τ_v. We call such arcs *shipping links*. In order to account for penalties associated with delayed shipments, a new node is created for each commodity and serves as its ultimate sink. For a given commodity, a link between a node representing its associated retailer at a specific time period and its corresponding sink node represents the penalty cost of delivering a specific shipment in that time period; it is called the *penalty link*. Similarly, to include production decisions in the network model, we add for each node i_t, corresponding to a production facility (supplier) i at a particular point in time t, a dummy node i_t' and an arc from i_t' to i_t whose cost represents the piecewise linear concave manufacturing costs. Observe that these *production links* have the same cost structure as the shipping links. Consequently, in our analysis of the network model, we will include them in the set of shipping links. Finally, we add links (i_l, i_{l+1}) for $l = 1, 2, \ldots, T - 1$, referred to as *inventory links*.

Let $G_T = (V, E)$ be the **expanded network**. Figure 14.1 illustrates the expanded network for a simple scenario where the shipping and inventory costs have to be balanced over a time horizon of just three periods and shortages are not allowed. For simplicity, we assume that travel times are zero.

Observe that, using the expanded network, we can formulate the shipper problem as a concave-cost multicommodity network flow problem.

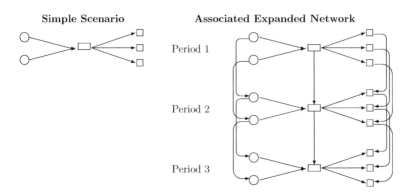

Simple Scenario   **Associated Expanded Network**

Period 1

Period 2

Period 3

FIGURE 14.1. Example of expanded network

14.2.2 A Set-Partitioning Approach

To describe our modeling approach, we introduce the following notation. Let $\mathcal{K} = \{1, 2, \ldots, K\}$ be the index set of all commodities, or different demands with fixed origin and destination, and let w_k, $k = 1, 2, \ldots, K$, be their corresponding size. For instance, commodity $k = 1$ may correspond to a demand of $w_1 = 100$ units that need to be shipped from a certain supplier to a certain retailer and must arrive by a particular period of time or incur delay penalties. Let the set of all possible paths for commodity k be P_k, and let c_{pk} be the sum of inventory and penalty costs incurred when commodity k is shipped along path $p \in P_k$. Observe that the shipping cost associated with a path will depend on the total quantity of all commodities being sent along each of its shipping links and, consequently, it can't be added to the path cost a priori. Thus, each shipping edge, whose cost must be globally computed, needs to be considered separately. Let the set of all shipping edges be SE, and for each edge $e \in SE$, let z_e be the total sum of weight of the commodities traveling on that edge.

We assume that the cost of a shipping edge e, $e \in SE$, of the expanded network $G_T(V, E)$, is $F_e(z_e)$, a **piecewise linear and concave cost function** that is nondecreasing in the total quantity, z_e, of the commodities sharing edge e. As presented in Balakrishnan and Graves (1989), this special cost structure allows for a formulation of the problem as a mixed integer linear program. For this purpose, the piecewise linear concave functions are modeled as follows. Let R be the number of different slopes in the cost function, which we assume, without loss of generality, is the same for all edges to avoid cumbersome notation. Let M_e^{r-1}, M_e^r, $r = 1, \ldots, R$, denote the lower and upper limits, respectively, on the interval of quantities corresponding to the rth slope of the cost function associated with edge e. Note that $M_e^0 = 0$ and M_e^R can be set to the total quantity of all commodities that may use arc e. We associate with each of these intervals, say r, a variable cost per unit, denoted by α_e^r, equal to the slope of the corresponding line segment, and a fixed cost, f_e^r, defined as the y-intercept of the linear prolongation of that segment. See Fig. 14.2 for a graphical representation. Observe that the cost incurred by any quantity on a certain range is the sum of its associated fixed cost plus the

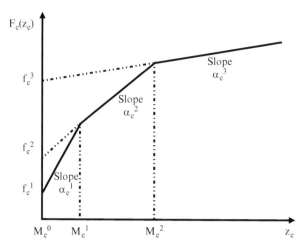

FIGURE 14.2. Piecewise linear and concave cost structure

cost of sending all units at its corresponding linear cost. That is, we can express
the arc flow cost function, $F_e(z_e)$, as

$$F_e(z_e) = f_e^r + \alpha_e^r z_e$$

if $z_e \in (M_e^{r-1}, M_e^r]$. Clearly,

Property 14.2.1 *The concavity and monotonicity of the function F_e imply that*

1. $\alpha_e^1 > \alpha_e^2 > \ldots > \alpha_e^R \geq 0$,

2. $0 \leq f_e^1 < f_e^2 < \ldots < f_e^R$,

3. $F_e(z_e) = \min_{r=1,\ldots,R} \left\{ f_e^r + \alpha_e^r z_e \right\}$. *The minimum is achieved at a unique index s, unless $z_e = M_e^s$, in which case the two consecutive indexes s and $s+1$ lead to the same minimum cost.*

We are now ready to introduce an integer linear programming formulation of the shipper problem for this special cost structure. Recall that z_e denotes the total flow on edge e, and let z_{ek} be the quantity of commodity k that is shipped along that edge. For all $e \in SE$ and $r = 1, \ldots, R$, define the *interval* variables,

$$x_e^r = \begin{cases} 1, & \text{if } z_e \in (M_e^{r-1}, M_e^r], \\ 0, & \text{otherwise,} \end{cases}$$

and, in addition, for every k, $k \in \mathcal{K}$, let the *quantity* variables be

$$z_{ek}^r = \begin{cases} z_{ek}, & \text{if } z_e \in (M_e^{r-1}, M_e^r], \\ 0, & \text{otherwise.} \end{cases}$$

In order to relate these edge flows to path flows, we define, for each $e \in SE$ and $p \in \bigcup_{k=1}^{K} P_k$,

$$\delta_p^e = \begin{cases} 1, & \text{if shipping link } e \text{ is in path } p, \\ 0, & \text{otherwise.} \end{cases}$$

Finally, let variables

$$y_{pk} = \begin{cases} 1, & \text{if commodity } k \text{ follows path } p \text{ in the optimal solution} \\ 0, & \text{otherwise,} \end{cases}$$

for each $k \in K$ and $p \in P_k$. These variables are referred to as *path flow* variables. Observe that defining these variables as binary variables implies that for every commodity k, only one of the variables y_{pk} takes a positive value. This reflects a common business practice in which each commodity, that is, items originated at the same source and destined to the same sink in the expanded network, is shipped along a single path. These integrality constraints are, however, not restrictive, as pointed out in Property 14.2.2 below, since the problem is uncapacitated and the cost functions concave.

In the *set-partitioning* formulation of the shipper problem, the objective is to select a minimum-cost set of feasible paths. Thus, we formulate the shipper problem for piecewise linear concave edge costs as the following mixed integer linear program, which we denote by Problem P.

$$\text{Problem } P: \quad Min \quad \sum_{k=1}^{K} \sum_{p \in P_k} y_{pk} c_{pk} + \sum_{e \in SE} \sum_{r=1}^{R} \left[f_e^r x_e^r + \alpha_e^r \left(\sum_{k=1}^{K} z_{ek}^r \right) \right]$$

s.t.

$$\sum_{p \in P_k} y_{pk} = 1, \quad \forall k = 1, 2, \dots, K, \tag{14.1}$$

$$\sum_{p \in P_k} \delta_p^e y_{pk} w_k = \sum_{r=1}^{R} z_{ek}^r, \quad \forall e \in SE, k = 1, \dots, K, \tag{14.2}$$

$$z_{ek}^r \leq w_k x_e^r \quad \forall e, r, k, \tag{14.3}$$

$$\sum_{k=1}^{K} z_{ek}^r \leq M_e^r x_e^r, \quad \forall e \in SE, r = 1, \dots, R, \tag{14.4}$$

$$\sum_{k=1}^{K} z_{ek}^r \geq M_e^{r-1} x_e^r, \quad \forall e \in SE, r = 1, \dots, R, \tag{14.5}$$

$$\sum_{r=1}^{R} x_e^r \leq 1 \quad \forall e \in SE, \tag{14.6}$$

$$y_{pk} \in \{0, 1\}, \quad \forall k = 1, 2, \dots, K, \text{ and } p \in P_k, \tag{14.7}$$

$$x_e^r \in \{0, 1\}, \quad \forall e \in SE, \text{ and } r = 1, 2, \dots, R, \tag{14.8}$$

$$z_{ek}^r \geq 0, \quad \forall e \in SE, \forall k = 1, 2, \dots, K,$$
$$\text{and } r = 1, 2, \dots, R.$$

In this formulation, constraints (14.1) ensure that exactly one path is selected for each commodity and constraints (14.2) set the total flow on an edge e to be equal to the total flow of all the paths that use that edge. Constraints (14.3)–(14.6) are used to model the piecewise linear concave function. Constraints (14.3) specify that if some commodity k is shipped on edge e using cost index r, the associated interval variable, x_e^r, must be 1. Constraints (14.4) and (14.5) make sure that if cost index r is used on edge e, then the total flow on that edge must fall in its associated interval, $[M_e^{r-1}, M_e^r]$. Finally, constraints (14.6) indicate that at most one cost range can be selected for each edge.

Let Z^* be the optimal solution to Problem P. Let Z_{R_x} and Z_{R_y} be the optimal solutions to relaxations of Problem P, where the integrality constraints of interval (x) and path flow (y) variables, respectively, are dropped. A consequence of Property 14.2.1 is the following result.

Property 14.2.2 *We have*

$$Z^* = Z_{R_x} = Z_{R_y}.$$

To find a robust and efficient heuristic algorithm for Problem P, we study the performance of a relaxation of Problem P that drops integrality and redundant constraints. Although constraints (14.3) are not required for a correct mixed integer programming formulation of the problem, we keep them because they significantly improve the performance of the linear programming relaxation of Problem P. In fact, Croxton et al. (2003), show that, without them, the linear programming relaxation of this model approximates the piecewise linear cost functions by their lower convex envelope. Furthermore, keeping these constraints makes constraints (14.4)–(14.6) redundant in the correct mixed integer programming formulation, as a direct consequence of Property 14.2.1, part 3, and in the linear programming relaxation of Problem P as well, as Lemma 14.2.3 below shows. This will be useful to considerably reduce the size of the formulation of the problem while preserving the tightness of its linear programming relaxation.

Let Problem P_{LP}^R be the linear program obtained from Problem P by relaxing the integrality constraints and constraints (14.4)–(14.6). That is,

$$\text{Problem } P_{LP}^R: \quad Min \quad \sum_{k=1}^{K} \sum_{p \in P_k} y_{pk} c_{pk} + \sum_{e \in SE} \sum_{r=1}^{R} \left[f_e^r x_e^r + \alpha_e^r \left(\sum_{k=1}^{K} z_{ek}^r \right) \right]$$

s.t.

$$(14.1) - (14.3)$$
$$y_{pk} \geq 0, \quad \forall k = 1, 2, \ldots, K, \text{ and } p \in P_k,$$
$$x_e^r \geq 0, \quad \forall e \in SE, \text{ and } r = 1, 2, \ldots, R,$$
$$z_{ek}^r \geq 0, \quad \forall e \in SE, \forall k = 1, 2, \ldots, K,$$
$$\text{and } r = 1, 2, \ldots, R.$$

Chan, Muriel and Simchi-Levi (1999), prove the following.

Lemma 14.2.3 *The optimal solution value to Problem P^R_{LP} is equal to the optimal solution value to the linear programming relaxation of Problem P.*

14.2.3   Structural Properties

To analyze the relaxed problem, we start by fixing the fractional path flows and study the behavior of the resulting linear program. Let $y = (y_{pk})$ be the vector of path flows in a feasible solution to the relaxed linear program, Problem P^R_{LP}.

Observe that, given the vector of path flows y, the amount of each commodity sent on each edge is known and, thus, Problem P^R_{LP} can be decomposed into multiple subproblems, one for every edge. Each subproblem determines the cost that the linear program associates with the corresponding edge flow. We refer to the subproblem associated with edge e as the fixed-flow subproblem on edge e, or Problem FF^e_y.

Let the proportion of commodity k shipped along edge e be

$$\gamma_{ek} = \sum_{p \in P_k} \delta^e_p y_{pk}.$$

We use Equation (14.2); the equality $\sum_{r=1}^R z^r_{ek} = w_k \gamma_{ek}$ must clearly hold; that is, the sum of all the flows of commodity k on the different cost intervals on edge e must be equal to the total quantity, $w_k \gamma_{ek}$, of commodity k that is shipped on that edge.

For each edge e, the total shipping cost on e, as well as the value of the corresponding variables z^r_{ek} and x^r_e, which Problem P^R_{LP} associates with the vector of path flows y, can be obtained by solving the fixed-flow subproblem on edge e:

Problem FF^e_y : $Min \quad \sum_{r=1}^R [f^r_e x^r_e + \alpha^r_e \sum_{k=1}^K z^r_{ek}]$

s.t.

$$z^r_{ek} \le w_k x^r_e \quad \forall k = 1, \ldots, K, \ and \ r = 1, \ldots, R, \quad (14.9)$$

$$\sum_{r=1}^R z^r_{ek} = w_k \gamma_{ek}, \quad \forall k = 1, \ldots, K, \quad (14.10)$$

$$z^r_{ek} \ge 0, \quad \forall k = 1, \ldots, K, \ and \ r = 1, \ldots, R,$$

$$x^r_e \ge 0, \quad \forall r = 1, \ldots, R.$$

Let $C^*_e(y) \equiv C^*_e(\gamma_{e1}, \ldots, \gamma_{eK})$ be the optimal solution to the fixed-flow subproblem on edge e for a given vector of path flows y or, equivalently, for given corresponding proportions $\gamma_{e1}, \ldots, \gamma_{eK}$ of the commodities shipped on that edge.

The following theorem determines the solution to the subproblem.

Theorem 14.2.4 *For any given edge $e \in SE$, let the proportion γ_{ek} of commodity k to be shipped on edge e be known and fixed, for $k = 1, 2, \ldots, K$, and let the* *commodities be indexed in nondecreasing order of their corresponding proportions, that is,*

$$\gamma_{e1} \leq \gamma_{e2} \leq \cdots \leq \gamma_{eK}.$$

Then the optimal solution to the fixed-flow subproblem on edge e is

$$C_e^*(\gamma_{e1}, \ldots, \gamma_{eK}) = \sum_{k=1}^{K} F_e\left(\sum_{i=k}^{K} w_i\right)[\gamma_{ek} - \gamma_{ek-1}], \qquad (14.11)$$

where $\gamma_{e0} := 0$.

Intuitively, the above theorem just says that in an optimal solution to the fixed-flow subproblem associated with any edge e, fractions of commodities are consolidated to be shipped at the cheapest possible cost per unit. At first, a fraction γ_{e1} of all commodities $1, 2, \ldots, K$ is available. Thus, these commodities get consolidated to achieve a cost per unit of $F_e(\sum_{k=1}^{K} w_k)/\sum_{k=1}^{K} w_k$, that is, the cost per unit associated with sending the full K commodities on that edge, and the available fraction γ_{e1} is sent incurring a cost of $\gamma_{e1} F_e(\sum_{k=1}^{K} w_k)$. At that point, none of commodity 1 is left and a fraction $(\gamma_{e2} - \gamma_{e1})$ is the maximum available simultaneously from all commodities $2, 3, \ldots, K$. These commodities are consolidated again, and that fraction, $(\gamma_{e2} - \gamma_{e1})$, from each commodity is sent at a cost $(\gamma_{e2} - \gamma_{e1}) F_e(\sum_{k=2}^{K} w_k)$. This process continues until the desired proportion of each commodity has been sent.

14.2.4 Solution Procedure

Theorem 14.2.4 provides a simple expression of the cost that the relaxed problem, Problem P_{LP}^R, assigns to any given fractional path flows, and thus it allows for the efficient computation of the impact of modifying the flow in a particular path. This is the key to the algorithm developed in this section. Indeed, the algorithm transforms an optimal fractional solution to the linear program P_{LP}^R into an integer solution by modifying path flows, choosing for each commodity the path that leads to the lowest increase in the objective of the linear program.

The Linear Programming-Based Heuristic:

Step 1: Solve the linear program, Problem P_{LP}^R. Initialize $k = 1$.

Step 2: For each arc, compute a *marginal cost,* which is the increase in cost incurred in the fixed-flow subproblem by augmenting the fractional flow of commodity k to 1. Note that this is easy to compute using Theorem 14.2.4.

Step 3: Determine a path for commodity k by finding the minimum-cost path on the expanded network with edge costs equal to the marginal costs.

Step 4: Update the flows and the costs on each link (again employing Theorem 14.2.4) to account for commodity k being sent along that path.

Step 5: Let $k = k + 1$, and repeat steps (2)–(5) until $k = K + 1$.

Evidently, the effectiveness of this heuristic depends on the tightness of the linear programming relaxation of Problem P. For this reason, we study the difference between integer and fractional solutions to Problem P. Chan et al. (1999), show that in some special cases an integer solution can be constructed from the optimal fractional solution of Problem P_{LP}^R without increasing its cost. In particular, using Theorem 14.2.4, they prove the following result.

Theorem 14.2.5 *In the following cases:*

1. *single period, multiple suppliers, multiple retailers, two warehouses,*

2. *two periods, single supplier, multiple retailers, single warehouse,*

3. *two periods, multiple supplier, multiple retailers, single warehouse using a cross-docking strategy,*

4. *multiple periods, single supplier, single retailer, single warehouse that uses a cross-docking strategy,*

the solution to the linear programming relaxation of Problem P is the optimal solution to the shipper problem. That is,

$$Z^* = Z^{\mathrm{LP}}.$$

Furthermore, in the first three cases, all extreme point solutions to the linear program are integer.

The **cross-docking** strategy referred to in the last two cases is a strategy in which the stores are supplied by central warehouses that do not keep any stock themselves. That is, in this strategy, the warehouses act as coordinators of the supply process and as transshipment points for incoming orders from outside vendors.

The theorem thus demonstrates the exceptional performance of the linear programming relaxation, and consequently of the heuristic, in some special cases. A natural question at this point is whether these results can be generalized. The answer is no in general. To show this, Chan et al. (1999) construct examples with a single supplier, a single warehouse, and multiple retailers and time periods, for which

$$\frac{Z^*}{Z^{\mathrm{LP}}} \to \infty,$$

as the number of retailers and time periods increases.

Lemma 14.2.6 *The linear programming relaxation of Problem P can be arbitrarily weak, even for a single-supplier, single-warehouse, multiretailer case in which demand for the retailers is constant over time.*

It is important to point out that the instances in which the heuristic solution is found to be arbitrarily bad are characterized by the unrealistic structure of the shipping cost. In these instances, the shipping cost between two facilities is a pure fixed charge (regardless of quantity shipped) in some periods, linear (with no fixed charges) in others, and yet prohibitively expensive so that nothing can be shipped in the remaining periods. The following examples illustrate this structure.

Example of weak linear programming solution: Consider a three-period, single-warehouse model in which a single supplier delivers goods to a warehouse that, in turn, replenishes inventory of three retailers over time. The warehouse uses a cross-docking strategy; thus, it does not keep any inventory. Let the transportation cost be a fixed charge of 100 for any shipment from the supplier to the warehouse at any period. Transportation from the warehouse to retailer i, $i = 1, 2, 3$, is very large for shipments made in period i (in other words, retailer i cannot be reached in period i) and negligible for periods $j \neq i$. Let the inventory cost be negligible for all retailers at all periods, and let the demand for each retailer be 0 units in periods 1 and 2 and 100 units in period 3.

Observe that in order to reach the three retailers, shipments need to be made in at least two different periods. Thus, the optimal integer solution is 200. However, in the solution to the linear program, 50 units are sent to each of the "reachable" retailers in each period, and a transportation cost of 50 is charged at each period (as stated in Theorem 14.2.4, since only a fraction of $1/2$ of the commodities is sent on any edge, exactly that fraction of the fixed cost is charged). Thus, the optimal fractional solution is 150 and the ratio of integer to fractional solutions is $3/2$.

In this instance, even if fractional and integer solutions are different, the linear programming-based heuristic generates the optimal integer solution. However, we can easily extend the above scenario to instances for which the difference between the solution generated by the heuristic and the optimal integer solution is arbitrarily large.

Example of weak heuristic solution: For that purpose, we add n new periods to the above setting. In period 4, the first of the new periods, the cost for shipping from supplier to warehouse is linear at a rate of $1/3$, and the cost for shipping from the warehouse to each of the three retailers is 0. In all the other $n - 1$ periods, the cost of shipping is very high, and thus no shipments will be made after period 4. Inventory costs at all retailers and all periods are negligible. Demand for each of the three retailers at each of the new n periods is 100, while demand during the first three periods is 0. It is easy to see that the optimal integer and fractional solutions are identical to those in the three-period case, with costs of 200 and 150, respectively. However, the heuristic algorithm will always choose to ship each commodity in period 4, since the increase in cost in the corresponding path would be $1/3 \times 100$, while it is at least 50 in any of the first three periods. Thus, the total cost of the heuristic solution is $1/3 \times 100 \times n$ and the gap with the optimal integer solution arbitrarily large.

The following section reports the performance of the algorithm on a set of randomly generated instances.

14.2.5 Computational Results

The computational tests carried out are divided into three categories:

1. single-period layered networks,

2. general networks,

3. multiperiod, single-warehouse distribution problems:

   - pure distribution instances.
   - production/distribution instances.

The first two categories are of special interest because they allow us to compare our results with those reported by Balakrishnan and Graves (1989), henceforth B&G. The third set of problems models practical situations in which each of the retailers is assigned to a single warehouse and production and transportation costs have to be balanced with inventory costs over time.

In the three categories, the tests were run on a Sun SPARC20 and CPLEX was used to solve the linear program, Problem P_{LP}^R, using an equivalent formulation where path flow variables are replaced by flow-balance constraints. During our computational work, we observed that the dual simplex method is more efficient than the primal simplex method in solving these highly degenerate problems, an observation also made by Melkote (1996). This is usually the case for programs with variable upper-bound constraints, such as our constraints $z_{ek}^r \leq w_k x_e^r$. We should also point out that most of the CPU time reported in our tests is used in solving the linear program. Thus, to enhance the computational performance of our algorithm and increase the size of the problems that it is capable of handling, we see the need for future research to focus on efficiently solving the linear program. For instance, the original set-partitioning formulation, Problem P_{LP}^R, could be solved faster using column-generation techniques. In these tests, however, we focused on evaluating the quality of the integer solutions provided by the heuristic and the tightness of the linear programming relaxation.

We now discuss each class of problems and the effectiveness of our algorithm.

Single-Period Layered Networks
B&G present exceptional computational results for single-period layered networks. In these instances, commodities flow from the manufacturing facilities to distribution centers, where they are consolidated with other shipments. These shipments are then sent to a number of warehouses, where they are split and shipped to their final destinations. Thus, every commodity must go through two layers of intermediate points: **consolidation points**, also referred to as distribution centers, and **breakbulk points**, or warehouses.

TABLE 14.1. Test problems generated as in Balakrishnan and Graves

Number of	Problem class				
nodes	LTL1	LTL2	LTL3	LTL4	LTL5
Source	4	5	6	8	10
Consolidn	5	10	12	15	20
Breakbulk	5	10	12	15	20
Destn	4	5	6	8	10
Arcs	42–47	131–141	190–207	309–312	358–372
Commodities	10	20	30	50	60

TABLE 14.2. Computational results for layered networks. Balakrishnan and Graves' results (B&G) vs. those of our linear programming-based heuristic (LPBH)

	B&G	LPBH	
Problem	LB/UB	LP/heuristic	Avg. CPU time
class	percentage	percentage	in seconds
LTL1	99.8	100	1.04
LTL2	100	100	7.94
LTL3	99.6	100	20.74
LTL4	99.1	100	55.72
LTL5	99.5	100	100.48

To test the performance of our algorithm and to compare it with that of B&G, we generated instances of the layered networks following the details given in their paper. In this computational work, we considered five different problem classes, referred to as LTL1–LTL5.

Table 14.1 shows the sizes of the different classes of problems. For each of these classes, the first column (B&G) of Table 14.2 presents the average ratio between the upper bounds generated by the heuristic proposed by B&G and a lower bound on the optimal solution, over five randomly generated instances. The numbers are taken from their paper. We do not include, though, their average CPU times because the machines they use are completely different than ours and, in addition, they do not report the total computational time for the entire algorithm. The second and third columns report the average deviation from optimality and the computational performance of the linear programming-based heuristic (LPBH) over 10 random instances, for each of the problem classes. In all of them, our algorithm **finds the optimal integer solution**; furthermore, the solution to the linear program in the first step of our algorithm is integer, providing the optimal solution to the problem.

Of course, since in all previous instances, the linear program provided the optimal integer solution, the performance of our procedure has not really been tested. In the following subsections, we present computational results for problem classes in which the solution to the linear program is not always integer.

TABLE 14.3. Computational results for general networks. Balakrishnan and Graves' results (B&G) vs. those of our linear programming-based heuristic (LPBH)

Problem class	Size			B&G	LPBH	
	No. of nodes	No. of arcs	No. of comm.	LB/UB percentage	LP/Heuristic percentage	Avg. CPU time in seconds
GEN1	10	47–54	10	99.9	100	2.18
GEN2	15	109–136	20	98.7	99.53	24.04
GEN3	20	196–235	30	98.4	99.88	139.83
GEN4	30	364–428	50	96.2	98.59	1313.06
GEN5	40	340–370	60	98.5	99.98	159.57

General Networks

In this subsection, we report on the performance of our algorithm on general networks, in which every node can be an origin and/or a destination, generated exactly as they are generated by B&G. These results, together with those of B&G, are reported in Table 14.3. In this category, B&G consider five different problem classes, referred to as GEN1,..., GEN5, and generate five random instances for each of them. We, in turn, solve 10 different randomly generated instances for each of the problem classes. Again, we do not include their average CPU times due to the reasons mentioned above.

Multiperiod, Single-Warehouse Distribution Problems

Here we consider a single-warehouse model where a set of suppliers replenishes the inventory of a number of retailers over time. We test two different types of instances: pure distribution instances in which the routing and timing of shipments are to be determined, and production/distribution instances in which the production schedule is also integrated with the transportation and inventory decisions.

A. Pure Distribution Instances

We assume that shortages are not allowed and analyze three different strategies:

1. *Classical inventory/distribution strategy:* Material always flows from the suppliers through a single warehouse, where it can be held as inventory.

2. *Cross-docking strategy:* All material flows through the warehouse, where shipments are reallocated and immediately sent to the retailers.

3. *A distribution strategy that allows for direct shipments:* Items may be sent either through the warehouse or directly to the retailer. The warehouse may keep inventory.

For each strategy, we analyze different situations where the number of suppliers is either 1, 2, or 5, the number of retailers is 10, 12, or 20, and the number of periods is 8 or 12. For each combination of the number of suppliers, retailers, and periods presented in Table 14.6, 10 instances are generated. The retailers and suppliers are randomly located on a 1000 × 1000 grid, while the warehouse is randomly

TABLE 14.4. Linear and setup costs used for all the test problems

Type of arc	α_e^1	α_e^2	α_e^3	Setup
Supplier–whs.	0.15	0.105	0.084	25
Whs.–retailer	0.25	0.20	0.16	10

TABLE 14.5. Inventory costs and different ranges for the different test problems

Problem Class	Inventory cost Warehouse	Inventory cost Retailer	Supplier–whs. cost Range 1	Supplier–whs. cost Range 2	Whs.–retailer cost Range 1	Whs.–retailer cost Range 2
I1					200	400
I2	5	10	800	1500	300	600
I3					300	600
I4					150	300
I5	10	20	1000	2000	200	400
I6					200	400
C1					200	400
C2	10	20	800	1500	300	600
C3					300	600
C4					150	300
C5	10	20	1000	2000	200	400
C6					200	400
D1					150	300
D2	10	20	500	1000	200	400
D3					200	400

assigned to the 400×400 subgrid at the center. Demand is generated for each retailer–supplier pair at each time period, except for the cases with five suppliers, in which each of these pairs has an associated demand with probability $1/3$. These demands are generated from a uniform distribution on the integers in the interval $[0, 100)$.

All suppliers and retailers are linked to the warehouse, and the distance associated is the corresponding Euclidean distance between the nodes of the grid. In the case of *a distribution strategy that allows for direct shipments*, shipping edges from each of the suppliers to each of the retailers are added. The holding costs per unit of inventory are different at the warehouses and retailer facilities and are presented in Table 14.5. All holding costs at the suppliers are set to zero. Two shipping cost functions, representing the cost per item per unit distance, are considered: The first is assigned to shipments from the suppliers to the warehouse. The second is incurred by the material flowing from the warehouse to the retailers. The cost function (dollars per mile per unit) associated with direct shipments is equal to that of shipments from the warehouse to a retailer. Both functions have an initial setup cost for using the link and three different linear rates depending on the quantity shipped; see Table 14.4. However, the ranges to which those linear costs

TABLE 14.6. Computational results for a single warehouse

Strategy	Problem class	Number of suppliers	Number of stores	Number of periods	LP/heuristic percentage	CPU time in seconds
	I1	1			100	65.21
Classical	I2	2	10	12	100	187.37
inventory/	I3	5			100	163.23
distribution	I4	1			99.946	83.5
strategy	I5	2	20	8	100	210.51
	I6	5			99.953	200.68
	C1	1			100	60.0
	C2	2	10	12	100	174.13
Cross-	C3	5			100	159.06
docking						
strategy	C4	1			100	79.73
	C5	2	20	8	100	202.83
	C6	5			100	186.0
Direct	D1	1			100	51.23
shipments	D2	2	12	8	100	165.83
allowed	D3	5			99.921	117.27

correspond are different for the different problem classes. This is done so that, in an optimal solution, shipments are consolidated, and thus the concave cost function plays an important role in the analysis. These ranges and the corresponding problem classes are presented in Table 14.5.

Observe (see Table 14.6) that in most of the instances tested, the linear program is tight and it provides the optimal integer solution. In only three of the 150 instances generated is the solution to the linear program not integer; in such cases, our algorithm finds a solution that is within 0.8 % from the optimal fractional solution.

B. Production/Distribution Instances

This section demonstrates the effectiveness of the algorithm when applied to production/distribution systems, that is, systems in which one needs to coordinate production planning, inventory control, and transportation strategies over time. For that purpose, we consider the same set of problems, I1–I3, as in the *classical inventory/distribution strategy* described in the previous section, and add production decisions at each of the supplier sites. This is incorporated into the model, as explained in Sect. 14.2.

We consider a fixed setup cost for producing at any period plus a certain cost per unit. The setup cost is varied in the set $\{50, 100, 500, 1000\}$, and the linear production cost is set to 1. The inventory holding rate at the supplier site (after production) is set to half of that at the warehouse. For the 60 different instances generated, the linear programming relaxation gave an **integer** solution every time.

14.3   Safety Stock Optimization

As observed earlier, the shipper model analyzed earlier is deterministic; safety stocks are determined exogenously and incorporated into the minimum inventory level that should be maintained at the beginning of each period. The objective of this section is to present a model for positioning and optimizing safety stock in the supply chain.

For this purpose, consider a single-product, single-facility, periodic-review inventory model. Let SI be the amount of time it takes from placing an order until the facility receives a shipment; this time is referred to as the **incoming service time**. Let S be the **committed service time** made by the facility to its own customers, and let T be the **processing time** at the facility. Of course, we must assume that $SI + T > S$; otherwise, no inventory is needed in the facility.

We assume that the facility manages its inventory following a periodic review policy and that demand is independent and identically distributed across time periods following a normal distribution. Given deterministic SI, S, and T, and with no setup costs, the level of safety stock that the facility needs to keep (see Exercise 14.1) is equal to

$$zh\sqrt{SI + T - S},$$

where z is the safety stock factor associated with a specified level of service and h is the inventory holding cost. The value $SI + T - S$ is referred to as the facility **net lead time**.

To understand our model, consider the following two-stage supply chain with facility 2 feeding facility 1, which serves the end customer (Fig. 14.3). Define SI_i, S_i, and T_i as before for $i = 1, 2$. Thus, S_1 is the committed service time to the end customer, S_2 is the commitment that facility 2 makes to facility 1, and hence $S_2 = SI_1$. Finally, SI_2 is the supplier commitment to facility 2.

Our objective is to minimize the total supply chain cost without affecting, or pushing, inventory to the supplier. Observe that if we reduce $S_2 = SI_1$, we will affect inventory at both facility 1 and facility 2. Indeed, by reducing the committed service time that facility 2 makes to facility 1, inventory at facility 1 is reduced, but inventory at the second facility is increased. Thus, our objective is to develop a model that selects the appropriate level of commitment that one facility makes to its downstream facility so as to minimize the total, or more precisely, systemwide safety stock cost.

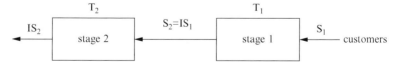

FIGURE 14.3. Illustration of the model

For this purpose, consider a supply network $G(N, A)$, which is acyclic, with N facilities and where A is the set of edges. Let $D \subset N$ be the set of customers, or demand points, with $\overline{S}_j$ being an upper bound on the commitment to be made to customer j, $j \in D$.

Following our discussion, we formulate the problem of setting commitments and safety stock levels as the following nonlinear optimization problem.

$$\text{Problem } SS: \quad Min \quad \sum_{j=1}^{N} z_j h_j \Phi(X_j)$$

s.t.

$$X_j = SI_j + T_j - S_j, \quad \forall j = 1, \ldots, N, \qquad (14.12)$$
$$X_j \geq 0, \quad \forall j = 1, \ldots, N, \qquad (14.13)$$
$$SI_j - S_i \geq 0, \quad \forall (i, j) \in A, \qquad (14.14)$$
$$S_j \leq \overline{S}_j, \quad \forall j \in D, \qquad (14.15)$$
$$S_j, SI_j \geq 0, \quad \forall j = 1, \ldots, N, \qquad (14.16)$$

where $\Phi(X_j) = \sqrt{X_j}$.

Observe that in this formulation, there are two sets of decision variables. The first is S_j, the commitment made by facility j to all its downstream facilities. The second is the implied incoming service time to facility j. This incoming service time is the maximum of the committed service time of all the upstream facilities feeding facility j.

Thus, constraint (14.12) defines the net lead time at facility j. Constraint (14.14) forces the incoming service time for facility j to be no smaller than the commitment that each facility i with $(i, j) \in A$ makes to facility j. Finally, constraint (14.15) forces the commitment to the end customer to be no larger than the target.

Of course, the challenge is to solve this formulation effectively. Graves and Willems (2000) propose a dynamic programming algorithm, while Magnanti et al. (2003) develop an efficient algorithm based on a similar approach to what is described earlier for the shipper problem.

14.4 Exercises

Exercise 14.1. Consider a single-product, single-facility, infinite-horizon, periodic-review model. Assume that the inventory is managed based on a stationary base stock policy and unsatisfied demand is backlogged. At each period, demand arrives according to a normal distribution $N(\mu, \sigma)$. (Let's assume that the probability of having negative demand is negligible.) Let SI be the incoming service time, S be the committed service time, and T be the processing time at the facility. Finally, assume that the initial inventory level equals the base stock level and assume the service level is α, which is defined as the probability that the demand can be fully satisfied from current on-hand inventories. Show that the base stock level is $(SI+T-S)\mu + \Psi^{-1}(\alpha)\sqrt{SI+T-S}\sigma$, where Ψ^{-1} is the inverse of the cumulative distribution function of the standard normal distribution.

Exercise 14.2. Assume, in Problem SS of Sect. 14.3, that the supply network reduces to a serial supply chain. Show that $S_i = 0$ or $S_i = S_{i+1} + T_i$, where stage $i+1$ serves stage i. Based on this observation, propose an algorithm to solve Problem SS.

15
Facility Location Models

15.1   Introduction

One of the most important aspects of logistics is deciding where to locate new facilities such as retailers, warehouses, or factories. These strategic decisions are a crucial determinant of whether materials will flow efficiently through the distribution system.

In this chapter, we consider several important warehouse location problems: the p-median problem; the single-source capacitated facility location problem; and a distribution system design problem. In each case, the problem is to locate a set of warehouses in a distribution network. We assume that the cost of locating a warehouse at a particular site includes a *fixed* cost (e.g., building costs, rental costs, etc.) and a *variable* cost for transportation. This variable cost includes the cost of transporting the product to the retailers as well as possibly the cost of moving the product from the plants to the warehouse. In general, the objective is to locate a set of facilities so that the total cost is minimized subject to a variety of constraints, which might include

- Each warehouse has a capacity, which limits the area it can supply.

- Each retailer receives shipments from *one and only one* warehouse.

- Each retailer must be within a fixed distance of the warehouse that supplies it, so that a reasonable delivery lead time is ensured.

D. Simchi-Levi et al., *The Logic of Logistics: Theory, Algorithms, and Applications for Logistics Management*, Springer Series in Operations Research and Financial Engineering, DOI 10.1007/978-1-4614-9149-1_15, © Springer Science+Business Media New York 2014

Location analysis has played a central role in the development of the operations research field. In this area lie some of the discipline's most elegant results and theories. We note here the paper of Cornuéjols et al. (1977) and the two excellent books devoted to the subject by Mirchandani and Francis (1990) and Daskin (1995). Location problems encompass a wide range of problems such as the location of emergency services (firehouses or ambulances), the location of hazardous materials, and problems in telecommunications network design, just to name a few.

In the next section, we present an exact algorithm for one of the simplest location problems, the p-median problem. We then generalize this model and algorithm to incorporate additional factors important to the design of the distribution network, such as warehouse capacities and fixed costs. In Sect. 15.4, we present a more general model where all levels of the distribution system (plants and retailers) are taken into account when deciding warehouse locations. We also present an efficient algorithm for its solution. All of the algorithms developed in this chapter are based on the Lagrangian relaxation technique described in Sect. 6.3, which has been applied successfully to a wide range of location problems. Finally, in Sect. 15.5, we describe the structure of the optimal solution to problems in the design of large-scale logistics systems.

15.2   An Algorithm for the p-Median Problem

Consider a set of retailers geographically dispersed in a region. The problem is to choose where in the region to locate a set of p identical warehouses. We assume there are $m \geq p$ sites that have been preselected as possible locations for these warehouses. Once the p warehouses have been located, each of n retailers will get its shipments from the warehouse closest to it. We assume the following:

- There is no fixed cost for locating at a particular site.

- There is no capacity constraint on the demand supplied by a warehouse.

Note that the first assumption also encompasses the case where the fixed cost is not site-dependent, and therefore the fixed setup cost for locating p warehouses is independent of where they are located.

Let the set of retailers be N, where $N = \{1, 2, \ldots, n\}$, and let the set of potential sites for warehouses be M, where $M = \{1, 2, \ldots, m\}$. Let w_i be the demand or flow between retailer i and its warehouse for each $i \in N$. We assume that the cost of transporting the w_i units of product from warehouse j to retailer i is c_{ij}, for each $i \in N$ and $j \in M$.

The problem is to choose p of the m sites where a warehouse will be located in such a way that the total transportation cost is minimized. This is the p-median problem.

The continuous version of this problem, where any point is a potential warehouse location, was first treated as early as 1909 by Weber. The discrete version was analyzed by Kuehn and Hamburger (1963) as well as Hakimi (1964), Manne (1964), Balinski (1965), and many others.

We present here a highly effective approach to the problem. Define the following decision variables:

$$Y_j = \begin{cases} 1, & \text{if a warehouse is located at site } j, \\ 0, & \text{otherwise,} \end{cases}$$

for $j \in M$, and

$$X_{ij} = \begin{cases} 1, & \text{if retailer } i \text{ is served by a warehouse at site } j, \\ 0, & \text{otherwise,} \end{cases}$$

for $i \in N$ and $j \in M$. The p-median problem is then

$$\text{Problem } P : Min \ \sum_{i=1}^{n} \sum_{j=1}^{m} c_{ij} X_{ij}$$

$$\text{s.t.} \ \sum_{j=1}^{m} X_{ij} = 1, \quad \forall i \in N, \tag{15.1}$$

$$\sum_{j=1}^{m} Y_j = p, \tag{15.2}$$

$$X_{ij} \leq Y_j, \quad \forall i \in N, \ j \in M, \tag{15.3}$$

$$X_{ij}, Y_j \in \{0,1\}, \quad \forall i \in N, \ j \in M. \tag{15.4}$$

Constraints (15.1) guarantee that each retailer is assigned to a warehouse. Constraint (15.2) ensures that p sites are chosen. Constraints (15.3) guarantee that a retailer selects a site only from among those that are chosen. Constraints (15.4) force the variables to be integer.

This formulation can easily handle several side constraints. If a handling fee is charged for each unit of product going through a warehouse, these costs can be added to the transportation cost along all arcs leaving the warehouse. Also, if a particular limit is placed on the length of any arc between retailer i and warehouse j, this can be incorporated by simply setting the per-unit shipping cost (c_{ij}) to $+\infty$. In addition, the model can be easily extended to cases where a set of facilities is already in place and the choice is whether to open new facilities or *expand* the existing facilities.

Let Z^* be an optimal solution to Problem P. One simple and effective technique to solve this problem is the method of Lagrangian relaxation described in Sect. 6.3.

As described in Sect. 6.3, Lagrangian relaxation involves relaxing a set of constraints and introducing them into the objective function with a multiplier vector. This provides a lower bound on the optimal solution to the overall problem. Then, using a subgradient search method, we iteratively update our multiplier vector in an attempt to increase the lower bound. At each step of the subgradient procedure (i.e., for each set of multipliers), we also attempt to construct a feasible solution to the location problem. This step usually consists of a simple and efficient subroutine. After a prespecified number of iterations, or when the solution found is within a fixed error tolerance of the lower bound, the algorithm is terminated.

To solve the p-median problem, we choose to relax constraints (15.1). We incorporate these constraints in the objective function with the multiplier vector $\lambda \in \mathbb{R}^n$. The resulting problem, call it P_λ, with optimal objective function value Z_λ, is

$$Min \ \sum_{i=1}^{n}\sum_{j=1}^{m} c_{ij}X_{ij} + \sum_{i=1}^{n}\lambda_i\left(\sum_{j=1}^{m} X_{ij} - 1\right)$$

$$\text{subject to } (15.2) - (15.4).$$

Disregarding constraint (15.2) for now, the problem decomposes by site; that is, each site can be considered separately. Let Subproblem P_λ^j, with optimal objective function value Z_λ^j, be the following:

$$Min \ \sum_{i=1}^{n}(c_{ij} + \lambda_i)X_{ij}$$

$$s.t. \ X_{ij} \leq Y_j, \quad \forall i \in N,$$

$$X_{ij} \in \{0,1\}, \quad \forall i \in N,$$

$$Y_j \in \{0,1\}.$$

Solving Subproblem P_λ^j

Assume λ is fixed. In Problem P_λ^j, site j is either selected ($Y_j = 1$) or not ($Y_j = 0$). If site j is not selected, then $X_{ij} = 0$ for all $i \in N$, and therefore, $Z_\lambda^j = 0$. If site j is selected, then we set $Y_j = 1$ and assign exactly those retailers i with $c_{ij} + \lambda_i < 0$ to site j. In this case:

$$Z_\lambda^j = \sum_{i=1}^{n} \min\{c_{ij} + \lambda_i, 0\}. \tag{15.5}$$

We see that P_λ^j is solved easily and its optimal objective function value is given by (15.5).

To solve P_λ, we must now reintroduce constraint (15.2). This constraint forces us to choose only p of the m sites. In P_λ, we can incorporate this constraint by choosing the p sites with smallest values Z_λ^j. To do this, let π be a permutation of the numbers $1, 2, \ldots, m$ such that

$$Z_\lambda^{\pi(1)} \leq Z_\lambda^{\pi(2)} \leq Z_\lambda^{\pi(3)} \leq \cdots \leq Z_\lambda^{\pi(m)}.$$

Then the optimal solution to P_λ has the objective function value:

$$Z_\lambda \doteq \sum_{j=1}^{p} Z_\lambda^{\pi(j)} - \sum_{j=1}^{n}\lambda_j.$$

The value Z_λ is a lower bound on the optimal solution of Problem P for any vector $\lambda \in \mathbb{R}^n$. To find the best such lower bound, we consider the Lagrangian dual:

$$\max_\lambda \{Z_\lambda\}.$$

Using the subgradient procedure (described in Sect. 6.3), we can iteratively improve this bound.

Upper Bounds

It is crucial to construct good upper bounds on the optimal solution value as the subgradient procedure advances. Clearly, solutions to P_λ will not necessarily be feasible to Problem P. This is due to the fact that the constraints (15.1) (that each retailer choose *one and only one* warehouse) may not be satisfied. The solution to P_λ may have facilities choosing a number of sites. If, in the solution to P_λ, each retailer chooses only *one* site, then this must be the optimal solution to P, and therefore, we stop. Otherwise, there are retailers that are assigned to several or no sites. A simple heuristic can be implemented that fixes those retailers that are assigned to only one site and assigns the remaining retailers to these and other sites by choosing the next site to open in the ordering defined by π. When p sites have been selected, we can do a simple check that each retailer is assigned to its closest site (of those selected); doing so can further improve the solution.

Computational Results

Below we give a table listing results of various computational experiments (Table 15.1). The retailer locations were chosen uniformly over the unit square. For simplicity, we made each retailer location a potential site for a warehouse; thus, $m = n$. The cost of assigning a retailer to a site was the Euclidean distance between the two locations. The values of w_i were chosen uniformly over the unit interval. We applied the algorithm mentioned above to many problems and recorded the relative error of the best solution found and the computational time required. The algorithm is terminated when the relative error is below 1% or when a prespecified number of iterations is reached. The numbers below "Error" are the relative errors averaged over 10 randomly generated problem instances. The numbers below "CPU time" is the CPU time averaged over the 10 problem instances. All computational times are on an IBM RISC 6000 Model 950.

15.3   An Algorithm for the Single-Source Capacitated Facility Location Problem

Consider the p-median problem, where we make the following two changes in our assumptions:

- The number of warehouses to locate (p) is not fixed beforehand.

TABLE 15.1. Computational results for the p-median algorithm

n	p	Error	CPU time
10	3	0.3%	0.2 s
20	4	1.7%	2.6 s
50	5	1.4%	20.7 s
100	7	1.3%	87.7 s
200	10	2.4%	715.4 s

- If a warehouse is located at site j:
  - A fixed cost f_j is incurred, and
  - There is a capacity q_j on the amount of demand it can serve.

The problem is to decide where to locate the warehouses and then how the retailers should be assigned to the open warehouses in such a way that the total cost is minimized. We see that the problem is considerably more complicated than the p-median problem. We now have capacity constraints on the warehouses, and therefore a retailer will not always be assigned to its nearest warehouse. Allowing the optimization to choose the appropriate number of warehouses also adds to the level of difficulty.

This problem is called the single-source capacitated facility location problem (CFLP), or sometimes the capacitated concentrator location problem (CCLP). This problem was successfully used in Chap. 17 as a framework for solving the capacitated vehicle routing problem.

Using the same decision variables as in the p-median problem, we formulate the single-source CFLP as the following integer linear program:

$$Min \quad \sum_{i=1}^{n}\sum_{j=1}^{m} c_{ij}X_{ij} + \sum_{j=1}^{m} f_j Y_j$$

$$s.t. \quad \sum_{j=1}^{m} X_{ij} = 1, \qquad \forall i \in N, \tag{15.6}$$

$$\sum_{i=1}^{n} w_i X_{ij} \leq q_j Y_j, \qquad \forall j \in M, \tag{15.7}$$

$$X_{ij}, Y_j \in \{0,1\}, \qquad \forall i \in N,\ j \in M. \tag{15.8}$$

Constraints (15.6) (along with the integrality conditions (15.8)) ensure that each retailer is assigned to exactly one warehouse. Constraints (15.7) ensure that the warehouse's capacity is not exceeded and also that if a warehouse is not located at site j, no retailer can be assigned to that site.

Let Z^* be the optimal solution value of single-source CFLP. Note we have restricted the assignment variables (X) to be integer. A related problem, where this assumption is relaxed, is simply called the (multiple-source) capacitated facility location problem. In that version, a retailer's demand can be *split* between any number of warehouses. In the single-source CFLP, it is required that each retailer have only *one* warehouse supplying it. In many logistics applications, this is a realistic assumption since without this restriction, optimal solutions might have a retailer receive many deliveries of the same product (each for, conceivably, a very small amount of the product). Clearly, from a managerial, marketing, and accounting point of view, restricting deliveries to come from only one warehouse is a more appropriate delivery strategy.

Several algorithms have been proposed to solve the CFLP in the literature; all are based on the Lagrangian relaxation technique. This includes Neebe and Rao (1983), Barcelo and Casanovas (1984), Klincewicz and Luss (1986), and Pirkul (1987). The one we derive here is similar to Pirkul's algorithm, which seems to be the most effective.

We apply the Lagrangian relaxation technique by including constraints (15.6) in the objective function. For any $\lambda \in I\!R^n$, consider the following problem P_λ:

$$Min \quad \sum_{i=1}^{n}\sum_{j=1}^{m} c_{ij}X_{ij} + \sum_{j=1}^{m} f_jY_j + \sum_{i=1}^{n}\lambda_i\left(\sum_{j=1}^{m} X_{ij} - 1\right)$$

$$\text{subject to} \quad (15.7) - (15.8).$$

Let Z_λ be its optimal solution and note that

$$Z_\lambda \leq Z^*, \quad \forall \lambda \in I\!R^n.$$

To solve P_λ, as in the p-median problem, we separate the problem by site. For a given $j \in M$, define the following problem P_λ^j, with the optimal objective function value Z_λ^j:

$$Min \quad \sum_{i=1}^{n}(c_{ij} + \lambda_i)X_{ij} + f_jY_j$$

$$s.t. \quad \sum_{i=1}^{n} w_iX_{ij} \leq q_jY_j,$$

$$X_{ij} \in \{0,1\}, \qquad \forall i \in N,$$

$$Y_j \in \{0,1\}.$$

Solving P_λ^j

Problem P_λ^j can be solved efficiently. In the optimal solution to P_λ^j, Y_j is either 0 or 1. If $Y_j = 0$, then $X_{ij} = 0$ for all $i \in N$. If $Y_j = 1$, then the problem is no more difficult than a single constraint 0–1 knapsack problem, for which efficient algorithms exist; see, for example, Nauss (1976). If the optimal knapsack solution is less than $-f_j$, then the corresponding optimal solution to P_λ^j is found by setting $Y_j = 1$ and X_{ij} according to the knapsack solution, indicating whether retailer i

is assigned to site j. If the optimal knapsack solution is more than $-f_j$, then the optimal solution to P_λ^j is found by setting $Y_j = 0$ and $X_{ij} = 0$ for all $i \in N$.

The solution to P_λ is then given by

$$Z_\lambda \doteq \sum_{j=1}^{m} Z_\lambda^j - \sum_{i=1}^{n} \lambda_i.$$

For any vector $\lambda \in \mathbb{R}^n$, this is a lower bound on the optimal solution Z^*. In order to find the best such lower bound, we use a subgradient procedure.

Note that if the problem has a constraint on the number of warehouses (facilities) that can be opened (chosen), this can be handled in essentially the same way as it was handled in the algorithm for the p-median problem.

Upper Bounds

For a given set of multipliers, if the values $\{X\}$ satisfy (15.6), then we have an optimal solution to Problem P, and we stop. Otherwise, we perform a simple subroutine to find a feasible solution to P. The procedure is based on the observation that the knapsack solutions found when solving P_λ give us some information concerning the benefit of setting up a warehouse at a site (relative to the current vector λ). If, for example, the knapsack solution corresponding to a given site is 0, that is, the optimal knapsack is empty, then this is most likely not a "good" site to select at this time. In contrast, if the knapsack solution has a very negative cost, then this is a "good" site. Given the values Z_λ^j for each $j \in M$, let π be a permutation of $1, 2, \ldots, m$ such that

$$Z_\lambda^{\pi(1)} \le Z_\lambda^{\pi(2)} \le \cdots \le Z_\lambda^{\pi(m)}.$$

The procedure we perform allocates retailers to sites in a myopic fashion. Let M be the minimum possible number of warehouses used in the optimal solution to CFLP. This can be found by solving the bin-packing problem defined on the values w_i with bin capacities Q_j; see Sect. 4.2. Starting with the "best" site, in this case site $\pi(1)$, assign the retailers in its optimal knapsack to this site. Then, following the indexing of the knapsack solutions, take the next-"best" site [say site $j \doteq \pi(2)$] and solve a new knapsack problem: one defined with costs $\bar{c}_{ij} \doteq c_{ij} + \lambda_i$ for each retailer i still unassigned. Assign all retailers in this knapsack solution to site j. If this optimal knapsack is empty, then a warehouse is not located at that site, and we go on to the next site. Continue in this manner until M warehouses are located.

The solution may still not be a feasible solution to P since some retailers may not be assigned to a site. In this case, unassigned retailers are assigned to sites that are already chosen, where they fit with minimum additional cost. If needed, additional warehouses may be opened following the ordering of π. A local improvement heuristic can be implemented to improve on this solution, using simple interchanges between retailers.

Computational Results

We now report on various computational experiments using this algorithm (Table 15.2). The retailer locations were chosen uniformly over the unit square. Again, for simplicity, we made each retailer location a potential site for a warehouse; thus, $m = n$. The fixed cost of a site was chosen uniformly between 0 and 1. The cost of assigning a retailer to a site was the Euclidean distance between the two locations. The values of w_i were chosen uniformly over the interval 0 to $\frac{1}{2}$ with warehouse capacity equal to 1. We applied the algorithm mentioned above to 10 problems and recorded the average relative error of the best solution found and the average computation time required. The algorithm is terminated when the relative error is below 1% or when a prespecified number of iterations is reached. The numbers below "Error" are the relative errors averaged over the 10 randomly generated problem instances. The numbers below "CPU Time" are the CPU time averaged over the 10 problem instances. All computational times are on an IBM RISC 6000 Model 950.

TABLE 15.2. Computational results for the single-source CFLP algorithm

n	Error	CPU time
10	1.2%	1.2 s
20	1.0%	8.1 s
50	1.1%	110.0 s
100	1.1%	558.3 s

15.4   A Distribution System Design Problem

So far, the location models we have considered have been concerned with minimizing the costs of transporting products between warehouses and retailers. We now present a more realistic model that considers the cost of transporting the product from manufacturing facilities to the warehouses as well.

Consider the following warehouse location problem. A set of plants and retailers are geographically dispersed in a region. Each retailer experiences demands for a variety of products that are manufactured at the plants. A set of warehouses must be located in the distribution network from a list of potential sites.

The cost of locating a warehouse includes the transportation cost per unit from warehouses to retailers but also the transportation cost from plants to warehouses. In addition, as in the CFLP, there is a site-dependent fixed cost for locating each warehouse.

The data for the problem are the following:

- L = number of plants; we will also let $L = \{1, 2, \ldots, L\}$;

- J = number of potential warehouse sites; also let $J = \{1, 2, \ldots, J\}$;

- I = number of retailers; also let $I = \{1, 2, \ldots, I\}$;

- K = number of products; also let $K = \{1, 2, \ldots, K\}$;

- W = number of warehouses to locate;

- $c_{\ell jk}$ = cost of shipping one unit of product k from plant ℓ to warehouse site j;

- d_{jik} = cost of shipping one unit of product k from warehouse site j to retailer i;

- f_j = fixed cost of locating a warehouse at site j;

- $v_{\ell k}$ = supply of product k at plant ℓ;

- w_{ik} = demand for product k at retailer i;

- s_k = volume of one unit of product k;

- q_j = capacity (in volume) of a warehouse at site j.

We make the additional assumption that a retailer gets delivery for a product from one warehouse only. This does not preclude solutions where a retailer gets shipments from different warehouses, but these shipments must be for different products. On the other hand, we assume that the warehouse can receive shipments from any plant and for any amount of product.

The problem is to determine where to locate the warehouses, how to ship the product from the plants to the warehouses, and also how to ship the product from the warehouses to the retailers. This problem is similar to that analyzed by Pirkul and Jayaraman (1996).

We again use a mathematical programming approach. Define the following decision variables:

$$Y_j = \begin{cases} 1, & \text{if a warehouse is located at site } j, \\ 0, & \text{otherwise,} \end{cases}$$

and

$$U_{\ell jk} = \text{amount of product } k \text{ shipped from plant } \ell \text{ to warehouse } j,$$

for each $\ell \in L$, $j \in J$, and $k \in K$. Also, define

$$X_{jik} = \begin{cases} 1, & \text{if retailer } i \text{ receives product } k \text{ from warehouse } j, \\ 0, & \text{otherwise,} \end{cases}$$

for each $j \in J$, $i \in I$, and $k \in K$.

Then the distribution system design problem can be formulated as the following integer program:

$$\text{Min} \quad \sum_{\ell=1}^{L}\sum_{j=1}^{J}\sum_{k=1}^{K} c_{\ell jk} U_{\ell jk} + \sum_{i=1}^{I}\sum_{j=1}^{J}\sum_{k=1}^{K} d_{jik} w_{ik} X_{jik} + \sum_{j=1}^{J} f_j Y_j$$

$$\text{s.t.} \quad \sum_{j=1}^{J} X_{jik} = 1, \qquad \forall i \in I, \ k \in K, \tag{15.9}$$

$$\sum_{i=1}^{I}\sum_{k=1}^{K} s_k w_{ik} X_{jik} \le q_j Y_j, \quad \forall j \in J, \tag{15.10}$$

$$\sum_{i=1}^{I} w_{ik} X_{jik} = \sum_{\ell=1}^{L} U_{\ell jk}, \quad \forall j \in J, \ k \in K, \tag{15.11}$$

$$\sum_{j=1}^{J} U_{\ell jk} \le v_{\ell k} \qquad \forall \ell \in L, \ k \in K, \tag{15.12}$$

$$\sum_{j=1}^{J} Y_j = W, \tag{15.13}$$

$$Y_j, X_{jik} \in \{0,1\}, \qquad \forall i \in I, j \in J, k \in K, \tag{15.14}$$

$$U_{\ell jk} \ge 0, \qquad \forall \ell \in L, j \in J, k \in K. \tag{15.15}$$

The objective function measures the transportation costs between plants and warehouses, those costs between warehouses and retailers, and also the fixed cost of locating the warehouses. Constraints (15.9) ensure that each retailer/product pair is assigned to one warehouse. Constraints (15.10) guarantee that the capacity of the warehouses is not exceeded. Constraints (15.11) ensure that there is a conservation of the flow of products at each warehouse; that is, the amount of each product arriving at a warehouse from the plants is equal to the amount being shipped from the warehouse to the retailers. Constraints (15.12) are the supply constraints. Constraints (15.13) ensure that we locate exactly W warehouses.

The model can handle several extensions, such as a warehouse handling fee or a limit on the distance of any link used just as in the p-median problem. Another interesting extension is when there is a fixed number of possible warehouse types from which to choose. Each type has a specific cost along with a specific capacity. The model can easily be extended to handle this situation (see Exercise 15.1).

As in the previous problems, we will use Lagrangian relaxation. We relax constraints (15.9) (with multipliers λ_{ik}) and constraints (15.11) (with multipliers θ_{jk}). The resulting problem is

$$Min \quad \sum_{\ell=1}^{L}\sum_{j=1}^{J}\sum_{k=1}^{K}c_{\ell jk}U_{\ell jk} + \sum_{j=1}^{J}\sum_{i=1}^{I}\sum_{k=1}^{K}d_{jik}w_{ik}X_{jik} + \sum_{j=1}^{J}f_jY_j$$

$$+ \sum_{j=1}^{J}\sum_{k=1}^{K}\theta_{jk}\left[\sum_{i=1}^{I}w_{ik}X_{jik} - \sum_{\ell=1}^{L}U_{\ell jk}\right] + \sum_{i=1}^{I}\sum_{k=1}^{K}\lambda_{ik}\left[1 - \sum_{j=1}^{J}X_{jik}\right],$$

subject to $(15.10), (15.12) - (15.15)$.

Let $Z_{\lambda,\theta}$ be the optimal solution to this problem. This problem can be decomposed into two separate problems P_1 and P_2:

$$\text{Problem } P_1 : Z_1 \doteq Min \quad \sum_{\ell=1}^{L}\sum_{j=1}^{J}\sum_{k=1}^{K}[c_{\ell jk} - \theta_{jk}]U_{\ell jk}$$

$$\text{s.t.} \quad \sum_{j=1}^{J}U_{\ell jk} \leq v_{\ell k}, \forall \ell \in L, \ k \in K, \tag{15.16}$$

$$U_{\ell jk} \geq 0, \quad \forall \ell \in L, j \in J, k \in K.$$

$$\text{Problem } P_2 : Z_2 \doteq Min \quad \sum_{j=1}^{J}\sum_{i=1}^{I}\sum_{k=1}^{K}[d_{jik}w_{ik} - \lambda_{ik} + \theta_{jk}w_{ik}]X_{jik} + \sum_{j=1}^{J}f_jY_j,$$

$$\text{s.t.} \quad \sum_{i=1}^{I}\sum_{k=1}^{K}s_kw_{ik}X_{jik} \leq q_jY_j, \quad \forall j \in J \tag{15.17}$$

$$\sum_{j=1}^{J}Y_j = P, \tag{15.18}$$

$$Y_j, X_{jik} \in \{0,1\}, \quad \forall i \in I, j \in J, k \in K.$$

Solving P_1

Problem P_1 can be solved separately for each plant/product pair. In fact, the objective functions of each of these subproblems can be improved (without loss in computation time) by adding the constraints

$$s_k\sum_{\ell=1}^{L}U_{\ell jk} \leq q_j, \quad \forall j \in J, \ k \in K. \tag{15.19}$$

For each plant/product combination, say plant ℓ and product k, sort the J values $\bar{c}_j \doteq c_{\ell jk} - \theta_{jk}$. Starting with the smallest value of $\bar{c}_j$, say $\bar{c}_{j'}$, if $\bar{c}_{j'} \geq 0$, then the solution is to ship none of this product from this plant. If $c_{\ell j'k} < 0$, then ship as much of this product as possible along arc (ℓ, j') subject to satisfying constraints (15.16) and (15.19). Then if the supply $v_{\ell k}$ has not been completely shipped, do the same for the next-cheapest arc, as long as it has negative reduced cost $(\bar{c})$. Continue in this manner until all of the product has been shipped or the reduced costs are no longer negative. Then proceed to the next plant/product combination, repeating this procedure. Continue until all the plant/product combinations have been scanned in this fashion.

Solving P_2

Solving Problem P_2 is similar to solving the subproblem in the CFLP. For now, we can ignore constraints (15.18). Then we separate the problem by warehouse. In the problem corresponding to warehouse j, either $Y_j = 0$ or $Y_j = 1$. If $Y_j = 0$, then $X_{jik} = 0$ for all $i \in N$ and $k \in K$. If $Y_j = 1$, then we get a knapsack problem with NK items, one for each retailer/product pair. Let Z_2^j be the objective function value when Y_j is set to 1 and the resulting knapsack problem is solved. After having solved each of these, let π be a permutation of the numbers $1, 2, \ldots, J$ such that

$$Z_2^{\pi(1)} \leq Z_2^{\pi(2)} \leq \cdots \leq Z_2^{\pi(J)}.$$

The optimal solution to P_2 is to choose the W smallest values:

$$Z_2 \doteq \sum_{j=1}^{W} Z_2^{\pi(j)}.$$

For fixed vectors λ and θ, the lower bound is

$$Z_{\lambda,\theta} \doteq Z_1 + Z_2 + \sum_{i=1}^{I} \sum_{k=1}^{K} \lambda_{ik}.$$

To maximize this bound, that is,

$$\max_{\lambda,\theta} \{Z_{\lambda,\theta}\},$$

we again use the subgradient optimization procedure.

Upper Bounds

At each iteration of the subgradient procedure, we attempt to construct a feasible solution to the problem. Consider Problem P_2. Its solution may have a retailer/product combination assigned to several warehouses. We determine the set of retailer/product combinations that are assigned to one and only one retailer and fix these. Other retailer/product combinations are assigned to warehouses using the following mechanism. For each retailer/product combination, we determine the cost of assigning it to a particular warehouse. After determining that this assignment is feasible (from a warehouse-capacity point of view), we calculate the assignment cost as the cost of shipping all of the demand for this retailer/product combination through the warehouse plus the cost of shipping the demand from the plants to the warehouse (along one or more arcs from the warehouse to the plants). For each retailer/product combination, we determine the penalty associated with assigning the shipment to its second-best warehouse instead of its best warehouse. We then assign the retailer/product combination with the highest such penalty and update all arc flows and remaining capacities. We continue in this manner until all retailer/product combinations have been assigned to warehouses.

Computational results for this problem appear at the end of Chap. 20.

15.5   The Structure of the Asymptotic Optimal Solution

In this section, we describe a region-partitioning scheme to solve large instances of the CFLP.

Assume there are n retailers located at points $\{x_1, x_2, \ldots, x_n\}$. Each retailer also serves as a potential site for a warehouse of fixed capacity q. The fixed cost of locating a warehouse at a site is assumed to be proportional to the distance the site is from a manufacturing facility located at x_0, which is assumed (without loss of generality) to be the origin $(0,0)$. Retailer i has a demand w_i, which is assumed to be less than or equal to q. Without loss of generality, we assume $q = 1$, and therefore, $w_i \in [0,1]$ for each $i \in N$. Let α be the per-unit cost of transportation between warehouses and the manufacturing facility, and let β be the per-unit cost of transportation between warehouses and retailers.

We assume the retailer locations are independently and identically distributed in a compact region $A \subset \mathbb{R}^2$ according to some distribution μ. Assume the retailer demands are independently and identically distributed according to a probability measure ϕ on $[0,1]$. The bin-packing constant associated with the distribution ϕ (denoted by γ_ϕ or simply γ) is the asymptotic number of bins used per item in an optimal packing of the retailer demands into unit size bins, when items are drawn randomly from the distribution ϕ (see Sect. 5.2).

The following theorem shows that if the retailer locations and demand sizes are random (from a general class of distributions), then as the problem size increases, the optimal solution has a very particular structure. This structure can be exploited using a region-partitioning scheme, as demonstrated below.

Theorem 15.5.1 *Let x_k, $k = 1, 2, \ldots, n$, be a sequence of independent random variables having a distribution μ with compact support in $\mathbb{R}^2$. Let $\|x\|$ be the Euclidean distance between the manufacturing facility and the point $x \in \mathbb{R}^2$, and let*

$$E(d) = \int \|x\| d\mu(x).$$

Let the demands w_k, $k = 1, 2, \ldots, n$, be a sequence of independent random variables having a distribution ϕ with bin-packing constant equal to γ. Then, almost surely,

$$\lim_{n \to \infty} \frac{1}{n} Z_n^* = \alpha \gamma E(d).$$

This analysis demonstrates that simple approaches that consider only the geography and the packing of the demands can be very efficient on large problem instances. Asymptotically, this is, in fact, the optimal strategy. This analysis also demonstrates that, asymptotically, the cost of transportation between retailers and warehouses becomes a very small fraction (eventually zero) of the total cost.

15.6   Exercises

Exercise 15.1. In the distribution system design problem, explain how the solution methodology changes when there is a fixed number of possible warehouse capacities. For example, at each site, if we decide to install a warehouse, we can install a *small*, *medium*, or *large* one.

Exercise 15.2. Prove Theorem 15.5.1.

Exercise 15.3. Show how any instance of the bin-packing problem (see Part I) can be formulated as an instance of the single-source CFLP.

Exercise 15.4. Consider Problem P_1 of Sect. 15.4.

(a) Show that this formulation can be strengthened by adding the constraints:

$$\sum_{\ell=1}^{L} \sum_{k=1}^{K} s_k U_{\ell jk} \leq q_j, \qquad \forall j \in J.$$

(b) Show that this new formulation can be transformed to a specialized kind of linear program called a transportation problem.

(c) Why might we not want to use this stronger formulation?

Exercise 15.5. (Mirchandani and Francis 1990) Define the uncapacitated facility location problem (UFLP) in the following way. Let F_j be the fixed charge of opening a facility at site j, for $j = 1, 2, \ldots, m$.

$$\text{Problem UFLP} \quad : Min \sum_{i=1}^{n} \sum_{j=1}^{m} c_{ij} X_{ij} + \sum_{j=1}^{m} F_j Y_j$$

$$\text{s.t.} \sum_{j=1}^{m} X_{ij} = 1, \quad \forall i \in N,$$

$$X_{ij} \leq Y_j, \quad \forall i \in N, \ j \in M,$$
$$X_{ij}, Y_j \in \{0, 1\}, \quad \forall i \in N, \ j \in M.$$

Show that UFLP is $\mathcal{NP}$-Hard by showing that any instance of the $\mathcal{NP}$-Hard node cover problem can be formulated as an instance of UFLP. The node cover problem is defined as follows: Given a graph G and an integer k, does there exist a subset of k nodes of G that cover all the arcs of G? (Node v is said to cover arc e if v is an endpoint of e.)

Exercise 15.6. (Mirchandani and Francis 1990) It appears that the p-median problem can be solved by solving the resulting problem UFLP (see Exercise 15.5) for different values of $F = F_j, \forall j$, until a value F^* is found where the UFLP opens exactly p facilities. Show that this method does not work by giving an instance of a 2-median problem for which no value of F provides an optimal solution to UFCLP with two open facilities.

Part IV

Vehicle Routing Models

16

The Capacitated VRP with Equal Demands

16.1 Introduction

A large part of many logistics systems involves the management of a fleet of vehicles used to serve warehouses, retailers, and/or customers. In order to control the costs of operating the fleet, a dispatcher must continuously make decisions on how much to load on each vehicle and where to send it. These types of problems fall under the general class of vehicle routing problems mentioned in Chap. 1.

The most basic vehicle routing problem (VRP) is the single-depot capacitated vehicle routing problem (CVRP). It can be described as follows: A set of customers has to be served by a fleet of identical vehicles of limited capacity. The vehicles are initially located at a given depot. The objective is to find a set of routes for the vehicles of minimal total length. Each route begins at the depot, visits a subset of the customers, and returns to the depot without violating the capacity constraint.

Consider the following scenario. A customer requests w units of product. If we allow this load to be *split* between more than one vehicle (i.e., the customer gets several deliveries, which together sum up to the total load requested), then we can view the demand for w units as w different customers each requesting *one* unit of product located at the same point. The capacity constraint can then be viewed as simply the maximum number of customers (in this new problem) that can be visited by a single vehicle. This is the capacity $Q \geq 1$. Therefore, if we allow this splitting of demands, and this may not be a desirable property (we investigate the unsplit-demand case in Chap. 17), there is no loss in generality in assuming that

D. Simchi-Levi et al., *The Logic of Logistics: Theory, Algorithms, and Applications for Logistics Management*, Springer Series in Operations Research and Financial Engineering, DOI 10.1007/978-1-4614-9149-1_16, © Springer Science+Business Media New York 2014

each customer has the same demand, namely, one unit, and the vehicle can visit at most Q of these customers on a route. Therefore, this model is sometimes called the CVRP with *splittable* demands or the ECVRP.

We denote the depot by x_0 and the set of customers by $N = \{x_1, x_2, \ldots, x_n\}$. The set $N_0 \doteq N \cup \{x_0\}$ designates all customers and the depot. The customers and the depot are represented by a set of nodes on an undirected graph $G = (N_0, E)$. We denote by d_i the distance between customer i and the depot, by $d_{\max} \doteq \max_{i \in N} d_i$ the distance from the depot to the furthest customer, and by d_{ij} the distance between customer i and customer j. The distance matrix $\{d_{ij}\}$ is assumed to be symmetric and to satisfy the triangle inequality; that is, $d_{ij} = d_{ji}$ for all i, j and $d_{ij} \leq d_{ik} + d_{kj}$ for all i, k, j. We denote the optimal solution value of the CVRP by Z^* and the solution provided by a heuristic H by Z^{H}.

In what follows, the optimal traveling salesman tour plays an important role. So, for any set $S \subseteq N_0$, let $L^*(S)$ be the length of the optimal traveling salesman tour through the set of points S. Also, let $L^{\alpha}(S)$ be the length of an α-optimal traveling salesman tour through S, that is, one whose length is bounded from above by $\alpha L^*(S)$, $\alpha \geq 1$.

The graph depicted in Fig. 16.1, which is denoted by $\mathcal{G}(t, s)$, also plays an important role in our worst-case analyses. It consists of s groups of Q nodes and another $s - 1$ nodes, called *white nodes*, separating the groups. The nodes within the same group have zero interdistance and each group is connected to the depot by an arc of unit length. The white nodes are of *zero distance apart* and t units' distance away from the depot. Each white node is connected to the two groups of nodes it separates by an arc of unit length. Note that when $0 \leq t \leq 2$, $\mathcal{G}(t, s)$ satisfies the triangle inequality [if an edge (i, j) is not shown in the graph, then the distance between node i and node j is defined as the length of the shortest path from i to j]. Also note that whenever $0 \leq t \leq 2$, the tour depicted in Fig. 16.2 is an optimal traveling salesman tour of length $2s$.

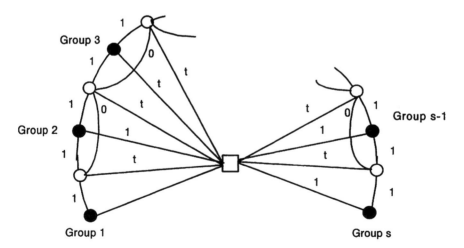

FIGURE 16.1. Every group contains Q customers with interdistance zero

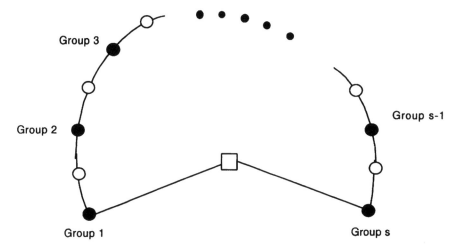

FIGURE 16.2. An optimal traveling salesman tour in $\mathcal{G}(t, s)$

In this chapter, we analyze this problem using the two tools developed earlier, worst-case and average-case analyses. Later, in Chap. 17, we will analyze a more general model of the CVRP.

16.2   Worst-Case Analysis of Heuristics

A simple heuristic for the CVRP, suggested by Haimovich and Rinnooy Kan (1985) and later modified by Altinkemer and Gavish (1990), is to partition a traveling salesman tour into segments, such that each segment of customers is served by a single vehicle; that is, each segment has no more than Q points. The heuristic, called the iterated tour-partitioning (ITP) heuristic, starts from a traveling sales-man tour through all $n = |N|$ customers and the depot. Starting at the depot and following the tour in an arbitrary orientation, the customers and the depot are numbered $x^{(0)}, x^{(1)}, x^{(2)}, \ldots, x^{(n)}$, where $x^{(0)}$ is the depot. We partition the path from $x^{(1)}$ to $x^{(n)}$ into $\lceil \frac{n}{Q} \rceil$ (or $\lceil \frac{n}{Q} \rceil + 1$) disjoint segments, such that each one contains no more than Q customers, and connect the endpoints of each segment to the depot. The first segment contains only customer $x^{(1)}$. All the other segments contain exactly Q customers, except maybe the last one. This defines one feasible solution to the problem. We can repeat the above construction by shifting the end-points of all but the first and last segments up by one position in the direction of the orientation. This can be repeated $Q - 1$ times, producing a total of Q different solutions. We then choose the best of the set of Q solutions generated.

It is easy to see that for a given traveling salesman tour, the running time of the ITP heuristic is $O(nQ)$. The performance of this heuristic clearly depends on the quality of the initial traveling salesman tour chosen in the first step of the algorithm. Hence, when the ITP heuristic partitions an α-optimal traveling salesman tour, it is denoted $\text{ITP}(\alpha)$. To establish the worst-case behavior of the algorithm, we first find a lower bound on Z^* and then calculate an upper bound on the cost of the solution produced by the $\text{ITP}(\alpha)$ heuristic.

Lemma 16.2.1 $Z^* \geq \max\{L^*(N_0), \frac{2}{Q}\sum_{i \in N} d_i\}$.

Proof. Clearly, $Z^* \geq L^*(N_0)$ by the triangle inequality. To prove $Z^* \geq \frac{2}{Q}\sum_{i \in N} d_i$, consider an optimal solution in which N is partitioned into subsets $\{N_1, N_2, \ldots, N_m\}$, where each set N_j is served by a single vehicle. Clearly,

$$Z^* = \sum_j L^*(N_j \cup \{x_0\}) \geq \sum_j 2 \max_{i \in N_j} d_i \geq \sum_j \frac{2}{|N_j|} \sum_{i \in N_j} d_i$$

$$\geq \sum_j \frac{2}{Q} \sum_{i \in N_j} d_i = \frac{2}{Q} \sum_{i \in N} d_i.$$

■

Lemma 16.2.2 $Z^{\text{ITP}(\alpha)} \leq \frac{2}{Q}\sum_{i \in N} d_i + (1 - \frac{1}{Q})\alpha L^*(N_0)$.

Proof. We prove the lemma by finding the cumulative length of the Q solutions generated by the ITP heuristic. The ith solution consists of the segments

$$\{x^{(1)}, x^{(2)}, \ldots, x^{(i)}\}, \{x^{(i+1)}, x^{(i+2)}, \ldots, x^{(i+Q)}\}, \ldots, \{x^{(i+1+\lfloor \frac{n-i}{Q} \rfloor Q)}, \ldots, x^{(n)}\}.$$

Thus, among the Q solutions generated, each customer $x^{(i)}$, $2 \leq i \leq n-1$ appears exactly once as the first point of a segment and exactly once as the last point. Therefore, in the cumulative length of the Q solutions, the term $2d_{x^{(i)}}$ is incurred for each i, $2 \leq i \leq n-1$. Customer $x^{(1)}$ is the first point of a segment in each of the Q solutions, and in the first one it is also the last point. Thus, the term $d_{x^{(1)}}$ appears $Q+1$ times in the cumulative length. Similarly, $x^{(n)}$ is always the last point of a segment in each of the Q solutions, and once the first point. Thus, the term $d_{x^{(n)}}$ appears $Q+1$ times in the cumulative length as well. Finally, each one of the arcs $(x^{(i)}, x^{(i+1)})$ for $1 \leq i \leq n-1$ appears in exactly $Q-1$ solutions since it is excluded from only one solution. These arcs, together with the $Q-1$ arcs connecting the depot to $x^{(1)}$ and $Q-1$ arcs connecting the depot to $x^{(n)}$, form $Q-1$ copies of the initial traveling salesman tour selected in the first step of the heuristic. Thus, if the initial traveling salesman tour is an α-optimal tour, the cumulative length of all Q tours is

$$2\sum_{i \in N} d_i + (Q-1)L^{\alpha}(N_0)$$

$$\leq 2\sum_{i \in N} d_i + (Q-1)\alpha L^*(N_0).$$

Hence,

$$Z^{\text{ITP}(\alpha)} \leq \frac{2}{Q} \sum_{i \in N} d_i + (1 - \frac{1}{Q})\alpha L^*(N_0).$$

∎

Combining upper and lower bounds, we obtain the following result.

Theorem 16.2.3

$$\frac{Z^{\text{ITP}(\alpha)}}{Z^*} \leq 1 + \left(1 - \frac{1}{Q}\right)\alpha. \tag{16.1}$$

For example, if Christofides' polynomial-time heuristic ($\alpha = 1.5$) is used to obtain the initial traveling salesman tour, we have

$$\frac{Z^{\text{ITP}(1.5)}}{Z^*} \leq \frac{5}{2} - \frac{3}{2Q}.$$

The proof of the worst-case result for the ITP(α) heuristic suggests that if we can improve the bound in (16.1) for $\alpha = 1$, then the bound can be improved for any $\alpha > 1$. However, the following theorem, proved by Li and Simchi-Levi (1990), says that this is impossible; that is, the bound

$$\frac{Z^{\text{ITP}(1)}}{Z^*} \leq 2 - \frac{1}{Q}$$

is sharp.

Theorem 16.2.4 *For any integer $Q \geq 1$, there exists a problem instance with $Z^{\text{ITP}(1)}/Z^* = 2 - \frac{1}{Q}$.*

Proof. Let us consider the graph $\mathcal{G}(0, q)$. A solution obtained by the ITP heuristic is shown in Fig. 16.3. In this solution,

$$Z^{\text{ITP}(1)} = 2 + 2 + \underbrace{4 + 4 + \cdots + 4}_{Q-2 \text{ times}} + 2 = 4Q - 2.$$

One can construct a solution that has Q vehicles serve the Q groups of customers and the $(Q + 1)$st vehicle serve the other $Q - 1$ nodes. Thus,

$$Z^* \leq 2Q.$$

Hence,

$$\frac{Z^{\text{ITP}(1)}}{Z^*} \geq 2 - \frac{1}{Q}.$$

This, together with the upper bound of (16.1), completes the proof. ∎

Another variant of the tour-partitioning heuristic is the optimal partitioning (OP) heuristic described by Beasley (1983). The algorithm takes a traveling salesman tour and optimally partitions it into a set of feasible routes; that is, each route contains at most Q customers.

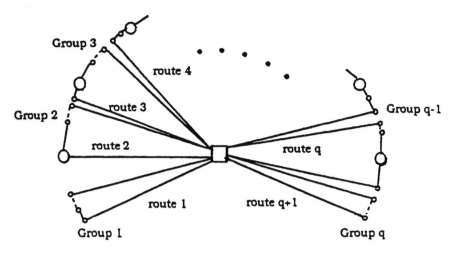

FIGURE 16.3. Solution obtained by the ITP heuristic

Given a traveling salesman tour through the customers and the depot, the heuristic numbers the points $x^{(0)}, x^{(1)}, \ldots, x^{(n)}$ in order of appearance on the tour, where $x^{(0)}$ is the depot. Define

$$
C_{jk} = \begin{cases} \text{the distance traveled by a vehicle that starts at } x^{(0)} \text{ visits,} \\ \text{customers } x^{(j+1)}, x^{(j+2)}, \ldots, x^{(k)} \text{ and returns to } x^{(0)}, & \text{if } k - j \le Q; \\ \infty, & \text{otherwise.} \end{cases}
$$

If we find the shortest path from $x^{(0)}$ to $x^{(n)}$ in the acyclic graph [with nodes $x^{(i)}$, $0 \le i \le n$, and arcs $(x^{(i)}, x^{(j)})$ for $0 \le i < j \le n$], where the distance between $x^{(j)}$ and $x^{(k)}$ is C_{jk}, we will have an optimal partition of the traveling salesman tour into feasible routes. For example, if the shortest path from $x^{(0)}$ to $x^{(n)}$ is $x^{(0)} \to x^{(t)} \to x^{(u)} \to x^{(n)}$, then three tours are formed, namely, $(x^{(0)}, x^{(1)}, \ldots, x^{(t)}, x^{(0)})$, $(x^{(0)}, x^{(t+1)}, x^{(t+2)}, \ldots, x^{(u)}, x^{(0)})$, and $(x^{(0)}, x^{(u+1)}, x^{(u+2)}, \ldots, x^{(n)}, x^{(0)})$.

For a given traveling salesman tour, the above shortest-path problem can be solved in $O(nQ)$ time, including the time required to evaluate the costs C_{jk}.

When the OP heuristic partitions an α-optimal traveling salesman tour, it is denoted OP(α). The partitions considered by the OP(α) heuristic include all Q of the partitions generated by the ITP(α) heuristic. Therefore, $Z^{OP(\alpha)} \le Z^{ITP(\alpha)}$, and hence, its worst-case bound is at least as good; that is,

$$
\frac{Z^{OP(\alpha)}}{Z^*} \le 1 + \left(1 - \frac{1}{Q}\right)\alpha.
$$

The next theorem implies that for $\alpha = 1$, this bound is asymptotically sharp; that is, $Z^{OP(1)}/Z^*$ tends to 2 when Q approaches infinity.

Theorem 16.2.5 *For any integer $Q \ge 1$, there exists a problem instance with $Z^{OP(1)}/Z^*$ arbitrarily close to $2 - \frac{2}{Q+1}$.*

Proof. Consider the graph $\mathcal{G}(1, Kq + 1)$, where K is a positive integer. It is easy to check that

$$Z^{\mathrm{OP}(1)} = 2(KQ + 1) + 2KQ.$$

On the other hand, consider the solution in which $KQ+1$ vehicles serve the $KQ+1$ groups of customers and another K vehicles serve the other nodes. Hence,

$$Z^* \leq 2(KQ + 1) + 2K,$$

and therefore,

$$\lim_{K \to \infty} \frac{Z^{\mathrm{OP}(1)}}{Z^*} \geq 2 - \frac{2}{Q + 1}.$$

∎

16.3 The Asymptotic Optimal Solution Value

In this and the following section, we assume that the customers are points in the plane and that the distance between any pair of customers is given by the Euclidean distance. Assume, without loss of generality, that the depot is the point $(0,0)$ and $||x||$ designates the distance from the depot to the point $x \in \mathbb{R}^2$. The results discussed in this section and the next are mainly based on Haimovich and Rinnooy Kan (1985).

The upper bound of Lemma 16.2.2 has two cost components; the first component is proportional to the total "radial" cost between the depot and the customers. The second component is proportional to the "circular" cost: the cost of traveling between customers. This cost is related to the cost of the optimal traveling sales-man tour. As discussed in Chap. 2, for large n, the cost of the optimal traveling salesman tour grows like $\sqrt{n}$, while the total radial cost between the depot and the customers grows like n. Therefore, it is intuitive that when the number of cus-tomers is large enough, the first cost component will dominate the second. This observation is now formally proven.

Theorem 16.3.1 *Let x_k, $k = 1, 2, \ldots, n$, be a sequence of independent random variables having a distribution μ with compact support in $\mathbb{R}^2$. Let*

$$E(d) = \int_{\mathbb{R}^2} ||x|| d\mu(x).$$

Then, with probability 1,

$$\lim_{n \to \infty} \frac{Z^*}{n} = \frac{2}{Q} E(d).$$

Proof. Lemma 16.2.1 and the strong law of large numbers tell us that

$$\lim_{n \to \infty} \frac{Z^*}{n} \geq \frac{2}{Q} E(d) \quad (a.s.). \tag{16.2}$$

On the other hand, from Lemma 16.2.2,

$$\frac{Z^*}{n} \le \frac{Z^{\text{ITP}(1)}}{n} \le \frac{2}{nQ}\sum_{i \in N} d_i + \left(1 - \frac{1}{Q}\right)\frac{L^*(N_0)}{n}.$$

From Chap. 5, we know that there exists a constant $\beta > 0$, independent of the distribution μ, such that with probability 1,

$$\lim_{n \to \infty} \frac{L^*(N_0)}{\sqrt{n}} = \beta \int_{I\!R^2} f^{1/2}(x)dx,$$

where f is the density of the absolutely continuous part of the distribution μ. Hence,

$$\lim_{n \to \infty} \frac{Z^*}{n} \le \frac{2}{Q}E(d) \qquad (a.s.).$$

This together with (16.2) proves the theorem. ∎

The following observation is in order. Haimovich and Rinnooy Kan prove Theorem 16.3.1 merely assuming $E(d)$ is finite rather than the stronger assumption of a compact support. However, the restriction to a compact support seems to be satisfactory for all practical purposes. The following is another important generalization of Theorem 16.3.1. Assume that a *cluster* of w_k customers (rather than a single customer) is located at point x_k, $k = 1, 2, \ldots, n$. The theorem then becomes

$$\lim_{n \to \infty} \frac{Z^*}{n} = \frac{2}{Q}E(w)E(d), \tag{16.3}$$

where $E(w)$ is the expected cluster size, provided that the cluster size is independent of the location. This follows from a straightforward adaptation of Lemmas 16.2.1 and 16.2.2.

16.4 Asymptotically Optimal Heuristics

The proof of the previous theorem (Theorem 16.3.1) reveals that the ITP(α) heuristic provides a solution whose cost approaches the optimal cost when n tends to infinity. Indeed, replacing ITP(1) by ITP(α) in the previous proof gives the following theorem.

Theorem 16.4.1 *Under the conditions of Theorem 16.3.1 and for any fixed $\alpha \ge 1$, the ITP(α) heuristic is asymptotically optimal.*

As is pointed out by Haimovich and Rinnooy Kan (1985), iterated tour-partitioning heuristics, although asymptotically optimal, hardly exploit the special topological structure of the Euclidean plane in which the points are located. It is therefore natural to consider *region-partitioning* (RP) heuristics that are more geometric in nature.

Haimovich and Rinnooy Kan consider three classes of regional partitioning schemes. In *rectangular region partitioning* (RRP), one starts with a rectangle containing the set of customers N and cuts it into smaller rectangles. In *polar region partitioning* (PRP) and *circular region partitioning* (CRP), one starts with a circle centered at the depot and partitions it by means of circular arcs and radial lines. We shall shortly discuss each one of these in detail.

In each case, the RP heuristics construct subregions of the plane, where subregion j contains a set of customers $N(j)$. These subregions are constructed so that each one of them has exactly Q customers except possibly one.

Since every subset $N(j)$ has no more than Q customers, each of these RP heuristics allocates one vehicle to each subregion. The vehicles then use the following routing strategy. The first customer visited is the one closest to the depot among all the customers in $N(j)$. The rest are visited in the order of an α-optimal traveling salesman tour through $N(j)$. After visiting all the customers in the subregion, the vehicle returns to the depot through the first (closest) customer. It is therefore natural to call these heuristics $RP(\alpha)$ heuristics. In particular, we have $RRP(\alpha)$, $PRP(\alpha)$, and $CRP(\alpha)$.

Lemma 16.4.2 $Z^{RP(\alpha)} \leq \frac{2}{Q} \sum_{i \in N} d_i + 2d_{\max} + \alpha \sum_j L^*(N(j))$.

Proof. We number the subsets $N(j)$ constructed by the $RP(\alpha)$ heuristic so that $|N(j)| = Q$ for every $j \geq 2$ and $|N(1)| \leq Q$. It follows that the total distance traveled by the vehicle that visits subset $N(j)$, for $j \geq 2$, is

$$\leq 2 \min_{i \in N(j)} d_i + \alpha L^*(N(j))$$

$$\leq \frac{2}{Q} \sum_{i \in N(j)} d_i + \alpha L^*(N(j)),$$

while the total distance traveled by the vehicle that visits $N(1)$ is no more than

$$2d_{\max} + \alpha L^*(N(1)).$$

Taking the sum over all subregions, we obtain the desired result. ∎

The quality of the upper bound of Lemma 16.4.2 depends, of course, on the quantity $\sum_j L^*(N(j))$. This value was analyzed in Chap. 5, where it was shown that for any RP heuristic,

$$\sum_j L^*(N(j)) \leq L^*(N) + \frac{3}{2} P^{RP}, \tag{16.4}$$

where P^{RP} is the sum of perimeters of the subregions generated by the RP heuristic. For this reason, we analyze the quantity P^{RP} in each of the three region-partitioning heuristics.

Rectangular Region Partitioning

This heuristic is identical to the one introduced for the traveling salesman problem in Sect. 5.3. The smallest rectangle with sides a and b containing the set of customers N is partitioned with horizontal and vertical lines. First, the region is subdivided by t vertical lines such that each subregion contains exactly $(h + 1)Q$ points except possibly the last one. Each of these $t + 1$ subregions is then partitioned with h horizontal lines into $h + 1$ smaller subregions such that each contains exactly Q points except possibly for the last one.

As before, h and t should satisfy

$$t = \left\lceil \frac{n}{(h + 1)Q} \right\rceil - 1$$

and

$$t(h + 1)Q < n \le (t + 1)(h + 1)Q.$$

The unique integer that satisfies these conditions is $h = \lceil \sqrt{\frac{n}{Q}} - 1 \rceil$. Note that the number of vertical lines added is $t \le \sqrt{\frac{n}{Q}}$, and each of these lines is counted twice in the quantity P^{RRP}.

In the second step of the RRP, we add h horizontal lines, where $h \le \sqrt{\frac{n}{Q}}$. These horizontal lines are also counted twice in P^{RRP}. It follows that

$$P^{\mathrm{RRP}} \le 2\sqrt{\frac{n}{Q}}(a + b) + 2(a + b) \le 8d_{\max}\sqrt{\frac{n}{Q}} + 8d_{\max}.$$

Polar Region Partitioning

The circle with radius $d_{\max}$ containing the set N and centered at the depot is partitioned in exactly the same way as in the previous partitioning scheme, with the exception that circular arcs and radial lines replace vertical and horizontal lines. Using the same analysis, one can show

$$P^{\mathrm{PRP}} \le 6\pi d_{\max}\sqrt{\frac{n}{Q}} + 2\pi d_{\max} + 2d_{\max}. \tag{16.5}$$

Circular Region Partitioning

This scheme partitions the circle centered at the depot with radius $d_{\max}$ into h equal sectors, where h is to be determined. Each sector is then partitioned into subregions via circular arcs, such that each subregion contains exactly Q customers except possibly the one closest to the depot. Thus, at most h subregions, each from one sector, have less than Q customers. These subregions (with the depot on their boundary) are then repartitioned with radial cuts such that at most $h - 1$ of them have exactly Q customers each except for possibly the last one.

The total length of the initial radial lines is $hd_{\max}$. The length of an inner circular arc bounding a subregion containing a set $N(j)$ is no more than

$$\frac{2\pi}{h} \min_{i \in N(j)} d_i \leq \frac{2\pi}{h} \frac{\sum_{i \in N(j)} d_i}{|N(j)|} = \frac{2\pi \sum_{i \in N(j)} d_i}{hQ},$$

while the length of the outer circle is $2\pi d_{\max}$. Finally, the repartitioning of the central subregions adds no more than $\frac{hd_{\max}}{2}$. Thus,

$$P^{CRP} \leq 2\left(hd_{\max} + \frac{2\pi \sum_{i \in N} d_i}{hQ} + \frac{hd_{\max}}{2}\right) + 2\pi d_{\max}.$$

Taking $h = \left\lceil \sqrt{\frac{4\pi \sum_{i \in N} d_i}{3Q d_{\max}}} \right\rceil$, we obtain the following upper bound on P^{CRP}:

$$P^{CRP} \leq 4\sqrt{3\pi d_{\max} \frac{1}{Q} \sum_{i \in N} d_i} + (3 + 2\pi)d_{\max}.$$

The reader should be aware that all of these partitioning schemes can be implemented in $O(n \log n)$ time. We now have all the necessary ingredients for an asymptotic analysis of the performance of these partitioning heuristics.

Theorem 16.4.3 *Under the conditions of Theorem 16.3.1 and for any fixed $\alpha \geq 1$, $RRP(\alpha)$, $PRP(\alpha)$, and $CRP(\alpha)$ are asymptotically optimal.*

Proof. Lemma 16.4.2, together with (16.4), provides the following upper bound on the total distance traveled by all vehicles in the solution produced by the above RP heuristics:

$$Z^{RP(\alpha)} \leq \frac{2}{Q} \sum_{i \in N} d_i + 2d_{\max} + \alpha L^*(N) + \frac{3}{2}\alpha P^{RP}.$$

By the strong law of large numbers and the fact that the distribution has compact support, $\frac{1}{n} \sum_{i \in N} d_i$ converges almost surely to $E(d)$, while $\frac{d_{\max}}{n}$ converges almost surely to 0. Furthermore, $\frac{L^*(N)}{n}$ converges to 0 almost surely; see the proof of Theorem 16.3.1. Finally, from the analysis of each of the region-partitioning heuristics and the fact that the points are in a compact region, $\frac{P^{RP}}{n}$ converges almost surely to zero as well. ∎

In conclusion, we see that the CVRP with equal demands is asymptotically solvable via several different region-partitioning schemes. In fact, since each customer has the same demand, the packing of the customers' demands into the vehicles is a trivial problem. Any Q customers can fit. The more difficult problem, when demands are of different sizes, presents complicating bin-packing features, which will prove to be more difficult.

16.5   Exercises

Exercise 16.1.   Consider the following version of the capacitated vehicle routing problem (CVRP). You are given a network $G = (V, A)$ with positive arc lengths. Assume that $E \subseteq A$ is a given set of edges that have to be "covered" by vehicles. The vehicles are initially located at a depot $p \in V$. Each vehicle has a "capacity" q; that is, each vehicle can cover no more than q edges from E. Once a vehicle starts an edge in E, it has to cover all of it. The objective is to design tours for vehicles so that all edges in E are covered, vehicles' capacities are not violated, and the total distance traveled is as small as possible.

(a) Suppose we want first to find a single tour that starts at the depot p, traverses all edges in E, and ends at p and whose total cost (length) is as small as possible. Generalize Christofides' heuristic for this case.

(b) Consider now the version of the CVRP described above and suggest two possible lower bounds on the optimal cost of the CVRP.

(c) Describe a heuristic algorithm based on a tour-partitioning approach using, as the initial tour, the tour you found in part (a). What is the worst-case bound of your algorithm?

Exercise 16.2.   Derive (16.3).

Exercise 16.3.   Consider an n-customer instance of the CVRP with equal demands. Assume there are m depots and at each depot is an unlimited number of vehicles of limited capacity. Suggest an asymptotically optimal region-partitioning scheme for this case.

Exercise 16.4.   Consider an n-customer instance of the CVRP with equal demands. There are K customer types: A customer is of type k with independent probability $p_k > 0$. Customers of different types cannot be served together in the same vehicle. Devise an asymptotically optimal heuristic for this problem. If K is a function of n, what conditions on $K(n)$ are necessary to ensure that this same heuristic is asymptotically optimal?

Exercise 16.5.   Derive (16.5).

17

The Capacitated VRP with Unequal Demands

17.1 Introduction

In this chapter, we consider the capacitated vehicle routing problem with unequal demands (UCVRP). In this version of the problem, each customer i has a demand w_i and the capacity constraint stipulates that the total amount delivered by a single vehicle cannot exceed Q. We let Z_u^* denote the optimal solution value of UCVRP, that is, the minimal total distance traveled by all vehicles.

In this version of the problem, the demand of a customer *cannot be split* over several vehicles; that is, each customer must be served by a single vehicle. This, more general, version of the model is sometimes called the CVRP with *unsplit* demands. The version where demands may be split is dealt with in Chap. 16. Splitting a customer's demand is often physically impossible or managerially undesirable due to customer service or accounting considerations.

17.2 Heuristics for the CVRP

A great deal of work has been devoted to the development of heuristics for the UCVRP; see, for example, Christofides (1985), Fisher (1995), and Federgruen and Simchi-Levi (1995), or Bertsimas and Simchi-Levi (1996). Following Christofides, we classify these heuristics into the four categories:

- constructive methods;

- route-first–cluster-second methods;

D. Simchi-Levi et al., *The Logic of Logistics: Theory, Algorithms, and Applications* 313
for Logistics Management, Springer Series in Operations Research and Financial Engineering,
DOI 10.1007/978-1-4614-9149-1_17, © Springer Science+Business Media New York 2014

- cluster-first–route-second methods;

- incomplete optimization methods.

We will describe the main characteristics of each of these classes and give examples of heuristics that fall into each.

Constructive Methods

The *savings algorithm* suggested by Clarke and Wright (1964) is the most important member of this class. This heuristic, which is the basis for a number of commercial vehicle routing packages, is one of the earliest heuristics designed for this problem and, without a doubt, the most widely known. The idea of the savings algorithm is very simple: Consider the depot and n demand points. Suppose that initially we assign a separate vehicle to each demand point. The total distance traveled by a vehicle that visits demand point i is $2d_i$, where d_i is the distance from the depot to demand point i. Therefore, the total distance traveled in this solution is $2\sum_{i=1}^{n} d_i$.

If we now combine two routes, say we serve i and j on a single trip (with the same vehicle), the total distance traveled by this vehicle is $d_i + d_{ij} + d_j$, where d_{ij} is the distance between demand points i and j. Thus, the *savings* obtained from combining demand points i and j, denoted s_{ij}, is

$$s_{ij} = 2d_i + 2d_j - (d_i + d_j + d_{ij}) = d_i + d_j - d_{ij}.$$

The larger the savings s_{ij}, the more desirable it is to combine demand points i and j. Based on this idea, Clarke and Wright suggest the following algorithm.

The Savings Algorithm

Step 1: Start with the solution that has each customer visited by a separate vehicle.

Step 2: Calculate the *savings* $s_{ij} = d_{0i} + d_{j0} - d_{ij} \geq 0$ for all pairs of customers i and j.

Step 3: Sort the savings in nonincreasing order.

Step 4: Find the first feasible arc (i, j) in the savings list, where
    (1) i and j are on different routes,
    (2) both i and j are either the first or last visited on their
        respective routes, and
    (3) the sum of demands of routes i and j is no more than Q.
  Add arc (i, j) to the current solution and delete arcs $(0, i)$ and $(j, 0)$. Delete arc (i, j) from the savings list.

Step 5: Repeat step 4 until no more arcs satisfy the conditions.

Additional constraints, which might be present, can easily be incorporated into step 4. Usually, a simple check can be performed to see whether combining the tours containing i and j violates any of these constraints.

Other examples of heuristics that fall into this class are the heuristics of Gaskel (1967), Yellow (1970), and Russell (1977). In particular, the first two are modifications of the savings algorithm.

Route-First–Cluster-Second Methods

Traditionally, this class has been defined as follows. The class consists of those heuristics that first construct a traveling salesman tour through all the customers (route first) and then partition the tour into segments (cluster second). One vehicle is assigned to each segment and visits the customers according to their appearance on the traveling salesman tour.

As we shall see in the next section, some strong statements can be made about the performance of this class's heuristics. For this purpose, we give a more precise definition of the class here.

Definition 17.2.1 *A heuristic is a route-first–cluster-second heuristic if it first orders the customers according to their locations, disregarding demand sizes, and then partitions this ordering to produce feasible clusters. These clusters consist of sets of customers that are consecutive in the initial order. Customers are then routed within their cluster depending on the specific heuristic.*

This definition of the class is more general than the traditional definition given above. The disadvantage of this class, of which we will give a rigorous analysis, can be highlighted by the following simple example. Consider a routing strategy that orders the demands in such a way that the sequence of demand sizes in the order is $(9, 2, 9, 2, 9, 2, 9, 2, \ldots)$. If the vehicle capacity is 10, then any partition of this tour *must* assign one vehicle to each customer. This solution would consist of half of the vehicles going to pick up two units (using 20 % of the vehicle capacity) and returning to the depot, not a very efficient strategy. By contrast, a routing strategy that looks at the demands at the same time as it looks at customer locations would clearly find a more intelligent ordering of the customers: one that sequences demands efficiently to decrease total distance traveled.

The route-first–cluster-second class includes classical heuristics such as the optimal partitioning heuristic introduced by Beasley (1983), and the sweep algorithm suggested by Gillett and Miller (1974).

In the optimal partitioning heuristic, one tries to find an optimal traveling salesman tour, or, if this is not possible, a tour that is close to optimal. This provides the initial ordering of the demand points. The ordering is then partitioned in an efficient way into segments. This step can be done by formulating a shortest-path problem. See Sect. 16.2 for details.

In the sweep algorithm, an arbitrary demand point is selected as the starting point. The other customers are ordered according to the angle made among them, the depot, and the starting point. Demands are then assigned to vehicles following this initial order. In effect, the points are "swept" in a clockwise direction around the depot and assigned to vehicles. Then efficient routes are designed for each vehicle. Specifically, the sweep algorithm is the following.

The Sweep Algorithm

Step 1: Calculate the polar coordinates of all customers, where the center is the depot and an arbitrary customer is chosen to be at angle 0. Reorder the customers so that

$$0 = \theta_1 \leq \theta_2 \leq \cdots \leq \theta_n.$$

Step 2: Starting from the unrouted customer i with smallest angle θ_i, construct a new cluster by sweeping consecutive customers $i + 1, i + 2 \ldots$ until the capacity constraint will not allow the next customer to be added.

Step 3: Continue step 2 until all customers are included in a cluster.

Step 4: For each cluster constructed, solve the TSP on the subset of customers and the depot.

In both of these methods, additional constraints can easily be incorporated into the algorithm.

We note that, traditionally, researchers have classified the sweep algorithm as a cluster-first–route-second method and *not* as a route-first–cluster-second method. Our opinion is that the essential part of any vehicle routing algorithm is the clustering phase of the algorithm, that is, how the customers are *clustered* into groups that can be served by individual vehicles. The specific sequencing within a cluster can and, for most problems, should be done once these clusters are determined. Therefore, a classification of algorithms for the CVRP should be solely based on how the clustering is performed. Thus, the sweep algorithm can be viewed as an algorithm of the route-first–cluster-second class since the clustering is performed on a fixed ordering of the nodes.

Cluster-First–Route-Second Methods

In this class of heuristics, the clustering is the most important phase. Customers are first clustered into feasible groups to be served by the same vehicle (cluster first) without regard to any preset ordering, and then efficient routes are designed for each cluster (route second).

Heuristics of this class are usually more technically sophisticated than the previous class, since determining the clusters is often based on a *mathematical programming* approach. This class includes the following three heuristics:

- the two-phase method (Christofides et al. 1978);

- the generalized assignment heuristic (Fisher and Jaikumar 1981);

- the location-based heuristic (Bramel and Simchi-Levi 1995).

The first two heuristics use, in a first step, the concept of *seed* customers. The seed customers are customers that will be in separate vehicles in the solution and around which tours are constructed. In both cases, the performance of the algorithm depends highly on the choice of these seeds. Placing the CVRP in the framework of a different combinatorial problem, the location-based heuristic selects the seeds in an *optimal* way and creates, at the same time, tours around these seeds. Thus, instead of decomposing the process into two steps, as is done in the two-phase method and the generalized assignment heuristic, the location-based heuristic simultaneously picks the seeds and designs tours around them. We will discuss this heuristic in detail in Sect. 17.7.

Incomplete Optimization Methods

These methods are optimization algorithms that, due to the prohibitive computing time involved in reaching an optimal solution, are terminated prematurely. Examples of these include

- cutting-plane methods (Cornuéjols and Harche 1993),

- minimum K-tree methods (Fisher 1994).

The disadvantage of incomplete optimization methods is that they still require large amounts of processing time, and they can handle problems with usually no more than 100 customers.

17.3   Worst-Case Analysis of Heuristics

In the worst-case analysis presented here, we assume that the customer demands $w_1, w_2, \ldots, w_n$ and the vehicle capacity Q are rationals. Hence, without loss of generality, we assume that Q and w_i are integers. Furthermore, we may assume that Q is even; otherwise, one can double Q as well as each w_i, $i = 1, 2, \ldots, n$, without affecting the problem. The following two-phase route-first–cluster-second heuristic was suggested by Altinkemer and Gavish (1987). In the first phase, we relax the requirement that the demand of a customer cannot be split. Each customer i is replaced by w_i-unit demand points that are zero distance apart. We then apply the ITP(α) heuristic (see Sect. 16.3) using a vehicle capacity of $\frac{Q}{2}$. In the second phase, we convert the solution obtained in Phase I to a feasible solution to the original problem without increasing the total cost. This heuristic is called the unequal-weight iterated tour-partitioning [UITP(α)] heuristic.

We now describe the second-phase procedure. Our notation follows the one suggested by Haimovich et al. (1988). Let $m = \sum_{i \in N} w_i$ be the number of demand points in the expanded problem. Recall that in the first phase an arbitrary orientation of the tour is chosen. The customers are then numbered $x^{(0)}, x^{(1)}, x^{(2)}, \ldots, x^{(n)}$, in order of their appearance on the tour, where $x^{(0)}$ is the depot. The ITP(α) heuristic partitions the path from $x^{(1)}$ to $x^{(n)}$ into $\lceil \frac{2m}{Q} \rceil$ (or $\lceil \frac{2m}{Q} \rceil + 1$) disjoint segments such that each one contains no more than $\frac{Q}{2}$ demand points and connects the endpoints of each segment to the depot. The segments are indexed by $j = 1, 2, \ldots, \lceil \frac{2m}{Q} \rceil$, such that the first customer of the jth segment is $x^{(b_j)}$ and the last customer is $x^{(e_j)}$. Hence, the jth segment, denoted by S_j, includes customers $\{x^{(b_j)}, \cdots, x^{(e_j)}\}$. Obviously, if $x^{(e_j)} = x^{(b_{j+1})}$ for some j, then the demand of customer $x^{(e_j)}$ is split between the jth and $(j+1)$th segments; therefore, these are not feasible routes. On the other hand, if $x^{(e_j)} \neq x^{(b_{j+1})}$ for all j, then the set of routes is feasible.

We now transform the solution obtained in the first phase into a feasible solution without increasing the total distance traveled. We use the following procedure.

The Phase 2 Procedure

Step 1: Set $S'_j = \emptyset$, for $j = 1, 2, \ldots, \lceil \frac{2m}{Q} \rceil$.

Step 2: For $j = 1$ to $\lceil \frac{2m}{Q} \rceil - 1$, do
$\quad$ If $x^{(e_j)} = x^{(b_{j+1})}$, then
$\qquad$ If $\sum_{i=b_j}^{b_{j+1}} w_{x^{(i)}} \leq Q$, then let $S'_j = \{x^{(b_j)}, \cdots, x^{(e_j)}\}$ and
$\qquad$ let $x^{(b_{j+1})} = x^{(b_{j+1}+1)}$;
$\qquad$ else let $S'_j = \{x^{(b_j)}, \cdots, x^{(e_j-1)}\}$ and $x^{(b_{j+1})} = x^{(e_j)}$
$\quad$ else, let $S'_j = \{x^{(b_j)}, \cdots, x^{(e_j)}\}$.

We argue that the procedure generates feasible sets S'_j for $j = 1, 2, \ldots, \lceil \frac{2m}{Q} \rceil$. Note that the jth set can be enlarged only in the $(j-1)$st and jth iterations (if at all). Moreover, if it is enlarged in the jth iteration, it is clearly done feasibly in view of the test $\sum_{i=b_j}^{b_{j+1}} w_{x^{(i)}} \leq Q$. On the other hand, if S_j is enlarged in the $(j-1)$st iteration, at most $\frac{Q}{2}$ demand points are added, thus ensuring feasibility. This can be verified as follows. Assume to the contrary that in the $(j-1)$st iteration more than $\frac{Q}{2}$ demand points are transferred from S'_{j-1} to S_j so that in the $(j-1)$st iteration $x^{(e_{j-1})} = x^{(b_j)}$. Since the original set S_{j-1} contains at most $\frac{Q}{2}$ demand points, we must have shifted demand points in the $(j-2)^{\text{nd}}$ iteration from S_{j-2} to S_{j-1} [and, in particular, $x^{(b_{j-1})} = x^{(e_{j-2})}$], part of which are now being transferred to S_j. This implies that $x^{(*)} \doteq x^{(e_{j-2})} = x^{(b_{j-1})} = x^{(e_{j-1})} = x^{(b_j)}$, where $e_{j-2}, b_{j-1}, e_{j-1},$ and b_j refer to the original sets $S_{j-2}, S_{j-1},$ and S_j. In other words, at the beginning of the $(j-1)$st iteration, the set S'_{j-1} contains a single customer $x^{(*)}$. But then, shifting $x^{(b_j)} = x^{(*)}$ backward to S'_{j-1} is feasible, contradicting the fact that more than $\frac{Q}{2}$ demand points need to be shifted forward from S'_{j-1} to S'_j. Therefore, the procedure generates feasible sets and we have the following worst-case bound.

Theorem 17.3.1 $\dfrac{Z^{\text{UITP}(\alpha)}}{Z^*_u} \leq 2 + (1 - \frac{2}{Q})\alpha.$

Proof. Recall that in the first phase, the vehicle capacity is set to $\frac{Q}{2}$. Hence, using the bound of Lemma 16.2.2, we obtain the following upper bound on the length of the tours generated in Phase I of the UITP(α) heuristic:

$$\frac{4}{Q}\sum_{i\in N} d_i w_i + \left(1 - \frac{2}{Q}\right)\alpha L^*(N_0). \tag{17.1}$$

In the second phase of the algorithm, the tour obtained in the first phase is converted into a feasible solution with total length no more than (17.1). To verify this, we need only to analyze those segments whose endpoints are modified by the procedure.

Suppose that S_j and S'_j differ in their starting point; then S'_j must start with $x^{(b_j+1)}$. This implies that arc $(x^{(b_j)}, x^{(b_j+1)})$, which is part of the Phase I solution, does not appear in the jth route. The triangle inequality ensures that the sum of the length of arcs $(x^{(0)}, x^{(b_j)})$ and $(x^{(b_j)}, x^{(b_j+1)})$ is no smaller than the length of arc $(x^{(0)}, x^{(b_j+1)})$. A similar argument can be applied if S_j and S'_j differ in their terminating point. Consequently, for every segment j, for $j = 1, 2, \ldots, \lceil \frac{2m}{Q} \rceil$, the length of the jth route according to the new partition is no longer than the length of the jth route according to the old partition. Hence,

$$Z^{\mathrm{UITP}(\alpha)} \leq \frac{4}{Q}\sum_{i\in N} d_i w_i + \left(1 - \frac{2}{Q}\right)\alpha L^*(N_0).$$

Clearly, $Z_u^* \geq Z^*$, and therefore using the lower bound on Z^* developed in Lemma 16.2.1 completes the proof. ∎

The UITP heuristic was divided into two phases to prove the above worst-case result. However, if the optimal partitioning heuristic is used in the unequal-weight model, the actual implementation is a one-step process. This is done as follows. Given a traveling salesman tour through the set of customers and the depot, we number the nodes $x^{(0)}, x^{(1)}, \ldots, x^{(n)}$ in order of their appearance on the tour, where $x^{(0)}$ is the depot. We then define a distance matrix with cost C_{jk}, where

$$C_{jk} = \begin{cases} \begin{array}{l}\text{the distance traveled by a vehicle that starts} \\ \text{at } x^{(0)}, \text{ visits customers } x^{(j+1)}, x^{(j+2)}, \ldots, x^{(k)}, \\ \text{and returns to } x^{(0)}, \end{array} & \text{if } \sum_{i=j+1}^{k} w_{x^{(i)}} \leq Q; \\ \\ \infty, & \text{otherwise.} \end{cases}$$

As in the equal-demand case (see Sect. 16.2), it follows that a shortest path from $x^{(0)}$ to $x^{(n)}$ in the directed graph with distance cost C_{jk} corresponds to an optimal partition of the traveling salesman tour. This version of the heuristic, developed by Beasley and called the unequal-weight optimal partitioning (UOP) heuristic, also has $Z^{\mathrm{UOP}(\alpha)}/Z^* \leq 2 + (1 - \frac{2}{Q})\alpha$. The following theorem, proved by Li and Simchi-Levi (1990), implies that when $\alpha = 1$, this bound is asymptotically tight as Q approaches infinity.

Theorem 17.3.2 *For any integer $Q \geq 1$, there exists a problem instance with $Z^{\mathrm{UOP}(1)}/Z_u^*$ (and therefore $Z^{\mathrm{UITP}(1)}/Z_u^*$) arbitrarily close to $3 - \frac{6}{Q+2}$.*

Proof. We modify the graph $\mathcal{G}(2, Kq+1)$, where K is a positive integer, as follows. Now, every group—instead of containing Q customers—contains only one customer with demand Q. The other KQ customers have unit demand. The optimal traveling salesman tour is again as shown in Fig. 16.2, and the solution obtained by the UOP(1) heuristic is to have $2KQ + 1$ vehicles, each one of them serving only one customer. Thus,

$$Z^{\text{UOP}(1)} = 2(KQ+1) + 4KQ.$$

The optimal solution to this problem has $KQ + 1$ vehicles serve those customers with demand Q,, and K other vehicles serve the unit demand customers. Hence,

$$Z_u^* = 2(KQ+1) + 4K.$$

Therefore,

$$\lim_{K \to \infty} \frac{Z^{\text{UOP}(1)}}{Z_u^*} = \lim_{K \to \infty} \frac{2(KQ+1) + 4KQ}{2(KQ+1) + 4K} = 3 - \frac{6}{Q+2}.$$

∎

17.4   The Asymptotic Optimal Solution Value

In the probabilistic analysis of the UCVRP, we assume, without loss of generality, that the vehicle's capacity Q equals 1, and the demand of each customer is no more than 1. Thus, vehicles and demands in a capacitated vehicle routing problem correspond to bins and item sizes (respectively) in a bin-packing problem. Hence, for every routing instance, there is a unique corresponding bin-packing instance.

Assume the demands $w_1, w_2, \ldots, w_n$ are drawn independently from a distribution Φ defined on $[0, 1]$. Assume customer locations are drawn independently from a probability measure μ with compact support in $\mathbb{R}^2$. We assume that $d_i > 0$ for each $i \in N$ since customers at the depot can be served at no cost. In this section, we find the asymptotic optimal solution value for *any* Φ and *any* μ. This is done by showing that an asymptotically optimal algorithm for the bin-packing problem, with item sizes distributed like Φ, can be used to solve, in an asymptotic sense, the UCVRP.

Given the demands $w_1, w_2, \ldots, w_n$, let b_n^* be the number of bins used in the optimal solution to the corresponding bin-packing problem. As demonstrated in Theorem 5.2.4, there exists a constant $\gamma > 0$ (depending only on Φ) such that

$$\lim_{n \to \infty} \frac{b_n^*}{n} = \gamma \qquad (a.s.). \qquad (17.2)$$

We shall refer to the constant γ as the *bin-packing constant* and omit the dependence of γ on Φ in the notation.

The following theorem was proved by Simchi-Levi and Bramel (1990). Recall, without loss of generality, the depot is positioned at $(0,0)$ and $||x||$ represents the distance from the point $x \in \mathbb{R}^2$ to the depot.

Theorem 17.4.1 *Let x_k, $k = 1, 2, \ldots, n$, be a sequence of independent random variables having a distribution μ with compact support in $\mathbb{R}^2$. Let*

$$E(d) = \int_{\mathbb{R}^2} ||x|| d\mu(x).$$

Let the demands w_k, $k = 1, 2, \ldots, n$, be a sequence of independent random variables having a distribution Φ with support on $[0, 1]$, and assume that the demands and the locations of the customers are independent of each other. Let γ be the bin-packing constant associated with the distribution Φ; then, almost surely,

$$\lim_{n \to \infty} \frac{1}{n} Z_u^* = 2\gamma E(d).$$

Thus, the theorem fully characterizes the asymptotic optimal solution value of the UCVRP, for any reasonable distributions Φ and μ. An interesting observation concerns the case where the distribution of the demands allows *perfect packing*, that is, when the wasted space in the bins tends to become a small fraction of the number of bins used. Formally, Φ is said to allow *perfect packing* if almost surely $\lim_{n \to \infty} \frac{b_n^*}{n} = E(w)$. Karmarkar (1982) proved that a nonincreasing probability density function (with some mild regularity conditions) allows perfect packing. Rhee (1988) completely characterizes the class of distribution functions Φ, which allow perfect packing. Clearly, in this case, $\gamma = E(w)$. Thus, Theorem 17.4.1 indicates that allowing the demands to be split or not does not change the asymptotic objective function value. That is, the UCVRP and the ECVRP can be said to be *asymptotically equivalent* when Φ allows perfect packing.

To prove Theorem 17.4.1, we start by presenting in Sect. 17.4.1 a lower bound on the optimal objective function value. In Sect. 17.4.2, we present a heuristic for the UCVRP based on a simple region-partitioning scheme. We show that the cost of the solution produced by the heuristic converges to our lower bound for any Φ and μ, thus proving the main theorem of the section.

17.4.1 A Lower Bound

We introduce a lower bound on the optimal objective function value Z_u^*. Let $A \subset \mathbb{R}^2$ be the compact support of μ and define $d_{\max} \doteq \sup_{x \in A} \{||x||\}$. For a given fixed positive integer $r \geq 1$, partition the circle with radius $d_{\max}$ centered at the depot into r rings of equal width. Let $\underline{d}_j \doteq (j-1) \frac{d_{\max}}{r}$ for $j = 1, 2, \ldots, r, r+1$, and construct the following $2r$ sets of customers:

$$S_j = \left\{ x_k \in N \middle| \underline{d}_j < d_k \leq \underline{d}_{j+1} \right\} \qquad \text{for } j = 1, \ldots, r,$$

and

$$F_j = \bigcup_{i=j}^{r} S_i \qquad \text{for } j = 1, 2, \ldots, r.$$

Note that $F_r \subseteq F_{r-1} \subseteq \cdots \subseteq F_1 = N$ since $d_k > 0$ for all $y_k \in N$.

In the lemma below, we show that $|F_r|$ grows to infinity almost surely as n grows to infinity. This implies that $|F_j|$ also grows to infinity almost surely for $j = 1, 2, \ldots, r$, since $|F_{j+1}| \leq |F_j|$, for $j = 1, 2, \ldots, r - 1$. The proof follows from the definitions of compact support and $d_{\max}$.

Lemma 17.4.2

$$\lim_{n \to \infty} \frac{|F_r|}{n} = p \ (a.s.) \ for \ some \ constant \ p > 0.$$

For any set of customers $T \subseteq N$, let $b^*(T)$ be the minimum number of vehicles needed to serve the customers in T; that is, $b^*(T)$ is the optimal solution to the bin-packing problem defined by item sizes equal to the demands of the customers in T. We can now present a family of lower bounds on Z_u^* that hold for different values of $r \geq 1$.

Lemma 17.4.3

$$Z_u^* > 2\frac{d_{\max}}{r} \sum_{j=2}^{r} b^*(F_j) \qquad for \ any \ r \geq 1.$$

Proof. Given an optimal solution to the UCVRP, let K_r^* be the number of vehicles in the optimal solution that serve at least one customer from S_r, and for $j = 1, 2, \ldots, r - 1$, let K_j^* be the number of vehicles in the optimal solution that serve at least one customer in the set S_j but do not serve any customers in F_{j+1}. Also, let V_j^* be the number of vehicles in the optimal solution that serve at least one customer in F_j. By these definitions, $V_j^* = \sum_{i=j}^{r} K_i^*$, for $j = 1, 2, \ldots, r$; hence, $K_j^* = V_j^* - V_{j+1}^*$ for $j = 1, 2, \ldots, r - 1$, and $K_r^* = V_r^*$.

Note that $V_j^* \geq b^*(F_j)$, for $j = 1, 2, \ldots, r$, since V_j^* represents the number of vehicles used in a feasible packing of the demands of customers in F_j, while $b^*(F_j)$ represents the number of bins used in an optimal packing.

By the definition of K_j^* and $\underline{d}_j$, $Z_u^* > 2 \sum_{j=1}^{r} \underline{d}_j K_j^*$ and therefore,

$$Z_u^* > 2\underline{d}_r V_r^* + \sum_{j=1}^{r-1} 2\underline{d}_j \left(V_j^* - V_{j+1}^* \right)$$

$$= 2\underline{d}_1 V_1^* + \sum_{j=2}^{r} 2(\underline{d}_j - \underline{d}_{j-1}) V_j^*$$

$$= 2 \sum_{j=2}^{r} (\underline{d}_j - \underline{d}_{j-1}) V_j^* \qquad \text{(since } \underline{d}_1 = 0\text{)}$$

$$\geq 2 \sum_{j=2}^{r} (\underline{d}_j - \underline{d}_{j-1}) b^*(F_j) \qquad \text{[since } V_j^* \geq b^*(F_j)]$$

$$= 2 \sum_{j=2}^{r} \frac{d_{\max}}{r} b^*(F_j).$$

∎

Note that Lemma 17.4.3 provides a deterministic lower bound; that is, no probabilistic assumptions are involved. Lemmas 17.4.2 and 17.4.3 are both required to provide a lower bound on $\frac{1}{n}Z_u^*$ that holds almost surely.

Lemma 17.4.4 *Under the conditions of Theorem 17.4.1, we have*

$$\lim_{n\to\infty} \frac{1}{n}Z_u^* \geq 2\gamma E(d) \qquad (a.s.).$$

Proof. Lemma 17.4.3 implies that

$$\lim_{n\to\infty} \frac{1}{n}Z_u^* \geq 2\frac{d_{\max}}{r} \lim_{n\to\infty} \sum_{j=2}^{r} \frac{b^*(F_j)}{n}$$

$$= 2\frac{d_{\max}}{r} \sum_{j=2}^{r} \lim_{n\to\infty} \frac{b^*(F_j)}{|F_j|} \lim_{n\to\infty} \frac{|F_j|}{n}.$$

From Lemma 17.4.2, $|F_j|$ grows to infinity almost surely as n grows to infinity, for $j = 1, 2, \ldots, r$. Moreover, since demands and locations are independent of each other, the demands in F_j, $j = 1, 2, \ldots, r$, are distributed like Φ. Therefore,

$$\lim_{n\to\infty} \frac{b^*(F_j)}{|F_j|} = \lim_{|F_j|\to\infty} \frac{b^*(F_j)}{|F_j|} = \gamma \qquad (a.s.).$$

Hence, almost surely,

$$\lim_{n\to\infty} \frac{1}{n}Z_u^* \geq 2\frac{d_{\max}}{r} \sum_{j=2}^{r} \gamma \lim_{n\to\infty} \frac{|F_j|}{n}$$

$$= 2\frac{d_{\max}}{r}\gamma \lim_{n\to\infty} \frac{1}{n} \sum_{j=2}^{r} |F_j|.$$

Since

$$F_j = \bigcup_{i=j}^{r} S_i \qquad \text{for } j = 1, 2, \ldots, r,$$

we have $|F_j| = \sum_{i=j}^{r} |S_i|$; hence, almost surely,

$$\lim_{n\to\infty} \frac{1}{n}Z_u^* \geq 2\frac{d_{\max}}{r}\gamma \lim_{n\to\infty} \frac{1}{n} \sum_{j=2}^{r} \sum_{i=j}^{r} |S_i|$$

$$= 2\frac{d_{\max}}{r}\gamma \lim_{n\to\infty} \frac{1}{n} \sum_{j=2}^{r} (j-1)|S_j|.$$

By the definition of $\underline{d}_j$,

$$\lim_{n\to\infty} \frac{1}{n} Z_u^* \geq 2\gamma \lim_{n\to\infty} \frac{1}{n} \sum_{j=2}^{r} \underline{d}_j |S_j| = 2\gamma \lim_{n\to\infty} \frac{1}{n} \sum_{j=1}^{r} \underline{d}_j |S_j|,$$

since $\underline{d}_1 = 0$ and $|S_1| \leq n$. By the definition of $\underline{d}_j$ and S_j, $\underline{d}_j \geq d_k - \frac{d_{\max}}{r}$, for all $x_k \in S_j$. Then almost surely,

$$\lim_{n\to\infty} \frac{1}{n} Z_u^* \geq 2\gamma \lim_{n\to\infty} \frac{1}{n} \sum_{x_k \in N} (d_k - \frac{d_{\max}}{r})$$

$$= 2\gamma \lim_{n\to\infty} \frac{1}{n} \sum_{x_k \in N} d_k - 2\gamma \frac{d_{\max}}{r}$$

$$= 2\gamma E(d) - 2\gamma \frac{d_{\max}}{r}.$$

This lower bound holds for arbitrarily large r; hence,

$$\lim_{n\to\infty} \frac{1}{n} Z_u^* \geq 2\gamma E(d) \qquad (a.s.).$$

■

In the next section, we show that this lower bound is tight by presenting an upper bound on the cost of the optimal solution that asymptotically approaches the same value.

17.4.2   An Upper Bound

We prove Theorem 17.4.1 by analyzing the cost of the following three-step heuristic that provides an upper bound on Z_u^*. In the first step, we partition the area A into subregions. Then, for each of these subregions, we find the optimal packing of the customers' demands in the subregion, into bins of unit size. Finally, for each subregion, we allocate one vehicle to serve the customers in each bin.

The Region-Partitioning Scheme

For a *fixed* $h > 0$, let $G(h)$ be an infinite grid of squares with side $\frac{h}{\sqrt{2}}$ and edges parallel to the system coordinates. Recall that A is the compact support of the distribution function μ, and let $A_1, A_2, \ldots, A_{t(h)}$ be the intersection of the squares of $G(h)$ with the compact support A that have $\mu(A_i) > 0$. Note $t(h) < \infty$ since A is compact and $t(h)$ is independent of n.

Let $N(i)$ be the set of customers located in subregion A_i, and define $n(i) \doteq |N(i)|$. For every $i = 1, 2, \ldots, t(h)$, let $b^*(i)$ be the minimum number of bins needed to pack the demands of customers in $N(i)$. Finally, for each subregion A_i, $i = 1, 2, \ldots, t(h)$, let $n_j(i)$ be the number of customers in the jth bin of this optimal packing, for each $j = 1, 2, \ldots, b^*(i)$.

We now proceed to find an upper bound on the value of our heuristic. Recall that for each bin produced by the heuristic, we send a single vehicle to serve all the customers in the bin. First, the vehicle visits the customer closest to the depot in the subregion to which the bin belongs, then serves all the customers in the bin in any order, and then returns to the depot through the closest customer again. Let $\underline{d}(i)$ be the distance from the depot to the closest customer in $N(i)$, that is, in subregion A_i. Note that since each subregion A_i is a subset of a square of side $\frac{h}{\sqrt{2}}$, the distance between any two customers in A_i is no more than h. Consequently, using the method just described, we calculate that the distance traveled by the vehicle that serves all the customers in the jth bin of subregion A_i is no more than

$$2\underline{d}(i) + h(n_j(i) + 1).$$

Therefore,

$$Z_u^* \le \sum_{i=1}^{t(h)} \sum_{j=1}^{b^*(i)} \left[2\underline{d}(i) + h(n_j(i) + 1) \right] \le 2 \sum_{i=1}^{t(h)} b^*(i)\underline{d}(i) + 2nh. \tag{17.3}$$

This inequality will be coupled with the following lemma to find an almost sure upper bound on the cost of this heuristic.

Lemma 17.4.5 *Under the conditions of Theorem 17.4.1, we have*

$$\varlimsup_{n\to\infty} \frac{1}{n} \sum_{i=1}^{t(h)} b^*(i)\underline{d}(i) \le \gamma E(d) \qquad (a.s.).$$

Proof. Let $p_i \doteq \mu(A_i)$ be the probability that a given customer x_k falls in subregion A_i. Since $p_i > 0$, by the strong law of large numbers, $\lim_{n\to\infty} \frac{n(i)}{n} = p_i$ almost surely, and therefore $n(i)$ grows to infinity almost surely as n grows to infinity. Thus, we have

$$\lim_{n\to\infty} \frac{b^*(i)}{n(i)} = \lim_{n(i)\to\infty} \frac{b^*(i)}{n(i)} = \gamma \qquad (a.s.).$$

Hence,

$$\varlimsup_{n\to\infty} \frac{1}{n} \sum_{i=1}^{t(h)} b^*(i)\underline{d}(i) = \varlimsup_{n\to\infty} \frac{1}{n} \sum_{i=1}^{t(h)} \frac{b^*(i)}{n(i)} n(i)\underline{d}(i)$$

$$\le \varlimsup_{n\to\infty} \frac{1}{n} \sum_{i=1}^{t(h)} \frac{b^*(i)}{n(i)} \sum_{x_k \in N(i)} d_k \quad [\text{since } \underline{d}(i) \le d_k, \forall x_k \in N(i)]$$

$$= \sum_{i=1}^{t(h)} \varlimsup_{n\to\infty} \frac{b^*(i)}{n(i)} \varlimsup_{n\to\infty} \frac{1}{n} \sum_{x_k \in N(i)} d_k$$

$$= \gamma \varlimsup_{n\to\infty} \frac{1}{n} \sum_{x_k \in N} d_k.$$

Using the strong law of large numbers, we have

$$\overline{\lim_{n\to\infty}} \frac{1}{n} \sum_{i=1}^{t(h)} b^*(i)\underline{d}(i) \le \gamma E(d) \qquad (a.s.),$$

which completes the proof of this lemma.   ∎

Remark: A simple modification of the proof of Lemma 17.4.5 shows that the inequality that appears in the statement of the lemma can be replaced by equality (see Exercise 17.5).

We can now finish the proof of the Theorem 17.4.1. From (17.3), we have

$$\frac{1}{n} Z_u^* \le \frac{2}{n} \sum_{i=1}^{t(h)} b^*(i)\underline{d}(i) + 2h.$$

Taking the limits and using Lemma 17.4.5, we obtain

$$\overline{\lim_{n\to\infty}} \frac{1}{n} Z_u^* \le 2\gamma E(d) + 2h \qquad (a.s.).$$

Since this inequality holds for arbitrarily small $h > 0$, we have

$$\overline{\lim_{n\to\infty}} \frac{1}{n} Z_u^* \le 2\gamma E(d) \qquad (a.s.).$$

This upper bound combined with the lower bound of Lemma 17.4.4 proves the main theorem.

17.5   Probabilistic Analysis of Classical Heuristics

Bienstock et al. (1993a) analyzed the average performance of heuristics that belong to the route cluster first–class second. Recall our definition of this class: all those heuristics that first order the customers according to their locations and then partition this ordering to produce feasible clusters.

It is clear that the UITP(α) and UOP(α) heuristics described in Sect. 17.3 belong to this class. As mentioned in Sect. 17.2, the sweep algorithm suggested by Gillett and Miller can also be viewed as a member of this class.

Bienstock et al. show that the performance of any heuristic in this class is strongly related to the performance of a nonefficient bin-packing heuristic called next-fit (NF). The next-fit bin-packing heuristic can be described in the following manner. Given a list of n items, start with item 1 and place it in bin 1. Suppose we are packing item j; let bin i be the highest indexed nonempty bin. If item j fits in bin i, then place it there; else place it in a new bin indexed $i + 1$. Thus, NF is an online heuristic; that is, it assigns items to bins according to the order in which they appear, without using any knowledge of subsequent items in the list.

The NF heuristic possesses some interesting properties that will be useful in the analysis of the class route first–cluster second. Assume the items are indexed $1, 2, \ldots, n$ and let a *consecutive* heuristic be one that assigns items to bins such that items in any bin appear consecutively in the sequence. The following is a simple observation.

Property 17.5.1 *Among all consecutive heuristics, NF uses the least number of bins.*

The next property is similar to a property developed in Sect. 5.2 for b_n^*, the optimal solution to the bin-packing problem.

Property 17.5.2 *Let the item sizes $w_1, w_2, \ldots, w_n, \ldots$ in the bin-packing problem be a sequence of independent random variables and let b_n^{NF} be the number of bins produced by NF on the items $1, 2, \ldots, n$. For every $t \geq 0$,*

$$Pr\left\{ |b_n^{\mathrm{NF}} - E(b_n^{\mathrm{NF}})| > t \right\} \leq 2 \exp(-t^2/8n). \qquad (17.4)$$

A direct result of this property is the following. The proof is left as an exercise (Exercise 17.2).

Corollary 17.5.3 *For any $n \geq 1$,*

$$b_n^{\mathrm{NF}} \leq E(b_n^{\mathrm{NF}}) + 4\sqrt{n \log n} \qquad (a.s.).$$

The next property is a simple consequence of the theory of subadditive processes (see Sect. 5.2) and the structure of solutions generated by NF.

Property 17.5.4 *For any distribution of item sizes, there exists a constant $\gamma^{\mathrm{NF}} > 0$ such that $\lim_{n \to \infty} \frac{b_n^{\mathrm{NF}}}{n} = \gamma^{\mathrm{NF}}$ almost surely, where b_n^{NF} is the number of bins produced by the NF packing and γ^{NF} depends only on the distribution of the item sizes.*

These properties are used to prove the following theorem, the main result of this section.

Theorem 17.5.5

(i) *Let H be a route-first–cluster-second heuristic. Then, under the assumptions of Theorem 17.4.1, we have*

$$\lim_{n \to \infty} \frac{1}{n} Z^{\mathrm{H}} \geq 2\gamma^{\mathrm{NF}} E(d) \qquad (a.s.).$$

(ii) *The $UOP(\alpha)$ heuristic is the best possible heuristic in this class; that is, for any fixed $\alpha \geq 1$, we have*

$$\lim_{n \to \infty} \frac{1}{n} Z^{\mathrm{UOP}(\alpha)} = 2\gamma^{\mathrm{NF}} E(d) \qquad (a.s.).$$

In view of Theorems 17.4.1 and 17.5.5, it is interesting to compare γ^{NF} to γ since the asymptotic error of any heuristic H in the class of route first–cluster second satisfies

$$\lim_{n\to\infty} Z^{\text{H}}/Z_u^* \geq \lim_{n\to\infty} Z^{\text{UOP}(\alpha)}/Z_u^* = \gamma^{\text{NF}}/\gamma.$$

Although in general the ratio is difficult to characterize, Karmarkarwas able to characterize it for the case when the item sizes are uniformly distributed on an interval $(0, a]$ for $0 < a \leq 1$. For instance, for a satisfying $\frac{1}{2} < a \leq 1$, we have

$$\gamma^{\text{NF}}/\gamma = \frac{2}{a}\left\{\frac{1}{12a^3}(15a^3 - 9a^2 + 3a - 1) + \sqrt{2}\left(\frac{1-a}{2a}\right)\tanh\left(\frac{1-a}{\sqrt{2a}}\right)\right\},$$

so that when the item sizes are uniform $(0, 1]$, the above ratio is $\frac{4}{3}$, which implies that UOP(α) converge to a value that is 33.3% more than the optimal cost, a very disappointing performance for the best heuristic currently available in terms of worst-case behavior.

Moreover, heuristics in the route-first–cluster-second class can never be asymptotically optimal for the UCVRP, except in some trivial cases (e.g., demands are all the same size). In fact, Theorem 17.5.5 clearly demonstrates that the route-first–cluster-second class suffers from misplaced priorities. The routing (in the first phase) is done without any regard to the customer demands, and thus this leads to a packing of demands into vehicles that is *at best* like the next-fit bin-packing heuristic. This is clearly suboptimal in all but trivial cases, one being when customers have equal demands, and thus we see the connection with the results of the previous chapter. Therefore, this theorem shows that an asymptotically optimal heuristic for the UCVRP must use an asymptotically optimal bin-packing heuristic to pack the customer demands into the vehicles.

In the next two subsections, we prove Theorem 17.5.5 by developing a lower bound (Sect. 17.5.1) on Z^{H} and an upper bound on $Z^{\text{UOP}(\alpha)}$ (Sect. 17.5.2).

17.5.1   A Lower Bound

In this section, we present a lower bound on the solution produced by these heuristics. Let H denote a route-first–cluster-second heuristic.

As in Sect. 17.4.1, let A be the compact support of the distribution μ, and define $d_{\max} \doteq \sup_{x \in A}\{\|x\|\}$. Given a fixed integer $r \geq 1$, define $\underline{d}_j = (j-1)\frac{d_{\max}}{r}$ for $j = 1, 2, \ldots, r$, and construct the following r sets of customers:

$$F_j = \left\{x_k \in N \,\middle|\, \underline{d}_j < d_k\right\} \qquad \text{for } j = 1, \ldots, r.$$

Note that $F_r \subseteq F_{r-1} \subseteq \ldots \subseteq F_1$, and $F_1 = N$ since, without loss of generality, $d_k > 0$ for all $x_k \in N$.

Let the customers be indexed $x_1, x_2, \ldots, x_n$ according to the order determined by the heuristic H in the route-first phase.

For any set of customers $T \subseteq N$, let $b^{\mathrm{NF}}(T)$ be the number of bins generated by the next-fit heuristic when applied to the bin-packing problem defined by item sizes equal to the demands of the customers in T, packed in the order of increasing index.

Lemma 17.5.6 *For any $r \geq 1$,*

$$Z_n^{\mathrm{H}} > 2\frac{d_{\max}}{r} \sum_{j=2}^{r} b^{\mathrm{NF}}(F_j).$$

Proof. For a given solution constructed by H, let $V(F_j)$ be the number of vehicles that serve at least one customer in F_j, for $j = 1, 2, \ldots, r$. By this definition, $V(F_j) - V(F_{j+1})$, $j = 1, 2, \ldots, r - 1$, is exactly the number of vehicles whose farthest customer visited is in F_j but not in F_{j+1}, and trivially $V(F_r)$ is the number of vehicles whose farthest customer visited is in F_r. Hence,

$$Z_n^{\mathrm{H}} > 2\underline{d}_r V(F_r) + \sum_{j=1}^{r-1} 2\underline{d}_j \left(V(F_j) - V(F_{j+1}) \right)$$

$$= 2\underline{d}_1 V(F_1) + \sum_{j=2}^{r} 2(\underline{d}_j - \underline{d}_{j-1}) V(F_j).$$

For a given subset of customers F_j, $j = 1, 2, \ldots, r$, the $V(F_j)$ vehicles that contain these customer demands (in the solution produced by H) can be ordered in such a way that the customer indices are in increasing order. Disregarding the demands of customers in these vehicles that are not in F_j, this represents the solution produced by a consecutive packing heuristic on the demands of customers in F_j. By Property 17.5.1, we must have $V(F_j) \geq b^{\mathrm{NF}}(F_j)$, for every $j = 1, 2, \ldots, r$. This, together with $\underline{d}_1 = 0$, $\underline{d}_j - \underline{d}_{j-1} = \frac{d_{\max}}{r}$, implies that

$$Z_n^{\mathrm{H}} > 2 \sum_{j=2}^{r} \frac{d_{\max}}{r} b^{\mathrm{NF}}(F_j).$$

∎

This lemma is used to derive an asymptotic lower bound on the cost of the solution produced by H that holds almost surely. The proof of the lemma is identical to the proof of Lemma 17.4.4.

Lemma 17.5.7 *Under the conditions of Theorem 17.4.1, we have*

$$\lim_{n \to \infty} \frac{1}{n} Z_n^{\mathrm{H}} \geq 2\gamma^{\mathrm{NF}} E(d) \qquad (a.s.).$$

In the next section, we show that this lower bound is asymptotically tight in the case of UOP(α) by presenting an upper bound that approaches the same value.

17.5.2   The UOP(α) Heuristic

We prove Theorem 17.5.5 by finding an upper bound on $Z_n^{\mathrm{UOP}(\alpha)}$. Let L^α be the length of the α-optimal tour selected by UOP(α). Starting at the depot and following the tour in an arbitrary orientation, the customers and the depot are numbered $x^{(0)}, x^{(1)}, x^{(2)}, \ldots, x^{(n)}$, where $x^{(0)}$ is the depot. Select an integer $m \doteq \lceil n^\beta \rceil$ for some fixed $\beta \in (\frac{1}{2}, 1)$, and note that for each such β, we have $\lim_{n\to\infty} \frac{m}{n} = 0$ [i.e., $m = o(n)$] and $\lim_{n\to\infty} \frac{\sqrt{n}}{m} = 0$ [i.e., $\sqrt{n} = o(m)$]. We partition the path from $x^{(1)}$ to $x^{(n)}$ into $m + 1$ segments, such that each one contains exactly $\lfloor \frac{n}{m} \rfloor$ customers, except possibly the last one.

Number the segments $1, 2, \ldots, m+1$ according to their appearance on the traveling salesman tour, where each segment has exactly $\lfloor \frac{n}{m} \rfloor$ customers except possibly segment $m + 1$. Let L_i (respectively, N_i) be the length of (respectively, subset of customers in) segment i, $1 \le i \le m + 1$. Finally, let $n_i = |N_i|$, $i = 1, 2, \ldots, m + 1$.

To obtain an upper bound on the cost of UOP(α), we apply the next-fit heuristic to each segment separately, where items are packed in bins in the same order they appear in the segment. This gives us a partition of the tour that must provide an upper bound on the cost produced by UOP(α). Let b_i^{NF} be the number of bins produced by the next-fit heuristic when applied to the customer demands in segment i. We assign a single vehicle to each bin produced by the above procedure, each of which starts at the depot, visits the customers assigned to its corresponding bin in the same order as they appear on the traveling salesman tour, and then returns to the depot. Let $\overline{d}_i$ be the distance from the depot to the farthest customer in N_i. Clearly, the total distance traveled by all the vehicles that serve the customers in segment i, $1 \le i \le m + 1$, is no more than

$$2b_i^{\mathrm{NF}} \, \overline{d}_i + L_i.$$

Hence,

$$Z^{\mathrm{UOP}(\alpha)} \le 2 \sum_{i=1}^{m+1} b_i^{\mathrm{NF}} \, \overline{d}_i + L^\alpha$$

$$\le 2 \sum_{i=1}^{m} b_i^{\mathrm{NF}} \, \overline{d}_i + 2 b_{m+1}^{\mathrm{NF}} d_{\max} + \alpha L^*. \qquad (17.5)$$

Lemma 17.5.8 *Under the conditions of Theorem 17.4.1, we have*

$$\overline{\lim_{n\to\infty}} \frac{1}{n} \sum_{i=1}^{m} b_i^{\mathrm{NF}} \, \overline{d}_i \le \gamma^{\mathrm{NF}} E(d) \qquad (a.s.).$$

Proof. Since the number of customers in every segment i, $1 \le i \le m$, is exactly $n_i = \lfloor \frac{n}{m} \rfloor$ and $\lim_{n\to\infty} \frac{m}{n} = 0$, we have for a given i, $1 \le i \le m$,

$$b_i^{\mathrm{NF}} \le E(b_i^{\mathrm{NF}}) + \sqrt{9 K n_i \log n_i} \qquad (a.s.),$$

for any $K \ge 2$.

We now show that, for sufficiently large n, these m inequalities hold simultaneously almost surely. To prove this, note that Property 17.5.2 tells us that, for n_i large enough, the probability that one such inequality does not hold is no more than $2\exp(-K \log n_i) = 2n_i^{-K}$. Thus, the probability that at least one of these inequalities is violated is no more than $2m(\frac{n}{m} - 1)^{-K}$. By the Borel–Cantelli lemma, these m inequalities hold almost surely if $\sum_n m(\frac{m}{n-m})^K < \infty$. Choosing $K > \frac{1+\beta}{1-\beta} > 3$ shows that this holds for any $m = \lceil n^\beta \rceil$, where $\frac{1}{2} < \beta < 1$.

Thus,

$$\overline{\lim_{n\to\infty}} \frac{1}{n} \sum_{i=1}^{m} b_i^{\mathrm{NF}} \, \overline{d}_i \leq \gamma^{\mathrm{NF}} \overline{\lim_{n\to\infty}} \sum_{i=1}^{m} \frac{1}{m} \overline{d}_i \qquad (a.s.).$$

Clearly, $\overline{d}_i \leq d_k + L_i$ for every $x_k \in N_i$ and every $i = 1, 2, \ldots, m$. Thus,

$$\overline{d}_i \leq \left(\left\lfloor \frac{n}{m} \right\rfloor\right)^{-1} \sum_{x_k \in N_i} d_k + L_i \qquad \text{for every } i = 1, 2 \ldots, m.$$

Hence,

$$\overline{\lim_{n\to\infty}} \sum_{i=1}^{m} \frac{1}{m} \overline{d}_i \leq \overline{\lim_{n\to\infty}} \frac{1}{n-m} \sum_{x_k \in N} d_k + \overline{\lim_{n\to\infty}} \frac{1}{m} L^\alpha$$

$$\leq \overline{\lim_{n\to\infty}} \frac{1}{n-m} \sum_{x_k \in N} d_k + \alpha \overline{\lim_{n\to\infty}} \frac{1}{m} L^*.$$

Applying the strong law of large numbers and using $\lim_{n\to\infty} \frac{m}{n} = 0$, we have

$$\overline{\lim_{n\to\infty}} \frac{1}{n-m} \sum_{x_k \in N} d_k = E(d) \qquad (a.s.).$$

Now from Chap. 5, we know that the length of the optimal traveling salesman tour through a set of k points independently and identically distributed in a given region grows almost surely like $\sqrt{k}$. This together with $\lim_{n\to\infty} \frac{\sqrt{n}}{m} = 0$ implies that

$$\lim_{n\to\infty} \frac{L^*}{m} = 0 \qquad (a.s.).$$

These facts complete the proof. ∎

We can now complete the proof of Theorem 17.4.1. From (17.5) and Lemma 5.2.1, we have

$$\overline{\lim_{n\to\infty}} \frac{1}{n} Z_n^{\mathrm{UOP}(\alpha)} \leq 2\gamma^{\mathrm{NF}} E(d) + 2d_{\max} \overline{\lim_{n\to\infty}} \frac{1}{n} b_{m+1}^{\mathrm{NF}} + \alpha \overline{\lim_{n\to\infty}} \frac{1}{n} L^* \qquad (a.s.).$$

Finally, using Beardwood et al.'s (1959) result (see Theorem 5.3.2) and the fact that the number of points in segment $m + 1$ is at most $\frac{n}{m}$, we obtain the desired result.

17.6   The Uniform Model

To our knowledge, no polynomial-time algorithm that is asymptotically optimal is known for the UCVRP for general Φ. We now describe such a heuristic for the case where Φ is uniform on the interval $[0, 1]$. In the unit interval, it is known that there exists an asymptotically optimal solution to the bin-packing problem with at most two items per bin. This forms the basis for the heuristic for the UCVRP, called optimal matching of pairs (OMP). It considers only feasible solutions in which each vehicle visits no more than two customers. Among all such feasible solutions, the heuristic finds the one with the minimum cost. This can be done by formulating the following integer linear program.

For every $x_k, x_l \in N$, let

$$
c_{kl} = \begin{cases}
d_k + d_{kl} + d_l, & \text{if } k \neq l \text{ and } w_k + w_l \leq 1; \\
2d_k, & \text{if } k = l; \\
\infty, & \text{otherwise.}
\end{cases}
$$

The integer program to solve is

$$
\text{Problem P}: \quad Min \quad \sum_{k \leq l} c_{kl} X_{kl}
$$

s.t.

$$
\sum_{l \geq k} X_{kl} + \sum_{l < k} X_{lk} = 1, \quad \forall k = 1, 2, \ldots, n, \qquad (17.6)
$$

$$
X_{kl} \in \{0, 1\}, \quad \forall k \leq l. \qquad (17.7)
$$

For $k < l$, X_{kl} is 1 if a vehicle delivers items to customers x_k and x_l and is 0 otherwise. Constraint (17.6) ensures that each customer is visited.

It is not hard to see that P can be solved in polynomial time since it is no more than a classical weighted matching problem defined on a specific graph. Define the following graph $\overline{G} = (\overline{N}, \overline{E})$, where each customer x_k is represented by two nodes v_k and v'_k, for $k = 1, 2, \ldots, n$. The set of edges of $\overline{G}$ is defined as follows:

$$
\overline{E} = \{(v_k, v'_k) | x_k \in N\}
$$
$$
\cup \{(v_k, v_l) | x_k \in N, x_l \in N, k \neq l, w_k + w_l \leq 1\}
$$
$$
\cup \{(v'_k, v'_l) | x_k \in N, x_l \in N, k \neq l, w_k + w_l \leq 1\}.
$$

Thus, $\overline{G}$ has $2n$ vertices. The length of edge (v_k, v_l), for $k \neq l$, is c_{kl}, of edge (v_k, v'_k) is c_{kk}, and of edge (v'_k, v'_l) is 0, for all k and l.

Note that any given feasible solution to P can be transformed into a feasible solution to the matching problem on $\overline{G}$ with the same cost. For any feasible solution to P, choose edge (v_k, v'_k) if customer k is served by a vehicle that does not serve any other customer and choose edges (v_k, v_l) and (v'_k, v'_l) if customers x_k and x_l

are visited together. Similarly, any feasible solution to the matching problem can be transformed into a feasible solution to P with the same cost. Hence, the two problems are equivalent.

An optimal matching in $\overline{G}$ can be found in $O(n^3)$ using Lawler's (1976) algorithm. The main result of this section is the following.

Theorem 17.6.1 *Let x_k, $k = 1, 2, \ldots, n$, be a sequence of independent random variables having a distribution μ with compact support in $\mathbb{R}^2$. Let*

$$E(d) = \int_{\mathbb{R}^2} ||x|| d\mu(x).$$

Let the demands w_k, $k = 1, 2, \ldots, n$, be a sequence of independent random variables having a uniform distribution on $[0, 1]$, and assume that the demands and the location of the customers are independent of each other. Then, the OMP heuristic is asymptotically optimal. That is, with probability 1,

$$\lim_{n \to \infty} \frac{Z_u^*}{n} = \lim_{n \to \infty} \frac{Z^{OMP}}{n} = E(d).$$

To prove that the *OMP* heuristic is asymptotically optimal, we approximate its performance by that of the sliced region-partitioning heuristic with parameters h and r $(SRP(h, r))$. For any *fixed* positive integer $r \geq 1$, the set N is partitioned into the following $2r$ disjoint subsets, some of which may be empty:

$$N_j = \left\{ x_k \in N \,\middle|\, \frac{1}{2}\left(1 - \frac{j+1}{r}\right) < w_k \leq \frac{1}{2}\left(1 - \frac{j}{r}\right) \right\}, \qquad j = 1, 2, \ldots, r - 1,$$

and

$$N^j = \left\{ x_k \in N \,\middle|\, \frac{1}{2}\left(1 + \frac{j-1}{r}\right) < w_k \leq \frac{1}{2}\left(1 + \frac{j}{r}\right) \right\}, \qquad j = 1, 2, \ldots, r - 1.$$

Also,

$$N_0 = \left\{ x_k \in N \,\middle|\, \frac{1}{2}\left(1 - \frac{1}{r}\right) < w_k \leq \frac{1}{2} \right\}$$

and

$$N^r = \left\{ x_k \in N \,\middle|\, \frac{1}{2}\left(1 + \frac{r-1}{r}\right) < w_k \right\}.$$

The number of customers in each N_j (respectively, N^j) is denoted by n_j (respectively, n^j) for all possible values of j.

Note that for any $j = 1, 2, \ldots, r - 1$, one vehicle can deliver the demand of a customer from N_j together with the demand of exactly one customer from N^j. The $SRP(h, r)$ heuristic generates pairs of customers, one customer from N_j and one from N^j, for every $j = 1, 2, \ldots, r-1$, using the same region-partitioning scheme used in the proof of Theorem 17.4.1 (Sect. 17.4.2). The customers in $N_0 \cup N^r$ are served separately; a single vehicle is assigned to each of these customers.

For every subregion A_i, $i = 1, 2, \ldots, t(h)$, generated by the grid $G(h)$ (see Sect. 17.4.2) and for every $j = 1, 2, \ldots, r-1$, let $N_j(i)$ [respectively, $N^j(i)$] be the subset of points in N_j (respectively, N^j) that fall in subregion A_i. Also, let $n_j(i) = |N_j(i)|$ and $n^j(i) = |N^j(i)|$.

In each subregion A_i, $i = 1, 2, \ldots, t(h)$, and for any $j = 1, 2, \ldots, r-1$, we arbitrarily match one customer from $N_j(i)$ with exactly one customer from $N^j(i)$; one vehicle serves each such pair. If $n_j(i) = n^j(i)$, then all customers in $N_j(i) \cup N^j(i)$ are matched and therefore visited in pairs. If, however, $n_j(i) \neq n^j(i)$, then we can match exactly $\min\{n_j(i), n^j(i)\}$ pairs of customers. The remaining $|n_j(i) - n^j(i)|$ customers in $N_j(i) \cup N^j(i)$ that have not yet been matched are each served by one vehicle. Thus, the total number of vehicles used in subregion A_i is

$$n_0(i) + n^r(i) + \sum_{j=1}^{r-1} \max\{n_j(i), n^j(i)\}.$$

The heuristic clearly generates a feasible solution to the UCVRP. Moreover, this solution is feasible for P, as each vehicle visits at most two customers. Thus,

$$Z^{OMP} \leq Z^{SRP(h,r)} \qquad \text{for any } r \geq 1 \text{ and } h > 0.$$

We now proceed by finding an upper bound on $Z^{SRP(h,r)}$. Essentially the same analysis as in Sect. 17.4.2 shows that the total distance traveled by all vehicles is no more than

$$2 \sum_{i=1}^{t(h)} \underline{d}(i) \left[n_0(i) + n^r(i) + \sum_{j=1}^{r-1} \max\{n_j(i), n^j(i)\} \right] + 2nh.$$

Since

$$\lim_{n(i) \to \infty} \frac{n_j(i)}{n(i)} = \lim_{n(i) \to \infty} \frac{n^j(i)}{n(i)} = \frac{1}{2r} \qquad (a.s.) \qquad \text{for all } j = 1, 2, \ldots, r,$$

we have

$$\lim_{n(i) \to \infty} \frac{1}{n(i)} \left[n_0(i) + n^r(i) + \sum_{j=1}^{r-1} \max\{n_j(i), n^j(i)\} \right] = \frac{1}{2} + \frac{1}{2r} \qquad (a.s.).$$

The remainder of the proof is identical to the proof of the upper bound of Theorem 17.4.1.

Therefore, the OMP is asymptotically optimal when demands are uniformly distributed between 0 and 1. In fact, the proof can be extended to a larger class of demand distributions. For example, for any demand distribution with symmetric density, one with $f(x) = f(1-x)$ for $x \in [0,1]$, one can show that the same result holds.

17.7   The Location-Based Heuristic

Bramel and Simchi-Levi (1995) used the insight obtained from the analysis of the asymptotic optimal solution value (see Theorem 17.4.1 above and the discussion that follows it) to develop a new and effective class of heuristics for the UCVRP called location-based heuristics. Specifically, this class of heuristics was motivated by the following observations.

A byproduct of the proof of Theorem 17.4.1 is that the region-partitioning scheme used to find an upper bound on Z_u^* is asymptotically optimal. Unfortunately, the scheme is not polynomial since it requires, among other things, optimally solving the bin-packing problem. But, the scheme suggests that, asymptotically, the tours in an optimal solution will be of a very simple structure consisting of two parts. The first is the round trip the vehicle makes from the depot to the subregion (where the customers are located); we call these the *simple tours*. The second is the additional distance (we call this *insertion cost*) accrued by visiting each of the customers it serves in the subregion. Our goal is therefore to construct a heuristic that assigns customers to vehicles so as to minimize the sum of the length of all simple tours plus the total insertion costs of customers into each simple tour. If done carefully, the solution obtained is asymptotically optimal.

To construct such a heuristic, we formulate the routing problem as another combinatorial problem commonly called [see, e.g., Pirkul (1987)] the single-source capacitated facility location problem (CFLP). This problem can be described as follows: Given m possible sites for facilities of fixed capacity Q, we would like to locate facilities at a subset of these m sites and assign n retailers, where retailer i demands w_i units of a facility's capacity, in such a way that each retailer is assigned to *exactly one* facility, the facility capacities are not exceeded, and the total cost is minimized. A site-dependent cost is incurred for locating each facility; that is, if a facility is located at site j, the *setup* cost is v_j, for $j = 1, 2, \ldots, m$. The cost of assigning retailer i to facility j is c_{ij} (the *assignment* cost), for $i = 1, 2, \ldots, n$ and $j = 1, 2, \ldots, m$.

The single-source CFLP can be formulated as the following integer linear program. Let

$$y_j = \begin{cases} 1, & \text{if a facility is located at site } j, \\ 0, & \text{otherwise,} \end{cases}$$

and let

$$x_{ij} = \begin{cases} 1, & \text{if retailer } i \text{ is assigned to a facility at site } j, \\ 0, & \text{otherwise.} \end{cases}$$

$$\text{Problem CFLP}: Min \quad \sum_{i=1}^{n}\sum_{j=1}^{m} c_{ij}x_{ij} + \sum_{j=1}^{m} v_j y_j$$

$$s.t. \quad \sum_{j=1}^{m} x_{ij} = 1, \qquad \forall i, \qquad (17.8)$$

$$\sum_{i=1}^{n} w_i x_{ij} \le Q, \qquad \forall j, \qquad (17.9)$$

$$x_{ij} \le y_j, \qquad \forall i, j, \qquad (17.10)$$

$$x_{ij} \in \{0, 1\}, \qquad \forall i, j, \qquad (17.11)$$

$$y_j \in \{0, 1\}, \qquad \forall j. \qquad (17.12)$$

Constraints (17.8) ensure that each retailer is assigned to exactly one facility, and constraints (17.9) ensure that the facility's capacity constraint is not violated. Constraints (17.10) guarantee that if a retailer is assigned to site j, then a facility is located at that site. Constraints (17.11) and (17.12) ensure the integrality of the variables.

In formulating the UCVRP as an instance of the CFLP, we set every customer x_j in the UCVRP as a possible facility site in the location problem. The length of the simple tour that starts at the depot visits customer x_j and then goes back to the depot is the setup cost in the location problem (i.e., $v_j = 2d_j$). Finally, the cost of inserting a customer into a simple tour in the UCVRP is the assignment cost in the location problem (i.e., $c_{ij} = d_i + d_{ij} - d_j$). This cost should represent the added cost of inserting customer i into a simple tour through the depot and customer j. Consequently, when i is added to a tour with j, the added cost is $c_{ij} = d_i + d_{ij} - d_j$, so that $v_j + c_{ij} = d_i + d_{ij} + d_j$. However, when a third customer is added, the calculation is not so simple, and therefore, the values of c_{ij} should, in fact, represent an *approximation* to the cost of adding i to a tour that goes through customer j and the depot. Hence, finding a solution for the CVRP is obtained by solving the CFLP with the data as described above. The solution obtained from the CFLP is transformed (in an obvious way) to a solution to the CVRP.

Although $\mathcal{NP}$-Hard the CFLP can efficiently, but approximately, be solved by the familiar Lagrangian relaxation technique (see Chap. 15), as described in Pirkul or Bramel and Simchi-Levi (1995), or by a cutting-plane algorithm, as described in Deng and Simchi-Levi (1992).

We can now describe the location-based heuristic (LBH):

The Location-Based Heuristic

Step 1: Formulate the UCVRP as an instance of the CFLP.

Step 2: Solve the CFLP.

Step 3: Transform the solution obtained in step 2 into a solution for the UCVRP.

Variations of the LBH can also be applied to other problems; we discuss this and related issues in the next chapter, where we consider a more general vehicle routing problem.

The LBH algorithm was tested on a set of 11 standard test problems taken from the literature. The problems are in the Euclidean plane, and they vary in size from 15 to 199 customers. The performance of the algorithm on these test problems was found to be comparable to the performance of most published heuristics. This includes both the running time of the algorithm as well as the quality (value) of the solutions found; see Bramel and Simchi-Levi (1995) for a detailed discussion.

One way to explain the excellent performance of the LBH is by analyzing its average performance. Indeed, a proof similar to the proof of Theorem 17.4.1 reveals [see also Bramel and Simchi-Levi (1995)] that

Theorem 17.7.1 *Under the assumptions of Theorem 17.4.1, there are versions of the LBH that are asymptotically optimal; that is,*

$$\lim_{n\to\infty} \frac{1}{n} Z^{\mathrm{LBH}} = 2\gamma E(d) \qquad (a.s.).$$

Finally, we observe that the generalized assignment heuristic due to Fisher and Jaikumar (1981) can be viewed as a special case of the LBH in which the seed customers are selected by a dispatcher. In the second step, customers are assigned to the seeds in an efficient way by solving a generalized assignment problem. The advantage of the LBH is that the selection of the seeds and the assignment of customers to seeds are done simultaneously, and not sequentially as in the generalized assignment heuristic. Note that neither of these heuristics (the LBH or the generalized assignment heuristic) requires that potential seed points be customer locations; both can be easily implemented to start with seed points that are simply points on the plane. A byproduct of the analysis, therefore, is that when the generalized assignment heuristic is carefully implemented (i.e., "good" seeds are selected), it is asymptotically optimal as well.

17.8   Rate of Convergence to the Asymptotic Value

While the results in the two previous sections completely characterize the asymptotic optimal solution value of the UCVRP, they do not say anything about the rate of convergence to the asymptotic solution value. See Psaraftis (1984) for an informal discussion of this issue.

To get some intuition on the rate of convergence, it is interesting to determine the expected difference between the optimal solution for a given number of customers n, and the asymptotic solution value (i.e., $2\gamma E[d]$). This can be done for the uniform model discussed in Sect. 17.6.

In this case, Bramel et al. (1991) and, independently, Rhee (1991) proved the following strong result.

Theorem 17.8.1 *Let x_k, $k = 1, 2, \ldots, n$, be a sequence of independent random variables uniformly distributed in the unit square $[0, 1]^2$. Let the demands w_k, $k = 1, 2, \ldots, n$, be drawn independently from a uniform distribution on $(0, 1]$. Then*

$$E[Z_n^*] = nE[d] + \Theta(n^{2/3}).$$

The proof of Theorem 17.8.1 relies heavily on the theory of three-dimensional stochastic matching, which is outside the scope of our survey. We refer the reader to Coffman and Lueker (1991, Chap. 3) for an excellent review of matching problems.

Rhee has also found an upper bound on the rate of convergence to the asymptotic solution value, for general distribution of the customers' locations and their demands. Using a new matching theorem developed together with Talagrand, she proved

Theorem 17.8.2 *Under the assumptions of Theorem 17.4.1, we have*

$$2n\gamma E[d] \leq E[Z_n^*] \leq 2n\gamma E[d] + O((n \log n)^{2/3}).$$

17.9 Exercises

Exercise 17.1. Consider the following heuristic for the CVRP with unequal demands. All customers of demand $w_i > \frac{1}{2}$ are served individually, one customer per vehicle. To serve the rest, apply the UITP heuristic with vehicle capacity Q. Prove that this solution can be transformed into a feasible solution to the CVRP with unequal demands. What is the worst-case bound of this heuristic?

Exercise 17.2. Prove Corollary 17.5.3.

Exercise 17.3. Given a seed point i, assume you must estimate the cost of the optimal traveling salesman tour through a set of points $S \cup \{i\}$ using the following cost approximation. Starting with $2d_i$, when each point j is added to the tour, add the cost $c_{ij} = d_j + d_{ij} - d_i$. That is, show that for any $r \geq 1$, there is an example where the approximation is r times the optimal cost.

Exercise 17.4. Construct an example of the single-source CFLP where each facility is a potential site (and vice versa) in which an optimal solution chooses a facility, but the demand of that facility is assigned to another chosen site.

Exercise 17.5. Show that Lemma 17.4.5 can be replaced by an equality instead of an inequality.

Exercise 17.6. Prove that the version of the LBH with setup costs $v_j = 2d_j$ and assignment costs $c_{ij} = d_i + d_{ij} - d_j$ is asymptotically optimal.

Exercise 17.7. Explain why the following constraints can or cannot be integrated into the savings algorithm.

(a) *Distance constraint.* Each route must be at most λ miles long.

(b) *Minimum route size.* Each route must pick up at least m points.

(c) *Mixing constraints.* Even indexed points cannot be on the same route as odd indexed points.

Exercise 17.8. Consider an instance of the CVRP with n customers. A customer is red with probability p and blue with probability $1 - p$, for some $p \in [0, 1]$. Red customers have loads of size $\frac{2}{3}$, while blue customers have loads of size $\frac{1}{3}$. What is $\lim_{n \to \infty} \frac{Z^*}{n}$ as a function of p?

18

The VRP with Time-Window Constraints

18.1 Introduction

In many distribution systems, in addition to the load that has to be delivered to it, each customer specifies a period of time, called a *time window*, in which this delivery must occur. The objective is to find a set of routes for the vehicles, where each route begins and ends at the depot, that serves a subset of the customers without violating the vehicle capacity and time-window constraints, while minimizing the total length of the routes. We call this model the vehicle routing problem with time windows (VRPTW).

Due to the wide applicability and the economic importance of the problem, variants of it have been extensively studied in the vehicle routing literature; for a review, see Solomon and Desrosiers (1988). Most of the work on the problem has focused on an empirical analysis, while very few papers have studied the problem from an analytical point of view. This is done in an attempt to characterize the theoretical behavior of heuristics and to use the insights obtained to construct effective algorithms. Some exceptions are the recent works of Federgruen and van Ryzin (1997) and Bramel and Simchi-Levi (1996). Here we describe the results of the latter paper.

18.2 The Model

To formally describe the model we analyze here, let the index set of the n customers be denoted $N = \{1, 2, \ldots, n\}$. Let $x_k \in I\!R^2$ be the location of customer $k \in N$. Assume, without loss of generality, that the depot is at the origin and, by rescaling,

D. Simchi-Levi et al., *The Logic of Logistics: Theory, Algorithms, and Applications for Logistics Management*, Springer Series in Operations Research and Financial Engineering, DOI 10.1007/978-1-4614-9149-1_18, © Springer Science+Business Media New York 2014

that the vehicle capacity is 1 and that the length of the working day is 1. We assume vehicles can leave and return to the depot at any time. Associated with customer k is a quadruplet (w_k, e_k, s_k, l_k), called the customer *parameters*, which represents, respectively, the load that must be picked up, the earliest starting time for service, the time required to complete the service, called the *service time*, and the latest time service can end. Clearly, feasibility requires that $e_k + s_k \leq l_k$ and $w_k, e_k, l_k \in [0, 1]$, for each $k \in N$.

For any point $x \in \mathbb{R}^2$, let $\|x\|$ denote the Euclidean distance between x and the depot. Let $d_k \doteq \|x_k\|$ be the distance between customer k and the depot. Also, let $d_{jk} \doteq \|x_j - x_k\|$ be the distance between customer j and customer k. Let Z_t^* be the total distance traveled in an optimal solution to the VRPTW, and let Z_t^H be the total distance traveled in the solution provided by a heuristic H.

Consider the customer locations to be distributed according to a distribution μ with compact support in $\mathbb{R}^2$. Let the customer parameters $\{(w_k, e_k, s_k, l_k) : k \in N\}$ be drawn from a joint distribution Φ with a continuous density ϕ. Let C be the support of ϕ; that is, C is a subset of $\{(a_1, a_2, a_3, a_4) \in [0, 1]^4 : a_2 + a_3 \leq a_4\}$. Each customer is therefore represented by its location in the Euclidean plane along with a point in C. Finally, we assume that a customer's location and its parameters are *independent* of each other.

In our analysis we associate a *job* with each customer. The parameters of job k are the parameters of customer k, that is, (w_k, e_k, s_k, l_k), where w_k is referred to as the *load* of job k and, using standard scheduling terminology, e_k represents the earliest time job k can begin processing, s_k represents the processing time and l_k denotes the latest time the processing of the job can end. The value of e_k can be thought of as the *release time* of job k, that is, the time it is available for processing. The value of l_k represents the *due date* for the job. Each job can be viewed abstractly as simply a point in C. Occasionally, we will refer to customers and jobs interchangeably; this convenience should cause no confusion.

To any set of customers $T \subseteq N$ with parameters $\{(w_k, e_k, s_k, l_k) : k \in T\}$, we associate a corresponding *machine scheduling problem* as follows. Consider the set of jobs T and an infinite sequence of *parallel* machines. Job k becomes available for processing at time e_k and must be finished processing by time l_k. The objective in this scheduling problem is to assign each job to a machine such that (i) each machine has at most one job being processed on it at a given time, (ii) the processing time of each job starts no earlier than its release time and ends no later than its due date, and (iii) the total load of all jobs assigned to a machine is no more than 1, and the number of machines used is minimized. In our discussion we refer to (ii) as the *job time-window constraint* and to (iii) as the *machine load constraint*.

Scheduling problems have been widely studied in the operations research literature; see Lawler et al. (1993) and Pinedo (1995). Unfortunately, no paper has considered the scheduling problem in its general form with the objective function of minimizing the number of machines used.

Observe that in the absence of time window constraints, the scheduling problem is no more than a bin-packing problem. Indeed, in that case, the VRPTW reduces to the model analyzed in the previous chapter, the CVRP. Thus, our strategy is to try to relate the machine scheduling problem to the VRPTW in much the same way as we used results obtained for the bin-packing problem in the analysis of the CVRP. As we shall shortly see, this is much more complex.

Let $M^*(S)$ be the minimum number of machines needed to schedule a set S of jobs. It is clear that this machine scheduling problem possesses the subadditivity property, described in Sect. 5.2. This implies that if M_n^* is the minimum number of machines needed to schedule a set of n jobs whose parameters are drawn independently from a distribution Φ, then there exists a constant $\gamma > 0$ (depending only on Φ) such that $\lim_{n\to\infty} M_n^*/n = \gamma$ (a.s.).

In this chapter, we relate the solution to the VRPTW to the solution to the scheduling problem defined by the customer parameters. That is, we show that asymptotically the VRPTW is no more difficult to solve than the corresponding scheduling problem. Our main result is the following.

Theorem 18.2.1 *Let $x_1, x_2, \ldots, x_n$ be independently and identically distributed according to a distribution μ with compact support in $\mathbb{R}^2$, and define*

$$E(d) = \int_{\mathbb{R}^2} \|x\| d\mu(x).$$

Let the customer parameters $\{(w_k, e_k, s_k, l_k) : k \in N\}$ be drawn independently from Φ. Let M_n^* be the minimum number of machines needed to feasibly schedule the n jobs corresponding to these parameters, and $\lim_{n\to\infty} \frac{M_n^*}{n} = \gamma$ (a.s.). Then

$$\lim_{n\to\infty} \frac{1}{n} Z_t^* = 2\gamma E(d) \quad (a.s.).$$

We prove this theorem (in Sect. 18.3) by introducing a lower bound on the optimal solution value and then developing an upper bound that converges to the same value. The lower bound uses a similar technique to the one developed in Chap. 17. The upper bound can be viewed as a *randomized* algorithm that is guaranteed to generate a feasible solution to the problem. That is, different runs of the algorithm on the same data may generate different feasible solutions. In Sect. 18.4, we show that the analysis leads, in a natural way, to the development of a new deterministic algorithm that is asymptotically optimal for the VRPTW. Though not polynomial, computational evidence shows that the algorithm works very well on a set of standard test problems.

18.3 The Asymptotic Optimal Solution Value

We start the analysis by introducing a lower bound on the optimal objective function value Z_t^*. First, let A be the compact support of μ, and define $d_{\max} \doteq \sup\{\|x\| : x \in A\}$. Pick a fixed integer $r \geq 1$, and define $\underline{d}_j \doteq (j-1)\frac{d_{\max}}{r}$, for $j = 1, 2, \ldots, r$. Now define the sets:

$$F_j = \left\{k \in N \mid \underline{d}_j < d_k\right\} \qquad \text{for } j = 1, 2, \ldots, r.$$

For any set $T \subseteq N$, let $M^*(T)$ be the minimum number of machines needed to feasibly schedule the set of jobs $\{(w_k, e_k, s_k, l_k) : k \in T\}$. The next lemma provides a deterministic lower bound on Z_t^* and is analogous to Lemma 17.4.3 developed for the VRP with capacity constraints.

Lemma 18.3.1

$$Z_t^* > 2\frac{d_{\max}}{r} \sum_{j=2}^{r} M^*(F_j).$$

Proof. Let V_j^* be the number of vehicles in an optimal solution to the VRPTW that serve a customer from F_j, for $j = 1, 2, \ldots, r$. By this definition, V_r^* is exactly the number of vehicles whose farthest customer visited is in F_r, and $V_j^* - V_{j+1}^*$ is exactly the number of vehicles whose farthest customer visited is in $F_j \setminus F_{j+1}$. Observe that if $V_j^* = V_{j+1}^*$, then there are no vehicles whose farthest customer visited is in $F_j \setminus F_{j+1}$. Consequently,

$$Z_t^* > 2\underline{d}_r V_r^* + \sum_{j=1}^{r-1} 2\underline{d}_j (V_j^* - V_{j+1}^*)$$

$$= 2\underline{d}_1 V_1^* + \sum_{j=2}^{r} 2(\underline{d}_j - \underline{d}_{j-1}) V_j^*$$

$$= 2\frac{d_{\max}}{r} \sum_{j=2}^{r} V_j^*.$$

We now claim that for each $j = 1, 2, \ldots, r$, $V_j^* \geq M^*(F_j)$. This should be clear from the fact that the set of jobs in F_j can be feasibly scheduled on V_j^* machines by scheduling the jobs at the times they are served in the VRPTW solution. $\blacksquare$

We can now determine the asymptotic value of this lower bound. This can be done in a similar manner to that of Chap. 17, and hence we omit the proof here.

Lemma 18.3.2 *Under the conditions of Theorem 18.2.1,*

$$\lim_{n \to \infty} \frac{1}{n} Z_t^* \geq 2\gamma E(d) \quad (a.s.).$$

We prove Theorem 18.2.1 by approximating the optimal cost from above by that of the following four-step heuristic. In the first step, we partition the region where the customers are distributed into subregions. In the second step, we randomly separate the customers of each subregion into two sets. Then for each subregion, we solve a machine scheduling problem defined on the customers in one of these sets. Finally, we use this schedule to specify how to serve all the customers in the subregion.

Pick an $\epsilon > 0$, and let δ be given by the definition of continuity of ϕ; that is, $\delta > 0$ is such that for all $x, y \in C$ with $||x - y|| < \delta$, we have $|\phi(x) - \phi(y)| < \epsilon$. Finally, pick a $\Delta < \min\{\frac{\delta}{\sqrt{2}}, \epsilon\}$.

Let $G(\Delta)$ be an infinite grid of squares of *diagonal* Δ, that is, of side $\frac{\Delta}{\sqrt{2}}$, with edges parallel to the system coordinates. Recall that A is the compact support of μ, and let $A_1, A_2, \ldots, A_{t(\Delta)}$ be the subregions of $G(\Delta)$ that intersect A and have $\mu(A_i) > 0$.

Let $N(i)$ be the indices of the customers located in subregion A_i, and define $n(i) = |N(i)|$. For each customer $k \in N(i)$, with parameters (w_k, e_k, s_k, l_k), we associate a job with parameters $(w_k, e_k, s_k + \Delta, l_k + \Delta)$. For any set $T \subseteq N$ of customers, let $M_\Delta^*(T)$ be the minimum number of machines needed to feasibly schedule the set of jobs $\{(w_k, e_k, s_k + \Delta, l_k + \Delta) : k \in T\}$. In addition, for any set T of customers, let $T(i) = N(i) \cap T$, for $i = 1, 2, \ldots, t(\Delta)$.

For the given grid partition and for any set $T \subseteq N$ of customers, the following is a feasible way to serve the customers in N. All subregions are served separately; that is, no customers from different subregions are served by the same vehicle. In subregion A_i, we solve the machine scheduling problem defined by the jobs $\{(w_k, e_k, s_k + \Delta, l_k + \Delta) : k \in T(i)\}$. Then, for each machine in this scheduling solution, we associate a vehicle that serves the customers corresponding to the jobs on that machine. The customers are visited in the exact order they are processed on the machine, and they are served in exactly the same interval of time as they are processed. This is repeated for each machine of the scheduling solution. The customers of the set $N(i) \setminus T(i)$ are served one vehicle per customer. This strategy is repeated for every subregion, thus providing a solution to the VRPTW.

We will show that for a suitable choice of the set T, this routing strategy is asymptotically optimal for the VRPTW. An interesting fact about the set T is that it is a randomly generated set; that is, each time the algorithm is run, it results in different sets T.

The first step is to show that, for any set $T \subseteq N$ (possibly empty), the solution produced by the above-mentioned strategy provides a feasible solution to the VRPTW. This should be clear from the fact that having an extra Δ units of time to travel between customers in a subregion is enough since all subregions have diagonal Δ. Therefore, any sets of customers scheduled on a machine together can be served together by one vehicle. Customers of $N(i) \setminus T$ can clearly be served within their time windows since they are served individually, one per vehicle.

We now proceed to find an upper bound on the value of this solution. For each subregion A_i, let $n_j(i)$ be the number of jobs on the jth machine in the optimal schedule of the jobs in $T(i)$, for each $j = 1, 2, \ldots, M_\Delta^*(T(i))$. Let $\underline{d}(i)$ be the distance from the depot to the closest customer in $N(i)$, that is, in subregion A_i. Using the routing strategy described above, we specify that the distance traveled by the vehicle serving the customers whose job was assigned to the jth machine of subregion A_i is no more than

$$2\underline{d}(i) + \Delta(n_j(i) + 1).$$

Therefore,

$$Z_t^* \leq \sum_{i=1}^{t(\Delta)} \sum_{j=1}^{M_\Delta^*(T(i))} \left[2\underline{d}(i) + \Delta(n_j(i) + 1) \right] + \sum_{k \notin T} 2d_k$$

$$\leq 2\sum_{i=1}^{t(\Delta)} M_\Delta^*(T(i))\underline{d}(i) + 2n\Delta + \sum_{k \notin T} 2d_k.$$

Dividing by n and taking the limit, we have

$$\overline{\lim_{n\to\infty}} \frac{1}{n} Z_t^* \leq 2\sum_{i=1}^{t(\Delta)} \overline{\lim_{n\to\infty}} \frac{1}{n} M_\Delta^*(T(i))\underline{d}(i) + 2\Delta + \overline{\lim_{n\to\infty}} \frac{1}{n} \sum_{k\notin T} 2d_k$$

$$= 2\sum_{i=1}^{t(\Delta)} \overline{\lim_{n\to\infty}} \frac{n(i)}{n} \frac{M_\Delta^*(T(i))}{n(i)} \underline{d}(i) + 2\Delta + \overline{\lim_{n\to\infty}} \frac{1}{n} \sum_{k\notin T} 2d_k$$

$$\leq 2\sum_{i=1}^{t(\Delta)} \overline{\lim_{n\to\infty}} \frac{n(i)}{n} \overline{\lim_{n\to\infty}} \frac{M_\Delta^*(T(i))}{n(i)} \underline{d}(i) +$$

$$2\Delta + \overline{\lim_{n\to\infty}} \frac{1}{n} \sum_{k\notin T} 2d_k. \tag{18.1}$$

In order to relate this quantity to the lower bound of Lemma 18.3.2, we must choose the set T appropriately. For this purpose, we make the following observation. Recall that ϕ is the continuous density associated with the distribution Φ. The customer parameters (w_k, e_k, s_k, l_k) of each of the customers of N are drawn randomly from the density ϕ. Associated with each customer is a job whose parameters are perturbed by Δ in the third and fourth coordinates, that is, $(w_k, e_k, s_k + \Delta, l_k + \Delta)$. This is equivalent to randomly drawing the job parameters from a density that we call ϕ'. The density ϕ' can be found simply by translating ϕ by Δ in the third and fourth coordinates, that is, for each $x = (\theta_1, \theta_2, \theta_3, \theta_4) \in \mathbb{R}^4$, $\phi'(x) = \phi'(\theta_1, \theta_2, \theta_3, \theta_4) = \phi(\theta_1, \theta_2, \theta_3 - \Delta, \theta_4 - \Delta)$. Finally, for each $x \in \mathbb{R}^4$, define $\psi(x) \doteq \min\{\phi(x), \phi'(x)\}$ and let $q \doteq \int_{\mathbb{R}^4} \psi < 1$.

The n jobs (or customer parameters) $\{y_k \doteq (w_k, e_k, s_k + \Delta, l_k + \Delta) : k \in N\}$ are drawn randomly from the density ϕ', and our task is to select the set $T \subseteq N$. To simplify presentation, we refer interchangeably to the index set of jobs and to the set of jobs itself; that is, $k \in N$ will have the same interpretation as $y_k \in N$, where $y_k \doteq (w_k, e_k, s_k + \Delta, l_k + \Delta)$.

For each job y_k, generate a random value, call it u_k, uniformly in $[0, \phi'(y_k)]$. The point $(y_k, u_k) \in \mathbb{R}^5$ is a point below the *graph* of ϕ'; that is, $u_k \leq \phi'(y_k)$. Define T as the set of indices of jobs whose u_k value falls below the graph of ϕ; that is, $T \doteq \{k \in N : u_k \leq \phi(y_k)\}$. Then the set of jobs $\{y_k : k \in T\}$ can be viewed as a random sample of $|T|$ jobs drawn randomly from the density $\frac{\psi}{q}$.

In order to relate this upper bound to the lower bound, we need to present the following lemma.

Lemma 18.3.3 *For T generated as above and for each subregion A_i, $i = 1, 2, \ldots, t(\Delta)$,*

$$\varlimsup_{n \to \infty} \frac{M_\Delta^*(T(i))}{n(i)} \leq \gamma, \qquad (a.s.).$$

Proof. To prove the result for a given subregion A_i, we construct a feasible schedule for the set of jobs $\{y_k = (w_k, e_k, s_k + \Delta, l_k + \Delta) : k \in T(i)\}$. Generate $n(i) - |T(i)|$ jobs randomly from the density

$$\frac{1}{1-q} [\phi - \psi].$$

Call this set of jobs D, for *dummy* jobs. From the construction of the sets D and $T(i)$, it is a simple exercise to show that the parameters of the jobs in $D \cup T(i)$ are distributed like ϕ.

A feasible schedule of the jobs in $T(i)$ is obtained by optimally scheduling the jobs in $D \cup T(i)$ using, say, M_i machines. The number of machines needed to schedule the jobs in $T(i)$ is obviously no more than M_i, since the jobs in D can simply be ignored. Thus, we have the bound

$$M_\Delta^*(T(i)) \leq M_i.$$

Now dividing by $n(i)$ and taking the limits, we get

$$\varlimsup_{n \to \infty} \frac{M_\Delta^*(T(i))}{n(i)} \leq \varlimsup_{n \to \infty} \frac{M_i}{n(i)} = \gamma, \ (a.s.),$$

since the set of jobs $D \cup T(i)$ is just a set of $n(i)$ jobs whose parameters are drawn independently from the density ϕ. ∎

Lemma 18.3.3 thus reduces (18.1) to

$$\varlimsup_{n \to \infty} \frac{1}{n} Z_t^* \leq 2 \sum_{i=1}^{t(\Delta)} \gamma \varlimsup_{n \to \infty} \frac{n(i)}{n} \underline{d}(i) + 2\Delta + \varlimsup_{n \to \infty} \frac{1}{n} \sum_{k \notin T} 2d_k$$

$$= 2\gamma \varlimsup_{n \to \infty} \frac{1}{n} \sum_{i=1}^{t(\Delta)} n(i) \underline{d}(i) + 2\Delta + \varlimsup_{n \to \infty} \frac{1}{n} \sum_{k \notin T} 2d_k$$

$$\leq 2\gamma \varlimsup_{n \to \infty} \frac{1}{n} \sum_{k \in N} d_k + 2\Delta + \varlimsup_{n \to \infty} \frac{1}{n} \sum_{k \notin T} 2d_k$$

$$= 2\gamma E(d) + 2\Delta + \varlimsup_{n \to \infty} \frac{1}{n} \sum_{k \notin T} 2d_k$$

$$\leq 2\gamma E(d) + 2\Delta + 2d_{\max} \varlimsup_{n \to \infty} \frac{1}{n} |N \setminus T|.$$

The next lemma determines an upper bound on $\varlimsup_{n \to \infty} \frac{1}{n} |N \setminus T|$.

Lemma 18.3.4 *Given $\epsilon > 0$ and T generated as above,*

$$\varlimsup_{n \to \infty} \frac{1}{n} |N \setminus T| < (1 + \epsilon)^2 \epsilon \qquad (a.s.).$$

Proof. By the strong law of large numbers, the limit is equal to the probability that a job of N is not in the set T. The probability of a particular job y_k *not* being in T is simply

$$\begin{cases} \frac{\phi'(y_k) - \phi(y_k)}{\phi'(y_k)}, & \text{if } \phi'(y_k) \geq \phi(y_k), \\ 0, & \text{otherwise.} \end{cases}$$

Hence, almost surely,

$$\begin{aligned}
\varlimsup_{n \to \infty} \frac{1}{n} |N \setminus T| &= \int_{\mathbb{R}^4} \max \left\{ \frac{\phi'(x) - \phi(x)}{\phi'(x)}, 0 \right\} \phi'(x) dx \\
&\leq \int_{\mathbb{R}^4} \left| \frac{\phi'(x) - \phi(x)}{\phi'(x)} \right| \phi'(x) dx \\
&= \int_{\mathbb{R}^4} |\phi'(x) - \phi(x)| dx \\
&= \int_{\mathbb{R}^4} |\phi'(\theta_1, \theta_2, \theta_3, \theta_4) - \phi(\theta_1, \theta_2, \theta_3, \theta_4)| d(\theta_1, \theta_2, \theta_3, \theta_4) \\
&= \int_{\mathbb{R}^4} |\phi(\theta_1, \theta_2, \theta_3 - \Delta, \theta_4 - \Delta) - \phi(\theta_1, \theta_2, \theta_3, \theta_4)| d(\theta_1, \theta_2, \theta_3, \theta_4) \\
&< (1 + \Delta)^2 \epsilon \\
&< (1 + \epsilon)^2 \epsilon,
\end{aligned}$$

where the second-to-last inequality follows from $\|(\theta_1, \theta_2, \theta_3 - \Delta, \theta_4 - \Delta) - (\theta_1, \theta_2, \theta_3, \theta_4)\| \leq \Delta \sqrt{2} < \delta$ and the continuity of ϕ. ∎

We now have all the necessary ingredients to finish the proof of Theorem 18.2.1; thus,

$$\varlimsup_{n \to \infty} \frac{1}{n} Z_t^* \leq 2\gamma E(d) + 2 d_{\max}(1 + \epsilon)^2 \epsilon + 2\Delta \qquad (a.s.).$$

Since ϵ was arbitrary, and recalling that $\Delta < \epsilon$, we have

$$\varlimsup_{n \to \infty} \frac{1}{n} Z_t^* \leq 2\gamma E(d) \qquad (a.s.).$$

This upper bound combined with the lower bound proves Theorem 18.2.1.

18.4   An Asymptotically Optimal Heuristic

In this section, we generalize the LBH heuristic developed for the CVRP (see Chap. 17) to handle time window constraints. Similar to the original LBH, we prove that the generalized version is asymptotically optimal for the VRPTW. We refer to this more general version of the heuristic also as the location-based heuristic; this should cause no confusion.

18.4.1   The Location-Based Heuristic

The LBH can be viewed as a three-step algorithm. In the first step, the parameters of the VRPTW are transformed into data for a location problem called the capacitated vehicle location problem with time windows (CVLPTW), described below. This location problem is solved in the second step. In the final step, we transform the solution to the CVLPTW into a feasible solution to the VRPTW.

The Capacitated Vehicle Location Problem with Time Windows

The capacitated vehicle location problem with time windows (CVLPTW) is a generalization of the single-source capacitated facility location problem (see Sect. 17.7) and can be described as follows: We are given m possible sites to locate vehicles of capacity Q. There are n customers geographically dispersed in a given region, where customer i has w_i units of product that must be picked up by a vehicle. The pickup of customer i takes s_i units of time and must occur in the time window between times e_i and l_i; that is, the service of customer i can start at any time $t \in [e_i, l_i - s_i]$. The objective is to select a subset of the possible sites, to locate one vehicle at each site, and to assign the customers to the vehicles. Each vehicle must leave its site, pick up the load of customers assigned to it in such a way that the vehicle capacity is not exceeded and all pickups occur within the customer's time window, and then return to its site. The costs are as follows: A site-dependent cost is incurred for locating each vehicle; that is, if a vehicle is located at site j, the *setup* cost is v_j, for $j = 1, 2, \ldots, m$. The cost of assigning customer i to the vehicle at site j is c_{ij} (the *assignment* cost), for $i = 1, 2, \ldots, n$ and $j = 1, 2, \ldots, m$. We assume that there are enough vehicles and sites so that a feasible solution exists.

The CVLPTW can be formulated as the following mathematical program. Let

$$y_j = \begin{cases} 1, & \text{if a vehicle is located at site } j, \\ 0, & \text{otherwise,} \end{cases}$$

and let

$$x_{ij} = \begin{cases} 1, & \text{if customer } i \text{ is assigned to the vehicle at site } j, \\ 0, & \text{otherwise.} \end{cases}$$

For any set $S \subseteq N$, let $f_j(S) = 1$ if the set of customers S can be feasibly served in their time windows by one vehicle that starts and ends at site j (disregarding the capacity constraint), and 0 otherwise.

$$\text{Problem } P : Min \quad \sum_{i=1}^{n} \sum_{j=1}^{m} c_{ij} x_{ij} + \sum_{j=1}^{m} v_j y_j$$

$$s.t. \quad \sum_{j=1}^{m} x_{ij} = 1, \qquad \forall i, \qquad (18.2)$$

$$\sum_{i=1}^{n} w_i x_{ij} \leq Q, \qquad \forall j, \qquad (18.3)$$

$$x_{ij} \leq y_j, \qquad \forall i, j, \qquad (18.4)$$

$$f_j(\{i : x_{ij} = 1\}) = 1, \quad \forall j, \qquad (18.5)$$

$$x_{ij}, y_j \in \{0, 1\}, \qquad \forall i, j. \qquad (18.6)$$

Constraints (18.2) ensure that each customer is assigned to exactly one vehicle, and constraints (18.3) ensure that the vehicle's capacity constraint is not violated. Constraints (18.4) guarantee that if a customer is assigned to the vehicle at site j, then a vehicle is located at that site. Constraints (18.5) ensure that the time-window constraints are not violated. Constraints (18.6) ensure the integrality of the variables.

The Heuristic

To relate the CVLPTW to the VRPTW, consider each customer in the VRPTW to be a potential site for a vehicle; that is, the set of potential sites is exactly the set of customers, and therefore, $m = n$. Picking a subset of the sites in the CVLPTW corresponds to picking a subset of the customers in the VRPTW; we call this set of selected customers the *seed* customers. These customers are those that will form simple tours with the depot.

In order for the LBH to perform well, the costs of the CVLPTW should approximate the costs of the VRPTW. The setup cost for locating a vehicle at site j (v_j), or, in other words, of picking customer j as a seed customer, should be the cost of sending a vehicle from the depot to customer j and back (i.e., the length of the simple tour). Hence, we set $v_j = 2d_j$ for each $j \in N$. The assignment cost c_{ij} is the cost of assigning customer i to the vehicle at site j. Therefore, this cost should represent the added cost of inserting customer i into the simple tour through the depot and customer j. Consequently, when i is added to a tour with j, the added cost is $c_{ij} = d_i + d_{ij} - d_j$, so that $v_j + c_{ij} = d_i + d_{ij} + d_j$. This cost is exact for two and sometimes three customers. However, as the number of customers increases, the values of c_{ij} in fact represent an *approximation* to the cost of adding i to a tour that goes through customer j and the depot. In Sect. 18.4.3, we present values of c_{ij} that we have found to work well in practice.

Once these costs are determined, the second step of the LBH consists of solving the CVLPTW. The solution provided is a set of sites (seed customers) and a set of customers assigned to each of these sites (to each seed). This solution can then be easily transformed into a solution to the VRPTW, since a set of customers that can be feasibly served starting from site j can also be feasibly served starting from the depot.

18.4.2 A Solution Method for CVLPTW

The computational efficiency of the LBH depends on the efficiency with which the CVLPTW can be solved. We therefore present a method to solve the CVLPTW. As discussed earlier, the CVLPTW without constraints (18.5) is simply the single-source capacitated facility location problem (CFLP) for which efficient solution methods exist based on the celebrated Lagrangian relaxation technique; see Sect. 6.3. For the CVLPTW, we use a similar method, although the specifics are more complex in view of the existence of these time window constraints.

In this case, for a given multiplier vector $\lambda \in \mathbb{R}^n$, constraints (18.2) are relaxed and put into the objective function with the multiplier vector. The resulting problem can be separated into n subproblems (one for each of the n sites), since constraints (18.2) are the only constraints that relate the sites to one another. The subproblem for site j is

$$\text{Problem } P_j : Min \quad \sum_{i=1}^{n} \bar{c}_{ij} x_{ij} + v_j y_j$$

$$s.t. \quad \sum_{i=1}^{n} w_i x_{ij} \le Q$$

$$x_{ij} \le y_j, \qquad \forall i,$$

$$f_j(\{i : x_{ij} = 1\}) = 1,$$

$$x_{ij} \in \{0,1\}, \; \forall i \text{ and } y_j \in \{0,1\},$$

where $\bar{c}_{ij} \doteq c_{ij} + \lambda_i$, for each $i \in N$.

In the optimal solution to Problem P_j, y_j is either 0 or 1. If $y_j = 0$, then $x_{ij} = 0$ for all $i \in N$, and the objective function value is 0. If $y_j = 1$, then the problem reduces to a different, but simpler, routing problem. Consider a vehicle of capacity Q initially located at site j. The driver gets a profit of $p_{ij} \doteq -\bar{c}_{ij}$ for picking up the w_i items at customer i in the time window (e_i, l_i). The pickup operation takes s_i units of time. The objective is to choose a subset of the customers, to pick up their loads in their time windows, without violating the capacity constraint, using a vehicle that must begin and end at site j, while maximizing the driver's profit. Let G_j^* be the maximum profit attainable at site j; that is, G_j^* is the optimal solution to the problem just described for site j. This implies that $v_j - G_j^*$ is the optimal solution value of Problem P_j given that $y_j = 1$. Therefore, we can write the optimal solution to Problem P_j as simply $\min\{0, v_j - G_j^*\}$.

Unfortunately, in general, determining the values G_j^* for $j \in N$ is $\mathcal{NP}$-Hard. We can, however, determine upper bounds on G_j^*; call them G_j. This provides a *lower bound* on the optimal solution to Problem P_j, which is equal to $\min\{0, v_j - G_j\}$. We use the simple bound given by $G_j \doteq \sum_{\{i:p_{ij}>0\}} p_{ij}$. Consequently, $\sum_{j=1}^{n} \min\{0, v_j - G_j\} - \sum_{i=1}^{n} \lambda_i$ is a lower bound on the optimal solution to the CVLPTW.

To generate a feasible solution to the VRPTW at each iteration of the procedure, we use information from the upper bounds on profit G_j for $j \in N$. After every iteration of the lower bound (for each multiplier), we renumber the sites so that $G_1 \geq G_2 \geq \cdots \geq G_n$. The upper bounds on profit are used as an estimate of the profitability of placing a vehicle at a particular site. For example, site 1 is considered to be a "good" site (or seed customer), since a large profit is possible there. A large profit for site j corresponds to a seed customer where neighboring customers can be feasibly served from it at *low cost*. Therefore, a site with large profit is selected as a seed customer since it will tend to have neighboring customers around it that can be feasibly served by a vehicle starting at that site.

To generate a feasible solution to the CVLPTW, we do the following: Starting with $j = 1$ in the new ordering of the sites (customers), we locate a vehicle at site j. For every customer still not assigned to a site, we first determine if this customer can be feasibly served with the customers that are currently assigned to site j. Then, of the customers that can be served from this site, we determine the one that will cause the least increase in cost, that is, the one with minimum c_{ij} over all customers i that can be served from this site. We then assign this customer to the site. We continue until no more customers can be assigned to site j, due to capacity or time constraints. We then increment j to 2 and continue with site 2. After all customers have been feasibly assigned to a site, we obtain a feasible solution whose cost is compared to the cost of the current best solution.

As we find solutions to the CVLPTW, we also generate feasible solutions to the VRPTW, using the information from the lower bound to the CVLPTW. Starting with $j = 1$, pick customer j as a seed customer. Then, for every customer that can be feasibly served with this seed, we determine the added distance this would entail; that is, we determine the best place to insert the customer into the current tour through the customers assigned to seed j. We choose the customer that causes the least increase in distance traveled as the one to assign to seed j. This idea is similar to the nearest-insertion heuristic discussed in Sect. 4.3.2. We then continue trying to add customers in this way to seed j. Once no more can be added to this tour (due to capacity or time constraints), we increment j to 2, select seed customer 2, and continue. Once every customer appears in a tour, that is, every customer is assigned to a seed, we have a feasible solution to the VRPTW corresponding to the current set of multipliers. The cost of this solution is compared to the cost of the current best solution.

Multipliers are updated using (6.6). The step size is initially set to 2 and halved after the lower bound has not improved in a series of 30 iterations. After the step size has reached a preset minimum (0.05), the heuristic is terminated.

18.4.3   Implementation

It is clear that many possible variations of the LBH can be implemented depending on the type of assignment costs (c_{ij}) used. In the computational results discussed below, the following have been implemented.

$$direct\ cost:\ c_{ij} = 2d_{ij}, \quad \text{and}$$

$$nearest\text{-}insertion\ cost:\ c_{ij} = d_i + d_{ij} - d_j.$$

Direct cost c_{ij} has the advantage that, when several customers are added to the seed, the resulting cost, which is the sum of the setup costs and these direct costs, is an upper bound on the length of any efficient route through the customers. On the other hand, the nearest-insertion cost works well because it is accurate at least for tours through two customers, and often for tours through three customers as well.

Several versions of the LBH have been implemented and tested. In the first, the star-tours (ST) heuristic, the direct assignment cost is used, while in the second, the seed-insertion (SI) heuristic, the nearest-insertion assignment cost is applied. Observe that the LBH is not a polynomial-time heuristic. However, as we shall shortly demonstrate, the running times reported on standard test problems are very reasonable and are comparable to the running times of many heuristics for the vehicle routing problem.

The ST heuristic is of particular interest because it is asymptotically optimal as demonstrated in the following lemma. The proof is similar to the previous proofs and is therefore omitted.

Lemma 18.4.1 *Let n customers, indexed by N, be independently and identically distributed according to a distribution μ with compact support in $\mathbb{R}^2$. Define*

$$E(d) = \int_{\mathbb{R}^2} ||x|| d\mu(x).$$

Let the customer parameters $\{(w_k, e_k, s_k, l_k) : k \in N\}$ be jointly distributed like Φ. In addition, let M_n^* be the minimum number of machines needed to feasibly schedule the jobs $\{(w_k, e_k, s_k, l_k) : k \in N\}$, and let $\lim_{n \to \infty} M_n^*/n = \gamma$, (a.s.). Then

$$\lim_{n \to \infty} \frac{1}{n} Z^{ST} = \lim_{n \to \infty} \frac{1}{n} Z_t^* = 2\gamma E(d) \quad (a.s.).$$

18.4.4   Numerical Study

Tables 18.1 and 18.2 summarize the computational experiments with the standard test problems of Solomon (1986). The problem set consists of 56 problems of various types. All problems consist of 100 customers and one depot, and the distances are Euclidean. Problems with the "R" prefix are problems where the customer locations are randomly generated according to a uniform distribution.

TABLE 18.1. Computational results: Part I

Problem	Alg. ST	CPU time (s)	Alg. SI	CPU time (s)	Solomon's best solution
C201	591.6	245.9	591.6	260.5	591
C202	*652.8	276.1	*640.8	262.7	731
C203	*692.2	309.2	*741.1	308.9	786
C204	*721.6	335.9	782.3	340.6	758
C205	713.8	250.8	699.9	258.8	606
C206	770.8	257.3	*722.8	283.3	730
C207	767.2	265.7	708.9	275.8	680
C208	736.2	287.7	660.2	272.4	607
R201	*1665.3	207.1	*1533.4	209.6	1,741
R202	*1485.3	276.4	*1484.3	248.5	1,730
R203	*1371.5	406.5	*1349.3	389.0	1,567
R204	1096.7	532.0	1077.0	538.2	1,059
R205	1472.3	287.0	*1329.4	312.6	1,471
R206	*1237.0	412.2	*1283.7	374.2	1,405
R207	*1217.7	484.8	*1162.9	453.9	1,241
R208	* 966.1	587.8	* 959.9	612.6	1,046
R209	*1276.1	394.8	*1262.8	355.7	1,418
R210	*1312.5	380.7	*1340.6	388.6	1,425
R211	1080.9	474.7	1141.3	488.7	1,016
RC201	*1873.8	203.5	*1841.7	185.8	1,880
RC202	*1742.1	227.8	*1705.1	241.0	1,799
RC203	*1417.5	331.5	*1471.1	300.1	1,550
RC204	*1139.6	437.7	*1190.3	411.5	1,208
RC205	*1830.5	233.0	*1878.9	214.0	2,080
RC206	1640.1	259.0	1607.5	248.2	1,582
RC207	*1566.4	294.2	*1557.3	272.3	1,632
RC208	1254.8	345.7	1298.7	317.3	1,194

* indicates that the LBH improves upon the best solution known

Problems with the "C" prefix are problems where the customer locations are clustered. Problems with the "RC" prefix are a mixture of both random and clustered. In addition, all of the problems have a constraint on the latest time T_0 at which a vehicle can return to the depot. For a full description of these problems, we refer the reader to Solomon.

We compare the performance of the LBH against the heuristics of Solomon and the column-generation approach of Desrochers et al. (1992). The latter method was able to solve effectively 7 of the 56 test problems; we describe this approach in the next chapter.

To compare the LBH to these solution methods, we implemented a time-window reduction phase before the start of the heuristic. Here, the earliest time for service e_k is replaced by $\max\{e_k, d_k\}$; in that way, vehicles leave the depot no earlier than

TABLE 18.2. Computational results: Part II

		CPU		CPU	Solomon's	DDS solution
Problem	Alg. ST	time (s)	Alg. SI	time (s)	best solution	value
C101	828.9	74.1	828.9	67.0	829	827.3
C102	982.8	82.9	1043.4	73.1	968	827.3
C103	*1015.1	95.9	1232.9	88.4	1,026	
C104	*980.9	105.4	*976.1	114.5	1,053	
C105	*828.9	79.7	860.8	67.3	829	
C106	852.9	82.8	880.1	66.7	834	827.3
C107	828.9	83.1	841.2	74.7	829	827.3
C108	852.9	88.6	853.6	80.9	829	827.3
C109	991.0	88.6	1014.5	83.1	829	
R101	1983.7	57.2	2071.2	39.9	1,873	1607.7
R102	1789.0	70.8	1821.4	57.4	1,843	1434.0
R103	1594.5	88.6	1599.1	67.9	1,484	
R104	1242.0	106.2	1237.3	81.0	1,188	
R105	1604.4	67.0	1696.2	52.0	1,502	
R106	1606.9	78.0	1589.2	70.0	1,460	
R107	*1324.9	92.4	1361.2	70.4	1,353	
R108	1202.6	107.5	1205.5	101.1	1,134	
R109	1504.7	78.5	1491.8	69.6	1,412	
R110	1380.9	92.0	1434.4	69.4	1,211	
R111	1422.1	91.7	1432.4	69.5	1,202	
R112	1248.1	105.2	1284.6	79.4	1,086	
RC101	2045.1	60.6	2014.4	45.0	1,867	
RC102	1806.6	68.7	1969.5	52.2	1,760	
RC103	1708.9	81.7	1716.3	69.6	1,641	
RC104	1372.1	93.5	1458.8	79.5	1,301	
RC105	*1826.3	68.9	2036.8	51.3	1,922	
RC106	1710.8	68.0	1804.8	50.5	1,611	
RC107	1593.2	76.4	1630.9	64.9	1,385	
RC108	1421.0	84.7	1493.8	65.5	1,253	

* indicates that the LBH improves upon the best solution known

time 0. In addition, the latest time service can end l_k is replaced by $\min\{l_k, T_0 - d_k\}$. The LBH can then be run as it is described in Sect. 18.4.1.

As can be seen in the tables, both the ST and SI heuristics have been implemented. CPU times are in seconds on a Sun SPARC Station II. In Tables 18.1 and 18.2, the column "Solomon's best solution" corresponds to the best solution found by Solomon. Solomon tested eight different heuristics on problem sets R1 and C1, and six heuristics on problems RC1, R2, C2, and RC2. We see that the ST heuristic provides a better solution than Solomon's heuristics in 25 of the 56 problems, while the SI heuristic provides a better solution in 21 of the 56 problems. In Table 18.2, the column "DDS solution value" corresponds to the value of the solution found using the column-generation approach of Desrochers et al.

18.5   Exercises

Exercise 18.1.   You are given a network $G = (V, A)$, where $|V| = n$, $d(i, j)$ is the length of edge (i, j) and a specified vertex $a \in V$. One service unit is located at a and has to visit each vertex in V so that the total waiting time of all vertices is as small as possible. Assume the waiting time of a vertex is proportional to the total distance traveled by the server from a to the vertex. The total waiting time (summed up over all customers) is then

$$(n - 1)d(a, 2) + (n - 2)d(2, 3) + (n - 3)d(3, 4) + \cdots + d(n - 1, n).$$

The delivery man problem (DMP) is the problem of determining the tour that minimizes the total waiting time.

Assume that G is a tree with $d(i, j) = 1$ for every $(i, j) \in A$. Show that any tour that follows a depth-first search starting from a is optimal.

Exercise 18.2.   Consider the delivery man problem described in Exercise 18.1. A delivery man currently located at the depot must visit each of n customers. Let Z^{DM} be the total waiting time in the optimal delivery man tour through the n points. Let Z^* be the total time required to travel the optimal traveling salesman tour through the n points.

(a) Prove that

$$Z^{DM} \leq \left(\frac{n}{2}\right) Z^*.$$

(b) One heuristic proposed for this problem is the nearest-neighbor (NN) heuristic. In this heuristic, the vehicle serves the closest unvisited customer next. Provide a family of examples to show that the heuristic does not have a fixed worst-case bound.

Exercise 18.3.   Consider the vehicle routing problem with distance constraints. Formally, a set of customers has to be served by vehicles that are all located at a common depot. The customers and the depot are presented as the nodes of an undirected graph $G = (N, E)$. Each customer has to be visited by a vehicle. The jth vehicle starts from the depot and returns to the depot after visiting a subset $N_j \subseteq N$. The total distance traveled by the jth vehicle is denoted by T_j. Each vehicle has a distance constraint λ: No vehicle can travel more than λ units of distance (i.e., $T_j \leq \lambda$). We assume that the distance matrix satisfies the triangle inequality assumption. Also, assume that the length of the optimal traveling salesman tour through all the customers and the depot is greater than λ.

(a) Suppose the objective function is to minimize the total distance traveled. Let K^* be the number of vehicles in an optimal solution to this problem. Show that there always exists an optimal solution with total distance traveled $> \frac{1}{2}K^*\lambda$. Does this lower bound hold for any optimal solution?

(b) Consider the following greedy heuristic: Start with the optimal traveling salesman tour through all the customers and the depot. In an arbitrary orientation of this tour, the nodes are numbered $(i_0, i_1, \ldots, i_n) \equiv S$ in order of appearance, where $n = $ the number of customers, i_0 is the depot, and $i_1, i_2, \ldots, i_n$ are the customers. We break the tour into K^H segments and connect the endpoints of each segment to the depot. This is done in the following way. Each vehicle j, $1 \le j < K^H$, starts by traveling from the depot to the first customer i_q not visited by the previous $j - 1$ vehicles and then visits the maximum number of customers according to S without violating the distance constraint upon returning to the depot.

Show that $K^H \le \min\{n, \lceil \frac{T - 2d_m}{\lambda - 2d_m} \rceil\}$, where T is the length of the optimal traveling salesman tour and d_m is the distance from the depot to the farthest customer.

Exercise 18.4. Consider the pickup and delivery problem. Here customers are pickup customers with probability p and delivery customers with probability $1 - p$. Assume a vehicle capacity of 1. If customer i is a pickup customer, then a load of size $w_i \le 1$ must be picked up at the customer and brought to the depot. If customer i is a delivery customer, then a load of size $w_i \le 1$ must be brought from the depot to the customer. Assume pickup sizes are drawn randomly from a distribution with bin-packing constant γ_P and delivery sizes are drawn randomly from a distribution with bin-packing constant γ_D. A pickup and a delivery can be in the vehicle at the same time.

(a) Develop a heuristic H for this problem and determine $\lim_{n \to \infty} \frac{Z^H}{n}$ as a function of p, γ_P, and γ_D.

(b) Assume all pickups are of size $\frac{1}{3}$ and deliveries are of size $\frac{2}{3}$. Suggest a better heuristic for this case. What is $\lim_{n \to \infty} \frac{Z^H}{n}$ as a function of p for this heuristic?

19

Solving the VRP Using a Column-Generation Approach

19.1   Introduction

A classical method, first suggested by Balinski and Quandt (1964) , for solving the VRP with capacity and time-window constraints, is based on formulating the problem as a set-partitioning problem. (See Chap. 6 for a general discussion of set partitioning.) The idea is as follows: Let the index set of all feasible routes be $\{1, 2, \ldots, R\}$ and let c_r be the length of route r. Define

$$\alpha_{ir} = \begin{cases} 1, & \text{if customer } i \text{ is served in route } r, \\ 0, & \text{otherwise,} \end{cases}$$

for each customer $i = 1, 2, \ldots, n$ and each route $r = 1, 2, \ldots, R$. Also, for every $r = 1, 2, \ldots, R$, let

$$y_r = \begin{cases} 1, & \text{if route } r \text{ is in the optimal solution,} \\ 0, & \text{otherwise.} \end{cases}$$

In the *set-partitioning formulation* of the VRP, the objective is to select a minimum-cost set of feasible routes such that each customer is included in some route. It is

D. Simchi-Levi et al., *The Logic of Logistics: Theory, Algorithms, and Applications for Logistics Management*, Springer Series in Operations Research and Financial Engineering, DOI 10.1007/978-1-4614-9149-1_19, © Springer Science+Business Media New York 2014

$$\text{Problem } S: \quad Min \quad \sum_{r=1}^{R} c_r y_r$$

$$\text{s.t.} \quad \sum_{r=1}^{R} \alpha_{ir} y_r \geq 1, \quad \forall i = 1, 2, \ldots, n, \qquad (19.1)$$

$$y_r \in \{0, 1\}, \quad \forall r = 1, 2, \ldots, R.$$

Observe that we have written constraints (19.1) as inequality constraints instead of equality constraints. The formulation with equality constraints is equivalent if we assume the distance matrix $\{d_{ij}\}$ satisfies the triangle inequality; therefore, each customer will be visited *exactly* once in the optimal solution. The formulation with inequality constraints will prove to be easier to work with from an implementation point of view.

Cullen et al. (1981) was the first to use this formulation successfully to design heuristic methods for the VRP. Later, Desrochers et al. (1992) used it in conjunction with a branch-and-bound method to generate optimal or near-optimal solutions to the VRP. Similar methods have been used to solve crew scheduling problems, such as Hoffman and Padberg (1993).

Of course, the set of all feasible routes is extremely large, and one cannot expect to generate it completely. Even if this set is given, it is not clear how to solve the set-partitioning problem since it is a large-scale integer program. Any method based on this formulation must overcome these two obstacles. We start here, in Sect. 19.2, by showing how the linear relaxation of the set-partitioning problem can be solved to optimality without enumerating all possible routes. In Sect. 19.3, we combine this method with a polyhedral approach that generates an optimal or near-optimal solution to the VRP. Finally, in Sect. 19.4, we provide a probabilistic analysis that helps explain why a method of this type will be effective.

To simplify the presentation, we assume *no time-window constraints exist*; the extension to the more general model is straightforward, for the most part. The interested reader can find some of these extensions in Desrochers et al. (1992).

19.2   Solving a Relaxation of the Set-Partitioning Formulation

To solve the linear relaxation of Problem S without enumerating all the routes, Desrochers et al. (1992) use the celebrated column-generation technique. A thorough explanation of this method is given below, but the general idea is as follows. A portion of all possible routes is enumerated, and the resulting linear relaxation with this partial route set is solved. The solution to this linear program is then used to determine if there are any routes not included that can reduce the objective function value. This is the *column-generation* step. Using the values of the optimal dual variables (with respect to the partial route set), the program generates a new route, and the linear relaxation is resolved. This is continued until one can show that an optimal solution to the linear program is found, one that is optimal for the complete route set.

Specifically, this is done by enumerating a partial set of routes, $1, 2, \ldots, R'$, and formulating the corresponding linear relaxation of the set-partitioning problem with respect to this set:

$$\text{Problem } S' : \quad Min \quad \sum_{r=1}^{R'} c_r y_r$$

$$\text{s.t.}$$

$$\sum_{r=1}^{R'} \alpha_{ir} y_r \geq 1, \quad \forall i = 1, 2, \ldots, n, \qquad (19.2)$$

$$y_r \geq 0, \quad \forall r = 1, 2, \ldots, R'.$$

Let $\bar{y}$ be the optimal solution to Problem S', and let $\bar{\pi}$ be the corresponding optimal dual variables. We would like to know whether $\bar{y}$ (or equivalently, $\bar{\pi}$) is optimal for the linear relaxation of Problem S (respectively, the dual of the linear relaxation of Problem S). To answer this question, observe that the dual of the linear relaxation of Problem S is

$$\text{Problem } S_D : \quad Max \quad \sum_{i=1}^{n} \pi_i$$

$$\text{s.t.}$$

$$\sum_{i=1}^{n} \alpha_{ir} \pi_i \leq c_r, \quad \forall r = 1, 2, \ldots, R, \qquad (19.3)$$

$$\pi_i \geq 0, \quad \forall i = 1, 2, \ldots, n.$$

Clearly, if $\bar{\pi}$ satisfies every constraint (19.3), then it is optimal for Problem S_D, and therefore, $\bar{y}$ is optimal for the linear programming relaxation of Problem S. How can we check whether $\bar{\pi}$ satisfies every constraint in Problem S_D? Observe that the vector $\bar{\pi}$ is not feasible in Problem S_D if we can identify a single constraint, r, such that

$$\sum_{i=1}^{n} \alpha_{ir} \bar{\pi}_i > c_r.$$

Consequently, if we can find a column r minimizing the quantity $c_r - \sum_i^n \alpha_{ir} \bar{\pi}_i$ and this quantity is negative, then a violated constraint is found. In that case, the current vector $\bar{\pi}$ is not optimal for Problem S_D. The corresponding column just found can be added to the formulation of Problem S_P, which is solved again. The process repeats itself until no violated constraint (column) is found; in this case, we have found the optimal solution to the linear relaxation of Problem S (the vector $\bar{y}$) and the optimal solution to Problem S_D (the vector $\bar{\pi}$).

Our task is then to find a column, or a route, r minimizing the quantity

$$c_r - \sum_i^n \alpha_{ir} \overline{\pi}_i. \tag{19.4}$$

We can look at this problem in a different way. Suppose we replace each distance d_{ij} with a new distance d'_{ij} defined by

$$d'_{ij} \doteq d_{ij} - \frac{\overline{\pi}_i}{2} - \frac{\overline{\pi}_j}{2}.$$

Then a tour $u_1 \to u_2 \to \ldots \to u_\ell$ whose length using $\{d_{ij}\}$ is $\sum_{i=1}^{\ell-1} d_{u_i u_{i+1}} + d_{u_\ell u_1}$ has, using $\{d'_{ij}\}$, a length

$$\sum_{i=1}^{\ell-1} d'_{u_i u_{i+1}} + d'_{u_\ell u_1} = \sum_{i=1}^{\ell-1} d_{u_i u_{i+1}} + d_{u_\ell u_1} - \sum_{i=1}^{\ell} \pi_{u_i}.$$

Hence, finding a route r that minimizes (19.4) is the same as using the distance matrix $\{d'_{ij}\}$ to find a minimum-length tour that starts and ends at the depot, visits a subset of the customers, and has a total load no more than Q. Unfortunately, this itself is an $\mathcal{NP}$-Hard problem, and so we are left with a method that is not attractive computationally.

To overcome this difficulty, the set-partitioning formulation, Problem S, is modified to allow routes to visit the same customer more than once. The purpose of this modification will be clear in a moment. This model, call it Problem S_M (where M stands for the "modified" formulation), is defined as follows: Enumerate all feasible routes, satisfying the capacity constraint, that may visit the same customer a number of times; each such visit increases the total load by the demand of that customer. Let the number of routes (columns) be R_M, and let c_r be the total distance traveled in route r. For each customer $i = 1, 2, \ldots, n$ and route $r = 1, 2, \ldots, R_M$, let

$$\xi_{ir} = \text{number of times customer } i \text{ is visited in route } r.$$

Also, for each $r = 1, 2, \ldots, R_M$, define

$$y_r = \begin{cases} 1, & \text{if route } r \text{ is in the optimal solution,} \\ 0, & \text{otherwise.} \end{cases}$$

The VRP can be formulated as

$$\text{Problem } S_M : \quad Min \quad \sum_{r=1}^{R_M} c_r y_r$$

$$\text{s.t.}$$

$$\sum_{r=1}^{R_M} \xi_{ir} y_r \geq 1, \quad \forall i = 1, 2, \ldots, n, \tag{19.5}$$

$$y_r \in \{0, 1\}, \quad \forall r = 1, 2, \ldots, R_M.$$

This is the set-partitioning problem solved by Desrochers et al. (1992), and therefore, it is not exactly Problem S. Clearly, the optimal integer solution to Problem S_M is the optimal solution to the VRP. However, the optimal solution values of the linear relaxations of Problem S_M and Problem S may be different. Of course, the linear relaxation of Problem S_M provides a lower bound on the linear relaxation of Problem S.

To solve the linear relaxation of Problem S_M, we use the method described above (for solving Problem S): We enumerate a partial set of R'_M routes; solve Problem S'_M, which is the linear relaxation of Problem S_M defined only on this partial list; use the dual variables to see whether a column not in the current partial list with $\sum_{i=1}^{n} \xi_{ir} \overline{\pi}_i > c_r$ exists. If such a column(s) exists, we add it (them) to the formulation and solve the resulting linear program again. Otherwise, we have the optimal solution to the linear relaxation of Problem S_M.

The modification we have made makes the column-generation step computationally easier. This can now be found in pseudopolynomial time using dynamic programming.

For this purpose, we need the following definitions. Given a path $P = \{0, u_1, u_2, \ldots, u_\ell\}$, where it is possible that $u_i = u_j$ for $i \neq j$, let the load of this path be $\sum_{i=1}^{\ell} w_{u_i}$. That is, the load of the path is the sum, over all customers in P, of the demand of a customer multiplied by the number of times that customer appears in P. Let $f_q(i)$ be the cost [using $\{d'_{ij}\}$] of the least-cost path that starts at the depot and terminates at vertex i with total load q. This can be calculated using the recursion

$$f_q(i) = \min_{j \neq i} \left\{ f_{q-w_i}(j) + d'_{ij} \right\},\tag{19.6}$$

with the initial conditions

$$f_q(i) = \begin{cases} d'_{0i} & \text{if } q = w_i, \\ +\infty & \text{otherwise.} \end{cases}$$

Finally, let $f^0_q(i) = f_q(i) + d'_{0i}$. Thus, $f^0_q(i)$ is the length of a least-cost tour that starts at the depot, visits a subset of the customers, of which customer i is the last to be visited, has a total load q, and terminates at the depot. Observe that finding $f^0_q(i)$ for every q, $1 \leq q \leq Q$, and every i, $i \in N$, requires $O(n^2 Q)$ calculations. The recursion chooses the predecessor of i to be a node $j \neq i$. This requires repeat visits to the same customer to be separated by at least one visit to another customer. In fact, expanding the state space of this recursion can eliminate *two-loops*: loops of the type $\ldots i, j, i \ldots$. This forces repeat visits to the same customer to be separated by visits to at least *two* other customers. This can lead to a stronger relaxation of the set-partitioning model. For a more detailed discussion of this recursion, see Christofides et al. (1981).

If there exist a q, $1 \leq q \leq Q$, and i, $i \in N$ with $f^0_q(i) < 0$, then the current vectors $\overline{y}$ and $\overline{\pi}$ are not optimal for the linear relaxation of Problem S_M. In such a case, we add the column corresponding to this tour [the one with negative $f^0_q(i)$] to the set of columns in Problem S'_M. If, on the other hand, $f^0_q(i) \geq 0$ for every q and i, then the current $\overline{y}$ and $\overline{\pi}$ are optimal for S_M.

To summarize, the column-generation algorithm can be described as follows.

The Column-Generation Procedure

Step 1: Generate an initial set of R'_M columns.

Step 2: Solve Problem S'_M and find $\bar{y}$ and $\bar{\pi}$.

Step 3: Construct the distance matrix $\{d'_{ij}\}$ and find $f_i^0(q)$ for all $i \in N$ and $1 \leq q \leq Q$.

Step 4: For every i and q with $f_i^0(q) < 0$, add the corresponding column to R'_M and go to *step 2*.

Step 5: If $f_i^0(q) \geq 0$ for all i and q, stop.

The procedure produces a vector $\bar{y}$ that is the optimal solution to the linear relaxation of Problem S_M. This is a *lower bound* on the optimal solution to the VRP.

19.3   Solving the Set-Partitioning Problem

In the previous section, we introduced an effective method for solving the linear relaxation of the set-partitioning formulation of the VRP, Problem S_M. How can we use this solution to the linear program to find an optimal or near-optimal integer solution?

We'll start with the set of columns present at the end of the column-generation step (the set E); one approach to generating an integer solution to the set-partitioning formulation is to use the *branch-and-bound* method. This method consists of splitting the problem into easier subproblems by fixing the value of a *branching* variable. The variable (in this case, a suitable choice is y_r for some route r) is set to either 1 or 0. Each of these subproblems is solved using the same method; that is, another variable is branched. At each step, tests are performed to see if the entire branch can be eliminated; that is, no better solution than the one currently known can be found in this branch. The solution found by this method will be the best integer solution among all the solutions in E. This solution will not necessarily be the optimal solution to the VRP, but it may be close.

Another approach that will generate the same integer solution as the branch-and-bound method is the following. Given a fractional solution to S_M, we can generate a set of constraints that will cut off this fractional solution. Then we can resolve this linear program; if it is integer, we have found the optimal integer solution (among the columns of E). If it is still fractional, then we can continue generating constraints and resolving the linear program until an integer solution is found. Again, the best integer solution found using this method may be close to optimal. This is the method successfully used by Hoffman and Padberg (1993) to solve crew scheduling problems.

Formally, the method is as follows.

The Cutting-Plane Algorithm

Step 1: Generate an initial set of R'_M columns.

Step 2: Use column generation to solve Problem S'_M.

Step 3: If the optimal solution to Problem S'_M is integer, stop.
  Else, generate cutting planes separating this solution.
  Add these cutting planes to the linear program S'_M.

Step 4: Solve the linear program S'_M. Go to *step 3*.

To illustrate this constraint-generation step (*step 3*), we make use of a number of observations. First, let E be the set of routes at the end of the column-generation procedure. Clearly, we can split E into two subsets. One subset E_m includes every column r for which there is at least one i with $\xi_{ir} \geq 2$; these columns are called *multiple-visit columns*. The second subset E_s includes the remaining columns; these columns are referred to as *single-visit columns*. It is evident that an optimal solution to the VRP will use no columns from E_m. That is, there always exists a single-visit column of at most the same cost that can be used instead. We therefore can immediately add the following constraint to the linear relaxation of Problem S_M:

$$\sum_{r \in E_m} y_r = 0. \tag{19.7}$$

To generate more constraints, construct the *intersection* graph G. The graph G has a node for each column in E_s. Two nodes in G are connected by an edge if the corresponding columns have at least one customer in common. Observe that a solution to the VRP where no customer is visited more than once can be represented by an *independent set* in this graph. That is, it is a collection of nodes on the graph G such that no two nodes are connected by an edge.

These observations give rise to two inequalities that can be added to the formulation:

1. We select a subset of the nodes of G, say K, such that every pair of nodes $i, j \in K$ is connected by an edge of G. Each set K, called a *clique*, must satisfy the following condition:

$$\sum_{r \in K} y_r \leq 1. \tag{19.8}$$

Clearly, if there is a node $j \notin K$ such that j is adjacent to every $i \in K$, then we can replace K with $K \cup \{j\}$ in inequality (19.8) to strengthen it (this is called *lifting*). In that sense, we would like to use inequality (19.8) when the set of nodes K is *maximal* in that sense.

2. Define a cycle $C = \{u_1, u_2, \ldots, u_\ell\}$ in G such that node u_i is adjacent to u_{i+1}, for each $i = 1, 2, \ldots, \ell - 1$, and node u_ℓ is adjacent to node u_1. A cycle C is called an odd cycle if the number of nodes in C, $|C| = \ell$, is odd. An odd cycle is called an *odd hole* if there is no arc connecting two nodes of the cycle except the ℓ arcs defining the cycle. It is easy to see that in any optimal solution to the VRP, each odd hole must satisfy the following property:

$$\sum_{r \in C} y_r \leq \frac{|C| - 1}{2}. \tag{19.9}$$

19.3.1   Identifying Violated Clique Constraints

Hoffman and Padberg suggest several procedures for clique identification, one of which is based on the fact that small problems can be solved quickly by enumeration. For this purpose, select v to be the node with minimum degree among all nodes of G. Clearly, every clique of G containing v is a subset of the neighbors of v, denoted by $neigh(v)$. Thus, starting with v as a temporary clique, that is, $K = \{v\}$, we add an arbitrary node w from $neigh(v)$ to K. We now delete from $neigh(v)$ all nodes that are not connected to a node of K, in this case either v or w. Continue adding nodes in this manner from the current set $neigh(v)$ to K until either there is no node in $neigh(v)$ connected to all nodes in K, or $neigh(v) = \emptyset$. In the end, K will be a maximal clique. We can then calculate the *weight* of this clique, that is, the sum of the values (in the linear program) of the columns in the clique. If the weight is more than 1, then the corresponding clique inequality is violated. If not, then we continue the procedure with a new starting node. The method can be improved computationally by, for example, always choosing the "heaviest" among those nodes eligible to enter the clique.

19.3.2   Identifying Violated Odd Hole Constraints

Hoffman and Padberg use the following procedure to identify violated odd hole constraints. Suppose $\bar{y}$ is the current optimal solution to the linear program and G is the corresponding intersection graph. Starting from an arbitrary node $v \in G$, construct a *layered graph* $G_\ell(v)$ as follows. The node set of $G_\ell(v)$ is the same as the node set of G. Every neighbor of v in G is connected to v by an edge in $G_\ell(v)$. We refer to v as the root, or level-0 node, and we refer to the neighbors of v as level-1 nodes. Similarly, nodes at level $k \geq 2$ are those nodes in G that are connected (in G) to a level-$k - 1$ node but are not connected to any node at level $< k - 1$. Finally, each edge (u_i, u_j) in $G_\ell(v)$ is assigned a length of $1 - \bar{y}_{u_i} - \bar{y}_{u_j} \geq 0$. Now pick a node u in $G_\ell(v)$ at level $k \geq 2$ and find the shortest path from u to v in $G_\ell(v)$. Delete all nodes at levels i $(1 \leq i < k)$ that are either on the shortest path or adjacent to nodes along this shortest path (other than nodes that are adjacent to v). Now pick another node w that is adjacent (in G) to u in level k. Find the shortest path from w to v in the current graph $G_\ell(v)$. Combining these two paths with the arc (u, w)

creates an odd hole. If the total length of this cycle is less than 1, then we have found a violated odd hole inequality. If not, we continue with another neighbor of u and repeat the process. We can then choose a node different from u at level k. If no violated odd hole inequality is found at level k, we proceed to level $k + 1$. This subroutine can be repeated for different starting nodes (v) as well.

19.4 The Effectiveness of the Set-Partitioning Formulation

The effectiveness of this algorithm depends crucially on the quality of the initial lower bound; this lower bound is the optimal solution to the linear relaxation of Problem S_M. If this lower bound is not very tight, then the branch-and-bound or the constraint-generation method will most likely not be computationally effective. On the other hand, when the gap between the lower bound and the best integer solution is small, the procedure will probably be effective.

Fortunately, many researchers have reported that the linear relaxation of the set-partitioning problem, Problem S_M, provides a solution close to the optimal integer solution [see, e.g., Desrochers et al. (1992)]. That is, the solution to the linear relaxation of Problem S_M provides a very tight lower bound on the solution of the VRP. For instance, in their paper, Desrochers et al. report an average relative gap between the optimal solution to the linear relaxation and the optimal integer solution of only 0.733%. A possible explanation for this observation is embodied in the following theorem, which states that asymptotically the relative error between the optimal solution to the linear relaxation of the set-partitioning model and the optimal integer solution goes to zero as the number of customers increases. Consider again the general VRP with capacity and time-window constraints.

Theorem 19.4.1 *Let the customer locations $x_1, x_2, \ldots, x_n$ be a sequence of independent random variables having a distribution μ with compact support in $\mathbb{R}^2$. Let the customer parameters (see Chap. 18) be independently and identically distributed like Φ. Let Z^{LP} be the value of the optimal fractional solution to S, and let Z^* be the value of the optimal integer solution to S, that is, the value of the optimal solution to the VRP. Then*

$$\lim_{n \to \infty} \frac{1}{n} Z^{\text{LP}} = \lim_{n \to \infty} \frac{1}{n} Z^* \quad (a.s.).$$

The theorem thus implies that the optimal solution value of the linear programming relaxation of Problem S tends to the optimal solution of the vehicle routing problem as the number of customers tends to infinity. This is important since, as shown by Bramel and Simchi-Levi (1997), other classical formulations of the VRP can lead to diverging linear and integer solution values (see Exercise 19.8).

In the next section, we motivate Theorem 19.4.1 by presenting a simplified model that captures the essential ideas of the proof. Finally, in Sect. 19.4.2, we provide a formal proof of the theorem. Again, to simplify the presentation, we assume no time-window constraints exist; for the general case, the interested reader is referred to Bramel and Simchi-Levi (1997).

19.4.1 Motivation

Define a customer type to be a location $x \in I\!\!R^2$ and a customer demand w; that is, a customer type defines the customer location and a value for the customer demand. Consider a *discretized* vehicle routing model in which there are a finite number W of customer types and a finite number m of distinct customer locations. Let n_i be the number of customers of type i, for $i = 1, 2, \ldots, W$, and let $n = \sum_{i=1}^{W} n_i$ be the total number of customers. Clearly, this discretized vehicle routing problem can be solved by formulating it as a set-partitioning problem. To obtain some intuition about the linear relaxation of S, we introduce another formulation of the vehicle routing problem closely related to S.

Let a *vehicle assignment* be a vector $(a_1, a_2, \ldots, a_W)$, where $a_i \geq 0$ are integers, and such that a single vehicle can feasibly serve a_1 customers of type 1, and a_2 customers of type 2, $\ldots$, and a_W customers of type W together without violating the vehicle capacity constraint. Index all the possible vehicle assignments $1, 2, \ldots, R_a$ and let c_r be the total length of the shortest feasible route serving the customers in vehicle assignment r. (Note that R_a is independent of n.) The vehicle routing problem can be formulated as follows. Let

$$A_{ir} = \text{number of customers of type } i \text{ in vehicle assignment } r,$$

for each $i = 1, 2, \ldots, W$ and $r = 1, 2, \ldots, R_a$. Let

$$y_r = \text{number of times vehicle assignment } r \text{ is used in the optimal solution.}$$

The new formulation of this discretized VRP is

$$\text{Problem } S_N : \quad Min \quad \sum_{r=1}^{R_a} y_r c_r$$

$$\text{s.t.}$$

$$\sum_{r=1}^{R_a} y_r A_{ir} \geq n_i, \quad \forall i = 1, 2, \ldots, W,$$

$$y_r \geq 0 \text{ and integer}, \quad \forall r = 1, 2, \ldots, R_a.$$

Let Z_N^* be the value of the optimal solution to Problem S_N and let Z_N^{LP} be the optimal solution to the linear relaxation of Problem S_N. Clearly, Problem S and Problem S_N have the same optimal solution values, that is, $Z^* = Z_N^*$, while their linear relaxations may be different. Define $\bar{c} \doteq \max_{r=1,2,\ldots,R_a}\{c_r\}$; that is, $\bar{c}$ is the length of the longest route among the R_a vehicle assignments. Using an analysis identical to the one in Sect. 6.2, we obtain

Lemma 19.4.2

$$Z^{\text{LP}} \leq Z^* \leq Z_N^{\text{LP}} + W\bar{c} \leq Z^{\text{LP}} + W\bar{c}.$$

Observe that the upper bound on Z^* obtained in Lemma 19.4.2 consists of two terms. The first, Z^{LP}, is a lower bound on Z^*, which clearly grows with the number of customers n. The second term $(W\bar{c})$ is the product of two numbers that are fixed and independent of n. Therefore, the upper bound on Z^* of Lemma 19.4.2 is dominated by Z^{LP}, and consequently, we see that for large n, $Z^* \approx Z^{\text{LP}}$, exactly what is implied by Theorem 19.4.1. Indeed, much of the proof of the following section is concerned with approximating the distributions μ (customer locations) and Φ (customer demands) with discrete distributions and forcing the number of different customer types to be independent of n.

19.4.2 Proof of Theorem 19.4.1

It is clear that $Z^{\text{LP}} \leq Z^*$; therefore, $\underline{\lim}_{n\to\infty} \frac{1}{n}(Z^* - Z^{\text{LP}}) \geq 0$. The interesting part is to find an upper bound on Z^* that involves Z^{LP} and use this upper bound to show that $\overline{\lim}_{n\to\infty} \frac{1}{n}(Z^* - Z^{\text{LP}}) \leq 0$. We do this in essentially the same way as in Sect. 19.4.1. We successively discretize the problem by introducing a sequence of vehicle routing problems whose optimal solutions are "relatively" close to Z^*. The last vehicle routing problem is a discrete problem, which, therefore, as in Sect. 19.4.1, can be directly related to the linear relaxation of its set-partitioning formulation. This linear program is also shown to have an optimal solution close to Z^{LP}.

To prove the upper bound, let N be the index set of customers, with $|N| = n$, and let Problem P be the original VRP. Let A be the compact support of the distribution of the customer locations (μ), and define $d_{\max} \doteq \sup\{\|x\| : x \in A\}$, where $\|x\|$ is the distance from point $x \in A$ to the depot. Finally, pick a fixed $k > 1$.

Discretization of the Locations

We start by constructing the following vehicle routing problem with discrete locations. Define $\Delta \doteq \frac{1}{k}$ and let $G(\Delta)$ be an infinite grid of squares of *diagonal* Δ, that is, of side $\frac{\Delta}{\sqrt{2}}$, with edges parallel to the system coordinates. Let $A_1, A_2, \ldots, A_{m(\Delta)}$ be the subregions of $G(\Delta)$ that intersect A and have $\mu(A_i) > 0$. Since A is bounded, $m(\Delta)$ is finite for each $\Delta > 0$. For convenience, we omit the dependence of m on Δ in the notation. For each subregion, let X_i be the *centroid* of subregion A_i, that is, the point at the center of the grid square containing A_i. This defines m points $X_1, X_2, \ldots, X_m$; note that a customer is at most $\frac{\Delta}{2}$ units from the centroid of the subregion in which it is located.

Construct a new VRP, called $P(m)$, defined on the customers of N. Each of the customers in N is moved to the centroid of the subregion in which it is located. Let $Z^*(m)$ be the optimal solution to $P(m)$. We clearly have

$$Z^* \leq Z^*(m) + n\Delta. \tag{19.10}$$

Discretization of the Customer Demands

We now describe a VRP where the customer demands are also discretized in much the same way as it is done in Sect. 6.2. Partition the interval $(0, 1]$ into subintervals of size $\Delta(= \frac{1}{k})$. This produces k segments and $I \doteq k - 1$ points in the interval $(0, 1)$, which we call *corners*.

We refer to each centroid–corner pair as a customer type; each centroid defines a customer location and each corner defines the customer demand. It is clear that there are mI possible customer types. An instance of a fully discretized vehicle routing problem is then defined by specifying the number of customers of each of the mI types.

For each centroid $j = 1, 2, \ldots, m$ and corner $i = 1, 2, \ldots, I$, let

$$N_{ji} = \left\{ h \in N : \frac{i-1}{k} < w_h \leq \frac{i}{k} \text{ and } x_h \in A_j \right\}.$$

Finally, for every $j = 1, 2, \ldots, m$ and $i = 1, 2, \ldots, I$, let $n_{ji} = |N_{ji}|$.

We now define a fully discretized vehicle routing problem $P_k(m)$, whose optimal solution value is denoted $Z_k^*(m)$. The vehicle routing problem $P_k(m)$ is defined as having $\min\{n_{ji}, n_{j,i+1}\}$ customers located at centroid j with customer demand equal to $\frac{i}{k}$, for each $i = 1, 2, \ldots, I$ and $j = 1, 2, \ldots, m$.

We have the following result.

Lemma 19.4.3

$$Z^*(m) \leq Z_k^*(m) + 2d_{\max} \sum_{j=1}^{m} \sum_{i=1}^{I} |n_{ji} - n_{j,i+1}|.$$

Proof. Observe:

(i) In $P_k(m)$, the number of customers at centroid j and with demand defined by corner i is $\min\{n_{ji}, n_{j,i+1}\}$.

(ii) In $P(m)$, each customer belongs to exactly one of the subsets N_{ji}, for $j = 1, 2, \ldots, m$ and $i = 1, 2, \ldots, I$.

(iii) In $P(m)$, the customers in N_{ji} have smaller loads than the customers of $P_k(m)$ at centroid j with demand defined by corner i.

Given an optimal solution to $P_k(m)$, let us construct a solution to $P(m)$. For each centroid $j = 1, 2, \ldots, m$ and corner $i = 1, 2, \ldots, I$, we pick any $\max\{n_{ji} - n_{j,i+1}, 0\}$ customers from N_{ji} and serve them in individual vehicles. The remaining $\min\{n_{ji}, n_{j,i+1}\}$ customers in N_{ji} can be served with exactly the same vehicle schedules as in $P_k(m)$. This can be done due to (iii); therefore, one can always serve customers with a demand of $P(m)$ in the same vehicles in which the customers of $P_k(m)$ are served. ∎

Now $P_k(m)$ is fully discrete, and we can apply results as in Sect. 19.4.1. Let $Z_k^{\mathrm{LP}}(m)$ be the optimal solution to the linear relaxation of the set-partitioning formulation of the routing problem $P_k(m)$. Let $\bar{c}$ be defined as in Sect. 19.4.1; that is, it is the cost of the most expensive tour among all the possible routes in $P_k(m)$.

Lemma 19.4.4

$$Z_k^*(m) \leq Z_k^{\mathrm{LP}}(m) + mI\bar{c}.$$

Proof. Since the number of customer types is at most mI, we can formulate $P_k(m)$ as the integer program, like Problem S_N, described in Sect. 19.4.1, with mI constraints. The bound then follows from Lemma 19.4.2. ∎

Recall that Z^{LP} is the optimal solution to the linear relaxation of the set-partitioning formulation of the VRP defined by Problem P. Then

Lemma 19.4.5

$$Z_k^{\mathrm{LP}}(m) \leq Z^{\mathrm{LP}} + n\Delta.$$

Proof. Let $\{\bar{y}_r : r = 1, 2, \ldots, R\}$ be the optimal solution to the linear relaxation of the set-partitioning formulation of Problem P. We can assume (see Exercise 19.3) that $\sum_{r=1}^{R} \bar{y}_r \alpha_{ir} = 1$, for each $i = 1, 2, \ldots, n$. We construct a feasible solution to the linear relaxation of the set-partitioning formulation of $P_k(m)$ using the values $\bar{y}_r$. Since every customer in $P_k(m)$ assigned to centroid j and corner i can be associated with a customer in P with $x_k \in A_j$ and whose demand is at least as large, each route r with $\bar{y}_r > 0$ can be used to construct a route r' feasible for $P_k(m)$. Since in $P_k(m)$ the customers are at the centroids instead of at their original locations, we modify the route so that the vehicle travels from the customer to its centroid and back. Thus, the length (cost) of route r' is at most the cost of route r in P plus $n_r\Delta$, where n_r is the number of customers in route r.

To create a feasible solution to the linear relaxation of the set-partitioning formulation to $P_k(m)$, we take the solution to the linear relaxation of P and create the routes r' as above. Therefore,

$$Z_k^{\mathrm{LP}}(m) \leq Z^{\mathrm{LP}} + \sum_{r=1}^{R} \bar{y}_r n_r \Delta \leq Z^{\mathrm{LP}} + n\Delta.$$

∎

We can now prove Theorem 19.4.1.

$$Z^* \leq Z^*(m) + n\Delta$$

$$\leq Z_k^*(m) + 2d_{\max} \sum_{j=1}^{m} \sum_{i=1}^{I} |n_{ji} - n_{j,i+1}| + n\Delta$$

$$\leq Z_k^{\mathrm{LP}}(m) + mI\bar{c} + 2d_{\max} \sum_{j=1}^{m} \sum_{i=1}^{I} |n_{ji} - n_{j,i+1}| + n\Delta$$

$$\leq Z^{\mathrm{LP}} + mI\bar{c} + 2d_{\max} \sum_{j=1}^{m} \sum_{i=1}^{I} |n_{ji} - n_{j,i+1}| + 2n\Delta.$$

We now need to show that Z^{LP} is the dominant part of the last upper bound. We do that using the following lemma.

Lemma 19.4.6 *There exists a constant K such that*

$$\varlimsup_{n\to\infty} \frac{1}{n} \sum_{j=1}^{m} \sum_{i=1}^{I} |n_{ji} - n_{j,i+1}| \leq \frac{2K}{k}.$$

Proof. In Sect. 6.2, we prove that given i and j, there exists a constant K such that

$$\varlimsup_{n\to\infty} \frac{1}{n} |n_{ji} - n_{j,i+1}| \leq \frac{2K}{k^2}.$$

Therefore, a similar analysis gives

$$\varlimsup_{n\to\infty} \frac{1}{n} \sum_{j=1}^{m} \sum_{i=1}^{I} |n_{ji} - n_{j,i+1}| \leq \sum_{j=1}^{m} \mu(A_j) \frac{2K}{k} = \frac{2K}{k}.$$

∎

Finally, observe that each tour in $P_k(m)$ has a total length no more than 1 since the truck travels at a unit speed and the length of each working day is 1. Hence, $mI\bar{c} = O(1)$, and therefore,

$$\varlimsup_{n\to\infty} \frac{1}{n}(Z^* - Z^{\mathrm{LP}}) \leq 4d_{\max}\frac{K}{k} + 2\Delta$$

$$= \frac{2}{k}(2Kd_{\max} + 1).$$

Since K is a constant and k was arbitrary, we see that the right-hand side can be made arbitrarily small. Therefore,

$$0 \leq \varliminf_{n\to\infty} \frac{1}{n}(Z^* - Z^{\mathrm{LP}}) \leq \varlimsup_{n\to\infty} \frac{1}{n}(Z^* - Z^{\mathrm{LP}}) \leq 0.$$

We conclude this chapter with the following observation. The proof of Theorem 19.4.1 also reveals an upper bound on the rate of convergence of Z^{LP} to its asymptotic value. Indeed (see Exercise 19.1), we have

$$E(Z^*) \leq E(Z^{\mathrm{LP}}) + O(n^{3/4}). \tag{19.11}$$

19.5 Exercises

Exercise 19.1. Prove the upper bound on the convergence rate (19.11).

Exercise 19.2. Consider an undirected graph $G = (V, E)$, where each edge (i, j) has a cost c_{ij} and each vertex $i \in V$ a nonnegative penalty π_i. In the prize-collecting traveling salesman problem (PCTSP), the objective is to find a tour that visits a subset of the vertices such that the length of the tour plus the sum of the penalties of all vertices not in the tour is as small as possible. Show that the problem can be formulated as a longest-path problem between two prespecified nodes of a new network.

Exercise 19.3.   Consider the bin-packing problem. Let w_i be the size of item i, $i = 1, \ldots, n$, and assume the bin capacity is 1. An important formulation of the bin-packing problem is as a set-covering problem. Let

$$F = \{S : \sum_{i \in S} w_i \leq 1\}.$$

Define

$$\alpha_{iS} = \begin{cases} 1, & \text{if item } i \text{ is in } S, \\ 0, & \text{otherwise,} \end{cases}$$

for each $i = 1, 2, \ldots, n$ and each $S \in F$. Finally, for any S, $S \in F$, let

$$y_S = \begin{cases} 1, & \text{if the items in } S \text{ are packed in a single bin with no other items,} \\ 0, & \text{otherwise.} \end{cases}$$

In the set-covering formulation of the bin-packing problem, the objective is to select a minimum number of feasible bins such that each item is included in some bin. It is the following integer program.

$$\text{Problem P} \quad : \quad Min \quad \sum_{S \in F} y_S$$

$$\text{s.t.}$$

$$\sum_{S \in F} y_S \alpha_{iS} \geq 1, \quad \forall i = 1, 2, \ldots, n, \quad (19.12)$$

$$y_S \in \{0, 1\}, \quad \forall S \in F.$$

Let Z^* be the optimal solution to Problem P, and let Z^{LP} be the optimal solution to the linear relaxation of Problem P. We want to prove that

$$Z^* \leq 2Z^{\mathrm{LP}}. \quad (19.13)$$

(a) Formulate the dual of the linear relaxation of Problem P.

(b) Show that $\sum_{i=1}^{n} w_i \leq Z^{\mathrm{LP}}$.

(c) Argue that $Z^* \leq 2 \sum_{i=1}^{n} w_i$. Conclude that (19.13) holds.

(d) An alternative formulation to Problem P is obtained by replacing constraints (19.12) with equality constraints. Call the new problem Problem PE. Show that the optimal solution value of the linear relaxation of Problem P equals the optimal solution value of the linear relaxation of Problem PE.

Exercise 19.4. Recall the dynamic program given by (19.6). Let

$$f = \min_{i \in N} \min_{w_i \le q \le Q} f_q(i).$$

Consider the function defined as follows:

$$g_q(i) = \min_{w_i \le q' \le q} \{f_{q'}(i) + f_{q-q'+w_i}(i)\},$$

for each $i \in N$ and $w_i \le q \le Q$. Now define $\underline{g} = \min_{i \in N} \min_{w_i \le q \le Q} g_q(i)$. Show that $\underline{f} = \underline{g}$.

Exercise 19.5. Develop a dynamic programming procedure for the column-generation step similar to $f_q(i)$ that avoids two-loops (loops of the type $\ldots i, j, i \ldots$). What is the complexity of this procedure?

Exercise 19.6. Develop a dynamic programming procedure for the column-generation step in the presence of time-window constraints. What is required of the time-window data in order for this to be possible? What is the complexity of your procedure?

Exercise 19.7. Develop a dynamic programming procedure for the column-generation step in the presence of a distance constraint on the length of any route. What is required of the distance data in order for this to be possible? What is the complexity of your procedure?

Exercise 19.8. Consider an instance of the VRPTW with n customers. Given a subset of the customers S, let $b^*(S)$ be the minimum number of vehicles required to carry the demands of customers in S; that is, $b^*(S)$ is the solution to the bin-packing problem defined on the demands of all customers in S. For $i = 1, 2, \ldots, n$ and $j = 1, 2, \ldots, n$, let

$$x_{ij} = \begin{cases} 1, & \text{if a vehicle travels directly between points } i \text{ and } j, \\ 0, & \text{otherwise.} \end{cases}$$

Let 0 denote the depot and define c_{ij} as the cost of traveling directly between points i and j, for $i, j = 0, 1, 2, \ldots, n$. Let t_i represent the time a vehicle arrives at the location of customer i; also, for every i and j, such that $i < j$, define $M_{ij} = \max\{l_i + d_{ij} - e_j, 0\}$, where $d_{ij} \equiv \|Y_i - Y_j\|$. Then the following is a valid formulation of the VRPTW:

Problem P' :   *Min* $\displaystyle\sum_{i<j} c_{ij} x_{ij}$

s.t. $\displaystyle\sum_{i<j} x_{ij} + \sum_{i>j} x_{ji} = 2, \quad \forall i = 1, 2, \ldots, n,$

$\displaystyle\sum_{i,j \in S} x_{ij} \le |S| - b^*(S), \quad \forall S \subset \{1, 2, \ldots, n\}, 2 \le |S| \le n - 1,$

$$e_i \le t_i \le l_i - s_i, \quad 1 \le i \le n,$$

$$t_i + s_i + d_{ij} - t_j \le M_{ij}(1 - x_{ij}), \quad 1 \le i < j \le n,$$

$$x_{ij} \in \{0,1\}, \quad 1 \le i < j \le n, \tag{19.14}$$

$$x_{0j} \in \{0,1,2\}, \quad j = 1,2,\ldots,n. \tag{19.15}$$

The case $x_{0j} = 2$ corresponds to a vehicle serving only customer j. The linear programming relaxation of P' is obtained by replacing constraints (19.14) and (19.15) by their linear equivalents.

Construct an instance of the VRPTW in which the fractional and integer solutions to the above linear program do not approach the same value asymptotically.

Part V

Logistics Algorithms
in Practice

20

Network Planning

20.1   Introduction

In this chapter, we present some of the issues involved in the practice of supply chain design and planning. These are issues that are often not dealt with in traditional operations research analyses. However, they are essential in transforming raw data and problem characteristics into modeling assumptions, input data, and decisions.

Our focus is on what we call network planning—the process by which the firm structures and manages the supply chain in order to

- find the right balance among inventory, transportation, and manufacturing costs,

- match supply and demand under uncertainty by positioning inventory effectively,

- utilize resources effectively in a dynamic environment.

Of course, this is a complex process, which requires a hierarchical approach in which decisions on network design, inventory positioning and management, as well as resource utilization are combined to reduce cost and increase service level. Thus, we divide the network planning process into three steps:

1. **Network design:** This includes decisions on the number, locations, and size of manufacturing plants and warehouses, the assignment of retail outlets to

D. Simchi-Levi et al., *The Logic of Logistics: Theory, Algorithms, and Applications for Logistics Management*, Springer Series in Operations Research and Financial Engineering, DOI 10.1007/978-1-4614-9149-1_20, © Springer Science+Business Media New York 2014

warehouses, and so on. Major sourcing decisions are also made at this point, and the typical planning horizon is a few years.

2. **Inventory positioning:** This includes identifying stocking points as well as selecting facilities that will produce to stock, and thus keep inventory, and facilities that will produce to order and hence keep no inventory.

3. **Resource allocation:** Given the structure of the logistics network and the location of stocking points, the objective in this step is to determine when and how much to produce or purchase and where and when to store inventory. These decisions require identifying the optimal tradeoff between setup costs and times, and inventory and transportation costs, taking into account production, sourcing, and warehousing capacities as well as other business rules and constraints.

In this chapter, we analyze each of these steps and provide examples of the processes involved.

20.2   Network Design

Network design determines the physical configuration and infrastructure of the supply chain. As explained in Chap. 1, network design is a strategic decision that has a long-lasting effect on the firm. Network design involves decisions related to plant and warehouse location as well as distribution and sourcing.

The supply chain infrastructure typically needs to be reevaluated due to changes in demand patterns, product mix, production processes, sourcing strategies, or the cost of running facilities. In addition, mergers and acquisitions may mandate the integration of different logistics networks.

In the discussion below, we concentrate on the following key strategic decisions:

1. determining the appropriate number of facilities such as plants and warehouses;

2. determining the location of each facility;

3. determining the size of each facility;

4. allocating space for products in each facility;

5. determining the production requirements in each plant;

6. determining sourcing requirements;

7. determining distribution.

The objective is to design or reconfigure the logistics network in order to minimize annual system-wide cost, including production and purchasing costs, inventory holding costs, facility costs (storage, handling, and fixed costs), and transportation costs, subject to a variety of service-level requirements. In this setting, the tradeoffs are clear. Increasing the number of warehouses typically yields

- an improvement in the service level due to the reduction in the average travel time to the customers,

- an increase in inventory costs due to increased safety stocks required to protect each warehouse against uncertainties in customer demands,

- an increase in overhead and setup costs,

- a reduction in outbound transportation costs: transportation costs from the warehouses to the customers,

- an increase in inbound transportation costs: transportation costs from the suppliers and/or manufacturers to the warehouses.

In essence, the firm must balance the costs of opening new warehouses with the advantages of being close to the customer. Thus, warehouse-location decisions are crucial determinants of whether the supply chain is an efficient channel for the distribution of products.

We describe below some of the issues related to data collection and the calculation of costs required for the optimization models. Some of the information provided is based on logistics textbooks such as Ballou (1992), Johnson and Wood (1986), and Robeson and Copacino (1994).

Figures 20.1 and 20.2 present two screens of a typical advance planning system (APS); the user would see these screens at different stages of optimization. One screen represents the network prior to optimization, and the other represents the optimized network.

Data Collection

A typical network configuration problem involves large amounts of data, including information on

1. the locations of customers, retailers, existing warehouses and distribution centers, manufacturing facilities, and suppliers,

2. all products, including volumes, and special transport modes (e.g., refrigerated),

3. annual demand for each product by customer location,

4. transportation rates by mode,

5. warehousing costs, including labor, inventory carrying charges, and fixed operating costs,

6. shipment sizes and frequencies for customer delivery,

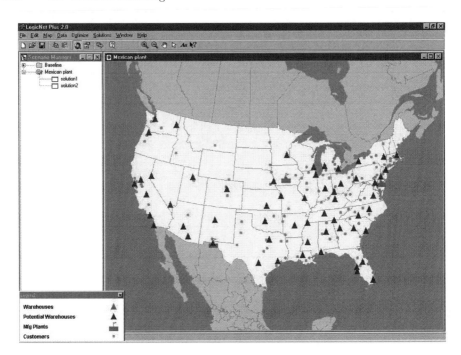

FIGURE 20.1. The APS screen representing data prior to optimization

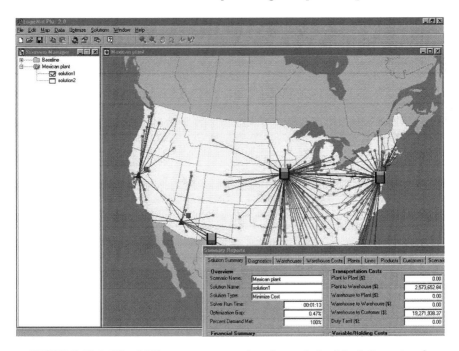

FIGURE 20.2. The APS screen representing the optimized logistics network

7. order processing costs,

8. customer service requirements and goals,

9. production and sourcing costs and capacities.

Data Aggregation

A quick look at the above list suggests that the amount of data involved in any optimization model for this problem is overwhelming. For instance, a typical soft drink distribution system has between 10,000 and 120,000 accounts (customers). Similarly, in a retail logistics network, such as Wal-Mart or JC Penney, the number of different products that flow through the network is in the thousands or even hundreds of thousands.

For that reason, an essential first step is data aggregation. This is carried out using the following procedure:

1. Customers located in close proximity to each other are aggregated using a grid network or other clustering technique. All customers within a single cell or a single cluster are replaced by a single customer located at the center of the cell or cluster. This cell or cluster is referred to as a customer zone. A very effective technique that is commonly used is to aggregate customers according to the five-digit or three-digit zip code. Observe that if customers are classified according to their service levels or frequency of delivery, they will be aggregated together by classes. That is, all customers within the same class are aggregated independently of the other classes.

2. Items are aggregated into a reasonable number of product groups, based on

   (a) Distribution pattern. All products picked up at the same source and destined to the same customers are aggregated together. Sometimes there is a need to aggregate not only by distribution pattern but also by logistics characteristics, such as weight and volume. That is, consider all products having the same distribution pattern. Within these products, we aggregate those SKUs with similar volume and weight into one product group.

   (b) Product type. In many cases, different products might simply be variations in product models or style or might differ only in the type of packaging. These products are typically aggregated together.

An important consideration, of course, is the impact on the model's effectiveness of replacing the original detailed data with the aggregated data. We address this question in two ways.

1. Even if the technology exists to solve the logistics network design problem with the original data, it may still be useful to aggregate data because our ability to forecast customer demand at the account and product levels

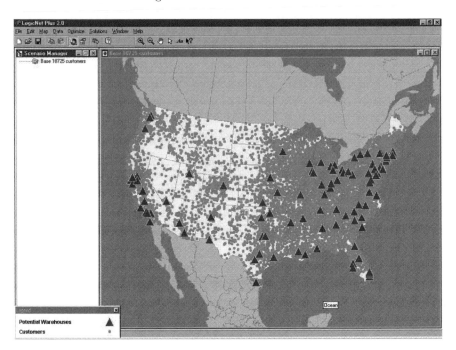

FIGURE 20.3. The APS screen representing data prior to aggregation

is usually poor. Because of the reduction in variability achieved through aggregation, forecast demand is significantly more accurate at the aggregated level.

2. Various researchers report that aggregating data into about 150–200 points usually results in no more than a 1 % error in the estimation of total transportation costs; see Ballou (1992) and House and Jamie (1981).

In practice, the following approach is typically used when aggregating the data:

- Aggregate demand points for 150–200 zones. If customers are classified into classes according to their service levels or frequency of delivery, each class will have 150–200 aggregated points.

- Make sure each zone has approximately an equal amount of total demand. This implies that the zones may be of different geographic sizes.

- Place the aggregated points at the center of the zone.

- Aggregate the products into 20–50 product groups.

Figure 20.3 presents information about 3,220 customers all located in North America, while Fig. 20.4 shows the same data after aggregation using a three-digit zip code, resulting in 217 aggregated points.

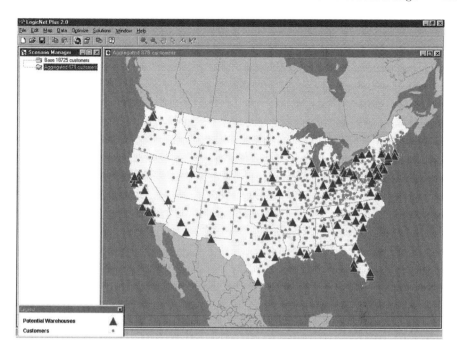

FIGURE 20.4. The APS screen representing data after aggregation

Transportation Rates

The next step in constructing an effective distribution-network design model is to estimate transportation costs. An important characteristic of most transportation rates, including truck, rail, and others, is that the rates are almost linear with distance but not with volume. We distinguish here between transportation costs associated with an internal and an external fleet.

Estimating transportation costs for company-owned trucks is typically quite simple. It involves annual costs per truck, annual mileage per truck, annual amount delivered, and the truck's effective capacity. All this information can be used to easily calculate the cost per mile per SKU.

Incorporating transportation rates for an external fleet into the model is more complex. We distinguish here between two modes of transportation: truckload, referred to as TL, and less than truckload, referred to as LTL.

In the United States, TL carriers subdivide the country into zones. Almost every state is a single zone, except for certain big states, such as Florida or New York, which are partitioned into two zones. The carriers then provide their clients with zone-to- zone table costs. This database provides the cost per mile per truckload between any two zones. For example, to calculate the TL cost from Chicago, Illinois, to Boston, Massachusetts, one needs to get the cost per mile for this pair and multiply it by the distance from Chicago to Boston. An important property of the TL cost structure is that it is not symmetric; that is, it is typically more expensive to ship a fully loaded truck from Illinois to New York than from New York to Illinois.

In the LTL industry, the rates typically belong to one of three basic types of freight rates: class, exception, and commodity. The class rates are standard rates that can be found for almost all products or commodities shipped. They are found with the help of a classification tariff that gives each shipment a rating or a class. For instance, the railroad classification includes 31 classes, ranging from 400 to 13, that are obtained from the widely used Uniform Freight Classification. The National Motor Freight Classification, on the other hand, includes only 23 classes, ranging from 500 to 35. In all cases, the higher the rating or class, the greater the relative charge for transporting the commodity. There are many factors involved in determining a product's specific class. These include product density, ease or difficulty of handling and transporting, and liability for damage.

Once the rating is established, it is necessary to identify the rate basis number. This number is the approximate distance between the load's origin and destination. With the commodity rating or class and the rate basis number, the specific rate per hundred pounds (hundred weight, or cwt) can be obtained from a carrier tariff table (i.e., a freight rate table).

The two other freight rates, namely, exception and commodity, are specialized rates used to provide either less expensive rates (exception), or commodity-specific rates (commodity). For an excellent discussion, see Johnson and Wood (1986) and Patton (1994). Most carriers provide a database file with all of their transportation rates; these databases are typically incorporated into advance planning systems.

The proliferation of LTL carrier rates and the highly fragmented nature of the trucking industry have created the need for sophisticated rating engines. An example of such a rating engine that is widely used is SMC3's RateWare (see www.smc3.com). This engine can work with various carrier tariff tables as well as SMC3's CzarLite, one of the most widely used and accepted forms of nationwide LTL zip code-based rates. Unlike an individual carrier's tariff, CzarLite offers a market-based price list derived from studies of LTL pricing on a regional, interregional, and national basis. This provides shippers with a fair pricing system and prevents any individual carrier's operational and marketing bias from overtly influencing the shipper choice. Consequently, CzarLite rates are often used as a base for negotiating LTL contracts among shippers, carriers, and third-party logistics providers.

In Fig. 20.5, we provide the LTL cost charged by one carrier for shipping 4,000 pounds as a function of the distance from Chicago. The cost is given for two classes, class 100 and class 150. As you can see, in this case, the transportation cost function is not linear with distance.

Warehouse Costs

Warehousing and distribution center costs include three main components:

1. Handling costs: These include labor and utility costs that are proportional to annual flow through the warehouse.

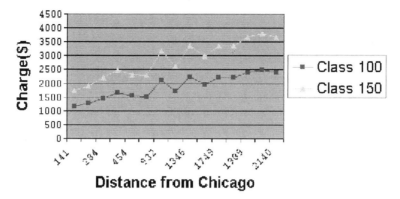

FIGURE 20.5. Transportation rates for shipping 4,000 lb

2. Fixed costs: These capture all cost components that are not proportional to the amount of material that flows through the warehouse. The fixed cost is typically proportional to warehouse size (capacity); the cost is a stepwise function of the warehouse capacity: That is, this cost is fixed in certain ranges of the warehouse size.

3. Storage costs. These represent inventory holding costs, which are proportional to average positive inventory levels.

Thus, estimating the warehouse handling costs is fairly easy, while estimating the other two cost values is quite difficult. To see this difference, suppose that during the entire year 1,000 units of product are required by a particular customer. These 1,000 units are not required to flow through the warehouse at the same time, so the average inventory level will likely be significantly lower than 1,000 units. Thus, when constructing the data for the APS, we need to convert these annual flows into actual inventory amounts over time. Similarly, annual flow and average inventory associated with this product tell us nothing about how much space is needed for the product in the warehouse. This is true because the amount of space that the warehouse needs is proportional to peak inventory, not annual flow or average inventory.

An effective way to overcome this difficulty is to utilize the inventory turnover ratio. This is defined as the annual sales divided by the average inventory level. Specifically, in our case, the inventory turnover ratio is the ratio of the total annual outflow from the warehouse to the average inventory level. Thus, the average inventory level is the total annual flow divided by the inventory turnover ratio. Multiplying the average inventory level by the inventory holding cost gives the annual storage costs. Finally, to calculate the fixed cost, we need to estimate the warehouse capacity. This is done in the next subsection.

Warehouse Capacities

Another important input to the distribution-network design model is the actual warehouse capacity. It is not immediately obvious, however, how to estimate the actual space required, given the specific annual flow of material through the warehouse. Again, the inventory turnover ratio suggests an appropriate approach. As before, the annual flow through a warehouse divided by the inventory turnover ratio allows us to calculate the average inventory level. Assuming a regular shipment and delivery schedule, such as that given in Fig. 7.1, it follows that the required storage space is approximately twice that amount. In practice, of course, every pallet stored in the warehouse requires an empty space to allow for access and handling; thus, considering this space as well as space for aisles, picking, sorting, and processing facilities, as well as automatic guided vehicles (AGVs), we typically multiply the required storage space by a factor (> 1). This factor depends on the specific application and allows us to assess the amount of space available in the warehouse more accurately. A typical factor used in practice is 3. This factor would be used in the following way. Consider a situation where the annual flow through the warehouse is 1,000 units and the inventory turnover ratio is 10.0. This implies that the average inventory level is about 100 units and, hence, if each unit takes $10\,\text{ft}^2$ of floor space, the required space for the products is $2,000\,\text{ft}^2$. Therefore, the total space required for the warehouse is about $6,000\,\text{ft}^2$.

Potential Warehouse Locations

It is also important to effectively identify potential locations for new warehouses. Typically, these locations must satisfy a variety of conditions:

- geographical and infrastructure conditions;
- natural resources and labor availability;
- local industry and tax regulations;
- public interest.

As a result, only a limited number of locations would meet all the requirements. These are the potential location sites for the new facilities.

Service-Level Requirements

There are various ways to define service levels in this context. For example, we might specify a maximum distance between each customer and the warehouse serving it. This ensures that a warehouse will be able to serve its customers within a reasonable time. Sometimes we must recognize that for some customers, such as those in rural or isolated areas, it is harder to provide the same level of service that most other customers receive. In this case, it is often helpful to define the service level as the proportion of customers whose distance to their assigned warehouse is no more than a given distance. For instance, we might require that 95 % of the customers be situated within 200 miles of the warehouses serving them.

Future Demand

As observed in Chap. 1, decisions at the strategic level, which include distribution-network design, have a long-lasting effect on the firm. In particular, decisions regarding the number, location, and size of warehouses have an impact on the firm for at least the next three to five years. This implies that changes in customer demand over the next few years should be taken into account when designing the network. This is most commonly addressed using a scenario-based approach incorporating net present value calculations. For example, various possible scenarios representing a variety of possible future demand patterns over the planning horizon can be generated. These scenarios can then be directly incorporated into the model to determine the best distribution strategy.

Model and Data Validation

The previous subsections document the difficulties in collecting, tabulating, and cleaning the data for a network-configuration model. Once this is done, how do we ensure that the data and model accurately reflect the network-design problem?

The process used to address this issue is known as model and data validation. This is typically done by reconstructing the existing network configuration using the model and collected data, and comparing the output of the model to existing data.

The importance of validation cannot be overstated. Valuable output of the model configured to duplicate current operating conditions includes all costs—warehousing, inventory, production, and transportation—generated under the current network configuration. These data can be compared to the company's accounting information. This is often the best way to identify errors in the data, problematic assumptions, modeling flaws, and so forth.

In one project we are aware of, for example, the transportation costs calculated during the validation process were consistently underestimating the costs suggested by the accounting data. After a careful review of the distribution practices, the consultants concluded that the effective truck capacity was only about 30 % of the truck's physical capacity; that is, trucks were being sent out with very little load. Thus, the validation process not only helped calibrate some of the parameters used in the model but also suggested potential improvements in the utilization of the existing network.

It is often also helpful to make local or small changes in the network configuration to see how the system estimates their impact on costs and service levels. Specifically, this step involves positing a variety of what-if questions. This includes estimating the impact of closing an existing warehouse on system performance. Or, to give another example, it allows the user to change the flow of material through the existing network and see the changes in the costs. Often, managers have good intuition about what the effect of these small-scale changes on the system should be, so they can more easily identify errors in the model. Intuition about the effect of radical redesign of the entire system is often much less reliable. To summarize, the model-validation process typically involves answering the following questions:

- Does the model make sense?

- Are the data consistent?

- Can the model results be fully explained?

- Did you perform sensitivity analysis?

Validation is critical for determining the validity of the model and data, but the process has other benefits. In particular, it helps the user make the connection between the current operations, which were modeled during the validation process, and possible improvements after optimization.

Key Features of a Network-Configuration APS

One of the key requirements of any advance planning system for network design is flexibility. In this context, we define flexibility as the ability of the system to incorporate a large set of preexisting network characteristics. Indeed, depending on the particular application, a whole spectrum of design options may be appropriate. At one end of this spectrum is the complete reoptimization of the existing network. This means that each warehouse can be either opened or closed and all transportation flows can be redirected. At the other end of the spectrum, it may be necessary to incorporate the following features in the optimization model:

1. Customer-specific service-level requirements.

2. Existing warehouses. In most cases, warehouses already exist and their leases have not yet expired. Therefore, the model should not permit the closing of these warehouses.

3. Expansion of existing warehouses. Existing warehouses may be expandable.

4. Specific flow patterns. In a variety of situations, specific flow patterns (e.g., from a particular warehouse to a set of customers) should not be changed, or perhaps more likely, a certain manufacturing location does not or cannot produce certain SKUs.

5. Warehouse-to-warehouse flow. In some cases, material may flow from one warehouse to another warehouse.

6. Production and bill of materials. In some cases, assembly is required and needs to be captured by the model. For this purpose, the user needs to provide information on the components used to assemble finished goods. In addition, production information down to the line level can be included in the model.

It is not enough for the advance planning system to incorporate all of the features described above. It also must have the capability to deal with all these issues with little or no reduction in its effectiveness. The latter requirement is directly related to the so-called robustness of the system. This stipulates that the relative quality of the solution generated by the system (i.e., cost and service level) should be independent of the specific environment, the variability of the data, or the particular setting. If a particular advance planning system is not robust, it is difficult to determine how effective it will be for a particular problem.

20.3 Strategic Safety Stock

Important questions when designing the logistics network and when managing inventory in a complex supply chain are where to keep safety stock and, similarly, which facilities should produce to stock and which should produce to order? The answers to these questions clearly depend on the desired service level, the supply network, lead times as well as a variety of operational issues and constraints. Thus, our focus is on a strategic model that allows the firm to position safety stock effectively in its supply chain.

To illustrate the tradeoffs and the impact of strategically positioning safety stock in the supply chain, consider the following example.

ElecComp Inc. is a large contract manufacturer of circuit boards and other high-tech parts. The company sells about 27,000 high-value products whose life cycle is relatively short. Competition in this industry forces ElecComp to commit to short lead times to its customers; this committed service time to the customers is typically much shorter than manufacturing lead time. Unfortunately, the manufacturing process is quite complex, including a complex sequence of assemblies at different stages.

Because of the long manufacturing lead time and the pressure to provide customers with a short response time, ElecComp kept inventory of finished products for many of its SKUs. Thus, the company managed its supply chain based on long-term forecast, the so-called push-based supply chain strategy. This make-to-stock environment required the company to build safety stock and resulted in huge financial and shortage risks.

Executives at ElecComp had long recognized that this push-based supply chain strategy was not the appropriate strategy for their supply chain. Unfortunately, because of the long lead time, a pull-based supply chain strategy, in which manufacturing and assembly are done based on realized demand, was not appropriate either.

Thus, ElecComp focused on developing a new supply chain strategy whose objectives are

1. reducing inventory and financial risks,

2. providing customers with competitive response times.

This could be achieved by

- determining the optimal location of inventory across the various stages of the manufacturing and assembly process,

- calculating the optimal quantity of safety stock for each component at each stage.

Thus, the focus of redesigning ElecComp's supply chain was on a hybrid strategy in which a portion of the supply chain is managed based on push, that is, a make-to-stock environment, while the remaining portion of the supply chain is managed

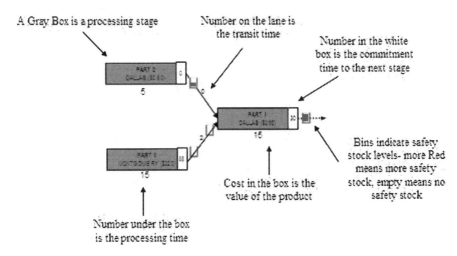

FIGURE 20.6. How to read the diagrams

based on pull, or a make-to-order strategy. Evidently, the supply chain stages that produce to stock will be the locations where the company keeps safety stock, while the make-to-order stages will keep no stock at all. Hence, the challenge was to identify the location in the supply chain in which the strategy is switched from a push-based, namely, a make-to-stock strategy, to a pull-based, that is, a make-to-order supply chain strategy. This location is referred to as the push–pull boundary.

ElecComp developed and implemented the new push–pull supply chain strategy, and the impact was dramatic! For the same customer lead times, safety stock was reduced by 40–60%, depending on product line. More importantly, with the new supply chain structure, ElecComp concluded that they could cut lead times to their customers by 50% and still enjoy a 30% reduction in safety stock.

Below we describe how this was achieved for a number of product lines.

An Illustrative Example

To understand the analysis and the benefit experienced by ElecComp, consider Fig. 20.6, in which a finished product (Part I) is assembled in a Dallas facility from two components, one produced in the Montgomery facility and one in a different facility in Dallas. Each box provides information about the value of the product produced by that facility; numbers under each box are the processing time at that stage; bins represent safety stock. Transit times between facilities are provided as well. Finally, each facility provides a committed response time to the downstream facilities. For instance, the assembly facility quotes 30 days' response time to its customers. This implies that any order can be satisfied in no more than 30 days. The Montgomery facility quotes an 88-day response time to the assembly facility. As a result, the assembly facility needs to keep inventory of finished products in order to satisfy customer orders within its 30 days' committed service time.

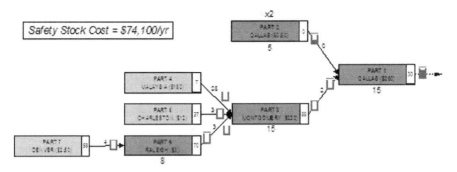

FIGURE 20.7. Current safety stock locations

Observe that if somehow ElecComp can reduce the committed service time from the Montgomery facility to the assembly facility from 88 days to say 50 or perhaps 40 days, the assembly facility will be able to reduce its finished goods inventory while the Montgomery facility will need to start building inventory. Of course, ElecComp's objective is to minimize system-wide inventory and manufacturing costs; this is precisely what Inventory Analyst^{TM} from LogicTools allows users to do. By looking at the entire supply chain, the tool determines the appropriate inventory level at each stage.

For instance, if the Montgomery facility reduces its committed lead time to 13 days, then the assembly facility does not need any inventory of finished goods. Any customer order will trigger an order for Parts II and III. Part II will be available immediately, since the facility producing Part II holds inventory, while Part III will be available at the assembly facility in 15 days: 13 days' committed response time by the manufacturing facility plus 2 days' transportation lead time. It takes another 15 days to process the order at the assembly facility; therefore, the order will be delivered to the customers within the committed service time. Thus, in this case, the assembly facility produces to order, that is, a pull-based strategy, while the Montgomery facility needs to keep inventory and hence is managed based on push, that is, a make-to-stock strategy.

Now that the tradeoffs are clear, consider the product structure depicted in Fig. 20.7. Light boxes (parts 4, 5, and 7) represent outside suppliers, whereas dark boxes represent internal stages within ElecComp's supply chain. Observe that the assembly facility commits a 30-day response time to the customers and keeps inventory of finished goods. More precisely, the assembly facility and the facility manufacturing Part II both produce to stock. All other stages produce to order.

Figure 20.8 depicts the optimized supply chain that provides customers with the same 30-day response time. Observe that by adjusting the committed service time of various internal facilities, the assembly system starts producing to order and keeps no finished goods inventory. On the other hand, the Raleigh and Montgomery facilities need to reduce their committed service time and hence keep inventory.

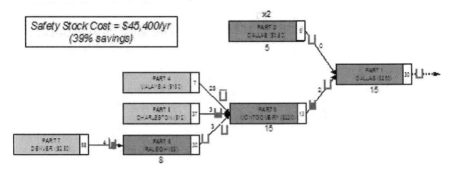

FIGURE 20.8. Optimized safety stock locations

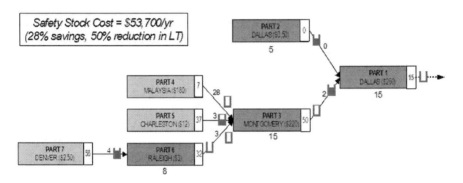

FIGURE 20.9. Optimized safety stock with reduced lead time

So where is the push and where is the pull in the optimized strategy? Evidently, the assembly facility and the Dallas facility that produces Part II both operate now in a make-to-order fashion—a pull strategy—while the Montgomery facility operates in a make-to-stock fashion, a push-based strategy. The impact on the supply chain is a 39 % reduction in safety stock!

At this point it was appropriate to analyze the impact of a more aggressive quoted lead time to the customers. That is, ElecComp executives considered reducing quoted lead times to the customers from 30 to 15 days. Figure 20.9 depicts the optimized supply chain strategy in this case. The impact was clear. Relative to the baseline (Fig. 20.7), inventory was down by 28 %, while response time to the customers is halved.

Finally, Figs. 20.10 and 20.11 present a more complex product structure. Figure 20.10 provides information about the supply chain strategy before optimization, and Fig. 20.11 depicts the supply chain strategy after optimizing the push–pull boundary as well as inventory levels at different stages in the supply chain. Again, the benefit is clear. By correctly selecting which stage is going to produce to order and which is producing to stock, ElecComp reduced the inventory cost by more than 60 % while maintaining the same quoted lead time to the customers.

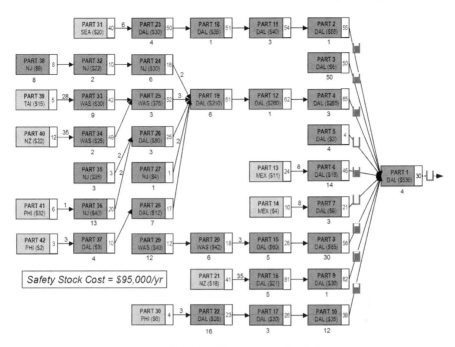

FIGURE 20.10. Current supply chain

Summary

Using a multistage inventory optimization technology (Inventory Analyst™ from LogicTools), ElecComp was able to significantly reduce inventory cost while maintaining and sometimes significantly decreasing quoted service times to the customers. This was achieved by

1. Identifying the push–pull boundary; that is, identifying supply chain stages that should operate in a make-to-stock fashion and hence keep safety stock. The remaining supply chain stages operate in a make-to-order fashion and thus keep no inventory. This is done by pushing inventory to less costly locations in the supply chain.

2. Taking advantage of the risk-pooling concept. This concept suggests that demand for a component used by a number of finished products has smaller variability and uncertainty than that of the finished goods.

3. Replacing traditional supply chain strategies that are typically referred to as sequential, or local, optimization by a globally optimized supply chain strategy. In a sequential, or local, optimization strategy, each stage tries to optimize its profit with very little regards to the impact of its decisions on other stages in the same supply chain. On the other hand, in a global supply chain strategy, one considers the entire supply chain and identifies strategies for each stage that will maximize the supply chain performance.

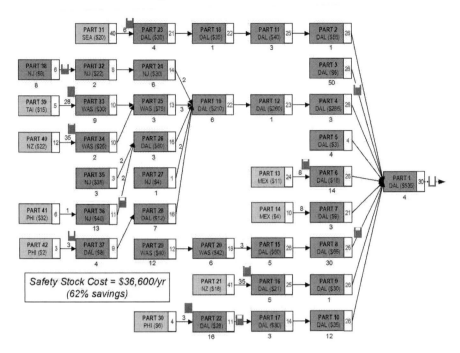

FIGURE 20.11. Optimized supply chain

To better understand the impact of the new supply chain paradigm employed by ElecComp, consider Fig. 20.12, where we plot the total inventory cost against the quoted lead time to the customers. The upper tradeoff curve represents the traditional relationship between cost and quoted lead time to the customers. This curve is a result of locally optimizing decisions at each stage in the supply chain. The lower tradeoff curve is the one obtained when the firm globally optimizes the supply chain by locating the push–pull boundary correctly.

Observe that this shift of the tradeoff curve, due to optimally locating the push–pull boundary, implies the following:

1. For the same quoted lead time, the company can significantly reduce cost, or

2. For the same cost, the firm can significantly reduce lead time.

Finally, notice that the curve representing the traditional relationship between cost and customer quoted lead time is smooth, while the new tradeoff curve representing the impact of optimally locating the push–pull boundary is not, with jumps in various places. These jumps represent situations in which the location of the push–pull boundary changes and significant cost savings are achieved.

Our experience is that those employing the new supply chain paradigm, like Elec-Comp, typically chose a supply chain strategy that both reduce cost andcustomer

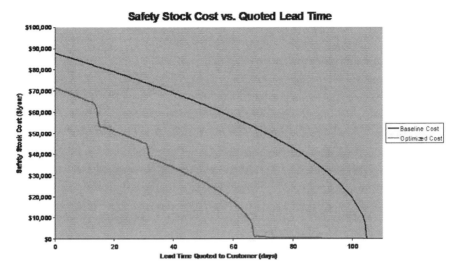

FIGURE 20.12. Global vs. local optimization

quoted lead time. This strategy allows ElecComp to satisfy demand much faster than their competitors and develop a cost structure that enables competitive pricing.

20.4   Resource Allocation

Supply chain master planning is defined as the process of coordinating and allocating production and distribution strategies and resources to maximize profit or minimize systemwide cost. In this process, the firm considers forecast demand for the entire planning horizon, such as the next 52 weeks, as well as safety stock requirements. The latter are determined, for instance, based on models similar to the one analyzed in the previous section.

The challenge of allocating production, transportation, and inventory resources in order to satisfy demand can be daunting. This is especially true when the firm is faced with seasonal demand, limited capacities, competitive promotions, or high volatility in forecasting. Indeed, decisions such as when and how much to produce, where to store inventory, and whether to lease additional warehouse space may have enormous impact on supply chain performance.

Traditionally, the supply chain planning process was performed manually with a spreadsheet and was done by each function in the company independently of other functions. That is, the production plan would be determined at the plant, independently from the inventory plan, and would typically require the two plans to be somehow coordinated at a later time. This implies that divisions typically end up "optimizing" just one parameter, usually production costs.

In modern supply chains, however, this sequential process is replaced by a process that takes into account the interaction between the various levels of the supply

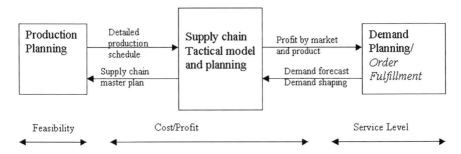

FIGURE 20.13. The extended supply chain: from manufacturing to order fulfillment

chain and identifies a strategy that maximizes supply chain performance. This is referred to as global optimization, and it necessitates the need for an optimization-based advance planning system. These systems, which model the supply chain as large-scale mixed integer linear programs, are analytical tools capable of considering the complexity and dynamic nature of the supply chain.

Typically, the output from the tool is an effective supply chain strategy that coordinates production, warehousing, transportation, and inventory decisions. The resulting plan provides information on production quantities, shipment sizes, and storage requirements by product, location, and time period. This is referred to as the supply chain master plan.

In some applications, the supply chain master plan serves as an input for a detailed production scheduling system. In this case, the production scheduling system employs information about production quantities and due dates received from the supply chain master plan. This information is used to propose a detailed manufacturing sequence and schedule. This allows the planner to integrate the back end of the supply chain—manufacturing and production—and the front end of the supply chain—demand planning and order replenishment; see Fig. 20.13. This diagram illustrates an important issue. The focus of order replenishment systems, which are part of the pull portion of the supply chain, is on the service level. Similarly, the focus of the tactical planning, that is, the process by which the firm generates a supply chain master plan, which is in the push portion of the supply chain, is on cost minimization or profit maximization. Finally, the focus in the detailed manufacturing scheduling portion of the supply chain is on feasibility. That is, the focus is on generating a detailed production schedule that satisfies all production constraints and meets all the due-date requirements generated by the supply chain master plan.

Of course, the output from the tactical planning process, namely, the supply chain master plan, is shared with supply chain participants to improve coordination and collaboration. For example, the distribution center managers can now better use this information to plan their labor and shipping needs. Distributors can share plans with their suppliers and customers in order to decrease costs for all partners in the supply chain and promote savings. Specifically, distributors can realign territories to better serve customers, store adequate amounts of inventory at the customer site, and coordinate overtime production with suppliers.

In addition, supply chain master planning tools can identify potential supply chain bottlenecks early in the planning process, allowing the planner to answer questions such as

- Will leased warehouse space alleviate capacity problems?

- When and where should the inventory for seasonal or promotional demand be built and stored?

- Can capacity problems be alleviated by rearranging warehouse territories?

- What impact do changes in the forecast have on the supply chain?

- What will be the impact of running overtime at the plants or outsourcing production?

- What plant should replenish each warehouse?

- Should the firm ship by sea or by air? Shipping by sea implies long lead times and therefore requires high inventory levels. On the other hand, using air carriers reduces lead times and hence inventory levels but significantly increases transportation cost.

- Should we rebalance inventory between warehouses or replenish from the plants to meet unexpected regional changes in demand?

Another important capability that tactical planning tools have is the ability to analyze demand plans and resource utilization to maximize profit. This enables balancing the effect of promotions, new product introductions, and other planned changes in demand patterns and supply chain costs. Planners now are able to analyze the impact of various pricing strategies as well as identify markets, stores, or customers that do not provide the desired profit margins.

A natural question is when should one focus on cost minimization and when on profit maximization? While the answer to this question may vary from instance to instance, it is clear that cost minimization is important when the structure of the supply chain is fixed or at times of a recession and therefore oversupply. In this case, the focus is on satisfying all demand at the lowest cost by allocating resources effectively. On the other hand, profit maximization is important at time of growth, namely, at the time when demand exceeds supply. In this case, capacity can be limited because of the use of limited natural resources or because of expensive manufacturing processes that are hard to expand, as is the case in the chemical and electronic industries. In these cases, deciding who to serve and for how much is more critical than costs savings.

Finally, an effective supply chain master planning tool must also be able to help the planners improve the accuracy of the supply chain model. This, of course, is counterintuitive since the accuracy of the supply chain master planning model depends on the accuracy of the demand forecast that is an input to the model. However, notice that the accuracy of the demand forecast is typically time-dependent. That is, the accuracy of forecast demand for the first few time periods, for instance, the first 10 weeks, is much higher than the accuracy of demand forecast for later time periods. This suggests that the planner should model the early portion of the demand forecast at a great level of detail, that is, apply weekly demand information. On the other hand, demand forecasts for later time periods are not as accurate, and hence the planner should model the later demand forecast month by month or by groups of 2–3 weeks each. This implies that later demand forecasts are aggregated into longer time buckets, and hence, due to the risk-pooling concept, the accuracy of the forecast improves.

In summary, supply chain master planning helps address fundamental tradeoffs in the supply chain such as setup cost versus holding costs or production lot sizes versus capacities. It takes into account supply chain costs such as production, supply, warehousing, transportation, taxes, and inventory, as well as capacities and changes in the parameters over time.

This example illustrates how supply chain master planning can be used dynamically and consistently to help a large food manufacturer manage the supply chain. The food manufacturer makes production and distribution decisions at the division level. Even at the division level, the problems tend to be large-scale. Indeed, a typical division may include hundreds of products, multiple plants, many production lines within a plant, multiple warehouses (including overflow facilities), bill-of-material structures to account for different packaging options, and a 52-week demand forecast for each product for each region. The forecast accounts for seasonality and planned promotions. The annual forecast is important because a promotion late in the year may require production resources relatively early in the year. Production and warehousing capacities are tight, and products have a limited shelf life that needs to be integrated into the analysis. Finally, the scope of the plan spans many functional areas, including purchasing, production, transportation, distribution, and inventory management. Traditionally, the supply chain planning process was performed manually with a spreadsheet and was done by each function in the company. That is, the production plan would be done at the plant, independently from the inventory plan, and would typically require the two plans to be somehow coordinated at a later time. This implies that divisions typically end up "optimizing" just one parameter, usually production costs. The tactical planning APS introduced in the company allows the planners to reduce systemwide cost and better utilize resources such as manufacturing and warehousing. Indeed, a detailed comparison of the plan generated by the tactical tool with the spreadsheet strategy suggests that the optimization-based tool is capable of reducing total costs across the entire supply chain. See Fig. 20.14 for illustrative results.

20.5 Summary

Optimizing supply chain performance is difficult because of conflicting objectives, demand and supply uncertainties, and supply chain dynamics. However, through network planning, which combines network design, inventory positioning, and

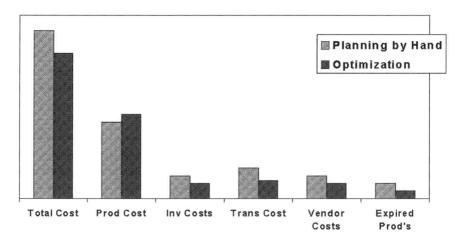

FIGURE 20.14. Comparison of manual vs. optimized scenarios

	Network design	Inventory positioning	Resource allocation
Decision focus	Infrastructure	Safety stock	Production distribution
Planning horizon	Years	Months	Months
Aggregation level	Family	Item	Classes
Frequency	Yearly	Monthly/ weekly	Monthly/ weekly
Return on investment	High	Medium	Medium
Implementation	Very short	Short	Short
User	Very few	few	few

TABLE 20.1. Network planning characteristics

resource allocation, the firm can globally optimize supply chain performance. This is achieved by considering the entire network, taking into account production, warehousing, transportation, and inventory costs, as well as service-level requirements.

Table 20.1 summarizes the key dimensions of each of the planning activities, network design, inventory positioning/management, and supply chain master planning. The table shows that network design involves long-term plans, typically over years, is done at a high level and can yield high returns. The planning horizon for supply chain master planning is months or weeks, the frequency of replanning is high (e.g., every week), and it typically delivers quick results as well. Inventory planning is focused on short-term uncertainty in demand, lead time, processing time, or supply. The frequency of replanning is high, such as monthly planning to determine appropriate safety stock based on the latest forecast and forecast error. Inventory planning can also be used more strategically to identify locations in the supply chain where the firm keeps inventory, as well as to identify stages that produce to stock and those that produce to order.

20.6   Exercises

Exercise 20.1. Consider n independent and identically distributed random variables, $X_1, X_2, \ldots, X_n$. Let $S_n = \frac{1}{n} \sum_{i=1}^{n} X_i$. Find the variance of the random variable S_n as a function of the variance of X_i.

21

A Case Study: School Bus Routing

21.1 Introduction

We now turn our attention to a case study in transportation logistics. We highlight particular issues that arise when implementing an optimization algorithm in a real-life routing situation. The case concerns the routing and scheduling of school buses in the five boroughs of New York City.

Many of the vehicle routing problems we have discussed so far (see Part II) have been simplified versions of the usually more complex problems that appear in practice. Typically, a vehicle routing problem will have many constraints on the types of routes that can be constructed, including multiple vehicle types, time and distance constraints, complex restrictions on what items can be in a vehicle together, and so forth. The problems that appear in the context of school bus routing and scheduling could be characterized as the most difficult types of vehicle routing problems since they have aspects of all these constraints. This is the problem we will consider here.

School bus routing and scheduling is an area where, in general, computerized algorithms can have a large impact. User-friendly software that call routing and scheduling algorithms at the click of a button and that result in workable solutions can greatly affect the day-to-day operations of a dispatching unit. With increasingly affordable high-speed computing power in desktop computers and the possibility of displaying geographic information on-screen, it is not surprising that many communities are using expert systems to perform the daunting task of routing and scheduling their school buses. In most cases, this has led to improved solutions in fractions of the time that was previously required.

D. Simchi-Levi et al., *The Logic of Logistics: Theory, Algorithms, and Applications for Logistics Management*, Springer Series in Operations Research and Financial Engineering, DOI 10.1007/978-1-4614-9149-1_21, © Springer Science+Business Media New York 2014

Unfortunately, providing workable solutions for such an application as this is not as simple as just "clicking" the right button. Anyone who has been involved in a real-life optimization application knows that much discussion is involved in determining what the problem *is* and how we are to "solve" it. In this chapter, we concern ourselves with some of the details that make it possible to put modeling assumptions and algorithms into action.

21.2   The Setting

The New York City school system is composed of 1,069 schools and approximately one million students. Most of these students either walk to school or are given public transportation passes. About 125,000 students ride school buses that are leased by the Board of Education. The majority, or about 83,000, of these are classified as General Education students. These students walk to their neighborhood bus stop in the morning and wait for a bus to take them to school. In the afternoon, a bus takes them from their school and drops them off back at their bus stop. The rest of these students with particular needs, classified as Special Education, are picked up and dropped off directly at their homes.

This is one distinction that makes the transportation policies governing Special Ed students fundamentally different from those of General Ed students. Another fundamental difference is that, in many cases, Special Ed students enroll in schools with specific services and therefore may be bused over long distances. General Ed students usually go to schools only a few miles from their homes and almost exclusively to schools within the same borough. In addition, Special Ed students, such as wheelchair-bound students, are transported in specially designed vehicles with much smaller carrying capacities.

For General Ed student transportation, currently the Board of Education leases approximately 1,150 buses a year. Many companies bid for the contract to transport the students; currently, the companies winning contracts are responsible for designing the routes. Independent of the company, the leasing cost to the Board is approximately $80,000 annually for each bus (and driver). The total yearly budget for General Education student transportation alone is therefore close to $100 million.

The routing of Special Education students is done differently. Using colored pins and large maps placed on walls, a team of inspectors/routers at the Board of Education Office of Pupil Transportation mark the students' homes and schools. Then, using their knowledge of the geography and street conditions acquired through their many years of work, they literally string pins together to form routes. Although the inspectors clearly do this well, this is very time-consuming. For example, a group of five people took approximately three months to manually generate routes just for the Borough of Manhattan.

Several years ago, the New York City Board of Education appropriated funds to develop a computerized system, called CATS (computer-assisted transportation system). This system is supposed to help in the design of routes for both the General and Special Ed students. The project consists of three phases, discussed next.

Phase I: Replicate the pinning-and-stringing approach on a computer. The purpose of this phase is to emulate on the computer screen what was previously done with maps, pins, and string. First, a database is needed to keep track of all relevant student and school information. The student data consist of address, bus stop, and school. For each school, the data consist of an address, as well as starting and ending times for all sessions. This makes data easily retrievable and updatable, and provides some of the basic information that is needed for routing and scheduling. In addition to the database, a method of generating maps on the computer is needed as well; this is the geographic information system (GIS). These systems, widely available only in the last few years, truly offer a new dimension to many decision-support systems. With this software, color-coded objects designating students or schools can easily be displayed on a computer screen. This enables the user to visualize the relative locations of important points. In addition, the user can "click and drag" with a mouse and get information about the area outlined. This information can include U.S. Census data such as number of households, median age, income, etc. More importantly, in this application, by designating two points, the GIS can calculate exact locations (latitude and longitude coordinates) and also the distance between the two points along the street network. By "stringing" together a series of points, the software can give the total distance traveled. When this phase is completed, inspectors currently designing Special Ed routes will be able to "click" on bus stops with a mouse and "string" them together on the computer screen. This is the method called "blocking and stringing."

Phase II: Extend the functionality developed in Phase I to the General Education stop-to-school service. The goal is to create a system whereby one could construct routes for the General Ed population on the computer screen. For example, by choosing a set of schools with a mouse, the pertinent bus stops (those with students going to the set of schools) are highlighted. The inspector can then string together the stops and schools to form a route directly on the computer screen, or again let the computer determine a good route through the stops. The immediate *visualization* of a possible solution (routes) along with relevant statistics (bus load, total travel time, total students picked up) makes it much easier to check the feasibility of the routes. This alone considerably simplifies the task of building efficient routes.

Phase III: Create an optimization module. The aim here is to build software that uses the student and school data and the GIS to generate efficient bus routes and schedules meeting existing transportation policies. The software should include subroutines that check the feasibility of suggested routes or design routes for any subset of the population, be it a school, a district, a borough, or the entire city. This is the phase in which we are the most interested.

We present here a range of issues related to the development of this optimization module (Phase III) and to the problem of routing buses through the New York City streets. We focus on routing the General Ed students; the routing of Special Ed students is currently being done at the Office of Pupil Transportation using the "blocking-and-stringing" approach.

In Sect. 21.3, we give a short summary of some of the important papers that have appeared in the literature in the area of school bus routing and scheduling and related vehicle routing problems. In Sect. 21.4, we present details of the school bus routing and scheduling problem in Manhattan. In Sect. 21.5, we give a brief overview of the methodologies we used to estimate distances, travel times, and the pickup and dropoff times.

When a computerized system for this problem is being designed, it is important to consider the following questions. First, is it possible to design an algorithm that will generate quality solutions in a reasonable amount of computing time? Second, are routes constructed by the computerized system truly *driveable*? Third, what is the best way to make these computerized algorithms of use to the people designing the routes? To answer the first two questions, we designed a school bus routing and scheduling algorithm and ran it on the Manhattan data. The algorithm is presented in Sect. 21.6. To answer the third question, in Sect. 21.8, we discuss some of the ways in which a computerized system for school bus routing can be made more interactive. In Sect. 21.9, we present results on the Manhattan data.

21.3   Literature Review

In the operations research literature, we find quite a few references to the problem as well as many different solution techniques. A standard way the school bus routing and scheduling problem has been analyzed is to decompose it into two problems: a route-generation problem, where simple routes are designed (usually with only one school); and a route scheduling problem, where these routes are linked to form longer routes (routes that visit more than one school).

As early as 1969, Newton and Thomas looked at a bus routing problem for a single school. Using some of the first local improvement procedures for vehicle routing problems, they designed a tour through all the bus stops and then partitioned it into smaller feasible routes that each could be covered by a bus.

In 1972, Angel et al. considered a clustering approach to generating routes. First, bus stops are grouped by their proximity using a clustering algorithm. Then an attempt is made to find minimum-length routes through these clusters in such a way that the constraints are satisfied. Finally, some clusters are merged if this is feasible. The algorithm was applied to an instance consisting of approximately 1,500 students and 5 schools in Indiana.

In 1972, Bennett and Gazis considered the problem of generating routes. They modified the savings algorithm of Clarke and Wright (1964) (see Sect. 17.2). They also experimented with different objective functions, such as minimizing total student-miles. The problem considered had 256 bus stops and approximately 30 routes in Toms River, New Jersey.

In 1979, Bodin and Berman used a 3-opt procedure to generate an initial traveling salesman tour that is then partitioned into feasible routes. This algorithm uses two additional components: a lookahead feature and a bus stop splitter. The lookahead feature allows the initial order to be changed slightly. The bus stop splitter allows a bus stop to be split into smaller bus stops. Two problems were studied. One dealt with a school district in a densely populated suburban area with 13,000 students requiring bus transportation each day and 25 schools. A second district, also in a suburban area, had 4,200 students transported.

In 1984, Swersey and Ballard addressed only the problem of scheduling a set of routes that had already been designed. Given a set of routes that delivered all students from their bus stops to their schools, the authors devised a method to find the minimum number of buses that could "cover" these routes. This scheduling problem can be formulated as a difficult integer program. The authors used some simple cutting planes to solve it heuristically. The size of the problem considered was approximately 30–38 buses and 100 routes.

Finally, in 1986, Desrosiers et al. studied a bus routing problem in Montréal, Canada. Using several techniques, depending on whether the stops were in rural or urban areas, they generated a set of routes. To schedule them, they formulated the problem as an integer program and solved it using a column-generation approach. The problem solved had 60 schools and 20,000 students.

21.4   The Problem in New York City

Depending on how generally it is formulated, the school bus routing and scheduling problem can take many forms. In its most general form, the problem consists of a set of students distributed in a region who have to be brought to and from their schools every school day. The problem consists of determining bus stop locations, assigning students to bus stops, and finally routing and scheduling the buses so as to minimize the total operating cost while following all transportation guidelines. The difficulty, of course, is that each of these subproblems is dependent and therefore should be looked at simultaneously. That is, any determination of bus stop locations, and who gets assigned to each, clearly has an impact on the routes and schedules of the buses. Hence, an integrated approach is required to avoid suboptimality. However, due to the complexity and the size of the problem, this has historically never been attempted. In addition, often it is not necessarily possible to reoptimize all aspects of the problem, such as bus stop locations or assignments.

To understand why this problem is so complex, consider, for instance, the bus stop location problem on its own. There are numerous constraints and requirements: No more than a certain number of students can be assigned to the same bus stop; bus stops cannot be within a certain distance of each other; each student must be within a short walk of the bus stop and must not cross a major thoroughfare; and so on.

In our case, the Board of Education decided that the bus stops that are currently being used will remain in use. Thus, the position of the bus stops as well as which students are assigned to each were assumed fixed. These stops satisfy all the requirements mentioned above. Our routing and scheduling problem thus starts with a set of bus stops, each with a particular number of students assigned to it destined for a particular school. Each school has starting and ending times for each session. In addition to bus stop and school data, it is assumed that the distance and travel time between any two points in the area are readily available. This issue will be discussed in more detail in Sect. 21.5.

We formally define a *route* as follows. A route is a sequence of stops and possibly several schools that can be feasibly driven by one bus. For example, routes for the morning problem always start with a pickup at a stop and end with dropoff at a school. In contrast, an afternoon route always starts with a pickup at a school and ends with a dropoff at a stop.

The goal is to design a set of minimum-cost routes satisfying all existing transportation guidelines. The major cost component to the Board of Education is the cost of leasing each bus and driver, and hence the objective is essentially to minimize the number of buses needed to feasibly transport the students. Clearly, safety is the first consideration, and it is the view of the Board of Education that bus routes that meet all transportation guidelines provide a high level of safety. The rest is up to the drivers.

Route feasibility is the most complex aspect of the problem. There are numerous side constraints. First, the bus can hold only a limited number of students at one time (*capacity constraint*). Second, each student must not be on the bus for more than a specific amount of time and/or distance (*time* or *distance constraint*). This is motivated by the simple observation that the less time spent on the bus, the safer and more desirable it is for the students. And finally, there are restrictions on the time a bus can arrive at a school in the morning, and on the time a bus can leave the school in the afternoon (*time-window constraints*). In many school bus routing and scheduling problems, transportation policies specify that students from different schools not be put on the same bus at the same time; that is, no *mixed loads* are allowed. Clearly, allowing mixed loads provides increased flexibility and therefore can lead to savings in cost. In New York City, for the most part, mixed loads are allowed. We list here the primary constraints. There are several other constraints, which we talk about in Sect. 21.7.

We will deal exclusively here with the problem of delivering the students to their school in the *morning*. Researchers have noted that this problem is usually more critical than the *afternoon* problem for two reasons. First, in the afternoon, the time windows are usually less constraining. For example, in Manhattan (in the morning), school starting times fall between 7:30 and 9:00 a.m. That gives roughly a one-and-a-half-hour time window to pick up all students and take them to their schools. In the afternoon, schools end at times over a wider range: anywhere between 1:00 and 4:15 p.m. Second, traffic congestion is usually higher in the morning hours than in the afternoon hours when the students are being bused. Therefore, it is very likely more buses will be needed in the morning than in the afternoon. Indeed, our computational experiments reported in Sect. 21.9 verify that this is true in Manhattan. Note that the solution found in the morning *cannot* be simply replicated in the afternoon, that is, having each bus travel the same route as in the morning but in the opposite direction. This is true since the sequencing of school ending times in the afternoon is different from the sequencing of school starting times; therefore, schools visited in one order in the morning cannot always be visited in the same or opposite order in the afternoon.

For the morning problem in Manhattan, the specific problem parameters are given below. During the 1992–1993 academic year, 4,619 students were transported by school buses from 838 bus stops to 73 schools. The constraints were as follows.

- *Vehicle capacity constraint.* At most 66 students can be on the bus at one time.

- *Distance constraint.* Each student cannot be on the bus for more than 5 miles.

- *Time-window constraints.* Buses must arrive at a school no earlier than 25 min before and no later than 5 min before the start of school.

- The earliest pickup must not be before 7:00 a.m.

- Mixed loads are allowed.

The 5-mile distance constraint is not applied uniformly to all students; students in District 6 (upper Manhattan) are often transported out of their district due to overcrowding. Therefore, since this involves longer trips, sometimes traversing most of the island, the 5-mile constraint is not applied to these students. Approximately 36% of the students in our application were in this group.

The Manhattan school bus routing problem presents many challenges. First of all, the number of bus stops and schools is much larger than those encountered in most vehicle routing applications. Second, there are many difficulties involved in calculating accurate distances and travel times in New York City. We now consider these two points.

21.5   Distance and Time Estimation

To accurately estimate distances, one needs a precise geographic representation of the area. This is achieved using a geographic information system (GIS,) which is based on data files built from satellite photographs. These files store geographic objects, such as streets, highways, parks, and rivers, that can be presented on a computer screen. An important feature is the ability to calculate the exact latitudes and longitudes of any point. When the GIS is given a street address, the process of *geocoding* returns the coordinates of the address with very high accuracy. Having these coordinates makes it easy to calculate "as the crow flies," or "Euclidean," distances. Some GISs also have the capability of calculating exact road network distances, that is, the distance between two points on the actual street network, sometimes even taking into account one-way streets.

The Office of Pupil Transportation at the Board of Education uses a GIS called MapInfo for Windows. The MapInfo version used by the City does not have a street network representation of New York City. However, such a network has been developed by a subcontractor; therefore, accurate shortest distances between any two points along the street network are readily available. The current version also takes into account one-way streets. Although incorporating one-way street information may seem like a trivial task, it turned out to be very difficult. We believe most current geographic information systems are highly inaccurate with regard to one-way streets and are probably unusable without substantial error checking. The New York City Department of Transportation does not keep the information in an easily retrievable format. We had to resort to checking the *one-way street sign* database at the NYC DOT to reconstruct accurate information about one-way streets. Inevitably, the data collection and error checking were extremely time-consuming.

Estimating accurate travel times in New York City is probably the trickiest part of the problem. As described above, a GIS with a street network representation simplifies the calculation of street distances. In addition, in the GIS, each data structure corresponding to a street segment has space to store the average travel speed and/or travel time along the segment. These estimates would make it possible to calculate travel times along any path. The difficulty lies, of course, in determining these travel speeds.

Most existing vehicle routing implementations that we are aware of use a fixed travel speed throughout the area of interest. Travel times are then determined by simply dividing the distance traveled by this universal speed. This method is most likely not satisfactory for New York City. Anyone who has driven in New York City knows the multitude of different street types and congestion levels that can produce a wide variety of different travel speeds. We decided to try to get some idea of the average speed in different parts of New York City.

In addition to performing various timing experiments, we obtained several reports from the New York City Department of Transportation. These include "Midtown Auto Speeds—Spring 1992" and "Midtown Auto Speeds—Fall 1992." These reports provide data on Midtown Manhattan average travel speeds as well as some data on the variance of these speeds. (Midtown Manhattan is defined as the rectangular area between First and Eighth Avenues and 30th and 60th Streets.) The data seem to suggest that speeds vary from an average of 6 miles per hour up to about 14 miles per hour, depending on street type, direction, and time of day.

Our approach was to choose an estimate of speed that would be specific to each district; thus, a district in the Bronx would not have the same speed estimate as one in Midtown Manhattan. These range from about 7 to 12 miles per hour. An important observation made when collecting data was that when a bus experienced below-average travel times (above-average speeds) along the beginning of the route, the bus driver will slow down or spend more time at the stops to get back on schedule. In addition, since the students have a scheduled pickup time, the bus cannot, as a rule, leave early. It must wait until a specific time before leaving the bus stop. If the bus experiences above-average travel times (below-average speeds), then the bus driver can speed up (slightly) and make sure to leave as soon as all students are on the bus. Consequently, the travel time is not quite as random as one might think.

To make sure that school buses meet the time-window constraints, simply having information about travel time along the streets of New York City is not sufficient. The time to pick up students from their bus stops and to drop off students at their schools must also be taken into account. By riding the buses, we collected data on the time it takes to pick up or drop off students at stops or at schools. A linear regression was performed on the data, providing the following model for the pickup time:

$$PTime = 19.0 + 2.6N,$$

where $PTime$ = pickup time (in s), and N = the number of students picked up at the bus stop. This regression was performed on 30 data points. The R^2 was 77.7 % and the p-value of the independent variable was very small (< 0.001). The regression performed on the dropoff times resulted in the equation

$$DTime = 29.0 + 1.9N,$$

where $DTime$ = dropoff time (in s), and N = the number of students dropped off at the school. This regression was performed on 30 data points. Here the R^2 was 41.9 % and the p-value of the independent variable was 0.01 %. In our implementation, we used these equations to determine approximate pickup and dropoff times.

Overall, the approximations and calculations made in testing the optimization module were designed with the goal of ensuring that a route constructed by the algorithm would be a driveable one. The next question is how to generate a good feasible solution to the school bus routing and scheduling problem.

21.6   The Routing Algorithm

There are many existing algorithms for school bus routing and scheduling. Numerous communities throughout the world have implemented computerized algorithms to perform these tasks. Overall, the success seems to be universally recognized. Almost all papers published in the literature mention cost savings of around 5–10%. We recognize that it may be useless to even contemplate the meaning of these savings numbers since the savings may come not only from a reduction in cost but also from increased control of the bus routes. The magnitude of the "savings" is also highly dependent on what methods were in use before the computerized system was put into place.

Transferability seems to be the critical factor. It is difficult to compare algorithms for this problem directly from the literature. Each problem has its own version of the constraints and even objectives. It is not always simple or even possible to take an existing algorithm in use in one community and simply apply it to another. Each problem has its peculiarities and may also have very different constraints. For instance, in an implementation in Montréal, the people designing the routes have the freedom to change existing school starting and ending times at their convenience. Clearly, this added flexibility can simplify the problem to some extent and can lead to additional savings in cost. In New York City, this was not possible.

Finally, this is all within the framework of an optimization problem, which we have seen is extremely difficult to solve. There is an absence of any strong lower bounds on the minimum number of buses required.

In determining what type of algorithm to apply to this large vehicle routing problem, we considered several important aspects of the problem and also the setting in which the algorithm would be used.

Efficiency This is an extremely large problem, so the solution method must be efficient in computation time and in space requirement. While we might want to optimize by district, the fact is that some districts have as many as 1,500 bus stops. Even though complete optimization of the solution might only be done once a year, the time involved in testing and experimenting with the problem parameters is reduced considerably if the algorithm is time- and space-efficient.

Transparency The algorithm would most likely need to be constructive in nature, thereby providing a dispatcher with the ability of viewing the algorithm progression in real time. This makes it possible to detect "problem routes" and correct errors without having to wait until the termination of the algorithm. That is, the approach should build routes in a sequential fashion and not, for example, work for hours and finally, in the last moments, provide a solution.

Flexibility The heuristic should be flexible enough to handle not only the constraints currently in place, but also additional constraints that might be imposed in the future.

Interactivity From our discussions with the inspectors, it is clear that the algorithm implemented must have an interactive component that would allow an experienced inspector to help construct routes using his or her prior knowledge. That is, the algorithm must be able to work in two different modes. First, it must be able to act like a black box, where data are input and a solution is output. Second, it must also serve as an interactive tool, where a starting solution can be presented along with a set of unrouted stops and the algorithm finds the best way to *add on* to this starting solution.

Multiple solutions The algorithm should be capable of producing a series of solutions, not simply one solution. This last point is important since each solution would have to be checked by an inspector, and it is possible that the inspector will rule out some solutions.

Finally, the urban nature of our application, in contrast to many of the problems seen in the literature, should also be taken into account. As many researchers have noted [see Bodin and Berman (1979) and Chapleau et al. (1983)], the vehicle capacity constraint tends to be the most binding constraint when routing in an urban area. This is due to the general rule that the bus will tend to "fill up" before the time constraints become an issue. Therefore, it seems as though algorithms developed for the capacitated vehicle routing problem (CVRP) (see Chap. 17) should be a good starting point. The difficulty is that the CVRP generally has a different objective function: Minimize the total distance traveled, not the number of vehicles used. Fortunately [see Chap. 17 or Bramel et al. (1991)], if the number of pickup points is very large and distances follow a general norm, when the distance is minimized, a byproduct of the solution is that the minimum number of vehicles will be used. Observe that distances in New York City come from the street network, not from a norm; however, since the blocks are short and somewhat uniform in size, the street network distance is fairly close to a norm distance, and similar results most likely hold.

For these reasons, our starting point for the algorithm for the school bus routing and scheduling problem was the location-based heuristic (LBH) (see Sect. 17.7) developed for the CVRP. This algorithm has the important property that it is *asymptotically optimal* for the CVRP (see Sect. 17.7); that is, the relative error between the value of the solution generated by the algorithm and the optimal solution value tends to zero as the number of pickup points increases.

Due to the size and complexity of the problem, we made several changes to the LBH. The algorithm is *serial* in nature, as it constructs one route at a time and not in parallel. To describe the algorithm, let the bus stops be indexed $1, 2, \ldots, n$. Let a route run by a single bus be denoted R_i. Let a full solution to the school bus routing and scheduling problem be written as a set of routes $\{R_1, R_2, \ldots, R_M\}$, where M is the number of buses used. For each bus stop j, let $school[j]$ be the index of the school to which the students at stop j are destined. Let U be the set of indices of all unvisited pickup points.

The following algorithm creates one solution to the school bus routing and scheduling problem. More solutions can be generated by starting the algorithm with different random seeds.

Randomized LBH:

    Let $U = \{1, 2, \ldots, n\}$ and $m = 0$.
    while $(U \neq \emptyset)$ do
        {
        Pick a seed stop from U using a selection criterion. Call it j.
        Let $U \leftarrow U \setminus \{j\}$.
        Let the current route be $R_m = \{j \rightarrow school[j]\}$.
        repeat
            {
            For each $i \in U$, calculate $c_i = routelength(i, R_m)$.
            Let $c_k = \min_{i \in U}\{c_i\}$.
            If $c_k < +\infty$, then

$$
\{
$$

Let $R_m \leftarrow buildroute(k, R_m)$.
Let $U \leftarrow U \setminus \{k\}$.
$$
\}
$$
$\}$ *until* $c_k = +\infty$.
$m \leftarrow m + 1$.
$\}$
$M \leftarrow m$.
The heuristic solution is $\{R_1, R_2, \ldots, R_M\}$.

The selection of the seed stops can be done in one of several different ways. One approach is to simply select these stops at random from the set of unvisited stops. Another approach is to select stops with large loads or stops that have tight delivery windows (i.e., the distance and time constraints force these stops to be delivered almost directly from the stop to the school with very few stops in between). Other criteria were used according to which constraints were binding at particular stops.

The function *routelength*(i, R) determines the approximate cost of inserting stop i into route R. Route R consists of a path through several stops and schools. While preserving the order of the stops and schools in route R, we determine the best insertion point for stop i. We check each consecutive pair of points (either stops or schools) along route R and check whether stop i can be inserted between these two. If *school*$[i]$ is not in route R, then we must find not only the best insertion point for stop i, but also the best insertion point for *school*$[i]$. It is possible that no insertion point(s) can be found that results in a feasible route. Checking whether a stop can be inserted requires checking all the constraints. If no feasible insertion point exists, then the value of *routelength*(i, R) is made $+\infty$. This indicates that it is not possible (while preserving the order of R) to insert stop i into route R. If an insertion is found that results in a feasible route, then the value of *routelength*(i, R) is made to be exactly the additional distance traveled.

To illustrate the difficulty of this step, consider simply the capacity constraint. In the case of the CVRP, all loads are dropped off at the same point (the final stop); therefore, the maximum load that is carried by the vehicle is when it picks up its last load. Therefore, it is easy to check whether a stop can be added to a route since we need only check that the maximum load is less than the vehicle capacity. This maximum load is always at the last stop, so the calculation is easy. By contrast, performing a similar calculation in the school bus routing and scheduling problem is much more complicated since there is more than one dropoff point. Checking feasibility when adding a stop to a route requires knowing *when* the student is getting on and off the bus, since this will affect whether there is room for a student at future points on the bus route. Therefore, checking whether the capacity constraint is violated in the school bus routing problem is much more complicated than in the CVRP.

The function *buildroute*(k, R) creates the route that results from the insertion of stop k into route R. Again, stop k is simply inserted between the two consecutive points (stops or schools) that result in the shortest total route. This route is guaranteed to be feasible since $c_k < +\infty$.

The algorithm satisfies the requirements that we described above. It runs efficiently for problems of a large size and builds routes sequentially. It is very flexible in the sense that constraints of almost any type can be included (e.g., disallowing mixed loads for some schools). Of course, each additional constraint causes the algorithm to take a little longer to find a solution. In terms of its interactivity (see the next section for details), the algorithm can be used in an interactive mode if this is desired. In this mode, a partial routing solution can be used as a starting point and unrouted stops can be added efficiently. The inspector can also have a major impact on the routes generated by the algorithm via the selection of the seed points (see Sect. 21.8 below for a further discussion on this point). Since the algorithm can be easily randomized (by randomizing the seed stop selection procedure), starting the algorithm with different random numbers makes it generate different solutions. Finally, the most important advantage of this heuristic is that it does not decompose the problem into subproblems, but solves the routing and scheduling components simultaneously.

21.7 Additional Constraints and Features

In the course of the implementation of our algorithm, several additional "soft" constraints came to our attention. These are subtle rules that inspectors used when constructing feasible routes, which were only determined once a set of routes were shown to the inspectors.

Limit on the number of buses to a particular school This is best explained with an example. Consider the situation where a school, say school A, has a late starting time relative to other schools, say 9:30 a.m., where all other schools start at 9 a.m., and assume only a dozen of the students from school A require bus service. Previously, if a solution required 20 buses to serve all schools, routers would take one of these and have it alone serve school A. That is, some time between 9 and 9:30 a.m., one bus would pick up the dozen students and deliver them to school A. Since 20 buses are used in the solution, this solution is equivalent to, for example, having 6 of the 20 buses each deliver 2 students to school A between 9 and 9:30 a.m. This, from a cost point of view, is just as good a solution. However, school A may only be able to handle one or two buses at a time due to limited driveway space. We therefore needed to add a constraint on the number of buses that could deliver students to each school. This constraint only became active for a few schools.

Multilevel relational distance constraints When a driver is delivering packages to warehouses or to customers, a distance constraint is usually set on the complete route and thus is limited to the driver's working day. When a bus driver is delivering students to schools, the distance constraint is really *student-specific*. That is, each student's trip is limited, not just the driver's. In the school bus routing and scheduling problem, the distance constraint also illustrates the difficulty of modeling, through simple constraints, a real-life problem. To illustrate this, consider the 5-mile distance constraint discussed earlier. We found that this simple constraint was actually unsatisfactory for this problem. For example, if a student was only 1 mile from school, then it was not considered desirable to have him or her end up traveling 5 miles on the bus. This student (and maybe more vociferously his or her parents) would not consider this an equitable solution. We therefore decided to implement what we call a relational distance constraint. That is, for a multiplier α, say $\alpha = 2$, a student could not travel on the bus for more than α times the distance the student's bus stop was from school. The question was then to what do we set α? We determined that the best rule was to divide the region around a particular school into concentric rings. For example, if the first ring was 3 miles in radius, then a stop that was $d \leq 3$ miles from the school would have a distance constraint (on the bus) of $\alpha_1 d$ miles. Ring i was assigned a multiplier α_i, and this was repeated for each ring. Although it took some time to determine appropriate multipliers, eventually this is the type of distance constraint that was implemented.

Waiting-time constraint Another constraint that did not come to our attention until we presented our routes to the inspectors was the waiting-time constraint. Again, this is something that is specific to the routing of people as opposed to packages. Consider a simple problem with two schools, school A starting at 8 a.m. and school B starting at 9 a.m. At 7:30 a bus picks up both students for schools A and B and then arrives at school A in the time window (say at 7:45) and drops off only those students who are going to school A. Since school B starts at 9 a.m., the bus waits for half an hour at school A until proceeding to pick up some more students for school B and then arriving at school B at 8:45 and dropping off all the students. A route of this type, where students wait on the bus for half an hour, was definitely not deemed acceptable. Therefore, we needed to add a constraint on the amount of time a bus (with students on it) can wait idle. Five minutes was the number that was eventually used.

Route balancing It is desirable that the routes in a solution be of similar duration and total distance. It does not seem fair if one driver serves morning routes from 7 to 7:30 a.m. while another works from 7 to 9:30 a.m. The balancing of the workloads is partially achieved by implementation of a *route-balance()* subroutine that is called once, at the end of the algorithm. This subroutine essentially moves stops and schools from heavily loaded routes to less heavily loaded routes while maintaining feasibility of the solution. This seemed to work very well.

Single-route optimization Once a solution is determined, we may (and should) optimize the sequencing of the stops and schools on each route individually. That is, given a set of stops and schools that can be feasibly served by one bus, in terms of service level, what is the "best" route to actually drive? An objective that guarantees a high service level is to minimize the total number of student-miles traveled (see, e.g., Bennett and Gazis (1972)). For each route created, we call a procedure called *route-opt()*, which minimizes the total number of student-miles while maintaining feasibility of the route.

21.8 The Interactive Mode

As we mentioned earlier, the complete rescheduling of all buses might only be done once a year (in August). However, throughout the course of the school year, there are quite a few small changes that must be made to the solution. These changes could be caused by, for example:

- A school that previously did not request bus service does request service in mid-year.

- A student changes address or school.

- A school's session time changes.

One option might be simply to reoptimize all routes that are affected by the changes. This might cause major disruptions in a large number of routes. These disruptions may translate to disruptions in the parents' morning schedules, which might overload the Office of Pupil Transportation telephone switchboard. In essence, it is desirable to implement the changes while making the fewest disruptions to other students' schedules.

This was the impetus for the development of the algorithm's *interactive mode*. Here it is possible to start the algorithm with a number of routes already created and to simply add stops to or delete stops from these routes. Let's consider what happens when a stop is added to an existing set of routes. The user has the ability to select from one of three options:

- *Complete reoptimization.* This corresponds to starting the reoptimization from scratch, that is, throwing away all previously created routes. Optimization then starts with all new stops added to the list of stops.

- *Single-route reoptimization.* This corresponds to selecting a route and checking whether a particular stop can be added to it. This is done through a simple *route-check()* subroutine. In this case, the route may be completely resequenced.

- *No reoptimization.* In this case, the stop is simply inserted between two stops on existing routes without any reoptimization.

Deleting a stop is somewhat easier to do; the user simply clicks the mouse on the stop in question and deletes it from the current solution. The fact that this may actually render the remaining route infeasible is a good illustration of the complexity of the bus routing and scheduling problem. This is due to the waiting-time constraint mentioned in the previous section. In either case, the user can specify whether a reoptimization of the route is desired.

These optimization tools proved quite useful as they provided simple ways to test what-if scenarios, tests that previously would have taken weeks, if not months.

21.9   Data, Implementation, and Results

To assess the effectiveness of our algorithm, we attempted to solve the problem using the Manhattan data given to us by the Office of Pupil Transportation, that is, to use our algorithm to generate a solution and to check it for actual drivability.

We solved both the morning and the afternoon problem. We first calculated the shortest-distance matrix between all 911 points of interest (838 stops and 73 schools) along the street network. In our implementation, we used a speed of 8 miles per hour for the entire borough. This was the lowest average speed in Midtown Manhattan along a street or avenue between 7 and 10 a.m. (the time interval that the bus would be traveling in the morning) reported by the Department of Transportation. We feel that this average speed is quite conservative and that, on average, a bus can travel more quickly. One reason for this is that the measurement was made in Midtown Manhattan, a location with very high congestion throughout the day.

The algorithm was run on a PC (486DX2/50 megahertz) under Windows over a period of several hours. To generate its first feasible solution, the algorithm takes about 40 min. We repeated the algorithm 40 times, keeping track of the best solution. The algorithm has as output a detailed schedule and directions for each bus.

In order to determine the sensitivity of the results to some of the assumptions we have made, we ran the algorithm with several settings for the average travel speed. We used 8, 10, and 12 mph. Note again that these speeds are conservative, as we have also taken into account the time to stop and pick up or drop off students. The following table lists the number of buses used in the best solutions found for each of these settings and for the morning and afternoon problems (Table 21.1).

TABLE 21.1. General Education routing

Universal speed (mph)	Number of buses used	
	Morning	Afternoon
8	74	67
10	64	60
12	59	56

As a comparison, these solutions use substantially fewer buses than are currently in use. We do not expect that the number of buses used will be as low as indicated by our preliminary results, due to the fact that the routes have not been checked by the inspectors. However, it is reasonable to assume that they will serve as a starting solution that can be modified by the inspectors.

References

Aggarwal, A., & Park, J. K. (1993). Improved algorithms for economic lot-size problems. *Operation Research, 41*, 549–571.

Agrawal, V., & Seshadri, S. (2000). Impact of Uncertainty and Risk Aversion on Price and Order Quantity in the Newsvendor Problem. *Manufacturing and Service Operations Management, 2*(4), 410–423.

Aho, A. V., Hopcroft, J. E., & Ullman, J. D. (1974). *The design and analysis of computer algorithms.* Reading, MA: Addison-Wesley.

Altinkemer, K., & Gavish, B. (1987). Heuristics for unequal weight delivery problems with a fixed error guarantee. *Operation Research Letter, 6*, 149–158.

Altinkemer, K., & Gavish, B. (1990). Heuristics for delivery problems with constant error guarantees. *Transportation Science, 24*, 294–297.

Angel, R. D., Caudle, W. L., Noonan, R., & Whinston, A. (1972). Computer-assisted school bus scheduling. *Management Science, 18*, 279–288.

Anily, S. (1991). Multi-item replenishment and storage problems (MIRSP): heuristics and bounds. *Operation Research, 39*, 233–239.

Anily, S., Bramel, J., & Simchi-Levi, D. (1994). Worst-case analysis of heuristics for the bin-packing problem with general cost structures. *Operation Research, 42*, 287–298.

D. Simchi-Levi et al., *The Logic of Logistics: Theory, Algorithms, and Applications for Logistics Management*, Springer Series in Operations Research and Financial Engineering, DOI 10.1007/978-1-4614-9149-1, © Springer Science+Business Media New York 2014

Archibald, B., & Silver, E. A. (1978). (s, S) policies under continuous review and discrete compound Poisson demand. *Management Science, 24*, 899–908.

Arkin, E., Joneja, D., & Roundy, R. (1989). Computational complexity of uncapacitated multi-echelon production planning problems. *Operation Research Letter, 8*, 61–66.

Arrow, K., Harris, T., & Marschak , J. (1951). Optimal inventory policy. *Econometrica, 19*, 250–272.

Atkins, D. R., & Iyogun, P. (1988). A heuristic with lower bound performance guarantee for the multi-product dynamic lot-size problem. *IIE Transactions, 20*, 369–373.

Azuma, K. (1967). Weighted sums of certain dependent random variables. *Tohoku Mathematical Journal, 19*, 357–367.

Baker, B. S. (1985). A new proof for the first-fit decreasing bin packing algorithm. *Journal of Algorithms, 6*, 49–70.

Baker, K. R., Dixon, P., Magazine, M. J., & Silver, E. A. (1978). An algorithm for the dynamic lot-size problem with time-varying production capacity constraints. *Management Science, 24*, 1710–1720.

Balakrishnan, A., & Graves, S. (1989). A composite algorithm for a concave-cost network flow problem. *Networks, 19*, 175–202.

Balinski, M. L. (1965). Integer programming: methods, uses, computation. *Management Science, 12*, 253–313.

Balinski, M. L., & Quandt, R. E. (1964). On an integer program for a delivery problem. *Operation Research, 12*, 300–304.

Ball, M. O., Magnanti, T. L., Monma, C. L., & Nemhauser, G. L. (Eds.) (1995). Network routing, *Handbooks in operations research and management science*. Amsterdam: North-Holland.

Ballou, R. H. (1992). *Business logistics management* (3rd ed.). Englewood Cliffs, NJ: Prentice-Hall.

Barcelo, J., & Casanovas, J. (1984). A heuristic Lagrangian algorithm for the capacitated plant location problem. *European Journal of Operational Research, 15*, 212–226.

Başar, T., & Olsder, G. J. (1999). *Dynamic noncooperative game theory (Classics in applied mathematics)*. Philadelphia: Society for Industrial and Applied Mathematics.

Beasley, J. (1983). Route first–cluster second methods for vehicle routing. *Omega, 11*, 403–408.

Beardwood, J., Halton, J. L., & Hammersley, J. M. (1959). The shortest path through many points. *Proceedings of the Cambridge Philosophical Society, 55*, 299–327.

Bell, C. (1970). Improved algorithms for inventory and replacement stock problems. *SIAM Journal on Applied Mathematics, 18*, 558–566.

Bennett, B., & Gazis, D. (1972). School bus routing by computer. *Transportation Research, 6*, 317–326.

Bertsekas, D. P. (1987). *Dynamic programming.* Englewood Cliffs, NJ: Prentice-Hall.

Bertsekas, D. P. (1995). *Nonlinear programming.* Boston: Athena Scientific.

Bernstein, F., & Federgruen, A. (2004). Dynamic inventory and pricing models for competing retailers. *Naval Research Logistics, 51*, 258–274.

Bertsimas, D., & Simchi-Levi, D. (1996). The new generation of vehicle routing research: robust algorithms addressing uncertainty. *Operation Research, 44*, 286–304.

Bienstock, D., & Simchi-Levi, D. (1993). A note on the prize collecting traveling salesman problem. Working Paper, Columbia University.

Bienstock, D., Bramel, J., & Simchi-Levi, D. (1993a). A probabilistic analysis of tour partitioning heuristics for the capacitated vehicle routing problem with unsplit demands. *Mathematics of Operations Research, 18*, 786–802.

Bienstock, D., Goemans, M., Simchi-Levi, D., & Williamson, D. (1993b). A note on the prize collecting traveling salesman problem. *Mathematical Programming, 59*, 413–420.

Bodin, L., & Berman, L. (1979). Routing and scheduling of school buses by computer. *Transportation Science, 13*, 113–129.

Bondareva, O. (1963). Some applications of linear programming methods to the theory of cooperative games (in Russian). *Problemy Kybernetiki, 10*, 119–139.

Bourakiz, M., & Sobel, M. J. (1992). Inventory control with an exponential utility criterion. *Operation Research, 40*, 603–608.

Braca, J., Bramel, J., Posner, B., & Simchi-Levi, D. (1997). A computerized approach to the New York City school bus routing problem. *IIE Transactions, 29*, 693–702.

Bramel, J., & Simchi-Levi, D. (1995). A location based heuristic for general routing problems. *Operation Research, 43*, 649–660.

Bramel, J., & Simchi-Levi, D. (1996). Probabilistic analysis and practical algorithms for the vehicle routing problem with time windows. *Operation Research, 44*, 501–509.

Bramel, J., & Simchi-Levi, D. (1997). On the effectiveness of set covering formulations for the vehicle routing problem. *Operation Research, 45*, 295–301.

Bramel, J., Coffman Jr., E. G., Shor, P., & Simchi-Levi, D. (1991). Probabilistic analysis of algorithms for the capacitated vehicle routing problem with unsplit demands. *Operation Research, 40*, 1095–1106.

Cachon, G. (2003). Supply chain coordination with contracts. In S. Graves, T. de Kok *Supply chain management: design, coordination and operation (Handbook of operations research and management science* (Vol. 11, pp. 229–340). Amsterdam: Elsevier.

Cachon, G., & Lariviere, M. (2005). Supply chain coordination with revenue sharing contracts: strengths and limitations. *Management Science, 51*, 30–44.

Cachon, G., & Netessine, S. (2004). Game theory in supply chain analysis. In D. Simchi-Levi, S. D. Wu, Z. J. Max Shen (Eds.) *Handbook of quantitative supply chain analysis: modeling in the eBusiness era*. Boston: Kluwer Academic.

Chapleau, L., Ferland, J. A., & Rousseau, J.-M. (1983). Clustering for routing in dense area. *European Journal of Operational Research, 20*, 48–57.

Chan, L. M. A., Simchi-Levi, D., & Bramel, J. (1998). Worst-case analyses, linear programming and the bin-packing problem. *Mathematical Programming, 83*, 213–227.

Chandra, B., Karloff, H., & Tovey, C. (1999). New results on the old k-opt algorithm for the traveling salesman problem. *SIAM Journal on Computing, 28*, 1998–2029.

Chan, L. M. A., Muriel, A., & Simchi-Levi, D. (1999). Production/distribution planning problems with piece-wise linear and concave cost structures. Northwestern University.

Chan, L. M. A., Max Shen, Z. J., Simchi-Levi, D., & Swann, J. (2004). Coordination of pricing and inventory (Chapter 3) In D. Simchi-Levi, S. D. Wu, Z. J. Max Shen (Eds.) *Handbook of quantitative supply chain analysis: modeling in the eBusiness era*. Boston: Kluwer Academic.

Chen, Y. F. (1996). On the optimality of (s, S) policies for quasiconvex loss functions. Working Paper, Northwestern University.

Chen, X. (2003). *Coordinating inventory control and pricing strategies* (Ph.D. Dissertation, Massachusetts Institute of Technology)

Chen, X. (2009), Inventory centralization games with price-dependent demand and quantity discount. *Operation Research, 57,* 1394–1406.

Chen, F., & Federgruen, A. (2000). Mean-variance analysis of basic inventory models. Working paper, Columbia University.

Chen, X., & Hu, P. (2012). Joint pricing and inventory management with deterministic demand and costly price adjustment. *Operation Research Letter, 40,* 385–389.

Chen, X., & Simchi-Levi, D. (2004a). Coordinating inventory control and pricing strategies with random demand and fixed ordering cost: the finite horizon case. *Operation Research, 52,* 887–896.

Chen, X., & Simchi-Levi, D. (2004b). Coordinating inventory control and pricing strategies with random demand and fixed ordering cost: the infinite horizon case. *Mathematics of Operations Research, 29,* 698–723.

Chen, X., & Simchi-Levi, D. (2009). A new approach for the stochastic cash balance problem with fixed costs. *Probability in the Engineering and Informational Sciences, 23,* 545–562.

Chen, X., & Simchi-Levi, D. (2012). Pricing and inventory management. P. Philips, Ö. Özer (Eds.) *Oxford handbook of pricing management* (pp. 784–822). Oxford: Oxford University Press.

Chen, X., & Sun, P. (2012). Optimal structural policies for ambiguity and risk averse inventory and pricing models. *SIAM Journal on Control and Optimization, 50,* 133–146.

Chen, X., & Zhang, J. (2009). A stochastic programming approach to inventory centralization games. *Operation Research, 57,* 840–851.

Chen, F., & Zheng, Y. S. (1994). Lower bounds for multi-echelon stochastic inventory systems. *Management Science, 40,* 1426–1443.

Chen, X., Sim, M., Simchi-Levi, D., & Sun, P. (2007). Risk aversion in inventory management. *Operation Research, 55,* 828–842.

Chen, X., Zhang, Y., & Zhou, S. (2010). Integration of inventory and pricing decisions with costly price adjustments. *Operation Research, 58,* 1012–1016.

Chen, X., Zhou, S., & Chen, F. (2011). Preservation of quasi-K-concavity and its application to joint inventory-pricing models with concave ordering costs. *Operation Research, 58,* 1012–1016.

Chen, X., Pang, Z., & Pan, L. (2012a). Coordinating inventory control and pricing strategies for perishable products. Working Paper, University of Illinois at Urbana-Champaign.

Chen, X., Hu, P., & He, S. (2012b). Preservation of supermodularity in two dimensional parametric optimization problems and its applications. This paper has been accepted by Operations Research.

Chen, X., Hu, P., Shum, S., & Zhang, Y. (2012c). Stochastic inventory model with reference price effects. Working Paper.

Chou, M., Teo, C., & Zheng, H. (2008). Process flexibility: design, evaluation, and applications. *Flexible Services and Manufacturing J., 20*(1), 59–94.

Chou, M., Chua, G., Teo, C., & Zheng, H. (2010). Design for process flexibility: efficiency of the long chain and sparse structure. *Operation Research, 58*, 43–58.

Chou, M., Chua, G., Teo, C., & Zheng, H. (2011). Processs flexibility revisited: the graph expander and its applications. *Operation Research, 59*, 1090–1105.

Chou, M., Chua, G., Teo, C., & Zheng, H. (2012). On the performance of sparse process structures in partial postponement production systems. Working Paper.

Christofides, N. (1976). Worst-case analysis of a new heuristic for the traveling salesman problem. Report 388, Graduate School of Industrial Administration, Carnegie-Mellon University, Pittsburgh, PA.

Christofides, N. (1985). Vehicle routing. In E. L. Lawler, J. K. Lenstra, A. H. G. Rinnooy Kan, D. B. Shmoys (Eds.) *The traveling salesman problem: a guided tour of combinatorial optimization* (pp. 431–448). New York: Wiley.

Christofides, N., Mingozzi, A., & Toth, P. (1978). The vehicle routing problem. In N. Christofides, A. Mingozzi, P. Toth, C. Sandi (Eds.) *Combinatorial optimization* (pp. 318–338). New York: Wiley.

Christofides, N., Mingozzi, A., & Toth, P. (1981). Exact algorithms for the vehicle routing problem based on spanning tree and shortest path relaxations. *Mathematical Programming, 20*, 255–282.

Churchman, C. W., Ackoff, R. L., & Arnoff, E. L. (1957). *Introduction to operations research.* New York: Wiley.

Clark, A. J., & Scarf, H. E. (1960). Optimal policies for a multi-echelon inventory problem. *Management Science, 6*, 475–490.

Clarke, G., & Wright, J. W. (1964). Scheduling of vehicles from a central depot to a number of delivery points. *Operation Research, 12*, 568–581.

Coffman, E. G. Jr., & Lueker, G. S. (1991). *Probabilistic analysis of packing and partitioning algorithms.* New York: Wiley.

Cornuéjols, G., & Harche, F. (1993). Polyhedral study of the capacitated vehicle routing problem. *Mathematical Programming, 60*, 21–52.

Cornuéjols, G., Fisher, M. L., & Nemhauser, G. L. (1977). Location of bank accounts to optimize float: an analytical study of exact and approximate algorithms. *Management Science, 23*, 789–810.

Council of Supply Chain Management Professionals: http://www.cscmp.org/.

Council on Logistics Management, mission statement, Council on Logistics Management Web Site: www.clm1.org/mission.html.

Croxton K. L., Gendron, B., Magnanti, T. L. (2003). A comparison of mixed-integer programming models for non-convex piecewise linear cost minimization problems. *Management Science, 49*, 1268–1273.

Cullen, F., Jarvis, J., & Ratliff, D. (1981). Set partitioning based heuristics for interactive routing. *Networks, 11*, 125–144.

Daskin, M. S. (1995). *Network and discrete location: models algorithms and applications.* New York: Wiley.

De Kok, A. G., & Graves, S. C. (Eds.) (2003). Supply chain management: design, coordination and operations, *Handbooks in operations research and management science.* Amsterdam: North-Holland.

Dematteis, J. J. (1968). An economic lot sizing technique: the part-period algorithm. *IBM Systems Journal, 7*, 30–38.

Denardo, E. V. (1996). Dynamic programming. In Avriel, M., Golany, B. (Eds.), *Mathematical programming for industrial engineers* (pp. 307–384). Englewood Cliffs, NJ: Marcel Dekker.

Deng, Q., & Simchi-Levi, D. (1992). Valid inequalities, facets and computational results for the capacitated concentrator location problem. Working Paper, Columbia University.

Deng, S., & Yano, C. (2006). Joint production and pricing decisions with setup costs and capacity constraints. *Management Science, 52*, 741–756.

Desrosiers, J., Ferland, J. A., Rousseau, J.–M., Lapalme, G., & Chapleau, L. (1986). TRANSCOL: A multi-period school bus routing and scheduling system. *TIMS Studies in the Management Sciences, 22*, 47–71.

Desrochers, M., Desrosiers, J., & Solomon, M. (1992). A new optimization algorithm for the vehicle routing problem with time windows. *Operation Research, 40*, 342–354.

Dobson, G. (1987). The economic lot scheduling problem: a resolution of feasibility using time varying lot sizes. *Operation Research, 35*, 764–771.

Dreyfus, S. E., & Law, A. M. (1977). *The art and theory of dynamic programming.* New York: Academic Press.

Edmonds, J. (1965). Maximum matching and a polyhedron with 0,1-vertices. *Journal of Research of the National Bureau of Standards B, 69B*, 125–130.

Edmonds, J. (1971). Matroids and the greedy algorithm. *Mathematical Programming, 1*, 127–136.

Eeckhoudt, L., Gollier, C., & Schlesinger, H. (1995). The risk-averse (and prudent) newsboy. *Management Science, 41*(5), 786–794.

Eliashberg, J., & Steinberg, R. (1991). Marketing-production joint decision making. In J. Eliashberg, J. D. Lilien (Eds.) *Management science in marketing*, Vol. 5 of *Handbooks in Operations Research and Management Science*. Amsterdam: North-Holland.

Elmaghraby, W., & Keskinocak, P. (2003). Dynamic pricing in the presence of inventory considerations: research overview, current practices, and future directions. *Management Science, 49*, 1287–1309.

Eppen, G., & Schrage, L. (1981). Centralized ordering policies in a multiwarehouse system with lead times and random demand. In L. Schwarz (Ed.) *Multi-level production/inventory control systems: theory and practice*. Amsterdam: North-Holland.

Erlenkotter, D. (1990). Ford Whitman Harris and the economic order quantity model. *Operation Research, 38*, 937–946.

Federgruen, A., & Heching, A. (1999). Combined pricing and inventory control under uncertainty. *Operation Research, 47*(3), 454–475.

Federgruen, A., & van Ryzin, G. (1997). Probabilistic analysis of a generalized bin packing problem with applications to vehicle routing and scheduling problems. *Operation Research, 45*, 596–609.

Federgruen, A., & Simchi-Levi, D. (1995). Analytical Analysis of Vehicle Routing and Inventory Routing problems. In M. O. Ball, T. L. Magnanti, C. L. Monma, G. L. Nemhauser (Eds.) *Handbooks in operations research and management science*, the volume on *Network routing* (pp. 297–373). Amsterdam: North-Holland.

Federgruen, A., & Tzur, M. (1991). A simple forward algorithm to solve general dynamic lot sizing models with n periods in $O(n \log n)$ or $O(n)$ time. *Management Science, 37*, 909–925.

Federgruen, A., & Zipkin, P. (1984a). Approximation of dynamic, multi-location production and inventory problems. *Management Science, 30*, 69–84.

Federgruen, A., & Zipkin, P. (1984b). Computational issues in the infinite horizon, multi-echelon inventory model. *Operation Research, 32*, 818–836.

Federgruen, A., & Zipkin, P. (1984c). Allocation policies and cost approximation for multi-location inventory systems. *Naval Research Logistic Quarterly, 31*, 97–131.

Few, L. (1955). The shortest path and the shortest road through n points. *Mathematika, 2*, 141–144.

Fisher, M. L. (1980). Worst-case analysis of algorithms. *Management Science, 26*, 1–17.

Fisher, M. L. (1981). The lagrangian relaxation method for solving integer programming problems. *Management Science, 27*, 1–18.

Fisher, M. L. (1994). Optimal solution of vehicle routing problems using minimum K-trees. *Operation Research, 42*, 626–642.

Fisher M. L. (1995). Vehicle routing. In M. O. Ball, T. L. Magnanti, C. L. Monma, G. L. Nemhauser (Eds.) *Handbooks in operations research and management science*, the volume on Network routing (pp. 1–33) Amsterdam: North-Holland.

Fisher, M. L., & Jaikumar, R. (1981). A generalized assignment heuristic for vehicle routing. *Networks, 11*, 109–124.

Florian, M., & Klein, M. (1971). Deterministic production planning with concave costs and capacity constraints. *Management Science, 18*, 12–20.

Florian, M., Lenstra, J. K., & Rinnooy Kan, A. H. G. (1980). Deterministic production planning: algorithms and complexity. *Management Science, 26*, 669–679.

Fudenberg, D., & Tirole, J. (1991). *Game theory*. Cambridge, MA: MIT Press.

Gale, D., & Politof, T. (1981). Substitutes and complements in network flow problems. *Discrete Applied Mathematics, 3*, 175–186.

Gallego, G., & van Ryzin, G. (1994). Optimal dynamic pricing of inventories with stochastic demand over finite horizons. *Management Science, 40*, 999–1020.

Gallego, G., Queyranne, M., & Simchi-Levi, D. (1996). Single resource multi-item inventory system. *Operation Research, 44*, 580–595.

Garey, M. R., & Johnson, D. S. (1979). *Computers and intractability*. New York: W. H. Freeman and Company.

Garey, M. R., Graham, R. L., Johnson, D. S., & Yao, A. C. (1976). Resource constrained scheduling as generalized bin packing. *Journal of Combinatorial Theory, Series A, 21*, 257–298.

Gaskel, T. J. (1967). Bases for vehicle fleet scheduling. *Operational Research Quarterly, 18*, 281–295.

Geunes, J., Romeijn, E., & Taaffe, K. (2006). Requirements planning with pricing and order selection flexibility. *Operation Research, 54*, 394–401.

Ghosh, A. (1994). *Retail management* (2nd ed.). New York, NY: Dryden Press Harcourt Brace College Publishers.

Gillett, B. E., & Miller, L. R. (1974). A heuristic algorithm for the vehicle dispatch problem. *Operation Research, 22*, 340–349.

Goemans M. X., & Bertsimas, D. J. (1993). Survivable networks, linear programming relaxations and the parsimonious property. *Mathematical Programming, 60*, 145–166.

Golden, B. L., & Stewart, W. R. (1985). Empirical analysis of heuristics. In E. L. Lawler, J. K. Lenstra, A. H. G. Rinnooy Kan, D. B. Shmoys (Eds.) *The traveling salesman problem: a guided tour of combinatorial optimization* (pp. 207–249). New York: Wiley.

Goyal, S. K. (1978). A note on "multi-product inventory situation with one restriction." *Journal of the Operational Research Society, 29*, 269–271.

Graves, S. C. (2008). Flexibility principles. In *Building intuition: insights from basic operations management models and principles* (Chapter 3, pp. 33–51) New York: Springer.

Graves, S. C., & Schwarz, L. B. (1977). Single cycle continuous review policies for arborescent production/inventory systems. *Management Science, 23*, 529–540.

Graves, S. C., & Willems, S. P. (2000). Optimizing strategic safety stock placement in supply chains." *Manufacturing Service Operation Managemant, 2*, 68–83.

Graves, S. C., Rinnooy Kan, A. H. G., & Zipkin, P. H. (Eds.) (1993). Logistics of production and inventory. In *Handbooks in operations research and management science*. Amsterdam: North-Holland.

Hadley, G., & Whitin, T. M. (1963). *Analysis of inventory systems.* Englewood Cliffs, NJ: Prentice-Hall.

Haimovich, M., & Rinnooy Kan, A. H. G. (1985). Bounds and heuristics for capacitated routing problems. *Mathematics of Operations Research, 10*, 527–542.

Haimovich, M., Rinnooy Kan, A. H. G., & Stougie, L. (1988). Analysis of heuristics for vehicle routing problems. In B. L. Golden, A. A. Assad (Eds.) *Vehicle routing: methods and studies* (pp. 47–61). New York, NY: Elsevier Science Publishers, B.V.

Hakimi, S. L. (1964) Optimum locations of switching centers and the absolute centers and medians of a graph. *Mathematics of Operations Research, 12*, 450–459.

Hall, N. G. (1988). A multi-item EOQ model with inventory cycle balancing. *Naval Research Logistics, 35*, 319–325.

Hariga, M. (1988). The warehouse scheduling problem (Ph. D. Thesis, School of Operations Research and Industrial Engineering, Cornell University).

Harris, F. (1915). *Operations and costs. Factory management series* (pp. 48–52). Chicago: A. W. Shaw Co.

Hartley, R., & Thomas, L. C. (1982). The deterministic, two-product, inventory system with capacity constraint. *Journal of the Operational Research Society, 33*, 1013–1020.

Hax, A. C., & Candea, D. (1984). *Production and inventory management.* Englewood Cliffs, NJ: Prentice-Hall.

Held, M., & Karp, R. M. (1962). A dynamic programming approach to sequencing problems. *SIAM Journal on Applied Mathematics, 10*, 196–210.

Held, M., & Karp, R. M. (1970). The traveling salesman problem and minimum spanning trees. *Mathematics of Operations Research, 18*, 1138–1162.

Held, M., & Karp, R. M. (1971). The traveling salesman problem and minimum spanning trees: part II. *Mathematical Programming, 1*, 6–25.

Hodgson, T. J., & Howe, T. J. (1982). Production lot sizing with material-handling cost considerations. *IIE Trans, 14*, 44–51.

Hoffman, K. L., & Padberg, M. (1993). Solving airline crew scheduling problems by branch-and-cut. *Management Science, 39*, 657–682.

Holt, C. C. (1958). Decision rules for allocating inventory to lots and cost foundations for making aggregate inventory decisions. *Journal of Industrial Engineering, 9*, 14–22.

Homer, E. D. (1966). Space-limited aggregate inventories with phased deliveries. *Journal of Industrial Engineering, 17*, 327–333.

Hopp, W., Tekin, E., & Van Oyen, M. (2004). Benefits of skill chaining in serial production lines with cross-trained workers. *Management Science, 50*, 83–98.

House, R. G., & Jamie, K. D. (1981). Measuring the impact of alternative market classification systems in distribution planning. *Journal of Business Logistics, 2*, 1–31.

Hu, P. (2011). Coordinated pricing and inventory management (Ph.D. Dissertation, University of Illinois at Urbana-Champaign)

Huh, T., & Janakiraman, G. (2008). (s, S) optimality in joint inventory-pricing control: an alternate approach. *Mathematics of Operations Research, 56*, 783–790.

Iglehart, D. (1963a). Optimality of (s, S) policies in the infinite horizon dynamic inventory problem. *Management Science, 9*, 259–267.

Iglehart, D. (1963b). Dynamic programming and stationary analysis in inventory problems. In H. Scarf, D. Guilford, M. Shelly (Eds.) *Multi-stage inventory models and techniques* (pp. 1–31). Stanford, CA: Stanford University Press.

Jaillet, P. (1985). Probabilistic traveling salesman problem (Ph.D. Dissertation, Operations Research Center, Massachusetts Institute of Technology, Cambridge, MA)

Johnson, D. S., & Papadimitriou, C. H. (1985). Performance guarantees for heuristics. In E. L. Lawler, J. K. Lenstra, A. H. G. Rinnooy Kan, D. B. Shmoys (Eds.) *The traveling salesman problem: a guided tour of combinatorial optimization* (pp. 145–180). New York: Wiley.

Johnson, J. C., & Wood, D. F. (1986). *Contemporary physical distribution and logistics*. New York: Macmillan.

Johnson, D. S., Demers, A., Ullman, J. D., Garey, M. R., & Graham, R. L. (1974). Worst-case performance bounds for simple one-dimensional packing algorithms. *SIAM Journal on Computing, 3*, 299–325.

Joneja, D. (1990). The joint replenishment problem: new heuristics and worst-case performance bounds. *Mathematics of Operations Research, 38*, 711–723.

Jones, P. C., & Inman, R. R. (1989). When is the economic lot scheduling problem easy? *IIE Trans, 21*, 11–20.

Jordan, W., & Graves, S. C. (1995). Principles on the benefits of manufacturing process flexibility. *Management Science, 41*, 577–594.

Karmarkar, N. (1982). Probabilistic analysis of some bin-packing algorithms. *Proceedings of 23rd Annual Symposium on Foundations of Computer Science*, 107–111.

Karlin, S., & Taylor, H. M. (1975). *A first course in stochastic processes*. San Diego, CA: Academic.

Karp, R. M. (1977). Probabilistic analysis of partitioning algorithms for the traveling salesman problem. *Mathematics of Operations Research, 2*, 209–224.

Karp, R. M., & Steele, J. M. (1985). Probabilistic analysis of heuristics. In E. L. Lawler, J. K. Lenstra, A. H. G. Rinnooy Kan, D. B. Shmoys (Eds.) *The traveling salesman problem: a guided tour of combinatorial optimization* (pp. 181–205). New York: Wiley.

Karp, R. M., Luby, M., & Marchetti-Spaccamela, A. (1984). A probabilistic analysis of multidimensional bin packing problems. *Proceedings of 16th Annual ACM Symposium on Theory of Computing*, 289–298.

Kimes, S. E. (1989). A tool for capacity-constrained service firms. *Journal of Operations Management, 8*(4), 348–363.

Kingman, J. F. C. (1976). Subadditive processes. *Lecture Notes in Mathematics 539*, 168–222. Berlin: Springer.

Klincewicz, J. G., & Luss, H. (1986). A lagrangian relaxation heuristic for capacitated facility location with single-source constraints. *Journal of the Operational Research Society, 37*, 495–500.

Kuehn, A. A., & Hamburger, M. J. (1963). A heuristic program for location warehouses. *Management Science, 9*, 643–666.

Lau, H. S. (1980) The newsboy problem under alternative optimization objectives. *Journal of the Operational Research Society, 31*, 525–535.

Lawler, E. L. (1976). *Combinatorial optimization: networks and matroids*. New York: Holt, Rinehart and Winston.

Lawler, E. L., Lenstra, J. K., Rinnooy Kan, A. H. G., & Shmoys, D. B. (Eds.) (1985). *The traveling salesman problem: a guided tour of combinatorial optimization*. New York: Wiley.

Lawler, E. L., Lenstra, J. K., Rinnooy Kan, A. H. G., & Shmoys, D. B. (1993). Sequencing and scheduling: algorithms and complexity. In S. C. Graves, A. H. G. Rinnooy Kan, P. H. Zipkin (Eds.) *Handbooks in operations research and management science*, the Volume on *Logistics of production and inventory* (pp. 445–522). Amsterdam: North-Holland.

Lee, H. L., & Nahmias, S. (1993). Single product, single location models. In S. C. Graves, A. H. G. Rinnooy Kan, P. H. Zipkin (Eds.) *Handbooks in operations research and management science*, the Volume on *Logistics of production and inventory* (pp. 3–55). Amsterdam: North-Holland.

Li, C. L., & Simchi-Levi, D. (1990). Worst-case analysis of heuristics for the multi-depot capacitated vehicle routing problems. *ORSA Journal on Computing, 2*, 64–73.

Lindsey (1996). A communication to the AGIS-L list server.

Lovasz, L. (1979). *Combinatorial problems and exercises*. Amsterdam: North-Holland.

Love, S. F. (1973). Bounded production and inventory models with piecewise concave costs. *Management Science, 20*, 313–318.

Magnanti, T. L., Shen, Z-J. M., Shu, J., Simchi-Levi, D., & Teo, C-P. (2003). Inventory placement in acyclic supply chain networks. Working Paper. *Operation Research Letter, 34*, 228–238.

Manne, A. S. (1964). Plant location under economies of scale—decentralization and computation. *Management Science, 11*, 213–235.

Martínez-de-Albéniz, V., & Simchi-Levi, D. (2005). A portfolio approach to procurement contracts. *Production and Operations Management, 14*, 90–114.

Martínez-de-Albéniz V., & Simchi-Levi, D. (2006). Mean-variance trde-offs in supply contracts. *Naval Research Logistics, 53*, 603–616.

Maxwell, W. L., & Singh, H. (1983). The effect of restricting cycle times in the economic lot scheduling problem. *IIE Trans, 15*, 235–241.

Melkote, S. (1996). Integrated models of facility location and network design (Ph.D. thesis, Northwestern University).

Mirchandani, P. B., & Francis, R. L. (1990). *Discrete location theory*. New York: Wiley.

Muckstadt, J. M., & Roundy, R. O. (1993). Analysis of multistage production systems. In S. C. Graves, A. H. G. Rinnooy Kan, P. H. Zipkin (Eds.) *Handbooks in operations research and management science*, the volume on *Logistics of production and inventory* (pp. 59–131). Amsterdam: North-Holland.

Murota, K. (2003). *Discrete convex analysis*. Philadelphia: Society for Industrial and Applied Mathematics.

Muriel, A., & Simchi-Levi, D. (2003). Supply chain design and planning – applications of optimization techniques for strategic and tactical models. In de. Kok, S. Graves (Eds.) *Handbooks in operations research and management science (Vol. 11): Supply chain management: design, coordiation and operation*. Boston: Elsevier.

Murota, K., & Shioura, A. (2004). Conjugacy relationship between M-convex and L-convex functions in continuous variables. *Mathematical Programming, 101*, 415–433.

Murota, K., & Shioura, A. (2005). Substitutes and complements in network flows viewed as discrete convexity. *Discrete Mathematics, 2*, 256–268.

Myerson, R. B. (1997). *Game theory: Analysis of Conflict*. Cambridge, MA: Harvard University Press.

Nagarajan, M., & Sošić, G. (2008). Game-theoretic analysis of cooperation among supply chain agents: review and extensions. *European Journal of Operational Research, 187*(3), 719–745.

Nauss, R. M. (1976). An efficient algorithm for the 0-1 knapsack problem. *Management Science, 23*, 27–31.

Neebe, A. W., & Rao, M. R. (1983). An Algorithm for the fixed-charged assigning users to sources problem. *Journal of the Operational Research Society, 34*, 1107–1113.

Newton, R. M., & Thomas, W. H. (1969). Design of school bus routes by computer. *Socio-Economic Planning Science, 3*, 75–85.

Osborne, M. J. (2003). *An introduction to game theory.* New York, Oxford: Oxford University Press.

Özer, Ö., & Phillips, R. (Eds.) (2012). *The oxford handbook of pricing management.* New York, Oxford: OUP Oxford.

Page, E., & Paul, R. J. (1976). Multi-product inventory situations with one restriction. *Journal of the Operational Research Society, 27*, 815–834.

Pang, Z. (2011). Optimal dynamic pricing and inventory control with stock deterioration and partial backordering. *Operation Research Letter, 39*, 375–379.

Pang, Z., Chen, Y. F., & Feng, Y. (2012). A note on the structure of joint inventory-pricing control with leadtimes. *Operation Research, 60*, 581–587.

Papadimitriou, C. H., & Stieglitz, K. (1982). *Combinatorial optimization: algorithms and complexity.* Englewood Cliffs, NJ: Prentice-Hall.

Park, K. S., & Yun, D. K. (1985). Optimal scheduling of periodic activities. *Operation Research, 33*, 690–695.

Patton, E. P. (1994). Carrier rates and tariffs. In J. A. Tompkins, D. Harmelink (Eds.) *The distribution management handbook* (Chapter 12). New York: McGraw-Hill.

Peleg, B., & P. Sudhölter (2007). *Introduction to the theory of cooperative games* (2nd ed.). Berlin: Springer.

Petruzzi, N. C., & Dada, M. (1999). Pricing and the newsvendor model: a review with extensions. *Operation Research, 47*, 183–194.

Pinedo, M. (1995). *Scheduling: theory, algorithms and systems.* Englewood Cliffs, NJ: Prentice-Hall.

Pirkul, H. (1987). Efficient algorithms for the capacitated concentrator location problem. *Computers and Operations Research, 14*, 197–208.

Pirkul, H., & Jayaraman, V. (1996). Production, transportation and distribution planning in a multi-commodity tri-echelon system. *Transportation Science, 30*, 291–302.

Polyak, B. T. (1967). A general method for solving extremum problems (in Russian). *Doklady Akademmi Nauk SSSR, 174* , 33–36.

Porteus, E. L. (1985). Investing in reduced setups in the EOQ model. *Management Science, 31*, 998–1010.

Porteus, E. L. (1990). Stochastic inventory theory. In D. P. Heyman, M. J. Sobel (Eds.) *Handbooks in operations research and management science*, the volume on *Stochastic models* (pp. 605–652). Amsterdam: North-Holland.

Psaraftis, H. N. (1984). On the practical importance of asymptotic optimality in certain heuristic algorithms. *Networks, 14*, 587–596.

Rhee, W. T. (1988). Optimal bin packing with items of random sizes. *Mathematics of Operations Research, 13*, 140–151.

Rhee, W. T. (1991). An asymptotic analysis of capacitated vehicle routing. Working Paper, The Ohio State University, Columbus, OH.

Rhee, W. T., & Talagrand, M. (1987). Martingale inequalities and NP-complete problems. *Mathematics of Operations Research, 12*, 177–181.

Robeson, J. F., & Copacino, W. C. (Eds.) (1994). *The logistics handbook.* New York: Free Press.

Rockafellar, R. T. (1970). *Convex analysis.* Princeton, NJ: Princeton University Press.

Rosen, J. B. (1965). Existence and uniqueness of equilibrium points for concave N-person games. *Econometrica, 33*, 520–534.

Ross, Sheldon M. (1983). Introduction to Stochastic Dynamic Programming. Academic Press, INC., London.

Rosenblatt, M., & Rothblum, U. (1990). On the single resource capacity problem for multi-item inventory systems. *Operation Research, 38*, 686–693.

Rosenkrantz, D. J., Stearns, R. E., & Lewis II, P. M. (1977). An analysis of several heuristics for the traveling salesman problem. *SIAM Journal on Computing, 6*, 563–581.

Ross, S. (1970). *Applied Probability models with optimization applications.* San Francisco: Holden-Day.

Rosling, K. (1989). Optimal inventory policies for assembly systems under random demand. *Operation Research, 37*, 565–579.

Roundy, R. (1985). 98%-effective integer-ratio lot-sizing for one-warehouse multi-retailer systems. *Management Science, 31*, 1416–1430.

Russell, R. A. (1977). An effective heuristic for the M-tour traveling salesman problem with some side constraints. *Operation Research, 25*, 521–524.

Sahni, S., & Gonzalez, T. (1976). P-complete approximation algorithms. *Journal of the Association for Computing Machinery, 23*, 555–565.

Scarf, H. E. (1960). The optimalities of (s, S) policies in the dynamic inventory problem. In K. Arrow, S. Karlin, P. Suppes (Eds.) *Mathematical methods in the social sciences* (pp. 196–202). Stanford, CA: Stanford University Press.

Schweitzer, M., & Chachon, G. (2000). Decision bias in the newsvendor problem with a known demand distribution: experimental evidence. *Management Science, 46*(3), 404–420.

Shapley, L. (1967), On balanced sets and cores. *Naval Research Logistics Quarterly, 14*, 453–460.

Shmoys, D., & Williamson, D. (1990). Analyzing the held-karp TSP bound: a monotonicity property with application. *Information Processing Letters, 35*, 281–285.

Silver, E. A. (1976). A simple method of determining order quantities in joint replenishments under deterministic demand. *Management Science, 22*, 1351–1361.

Silver, E. A., & Meal, H. C. (1973). A heuristic for selecting lot size quantities for the case of a deterministic time-varying demand rate and discrete opportunities for replenishment. *Production and Inventory Management, 14*, 64–74.

Silver, E. A., & Peterson, R. (1985). *Decision systems for inventory management and production planning.* New York: Wiley.

Simchi-Levi, D. (1994), New worst case results for the bin-packing problem. *Naval Research Logistics, 41*, 579–585.

Simchi-Levi, D. (2010). *Operations rules: delivering customer value through flexible operations.* Cambridge, MA: MIT Press.

Simchi-Levi, D., & Bramel, J. (1990). On the optimal solution value of the capacitated vehicle routing problem with unsplit demands. Working Paper, Department of IE&OR, Columbia University, New York.

Simchi-Levi, D., Kaminsky, P., & Simchi-Levi, E. (2003). *Designing and managing the supply chain* (2nd ed.). Burr Ridge, IL: McGraw-Hill.

Simchi-Levi, D., Kaminsky, P., & Simchi-Levi, E. (2007). *Designing and managing the supply chain* (3rd ed.). concepts, Strategies and Case Studies. McGraw-Hill.

Simchi-Levi, D., Wu, D., & Shen, Z. J. (Eds). (2004). *Handbook of quantitative supply chain analysis: modeling in the E-business era.* New York: Springer.

Simchi-Levi, D., Wei, Y. (2012) Understanding the Performance of the Long Chain and Sparse Designs in Process Flexibility. *Operations Research, 60*(5), 1125–1141

Solomon, M. M. (1986). On the worst-case performance of some heuristics for the vehicle routing and scheduling problem with time window constraints. *Networks*, *16*, 161–174.

Solomon, M. M., & Desrosiers, J. (1988). Time window constrained routing and scheduling problems: a survey. *Transportation Science*, *22*, 1–13.

Stankevich, D. (1996). Ace of diamonds. *Discount Merchandiser*, *38*(8), 28–37.

Steele, J. M. (1981). Subadditive euclidean functionals and nonlinear growth geometric probability. *Annals of Probability*, *9*, 365–375.

Steele, J. M. (1990). Lecture notes on "probabilistic analysis of algorithms."

Stout, W. F. (1974). *Almost sure convergence*. New York: Academic.

Strassen, V. (1969). Gaussian elimination is not optimal. *Nmerische Mathematik*, *13*, 354–356.

Swersey, A. J., & Ballard, W. (1984). Scheduling school buses. *Management Science*, *30*, 844–853.

Talluri, K., & van Ryzin, G. (2004). *The theory and practice of revenue management*. New York: Springer.

Tarski, A. (1955). A lattice-theorectical fixpoint theorem and its applications. *Pacific Journal of Mathematics*, *5*, 285–309.

Thomas, L. C., & Hartley, R. (1983). An algorithm for limited capacity inventory problem with staggering. *Journal of the Operational Research Society*, *34*, 81–85.

Topkis, D. M. (1998). *Supermodularity and complementarity*. Princeton, NJ: Princeton University Press.

van Ryzin, G. (2012). Models of demand. In P. Philips, Ö. Özer (Eds.) *Oxford handbook of pricing management* (pp. 340–380). Oxford: Oxford University Press.

Veinott, A. (1966). On the optimality of (s, S) inventory policies: new condition and a new proof. *Journal SIAM Applied Mathematics*, *14*, 1067–1083.

Veinott, A., & Wagner, H. (1965). Computing optimal (s, S) inventory policies. *Management Science*, *11*, 525–552.

Vives, X. (2000). *Oligopoly pricing: old ideas and new tools*. Cambridge, MA: MIT Press.

Wagner, H. M., & Whitin, T. M. (1958a). Dynamic problems in the theory of the firm. *Naval Research Logistics Quarterly*, *5*, 53–74.

Wagner, H. M., & Whitin, T. M. (1958b). Dynamic version of the economic lot size model. *Management Science*, *5*, 89–96.

Wagelmans, A., Van Hoesel, S., & Kolen, A. (1992). Economic lot sizing: an $O(n \log n)$ algorithm that runs in linear time in the Wagner–Whitin case. *Operation Research, 40*(Suppl 1), S145–S156.

Weber, A. (1909). In C. J. Friedrich (Ed. and transl.) *Theory of the location of industries*. Chicago: Chicago University Press.

Whitin, T. M. (1955). Inventory control and price theory. *Management Science, 2*, 61–80.

Wolsey, L. (1980). Heuristic analysis, linear programming and branch and bound. *Mathematical Programming Study, 13*, 121–134.

Yano, C., & Gilbert, S. (2002). Coordinated pricing and production/procurement decisions: a review. In A. Chakravarty, J. Eliashberg (Eds.) *Managing business interfaces: marketing, engineering and manufacturing perspectives*. Boston: Kluwer Academic.

Ye, Q., & Duenyas, I. (2007). Optimal capacity investment decisions with two-sided fixed-capacity adjustment costs. *Operation Research, 55*, 272–283.

Yellow, P. (1970). A computational modification to the savings method of vehicle scheduling. *Operational Research Quarterly, 21*, 281–283.

Zangwill, W. I. (1966). A deterministic multi-period production scheduling model with backlogging. *Management Science, 13*(1), 105–199.

Zavi, A. (1976). *Introduction to operations research, part II: dynamic programming and inventory theory* (in Hebrew). Tel-Aviv: Dekel.

Zheng, Y. S. (1991). A simple proof for the optimality of (s, S) policies for infinite horizon inventory problems. *Journal of Applied Probability, 28*, 802–810.

Zheng, Y. S., & Federgruen, A. (1991). Finding optimal (s, S) policies is about as simple as evaluating a single policy. *Operation Research, 39*, 654–665.

Zipkin, P. H. (2000). *Foundations of inventory management*. Burr Ridge, IL: Irwin.

Zipkin, P. H. (2008). On the structure of lost-sales inventory models. *Operation Research, 56*, 937–944.

Zoller, K. (1977). Deterministic multi-item inventory systems with limited capacity. *Management Science, 24*, 451–455.

Index

Manufactured by Amazon.ca
Bolton, ON

16262859R00258